Spon's
Landscape and
External Works
Price Book

2006

Spon's Landscape and External Works Price Book

Edited by

DAVIS LANGDON

in association with
LandPro Ltd
Landscape Surveyors

2006

Twenty-fifth edition

Taylor & Francis
Taylor & Francis Group

LONDON AND NEW YORK

First edition 1978
Twenty-fifth edition published 2006
by Taylor & Francis
2 Park Square, Milton Park, Abingdon, Oxon OX14 4RN

Simultaneously published in the USA and Canada
by Taylor & Francis
270 Madison Avenue, New York, NY 10016

Taylor & Francis is an imprint of the Taylor & Francis Group

Printed and bound in Great Britain by
TJ International Ltd, Padstow, Cornwall

Publisher's note
This book has been produced from camera-ready copy supplied by the authors.

British Library Cataloguing in Publication Data
A catalogue record for this book is available from the British Library

ISBN 0-415-37036-1
ISSN 0267-4181

Contents

4th Edition
Dictionary of Architecture and Building Technology

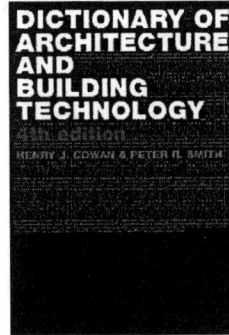

Edited by
Henry J. Cowan and Peter Smith

This dictionary of over 6000 terms is compiled specifically for the professional. The explanation of terms addresses technical issues in architecture and building.

The terms have been fully updated with many new entries for this new edition.

This is a valuable addition to all construction reference libraries, which will enable students and professionals from technical, management and professional fields to grasp vocabulary from outside their areas as they contribute towards the built environment.

June 2004: 234x156 mm: 352 pages
HB: 0-415-31233-7: £80.00
PB: 0-415-31234-5: £25.99

To Order: Tel: +44 (0) 1264 343071 Fax: +44 (0) 1264 343005, or
Post: Taylor and Francis Customer Services, Thomson Publishing Services, Cheriton House, Andover, Hants, SP10 5BE, UK Email: book.orders@tandf.co.uk

For a complete listing of all our titles visit:
www.sponpress.com

Taylor & Francis
Taylor & Francis Group plc

Professional Negligence in Construction

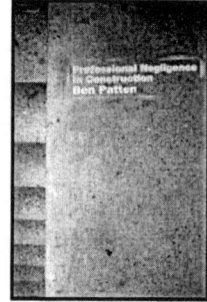

Ben Patten

What is professional negligence? Who, for that matter, is a "construction professional" and what are their obligations in contract and their obligations in tort? In what circumstances might the difference between the obligations be important?

These questions are of crucial importance to contractors, architects, quantlty surveyors, engineers, project managers, and multi disciplinary practices, and not simply to construction lawyers. Ben Pattern's guide is written for them. It is brief and simple, and yet solidly authoritative. It should also be helpful to lawyers and to students in construction and in law. However, the emphasis in the book is on the practical aspects of professional negligence in the construction industry and it is not primarily intended to be a 'legal practitioner' work.

Contents: 1. Legal Framework. 2. Insurance. 3. Architects and Engineers. 4. Quantity Surveyors. 5. Project Managers. 6. Claims Consultants and Expert Witnesses. 7. Dispute Resolution.

July 2003: 234x156mm: 192pp.
HB: 0-415-29066-X: £45.00

To Order: Tel: +44 (0) 1264 343071 Fax: +44 (0) 1264 343005, or
Post: Taylor and Francis Customer Services, Thomson Publishing Services, Cheriton House, Andover, Hants, SP10 5BE, UK Email: book.orders@tandf.co.uk

For a complete listing of all our titles visit:
www.sponpress.com

Taylor & Francis
Taylor & Francis Group plc

Preface to the Twenty Fifth Edition

Landscape contractors report a weakening of tender opportunities especially on the domestic and smaller commercial sectors at the time of going to press. Long term projects on the larger commercial or civil sector still have to run their course and the opportunity for landscape or external works contracts will as usual be largely dependent on the investment in new construction projects.

Tender prices in Greater London, measured by Davis Langdon's Tender Price Index, showed a rise of 1.5% in the first quarter of 2005, continuing an increase of 4% since mid 2004. Over the past twelve months, there has been a 6% increase in tender prices in London, compared with other regions, where increases of up to 9% have been observed.

Davis Langdon's Tender Price Index is expected to rise by between 4% and 5% over the next 12 months.

Market Conditions
Prices for labour within the Landscape sector of the construction industry have stabilized. During the course of this annual update of this book we have noticed the following trends:

Specialist materials and services suppliers have held their prices in many cases. Larger national suppliers have applied increases of 3.5% over last years prices. This reflects the same trend the previous year.

Steel prices have stabilized on last years prices but remain volatile. Steel suppliers report that whilst high prices generally persist, hard bargaining by buyers of steel often result in very keen prices to buyers. The range is from £350 -£650.00 per tonne dependent of quantity.

Labour Rates
Although BCE national wage rates increased by 9.5% from 27 June 2005, the labour rate used in this book, in line with market trends, has increased by only £0.25 (1.35%) over the rate used last year.

New Items
This edition includes:-
- new items for the comparative costs of excavation and earth moving machinery intended to aid contractors with cost of machine selection.
- construction of different formats of sleeper retaining walls.
- new cost tables for the fixing of various sizes of simple reinforcement to walls and slabs.
- aluminium edgings for soft and hard landscape applications.
- containerised large trees which are growing in popularity in the industry now that they are becoming more available due to the establishment of specialist growers.
- irrigation costs for typical gardens and landscape situations.

Additionally readers will find augmentation to many existing sections of the book.

New Approximate Estimates
There has been a major overhaul on the existing approximate estimates. Readers will find that many items have been revised in terms of updated specification and practise as well as the addition of numerous new items in this section.

Tables and Memoranda

We remind readers of the information contained in this section at the back of the book. It is often overlooked but contains a wealth of information useful to External works consultants and contractors.

Omitted sections

We have this year omitted the sections dealing with fees for Quantity surveyors, the Landscape consultants appointment and the Aggregates tax. We have revised the section on Preliminaries to tie in with the 2002 revision of the JCLI Landscape contract and included a list of other preliminary heads which may be addressed as part of a landscape contract. Readers should note however that the new family of JCT contracts have now been published (May 2005) and will be addressed in next years editions of the book.

Profit and Overhead

Spon's Landscape and External Works prices do not allow for profit or site overhead. Readers should evaluate the market conditions in the sector or environment in which they operate and apply a percentage to these rates. Company overhead is allowed for within the labour rates used. Please refer to the Cost information section at the front of the book for an explanation of the build-up of this years labour rate

Prices for Suppliers and Services

We acknowledge the support provided to us by the suppliers of products and services who provide us with the base information used and whose contact details are published in the directory section at the front of this book. We advise that readers wishing to evaluate the cost of a product or service should approach these suppliers directly should they wish to confirm prices prior to submission of tenders or quotations.

Whilst every effort is made to ensure the accuracy of the information given in this publication, neither the editors nor the publishers in any way accept liability for loss of any kind resulting from the use made by any person of such information.

We remind readers that we would be grateful to hear from them during the course of the year with suggestions or comments on any aspect of the contents of this years book and suggestions for improvements. Our contact details are shown below.

DAVIS LANGDON
MidCity Place
71 High Holborn,
London
WC1V 6QS

Tel: 020 7061 7000
e-mail: data.enquiry@davislangdon.com

SAM HASSALL
LANDPRO LTD
Landscape Surveyors
25b The Borough
Farnham
Surrey
GU9 7NJ

Tel: 01252 725513
e-mail: info@landpro.co.uk

List of Manufacturers and Suppliers

This list has been compiled from the latest information available but as firms frequently change their names, addresses and telephone numbers due to re-organization, users are advised to check this information before placing orders.

A Plant Acrow
Colnbrook By-Pass
Colnbrook
Slough
Berkshire
Plant hire
Tel: 01753 680420
Fax: 01753 681338

A Plant Hire Co Ltd
102 Dalton Avenue
Birchwood Park
Warrington
Cheshire WA3 6YE
Plant hire
Tel: 01925 281000
Fax: 01925 281001
Website: www.aplant.com
E-mail: johnburfoot@aplant.com

ABG Ltd
Unit E7
Meltham Mills Road
Meltham
West Yorkshire HD9 4DS
Geosynthetics
Tel: 01484 852096
Fax: 01484 851562
Website: www.abg-geosynthetics.com
E-mail: goran@abg-geosynthetics.com

Addagrip Surface Treatments UK Ltd
Addagrip House
Bell Lane Industrial Estate
Uckfield
East Sussex TN22 1QL
Epoxy bound surfaces
Tel: 01825 761333
Fax: 01825 768566
Website: www.addagrip.co.uk
E-mail: sales @addagrip.co.uk

Alumasc Exterior Building Products Ltd
White House Works
Bold Road
Sutton, St.Helens
Merseyside WA9 4JG
Green roof systems
Tel: 01744 648400
Fax: 01744 648401
Website: www.alumasc-exterior-building-
products.co.uk
E-mail: info@alumasc-exteriors.co.uk

Amberol Ltd
The Plantation
Spencer Road
Belper
Derbyshire DE56 1JW
Street furniture
Tel: 01773 830930
Fax: 01773 834191
Website: www.amberol.co.uk
E-mail: sales@amberol.co.uk

Amenity and Horticultural Services Ltd
Coppards Lane
Northiam
East Sussex TN31 6QP
Horticultural composts & fertilizers
Tel: 01797 252728
Fax: 01797 252724
E-mail: ahs@talk21.com

Amenity Land Services Ltd (ALS)
Long Lane
Willington
Telford
Shropshire TF6 6HA
Horticultural products supplier
Tel: 01952 641949
Fax: 01952 247369
Website: www.amenity.co.uk
E-mail: sales@amenity.co.uk

Ancon Ltd
President Way
President Park
Sheffield S4 7UR
Fixings
Tel: 0114 275 5224
Fax: 0114 276 8543
Website: www.ancon.co.uk
E-mail: info@ancon.co.uk

Anderton Concrete Products Ltd
Anderton Wharf
Soot Hill
Anderton
Northwich
Cheshire CW9 6AA
Concrete fencing
Tel: 01606 79436
Fax: 01606 871590
Website: www.andertonconcrete.co.uk

Anglo Aquarium Plant Co Ltd
Strayfield Road
Enfield
Middlesex EN2 9JE
Aquatic plants
Tel: 020 8363 8548
Fax: 020 8363 8547
Website: www.anglo-aquarium.co.uk
E-mail: sales@anglo-aquarium.co.uk

Architectural Heritage Ltd
Taddington Manor
Nr. Cutsdean
Cheltenham
Gloucestershire GL54 5RY
Ornamental stone buildings & water features
Tel: 01386 584414
Fax: 01386 584236
Website: www.architectural-heritage.co.uk
E-mail: puddy@architectural-heritage.co.uk

Artificial Lawn Co.
Hartshill Nursery
Thong Lane
Gravesend
Kent DA12 4AD
Artificial grass suppliers
Tel: 01474 364320
Fax: 01474 321587
Website: www.artificiallawn.co.uk
E-mail: sales@artificiallawn.co.uk

Autopa Ltd
Cottage Leap off Butlers Leap
Rugby
Warwickshire CV21 3XP
Street furniture
Tel: 01788 550556
Fax: 01788 550265
Website: www.autopa.co.uk
E-mail: info@autopa.co.uk

AVS Fencing Supplies Ltd
Unit 1 AVS Trading Park
Chapel Lane
Milford
Surrey GU8 5HU
Fencing
Tel: 01483 410960
Fax: 01483 860867
Website: www.avsfencing.co.uk
E-mail: sales@avsfencing.co.uk

Bauder Ltd
Broughton House
Broughton Road
Ipswich
Suffolk IP1 3QR
Green roof systems
Tel: 01473 257671
Fax: 01473 230761
Website: www.bauder.co.uk
E-mail: systems@bauder.co.uk

Baylis Landscape Contractors
Hartshill Nursery
Thong Lane
Gravesend
Kent DA12 4AD
Sportsfield constructon
Tel: 01474 569576
Fax: 01474 321587
E-mail: landscape@baylis.freeserve.co.uk

Belwood Trees Ltd
Brigton of Ruthven
Meigle
Perthshire PH12 8RQ
Mature tree grower
Tel: 01828 640219
Fax: 01828 640623
Website: www.belwoodtrees.co.uk
E-mail: belwood@belwoodtrees.co.uk

Blanc De Bierges
Eastrea Road
Whittlesey
Peterborough
Cambridgeshire PE7 2AG
Concrete paving supplier
Tel: 01733 202566
Fax: 01733 205405
Website: www.blancedebierges.com
E-mail: sales@blancedebierges.com

BSI Metro Sand
Unit 2 Vicarage Farm
Fair Oak
Southampton
Hampshire SO50 7HD
Structural tree soil
Tel: 02380 696957
Fax: 02380 696958
Website: www.metrosand.co.uk
E-mail: sales@metrosand.co.uk

Boughton Loam Ltd
Telford Way
Telford Way Industrial Estate
Kettering
Northants. NN16 8UN
Loams and topsoils
Tel: 01536 510515
Fax: 01536 510691
Website: www.boughton-loam.co.uk
E-mail: enquiries@boughton-loam.co.uk

Capital Garden Products Ltd
Gibbs Reed Barn
Pashley Road
Ticehurst
East Sussex TN5 7HE
Plant containers
Tel: 01580 201092
Fax: 01580 201093
Website: www.capital-garden.com
E-mail: sales@capital-garden.com

Brett Landscaping & Building Products
Salt Lane
Cliffe
Rochester
Kent ME3 7SZ
Landscaping and building products
Tel: 01634 222188
Fax: 01634 222001
Website: www.brett.co.uk/landscaping

CED Ltd
728 London Road
West Thurrock
Grays
Essex RM20 3LU
Natural stone
Tel: 01708 867237
Fax: 01708 867230
Website: www.ced.ltd.uk
E-mail: sales@ced.ltd.uk

British Seed Houses Ltd
Camp Road
Witham St Hughs
Lincoln LN6 9QJ
Grass and wildflower seed
Tel: 01522 868714
Fax: 01522 868095
Website: www.britishseedhouses.com
E-mail: simontaylor@bshlincoln.co.uk

Cemex Ltd
Dale Road
Dove Holes
Buxton
Derbyshire SK17 8EH
Tel: 01298 22324
Fax: 01298 815 221
Website: www.cemex.co.uk
Email: webmanager.hbm@rmc.co.uk

Broxap Streetscene
Rowhurst Industrial Estate
Chesterton
Newcastle-under-Lyme
Staffordshire ST5 6BD
Street furniture
Tel: 01782 564411
Fax: 01782 565357
Website: www.broxap.com
E-mail: sales@broxap.com

Charcon Hard Landscaping
Hulland Ward
Ashbourne
Derbyshire DE6 3ET
Paving and street furniture
Tel: 01335 372222
Fax: 01335 370074
Website: www.aggregate.com
E-mail: enquiries@aggregate.com

Charles Morris (Fertilizers) Ltd
Longford House
Long Lane
Stanwell
Middlesex TW19 7AT
Topsoil
Tel: 01784 449144
Fax: 01784 449133

Coblands Nurseries Ltd
Trench Road
Tonbridge
Kent TN11 9NG
Nursery stock supplier
Tel: 01732 770999
Fax: 01732 770271
Website: www.coblands.co.uk
E-mail: plants@coblands.co.uk

Colourpave Ltd
Laymore
Forest Vale Industrial Estate
Cinderford
Gloucestershire GL14 2PH
Macadam contractor
Tel: 01594 826768
Fax: 01594 826598

Columbia Cascade Ltd
Plasycoed
Wernoleu Road
Ammanford
Carmarthenshire SA18 2JL
Street furniture
Tel: 01269 596396
Fax: 01269 596395
Website: www.timberform.com
E-mail: columbiacascade@aol.com

Cooper Clarke Group Ltd
Special Products Division
Bloomfield Road
Farnworth
Bolton BL4 9LP
Wall drainage and erosion control
Tel: 01204 862222
Fax: 01204 575472
Website: www.heitonuk.com
E-mail: marketing@cooperclarke.co.uk

Coverite Ltd
Palace Gates
Bridge Road
Wood Green
London N22 4SP
Waterproofing specialist
Tel: 020 8888 7821
Fax: 020 8889 0731
Website: www.coverite.com

Crowders Nurseries
Lincoln Road
Horncastle
Lincolnshire LN9 5LZ
Plant protection
Tel: 01507 525000
Fax: 01507 524000
Website: www.crowders.co.uk
E-mail: sales@crowders.co.uk

CU Phosco Ltd
Charles House
Lower Road
Great Amwell
Ware
Hertfordshire SG12 9TA
Lighting
Tel: 01920 860600
Fax: 01920 860635
Website: www.cuphosco.co.uk
E-mail: sales@cuphosco.co.uk

Dee-Organ Ltd
5 Sandyford Road
Paisley
Renfrewshire PA3 4HP
Street furniture
Tel: 0141 889 7000
Fax: 0141 889 7764
Website: www.dee-organ.co.uk
E-mail: signs@dee-organ.co.uk

Deepdale Trees Ltd
Tithe Farm
Hatley Road
Potton
Sandy
Bedfordshire SG19 2DX
Trees, multi-stem and hedging
Tel: 01767 262636
Fax: 01767 262288
Website: www.deepdale-trees.co.uk
E-mail: mail@deepdale-trees.co.uk

DLF Perryfields Ltd
Thorn Farm
Evesham Road
Inkberrow
Worcestershire WR7 4LJ
Grass and wild flower seed
Tel: 01386 793135
Fax: 01386 792715
Website: www.perryfields.co.uk
E-mail: amenity@perryfields.co.uk

Duracourt (Spadeoak) Ltd
Town Lane
Wooburn Green
High Wycombe
Buckinghamshire HP10 0PD
Tennis courts
Tel: 01628 528421
Fax: 01628 810509
Website: www.duracourt.co.uk
E-mail: info@duracourt.co.uk

E.T. Clay Products
7 Fowler Road
Hainault
Essex IT6 3UT
Brick supplier
Tel: 020 8501 2100
Fax: 020 8500 9990
E-mail: eddie@etclay.freeserve.co.uk

Earth Anchors Ltd
15 Campbell Road
Croydon
Surrey CR0 2SQ
Anchors for site furniture
Tel: 020 8684 9601
Fax: 020 8684 2230
Website: www.earth-anchors.com

Edwards Sports Products Ltd
North Mills
Bridport
Dorset DT6 3AH
Sports equipment
Tel: 01308 424111
Fax: 01308 455800
Website: www.edsports.co.uk
E-mail: sales@edsports.co.uk

Elliotthire
The Fen
Baston
Peterborough
Cambridgeshire PE6 9BR
Site offices
Tel: 0800 454962
Website: www.elliotthire.co.uk
E-mail: hirediv@elliott-group.co.uk

English Woodlands
Burrow Nursery
Cross in Hand
Heathfield
East Sussex TN21 0UG
Plant protection
Tel: 01435 862992
Fax: 01435 867742
Website: www.ewburrownursery.co.uk
E-mail: sales@ewburrownursery.co.uk

Ennstone Breedon
Breedon-on-the-Hill
Nr. Melbourne
Derbyshire DE73 8AP
Specialist Gravels
Tel: 01332 862254
Fax: 01332 863149
Website: www.ennstone.co.uk
E-mail: breedon@breedon.co.uk

Ensor Building Products
Blackamoor Road
Guide
Blackburn
Lancashire BB1 2LQ
Drainage suppliers
Tel: 0870 7700484
Fax: 0870 7700485
Website: www.ensorbuilding.com
E-mail: sales@ensorbuilding.com

Eve Trakway
Bramley Vale
Chesterfield
Derbyshire S44 5GA
Portable roads
Tel: 08700 767676
Fax: 08700 737373
Website: www.evetrakway.co.uk
E-mail: marketing@evetrakway.co.uk

Exclusive Leisure Ltd
28 Cannock Street
Leicester LE4 7HR
Artificial sports surfaces
Tel: 0116 233 2255
Fax: 0116 246 1561
Website: www.exclusiveleisure.co.uk
E-mail: info@exclusiveleisure.co.uk

Exxon Chemical Geopolymers Ltd
Mamhilad Park Industrial Estate
Pontypool
Gwent NP4 0YR
Geofabrics
Tel: 01495 757722
Fax: 01495 762383

Fairwater Ltd
Lodge Farm
Malthouse Lane
Ashington
West Sussex RH20 3BU
Water feature contractors
Tel: 01903 892228
Fax: 01903 892522
Website: www.fairwater.co.uk
E-mail: info@fairwater.co.uk

Farmura Environmental Ltd
Stone Hill
Egerton
Ashford
Kent TN27 9DU
Organic fertilizer suppliers
Tel: 01233 756241
Fax: 01233 756419
Website: www.farmura.com
        www.alginure.co.uk
E-mail: enquiries@farmura.com

Fleet Line Markers Limited
Fleet House
Spring Lane Industrial Estate
Malvern
Worcestershire WR14 1AT
Sports line marking
Tel: 01684 573535
Fax: 01684 892784
Website: www.fleetlinemarkers.com
E-mail: sales@fleetlinemarkers.com

Forticrete Ltd
Hillhead Quarry
Harpur Hill
Buxton
Derbyshire SK17 9PS
Retaining wall systems
Tel: 01298 23333
Fax: 01298 23000
Website: www.forticrete.co.uk
E-mail: info@forticrete.com

Furnitubes International Ltd
Meridian House
Royal Hill
Greenwich
London SE10 8RT
Street furniture
Tel: 020 8378 3200
Fax: 020 8378 3250
Website: www.furnitubes.com
E-mail: sales@furnitubes.com

Geometric Furniture Ltd
Birch Mill
Heywood Old Road
Heywood
Lancashire OL10 2QQ
Street furniture
Tel: 0161 653 2233
Fax: 0161 653 2299
Website: www.geometric-furniture.co.uk
E-Mail: info@geometric-furniture.co.uk

Grace Construction Products
Ajax Avenue
Slough
Berkshire SL1 4BH
Bitu-thene tanking
Tel: 01753 692929
Fax: 01753 691623
Website: www.uk.graceconstruction.com
E-mail: uksales@grace.com

Grass Concrete Ltd
Walker House
22 Bond Street
Wakefield
West Yorkshire WF1 2QP
Grass block paving
Tel: 01924 379443
Fax: 01924 290289
Website: www.grasscrete.com
E-mail: info@grasscrete.com

Greenfix Ltd
Evesham Road
Bishops Cleeve
Cheltenham
Glos. GL52 8SA
Seeded erosion control mats
Tel: 01242 679555
Fax: 01242 679855
Website: www.greenfix.co.uk
E-mail: cheltenham@greenfix.co.uk

Greenleaf Horticulture
Ivyhouse Industrial Estate
Haywood Way
Hastings TN35 4PL
Root directors
Tel: 01424 717797
Fax: 01424 205240

H. Langdon & Son
51-53 Second Avenue
Chatham
Kent ME4 5BA
Railings
Tel: 01634 842485
Fax: 01634 831037

H.S.S. Hire
25 Willow Lane
Mitcham
Surrey CR4 4TS
Tool and plant hire
Tel: 08457 231141
Website: www.hss.com
E-mail: hire@hss.com

Haddonstone Ltd
The Forge House
East Haddon
Northampton NN6 8DB
Architectural stonework
Tel: 01604 770711
Fax: 01604 770027
Website: www.haddonstone.co.uk
E-mail: info@haddonstone.co.uk

Hanson Brick
Birchwood Way
Cotes Park Industrial Estate
Sumercotes
Alfreton
Derbyshire DE55 4NH
Brick manufacturer
Tel: 08705 258258
Fax: 01773 514041
Website: www.hansonbrick.com
E-mail: uksales@hansonplc.com

Harrison External Display Systems
Borough Road
Darlington
Co Durham DL1 1SW
Flagpoles
Tel: 01325 355433
Fax: 01325 461726
Website: www.harrisoneds.com
E-mail: sales@harrisoneds.com

Hepworth Plc
Hazlehead
Stocksbridge
Sheffield S30 5HG
Drainage
Tel: 0870 443 6000
Fax: 0870 443 8000
Website: www.hepworthdrainage.co.uk
E-mail: customerservices@hepworthdrainage.co.uk

Hill & Smith Ltd
Springvale Business & Industrial Park
Bilston
Wolverhampton WV14 0QL
Safety barriers
Tel: 01902 499400
Fax: 01902 499419
Website: www.hill-smith.co.uk
E-mail: barrier@hill-smith.co.uk

Hills Industries Ltd
Pontygwindy Industrial Estate
Caerphilly
Mid Glamorgan CF8 3HU
Industrial washing lines
Tel: 029 2088 3951
Fax: 029 2088 6102
Website: www.hills-industries.co.uk
E-mail: info@hills-industries.co.uk

Hodkin & Jones (Sheffield) Ltd
Callywhite Lane
Dronfield
Sheffield S18 6XP
Drainage
Tel: 01246 290890
Fax: 01246 290292
Website: www.plastermouldingsonline.com
E-mail: info@hodkin-jones.co.uk

Inturf
The Chestnuts
Wilberfoss
York YO41 5NT
Turf
Tel: 01759 321000
Fax: 01759 380130
Website: www.inturf.co.uk
E-mail: info@inturf.co.uk

J Toms Ltd
7 Marley Farm
Headcorn Road
Smarden
Ashford
Kent TN27 8PJ
Plant protection
Tel: 01233 770066
Fax: 01233 770055
Website: www.jtoms.co.uk
E-mail: jtoms@btopenworld.com

Jacksons Fencing
Stowting Common
Ashford
Kent TN25 6BN
Fencing
Tel: 01233 750393
Fax: 01233 750403
Website: www.jacksons-fencing.co.uk
E-mail: sales@jacksons-fencing.co.uk

John Anderson Hire
Unit 5 Smallford Works
Smallford Lane
St. Albans
Herts. AL4 0SA
Mobile toilet facilities
Tel: 01727 822485
Fax: 01727 822886
Website: www.superloo.co.uk
E-mail: sales@superloo.co.uk

Johnsons Wellfield Quarries Ltd
Crosland Hill
Huddersfield
West Yorkshire HD4 7AB
Natural Yorkstone pavings
Tel: 01484 652311
Fax: 01484 460007
Website: www.johnsons-wellfield.co.uk
E-mail: sales@johnsons-wellfield.co.uk

Jones of Oswestry
Whittington Road
Oswestry
Shropshire SY11 1HZ
Channels, gulleys, manhole covers
Tel: 01691 653251
Fax: 01691 658222
Website: www.jonesofoswestry.com
E-mail: sales@jonesofoswestry.com

Keller Comtec
Mereworth Business Centre
Danns Lane
Mereworth
Kent ME18 5LW
Hydro seeding specialist
Tel: 01622 816780
Fax: 01622 816791
Website: www.keller-ge.co.uk
E-mail: comtec@comtec-keller.co.uk

Kompan Ltd
Unit 20 Denbigh Hall
Bletchley
Milton Keynes
Buckinghamshire MK3 7QT
Play equipment
Tel: 01908 642466
Fax: 01908 270137
Website: www.kompan.com
E-mail: kompan.uk@kompan.com

Land and Water Services
3 Weston Yard
The Street
Albury
Guildford
Surrey GU5 9AF
Specialist plant hire
Tel: 01483 202733
Fax: 01483 202510
Website: www.land-water.co.uk
E-mail: enquiries@land-water.co.uk

Landline Ltd
1 Bluebridge Industrial Estate
Halstead
Essex CO9 2EX
Pond and lake installation
Tel: 01787 476699
Fax: 01787 472507
Website: www.landline.co.uk
E-mail: sales@landline.co.uk

Landscapes By Design
The Studio
17 Church Road
Farnborough
Kent BR6 7DB
Trellis
Tel: 01689 851570
Fax: 01689 861151

Lappset UK Ltd
Lappset House
Henson Way
Kettering
Northamptonshire NN16 8PX
Play equipment
Tel: 01536 412612
Fax: 01536 521703
Website: www.lappset.co.uk
E-mail: uk@lappset.com

LDC Limited
Loampits Farm
99 Westfield Road
Woking
Surrey GU22 9QR
Willow walling
Tel: 01483 767488
Fax: 01483 764293
Website: www.ldclandscape.co.uk
E-mail: info@ldc.co.uk

Leaky Pipe Systems Ltd
Frith Farm
Dean Street
East Farleigh
Maidstone
Kent ME15 0PR
Irrigation systems
Tel: 01622 746495
Fax: 01622 745118
Website: www.leakypipe.co.uk
E-mail: sales@leakypipe.co.uk

Lister Lutyens Co Ltd
6 Alder Close
Eastbourne
East Sussex BN23 6QF
Street furniture
Tel: 01323 431177
Fax: 01323 639314
Website: www.listerteak.com
E-mail: sales@listerteak.com

Longlyf Timber Products Ltd
Grange Road
Tilford
Farnham
Surrey GU10 2DQ
Treated timber products
Tel: 01252 795042
Fax: 01252 795043
Website: www.longlyftimber.co.uk
E-mail: enquiries@longlyftimber.co.uk

Lorenz von Ehren
Maldfeldstrasse 4
21077 Hamburg (Marmstorf)
Germany
Topiary
Tel: 0049 40 761080
Fax: 0049 40 761081
Website: lve@lve.de
E-mail: www.lve.de

Louis Poulsen UK Ltd
Unit C 44 Barwell Business Park
Leatherhead Road
Chessington KT9 2NY
Outdoor lighting
Tel: 0208 3974400
Fax: 0208 3974455
Website: www.louispoulsen.co.uk
E-mail: louis.poulsen.uk@lpmail.com

Maccaferri Ltd
7400 The Quorum
Oxford Business Park North
Garsington Road
Oxford OX4 2JZ
Gabions
Tel: 01865 770555
Fax: 01865 774550
Website: www.maccaferri.co.uk
E-mail: oxford@maccaferri.co.uk

Marlin Lighting
Hanworth Trading Estate
Hampton Road West
Feltham TW13 6DR
Lighting
Tel: 0870 6062030
Fax: 02088940911
Website: www.concordmarlin.com

Marshalls Plc
Hall Ings
Southowram
Halifax
West Yorkshire HX3 9TW
Hard landscape materials and street furniture
Tel: 01422 306400
Fax: 01422 330185
Website: www.marshalls.co.uk

McArthur Group
Geddings Road
Hoddeston
Herts EN11 0NZ
Security fencing
Tel: 01992 467111
Fax: 01992 469008
Website: www.mcarthur-group.com
E-mail: marketing@mcarthur-group.com

Melcourt Industries Limited
Boldridge Brake
Long Newnton
Tetbury
Gloucestershire GL8 8RT
Mulch and compost
Tel: 01666 502711
Fax: 01666 504398
Website: www.melcourt.co.uk
E-mail: mail@melcourt.co.uk

Milton Pipes Ltd
Milton Regis
Sittingbourne
Kent ME10 2QF
Soakaway rings
Tel: 01795 425191
Fax: 01795 478232
Website: www.miltonpipes.com
E-mail: sales@miltonpipes.com

Monarflex Ltd
Lyon Way
St Albans
Hertfordshire AL4 0LB
Lake liners
Tel: 01727 830116
Fax: 01727 868045
Website: monarflex.icopal.co.uk
E-mail: enq@monarflex.co.uk

Neptune Outdoor Furniture Ltd
Thompsons Lane
Marwell
Winchester
Hampshire SO21 1JH
Street furniture
Tel: 01962 777799
Fax: 01962 777723
Website: www.nofl.co.uk
E-mail: info@nofl.co.uk

Netlon Turf Systems
New Wellington Street
Blackburn
Lancashire BB2 4PJ
Erosion control, soil stabilisation, plant protection
Tel: 01254 266833
Fax: 01254 266868
Website: www.netlon.co.uk
E-mail: turf@netlon.co.uk

Notcutts Nurseries Ltd
Woodbridge
Suffolk IP12 4AF
Nursery stock supplier
Tel: 01394 383344
Fax: 01394 445307
Website: www.notcutts.co.uk

Orchard Street Furniture Ltd
119 The Street
Crowmarsh Gifford
Wallingford
Oxfordshire OX10 8EF
Street furniture
Tel: 01491 642123
Fax: 01491 642126
Website: www.orchardstreet.co.uk
E-mail: sales@orchardstreet.co.uk

Permaloc
Springfield House
41-45 Chapel Brow
Leyland
Lancashire PR25 3NH
Tel: 01772 454279
Fax: 01772 622696
Aluminium edging supplier
Website: www.permaloc.com
E-mail: info@permaloc.com

Platipus Anchors Ltd
Kingsfield Business Centre
Philanthropic Road
Redhill
Surrey RH1 4DP
Tree anchors
Tel: 01737 762300
Fax: 01737 773395
Website: www.platipus-anchors.com
E-mail: info@platipus-anchors.com

Quality Irrigation Ltd
309 Vale Road
Ash Vale
Surrey GU12 5LN
Irrigation
Tel: 01252 328017
Fax: 01252 328017
Website: www.qualityirrigation.co.uk
E-mail: alanaustin@qualityirrigation.co.uk

Rawell Water Control Systems Ltd
Carr Lane
Hoylake
Wirral
Merseyside CH47 4FE
Lake liners
Tel: 0151 632 5771
Fax: 0151 632 4363
E-mail: postmaster@rawell.com

RCC
Barholm Road
Tallington
Stamford
Lincolnshire PE9 4RL
Precast concrete retaining units
Tel: 01778 344460
Fax: 01778 345949
Website: www.tarmacprecast.co.uk
E-mail: rcc@tarmac.co.uk

Recycled Materials Ltd
PO Box 519
Surbiton
Surrey KT6 4YL
Disposal contractors
Tel: 020 8390 7010
Fax: 020 8390 7020
Website: www.recycledmaterials.co.uk
E-mail: mail@recycledmaterials.co.uk

Rigby Taylor Ltd
The Riverway Estate
Portsmouth Road
Peasmarsh
Guildford
Surrey GU3 1LZ
Horticultural supply
Tel: 01483 446900
Fax: 01483 534058
Website: www.rigbytaylor.com
E-mail: sales@rigbytaylor.com

RIW Ltd
Arc House
Terrace Road South
Binfield
Bracknell
Berkshire RG42 4PZ
Waterproofing products
Tel: 01344 861988
Fax: 01344 862010
Website: www.riw.co.uk
E-mail: enquiries@riw.co.uk

Road Equipment Ltd
28/34 Feltham Road
Ashford
Middlesex TW15 1DL
Plant hire
Tel: 01784 256565
Fax: 01784 240398
E-mail: roadequipment@aol.com

Rolawn Ltd
Elvington
York YO41 4XR
Industrial turf
Tel: 01904 608661
Fax: 01904 608272
Website: www.rolawn.co.uk
E-mail: ian.elwick@rolawn.co.uk

Rom Ltd
Wheaton Road
Witham
Essex CM8 3BU
Reinforcement steel bars
Tel: 01376 514321
Fax: 01376 518628

Scotts UK
Paper Mill Lane
Bramford
Ipswich
Suffolk IP8 4BZ
Fertilizers and chemicals
Tel: 01473 830492
Fax: 01473 830386
Website: www.scottsukonline.com

Sleeper Supplies Ltd
PO Box 1377
Kirk Sandall
Doncaster DN3 1XT
Sleeper supplier
Tel: 01302 888676
Fax: 01302 880547
Website: www.sleeper-supplies.co.uk
E-mail: sales@sleeper-supplies.co.uk

SMP Playgrounds Ltd
Ten Acre Lane
Thorpe
Egham
Surrey TW20 8RJ
Playground equipment
Tel: 01784 489100
Fax: 01784 431079
Website: www.smp.co.uk
E-mail: sales@smp.co.uk

Southern Conveyors
Unit 2 Denton Slipways Site
Wharf Road
Gravesend DAI2 2RU
Conveyors
Tel: 01474 564145
Fax: 01474 568036

Spadeoak Construction Co Ltd
Town Lane
Wooburn Green
High Wycombe
Bucks. HP10 0PD
Macadam contractors
Tel: 01628 529421
Fax: 01628 810509
Website: www.spadeoak.co.uk
E-mail: email@spadeoak.co.uk

St. Gobain Pipelines Plc
Holwell Works
Asfordby Hill
Melton Mowbray
Leicestershire LE14 3RE
Ductile iron access covers, gullies & grates
Tel: 0115 9305000
Fax: 0115 9329513
Website: www.saint-gobain-pipelines.co.uk
E-mail:
    technical.covers@saint-gobain-pipelines.co.uk

Steelway-Fensecure Ltd
Parkside Industrial Estate
Hickman Avenue
Wolverhampton
West Midlands VV1 2EN
Fencing
Tel: 01902 490919
Fax: 01902 490929
Website: www.steelway.co.uk
E-mail: sales@fensecure.co.uk

Sugg Lighting Ltd
Sussex Manor Business Park
Gatwick Road
Crawley
West Sussex RH10 9GD
Lighting
Tel: 01293 540111
Fax: 01293 540114
Website: www.sugglighting.co.uk
E-mail: sales@sugglighting.co.uk

Tarmac Southern
London Sales Office
24 Park Royal Road
London NW10 7JW
Ready mixed concrete
Tel: 020 8965 1864
Fax: 020 8965 1863

Tensar International (Netlon Group)
New Wellington Street
Blackburn
Lancashire BB2 4PJ
Erosion control, soil stabilisation
Tel: 01254 262431
Fax: 01254 266868
Website: www.tensar.co.uk
E-mail: sales@tensar.co.uk

Terram Limited
Mamhilad Park Industrial Estate
Pontypool
Gwent NP4 0YR
Geofabrics
Tel: 01495 757722
Fax: 01495 762383
Website: www.terram.com
E-mail: info@terram.co.uk

Terranova Lifting Ltd
Terranova House
Bennett Road
Reading
Berks RG2 OQX
Crane hire
Tel: 0118 931 2345
Fax: 0118 931 4114
Website: www.terranova-lifting.co.uk
E-mail: cranes@terranovagroup.co.uk

Townscape Products Ltd
Fulwood Road South
Sutton-in-Ashfield
Nottinghamshire NG17 2JZ
Hard landscaping
Tel: 01623 513355
Fax: 01623 440267
Website: www.townscape-products.co.uk
E-mail: sales@townscape-products.co.uk

Tubex Ltd
Aberaman Park
Aberdare
Mid Glamorgan CF44 6DA
Plant protection, tree guards
Tel: 01685 888000
Fax: 01685 888001
Website: www.tubex.com
E-mail: plantcare@tubex.com

Turf Management Systems
Dromenach Farm
Seven Hills Road
Iver Heath
Buckinghamshire SL0 0PA
Specialist turf systems
Tel: 01895 834411
Fax: 01895 834892

Wavin Plastics Ltd
Parsonage Way
Chippenham
Wiltshire SN15 5PN
Drainage products
Tel: 01249 766600
Fax: 01249 443286
Website: www.wavin.co.uk
E-mail: info@wavin.co.uk

White Horse Contractors Ltd
Blakes Oak Farm
Lodge Hill
Abingdon
Oxfordshire OX14 2JD
Drainage
Tel: 01865 736272
Fax: 01865 326176
Website: www.whitehorsecontractors.co.uk
E-mail: whc@whitehorsecontractors.co.uk

Wicksteed Leisure Ltd
Digby Street
Kettering
Northamptonshire NN16 8YJ
Play equipment
Tel: 01536 517028
Fax: 01536 410633
Website: www.wicksteed.co.uk
E-mail: sales@wicksteed.co.uk

Woodscape Ltd
Church Works
Church Street
Church
Lancashire BB5 4JT
Street furniture
Tel: 01254 383322
Fax: 01254 381166
Website: www.woodscape.co.uk
E-mail: sales@woodscape.co.uk

Wybone Ltd
Mason Way
Platts Common Industrial Estate
Hoyland
Barnsley
South Yorkshire S74 9TF
Street furniture
Tel: 01226 744010
Fax: 01226 350105
Website: www.wybone.co.uk
E-mail: sales@wybone.co.uk

Wyckham Blackwell
Old Station Road
Hampton in Arden
Solihull
West Midlands B92 0HB
Timber decking
Tel: 01675 442233
Fax: 01675 442227
Website: www.wyckham-blackwell.co.uk
E-mail: info@wyckham-blackwell.co.uk

Yeoman Aggregates Ltd
Stone Terminal
Horn Lane
Acton
London W3 9EH
Aggregates
Tel: 020 8896 6820
Fax: 020 8896 6811
E-mail: yeoman.sales@ukonline.co.uk

Common Arrangement of Work Sections

The main work sections relevant to landscape work and their grouping:

A   **Preliminaries/General Conditions**

A10	Project particulars
A11	Tender and contract documents
A12	The site/existing buildings
A13	Description of the work
A20	The contract/sub-contract
A30	Employers requirements: tendering/sub-letting/supply
A31	Employers requirements: provision, content and use of documents
A32	Employers requirements: management of the works
A33	Employers requirements: quality standards/control
A34	Employers requirements: security/safety/protection
A35	Employers requirements: specific limitations on method/sequence/timing/use of site
A36	Employers requirements: facilities/temporary works/services
A37	Employers requirements: operation/maintenance of the finished building
A40	Contractors general cost items: management and staff
A41	Contractors general cost items: site accommodation
A42	Contractors general cost items: services and facilities
A43	Contractors general cost items: mechanical plant
A44	Contractors general cost items: temporary works
A50	Works/products by/on behalf of the employer
A51	Nominated sub-contractors
A52	Nominated suppliers
A53	Work by statutory authorities/undertakers
A54	Provisional work
A55	Dayworks
A60	Preliminaries/general conditions for demolition contract
A61	Preliminaries/general conditions for investigation/survey contract
A62	Preliminaries/general conditions for piling/embedded retaining wall contract
A63	Preliminaries/general conditions for landscape contract
A70	General specification requirements for work package

B   **Complete buildings/structures/units**

B10	Prefabricated buildings/structures
B11	Prefabricated building units

C   **Existing site/buildings/services**

C10	Site survey
C11	Ground investigation
C12	Underground services survey

D   **Groundwork**

D11	Soil stabilisation
D20	Excavating and filling
D41	Crib walls/gabions/reinforced earth

E   **In situ concrete/large precast concrete**

E05	In situ concrete construction generally
E10	Mixing/casting/curing in situ concrete
E20	Formwork for in situ concrete
E30	Reinforcement for in situ concrete

F   Masonry

F10	Brick/block walling
F20	Natural stone rubble walling
F21	Natural stone ashlar walling/dressings
F22	Cast stone ashlar walling/dressings
F30	Accessories/sundry items for brick/block/stone walling
F31	Precast concrete sills/lintels/copings/features

G   Structural/carcassing metal/timber

G31	Prefabricated timber unit decking

H   Cladding/covering

H51	Natural stone slab cladding/features
H52	Cast stone slab cladding/features

J   Waterproofing

J10	Specialist waterproof rendering
J20	Mastic asphalt tanking/damp proofing
J21	Mastic asphalt roofing/insulation/finishes
J22	Proprietary roof decking with asphalt finish
J30	Liquid applied tanking/damp proofing
J31	Liquid applied waterproof roof coatings
J40	Flexible sheet tanking/damp proofing
J44	Sheet linings for pools/lakes/waterways

M   Surface finishes

M10	Cement:sand/Concrete screeds
M20	Plastered/rendered/roughcast coatings
M40	Stone/concrete/quarry/ceramic tiling/mosaic
M60	Painting/clear finishing

P   Building fabric sundries

P30	Trenches/pipeways/pits for buried engineering services

Q   Paving/planting/fencing/site furniture

Q10	Kerbs/edgings/channels/paving accessories
Q20	Granular sub-bases to roads/pavings
Q21	In situ concrete roads/pavings
Q22	Coated macadam/asphalt roads/pavings
Q23	Gravel/hoggin/woodchip roads/pavings
Q24	Interlocking brick/block roads/pavings
Q25	Slab/brick/sett/cobble pavings
Q26	Special surfacings/pavings for sport/general amenity
Q30	Seeding/turfing
Q31	Planting
Q32	Planting in special environments
Q35	Landscape maintenance
Q40	Fencing
Q50	Site/street furniture/equipment

R   Disposal systems

R12	Drainage below ground
R13	Land drainage

S   Piped supply systems

S10	Cold water
S14	Irrigation
S15	Fountains/water features

V   Electrical supply/power/lighting systems

V41	Street/area floodlighting

Architects' Guide to Fee Bidding

M. Paul Nicholson

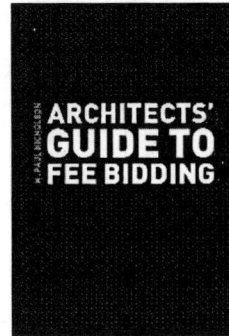

The architectural profession is renowned for poor levels of income, often exacerbated by the fact that practitioners most generally have to rely on guesswork when it comes to estimating for fees. This book actually introduces practicing architects, architectural managers and senior students, however, to the philosophy and practice of analytical estimating for fees. By means of a detailed case study it illustrates the many problems that may be encountered and is presented as a step-by-step guide and source of reference to successful bidding. A detailed discussion of the philosophy of design management and architectural management is developed as a backdrop to the preparation of a bid as it leads the reader through the mysteries of converting the calculation of a bid into a serious tender. This unique text is an essential guide for all practitioners, particularly those at the commencement of their careers and RIBA (Royal Institute of British Architects) Part 3 students. Indeed it will be of importance to all constructional professionals who operate within a highly competitive market.

September 2002: 246x189 mm: 176 pages
HB: 0-415-27335-8: £80.00
PB: 0-415-27336-6: £28.99

To Order: Tel: +44 (0) 1264 343071 Fax: +44 (0) 1264 343005, or
Post: Taylor and Francis Customer Services, Thomson Publishing Services, Cheriton House, Andover, Hants, SP10 5BE, UK Email: book.orders@tandf.co.uk

For a complete listing of all our titles visit:
www.sponpress.com

Taylor & Francis
Taylor & Francis Group plc

The Construction Sector in the Asian Economies

Michael Anson, Y. H. Chiang and John Raftery

A collection of essential data on 11 Asian economies, outlining new trends and highlighting increasing differences between developed and developing countries. Features a detailed analysis of the state of the construction industry and its economic effects in Australia, China Mainland, China Hong Kong, India, Indonesia, Japan, South Korea, Philippines, Singapore, Sri Lanka and Vietnam.

Foreword 1. Regional Overview 2. Australia 3. China Mainland 4. China Hong Kong 5. India 6. Indonesia 7. Japan 8. Singapore 9. South Korea 10. Sri Lanka 11. Vietnam.

November 2004: 246x174 mm: 512 pages
HB: 0-415-28613-1: £85.00

To Order: Tel: +44 (0) 1264 343071 Fax: +44 (0) 1264 343005, or
Post: Taylor and Francis Customer Services, Thomson Publishing Services, Cheriton House, Andover, Hants, SP10 5BE, UK Email: book.orders@tandf.co.uk

Part 1

How to Use This Book

1.1   INTRODUCTION

First-time users of *Spon's Landscape and External Works Price Book* and others who may not be familiar with the way in which prices are compiled may find it helpful to read this section before starting to calculate the costs of landscape works.

The cost of an item of landscape construction or planting is made up of many components:

- the cost of the product;
- the labour and additional materials needed to carry out the job;
- the cost of running the contractor's business.

These are described more fully below.

1.2   IMPORTANT NOTES ON THE PROFIT ELEMENT OF RATES IN THIS BOOK

The rates shown in the Measured Works and Approximate Estimates sections of this book do not generally contain a profit element unless the rate has been provided by a sub-contractor.

Analysed Rates Versus Sub-contractor Rates

As a general rule if a rate is shown as an analysed rate in the Measured Works section, i.e. it has figures shown in the columns other than the "Unit" and "Total Rate" column, it can be assumed that this has no profit or overhead element and that this calculation is shown as a direct labour/material supply/plant, task being performed by a contractor.

On the other hand if a rate is shown as a Total Rate only, this would normally be a sub-contractors rate and would contain the profit and overhead for the sub-contractor.

The foregoing applies for the most part to the Approximate Estimates section. However in some items there may be an element of sub-contractor rates within direct works build-ups.

As an example of this, to excavate, lay a base and place a macadam surface; a general landscape contractor would normally perform the earthworks and base installation. The macadam surfacing would be performed by a specialist sub-contractor to the landscape contractor.

The Approximate Estimate for this item uses rates from the Measured Works section to combine the excavation, disposal, base material supply and installation with all associated plant. There is no profit included on these elements. The cost of the surfacing, however, as supplied to us in the course of our annual price enquiry from the macadam sub-contractor, would include the sub-contractors profit element.

The landscape contractor would add this to his section of the work but would normally apply a mark up at a lower percentage to the sub-contract element of the composite task.

Users of this book should therefore allow for the profit element at the prevailing rates. Please see the worked examples below and the notes on overheads for further clarification.

1.3   INTRODUCTORY NOTES ON COSTING IN THIS BOOK

The Prices for Measured Work are intended to apply to a medium-sized contract of about 6000 m2 and a value of between £25 000.00 – £500 000.00 in the outer London area, and assume that 50% is hard surfacing and 50% is soft landscaping and planting. Similarly it has been necessary to assume that the work is undertaken as a main contract, but if the work is let as a sub-contract, consideration should be given to the need for the addition of main contractor's discount and profit. Additionally for purposes of this book it is assumed that all materials are delivered in full loads.

As explained in more detail later the prices are generally based on wage rates and material costs current at Spring 2005. They do not allow for preliminary items, which are dealt with in the Preliminaries section of this book, or for any Value Added Tax which may be payable.

Adjustments should be made to standard rates for time, location, local conditions, site constraints and any other factors likely to affect the costs of a specific scheme.

Term contracts for general maintenance of large areas should be executed at rates somewhat lower than those given in this section.

There is now a facility available to readers that enables a comparison to be made between the level of prices given and those for projects carried out in regions other than outer London; this is dealt with in Section 2.6.

The units of measurement have been varied to suit the type of work and care should be taken when using any prices to ascertain the basis of measurement adopted.

The prices per unit of area for executing various mechanical operations are for work in the following areas and under the following conditions.

Prices per m^2 relate to areas not exceeding 100 m^2
(any plan configuration)

Prices per 100 m^2 relate to areas exceeding 100 m^2 but not exceeding 1/4 ha
(generally clear areas but with some sub-division)

Prices per ha   relate to areas over 1/4 ha
(clear areas suitable for the use of tractors and tractor-operated equipment)

The prices per unit area for executing various operations by hand generally vary in direct proportion to the change in unit area.

Measured Works

Prime Cost: Commonly known as the 'PC'. Prime Cost is the actual price of the material item being addressed such as paving, shrubs, bollards or turves, as sold by the supplier. Prime Cost is given 'per square metre', 'per 100 bags' or 'each' according to the way the supplier sells his product. In researching the material prices for the book we requested that the suppliers price for full loads of their product delivered to a site close to the M25 in London. Spon's rates do not include VAT. Some companies may be able to obtain greater discounts on the list prices than those shown in the book. Prime Cost prices for those products and plants which have a wide cost range will be found under the heading of Market Prices in the main sections of this book, so that the user may select the product most closely related to his specification.

Materials: The PC material plus the additional materials required to fix the PC material.
Every job needs materials for its completion besides the product bought from the supplier. Paving needs sand for bedding, expansion joint strips and cement pointing; fencing needs concrete for post setting and nails or bolts; tree planting needs manure or fertilizer in the pit, tree stakes, guards and ties. If these items were to be priced out separately, Spon's Landscape and External Works Price Book (and the Bill of Quantities) would be impossibly unwieldy, so they are put together under the heading of Materials.

Free Updates

with three easy steps…

1. Register today on
   www.pricebooks.co.uk/updates

2. We'll alert you by email when new
   updates are posted on our website

3. Then go to
   www.pricebooks.co.uk/updates
   and download.

All four Spon Price Books – *Architects' and Builders'*, *Civil Engineering and Highway Works*, *Landscape and External Works* and *Mechanical and Electrical Services* – are supported by an updating service. Three updates are loaded on our website during the year, in November, February and May. Each gives details of changes in prices of materials, wage rates and other significant items, with regional price level adjustments for Northern Ireland, Scotland and Wales and regions of England. The updates terminate with the publication of the next annual edition.

As a purchaser of a Spon Price Book you are entitled to this updating service for the 2006 edition – free of charge. Simply register via the website www.pricebooks.co.uk/updates and we will send you an email when each update becomes available.

If you haven't got internet access we can supply the updates by an alternative means. Please write to us for details: Spon Price Book Updates, Spon Press, 2 Park Square, Milton Park, Abingdon, Oxfordshire, OX14 4RN.

Find out more about spon books
Visit www.sponpress.com for more details.

New books from Spon

Labour: This figure covers the cost of planting shrubs or trees, laying paving, erecting fencing etc. and is calculated on the wage rate (skilled or unskilled) and the time needed for the job. Extras such as highly skilled craft work, difficult access, intermittent working and the need for labourers to back up the craftsman all add to the cost. Large regular areas of planting or paving are cheaper to install than smaller intricate areas, since less labour time is wasted moving from one area to another.

Labour Rates Used in this Edition

The rates for labour used in this edition have been based on surveys carried out on a cross section of external works contractors.
These rates include for company overheads such as employee administration, transport insurance, and on costs such as National Insurance. The rates do not include for profit.

Plant, consumable stores and services: This rather impressive heading covers all the work required to carry out the job which cannot be attributed exactly to any one item. It covers the use of machinery ranging from JCB's to shovels and compactors, fuel, static plant, water supply (which is metered on a construction site), electricity and rubbish disposal. The cost of transport to site is deemed to be included elsewhere and should be allowed for elsewhere or as a preliminary item. Hired plant is calculated on an average of 36 hours working time per week

Overheads: An allowance for this is included in the labour rates which are described above. The general overheads of the contract such as insurance, site huts, security, temporary roads and the statutory health and welfare of the labour force are not directly assignable to each item, so they are distributed as a percentage on each, or as a separate preliminary cost item. The contractor's and sub-contractor's profits are not included in this group of costs. Site overheads, which will vary from contract to contract according to the difficulties of the site, labour shortages, inclement weather or involvement with other contractors, have not been taken into account in the build up of these rates, while Overhead (or profit) may have to take into account losses on other jobs and the cost to the contractor of unsuccessful tendering.

Sub-contract rates: Where there is no analysis against an item, this is deemed to be a rate supplied by a sub-contractor. In most cases these are specialist items where most external works contractors would not have the expertise or the equipment to carry out the task described. It should be assumed that sub-contractor rates include for the sub-contractors profit.
An example of this may be found for the Tennis court rates in section Q26.

1.4   ADJUSTMENT AND VARIATION OF THE RATES IN THIS BOOK

It will be appreciated that a variation in any one item in any group will affect the final Measured Work price.
Any cost variation must be weighed against the total cost of the contract and a small variation in Prime Cost where the items are ordered in thousands may have more effect on the total cost than a large variation on a few items, while a change in design that necessitates the use of earth-moving equipment which must be brought to the site for that one job will cause a dramatic rise in the contract cost. Similarly, a small saving on multiple items will provide a useful reserve to cover unforeseen extras.

Worked Examples

A variation in the Prime Cost of an item can arise from a specific quotation.

For example:
The PC of 900 x 600 x 50 mm precast concrete flags is given as £2.93 each which equates to £5.43/m^2; and to which the costs of bedding and pointing are added to give a total material cost of £9.48/m^2.
Further costs for labour and mechanical plant give a resultant price of £16.98/m^2.
If a quotation of £6.00/m^2 was received from a supplier of the flags the resultant price would be calculated as £18.91 less the original cost (£5.43) plus the revised cost (£6.00) to give £17.55/m^2.

A variation will also occur if, for example, the specification changes from 50 light standard trees to 50 extra heavy standard trees. In this case the Prime Cost will increase due to having to buy extra heavy standard trees instead of light standard stock.

Original price: The prime cost of an Acer platanoides bare root standard is given as £11.25, to which the cost of handling and planting, etc is added to give an overall cost of £18.75. Further costs for labour and mechanical plant for mechanical excavation of a 600 x 600 x 600 mm pit (£3.42), a single tree stake (£7.59), importing "Topgrow" compost in 75 litre bags (£1.38) backfilling with imported topsoil (£4.48) and disposing of the excavated material,(£3.64) give a resultant price of £39.26. Therefore the total cost of 50 light standard trees is £1963.00.

Revised price: The prime cost of an Acer platanoides extra heavy standard is given as £40.00, however, taking into account the additional staking, extra labour for handling and planting, a larger tree pit, a larger stake and more topsoil, and this cost gives a total unit price of £95.99. This is significantly more than the light standard version. Therefore the total cost of 50 extra heavy trees is £4799.64 and the additional cost is £2836.64

This example of the effects of changing the tree size, illustrates that caution is needed when revising tender prices as merely altering the prime cost of an item is will not accurately reflect the total cost of the revised item.

1.5   APPROXIMATE ESTIMATES

These are combined Measured Work prices which give an approximate cost for a complete section of landscape work. For example, the construction of a car park comprises excavation, levelling, road-base, surfacing and marking parking bays. Each of these jobs is priced separately in the Measured Works section, but a comprehensive price is given in the Approximate Estimates section, which is intended to provide a quick guide to the cost of the job. It will be seen that the more items that go to make up an approximate estimate, the more possibilities there are for variations in the PC prices and the user should ensure that any PC price included in the estimate corresponds to his specification.

Worked Example

In many instances a modular format has been use in order to enable readers to build up a rate for a required task. The following table describes the trench excavation pipe laying and backfilling operations as contained in section R12 of this book.

Pipe laying 100 m

	£
Excavate for drain 150 mm wide x 450 mm deep inclusive of disposal on site; by machine	94.88
Lay flexible plastic pipe 100 mm wide	186.51
Backfilling with gravel rejects, blinding with sand and topping with 150 mm topsoil	203.65
Total	**485.04**

The figures given in Spon's Landscape and External Works Price Book are intended for general guidance, and if a significant variation appears in any one of the cost groups, the Price for Measured Work should be re-calculated.

1.6   HOW THIS BOOK IS UPDATED EACH YEAR

The basis for this book is a database of Material, Labour, Plant and Sub-contractor resources each with its own area of the database.

Material, Plant and Sub-contract Resources

Each year the suppliers of each material, plant or subcontract item are approached and asked to update their prices to those that will prevail in September of that year.

These resource prices are individually updated in the database. Each resource is then linked to one or many tasks. The tasks in the task library section of the database is automatically updated by changes in the resource library. A quantity of the resource is calculated against the task.

The calculation is generally performed once and the links remain in the database.

On occasions where new information or method or technology are discovered or suggested, these calculations would be revisited. A further source of information is simple time and production observations made during the course of the last year.

Labour Resource Update

Most tasks except those shown as sub-contractor rates (see above) employ an element of labour. The Data Department at Davis Langdon conducts ongoing research into the costs of labour in various parts of the country.

Tasks or entire sections would then be re-examined and recalculated. Comments on the rates published in this book are welcomed and may be submitted to the contact addresses shown in the preface.

Spon's Manual for Educational Premises

Derek Barnsley

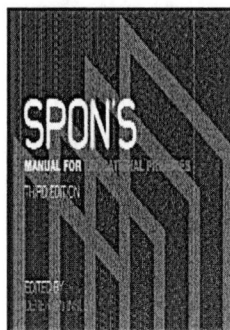

Spon's Manual for Educational Premises is an essential reference book for planning, building, remodelling, repairing, maintaining and managing a wide range of both modern and historic educational premises.

It contains clear, practical guidance for local authorities, head teachers and governing boards responsible for planning, costing, commissioning, supervising and paying for building work and maintenance services. The book is equally essential for architects, surveyors and other professionals and contractors carrying out work in the educational sector.

The third edition has been revised to respond to the evolution of new teaching methods and technology, the development of vocational training and to meet new mandatory regulations relating to children who require special care. The book focuses on cost and performance, including PFI and PPP criteria, but with added emphasis on economic planning to improve standards in design and performance to meet detailed educational needs.

January 2004: 297x210 mm: 464 pages
HB: 0-415-28077-X: £200.00

To Order: Tel: +44 (0) 1264 343071 Fax: +44 (0) 1264 343005, or
Post: Taylor and Francis Customer Services, Thomson Publishing Services, Cheriton House, Andover, Hants, SP10 5BE, UK Email: book.orders@tandf.co.uk

For a complete listing of all our titles visit:
www.sponpress.com

Taylor & Francis
Taylor & Francis Group plc

Understanding JCT Standard Building Contracts
7th Edition

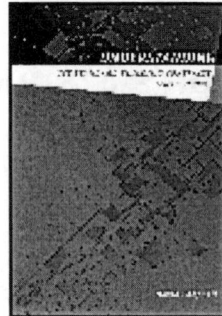

David Chappell

This latest edition of David Chappell's bestselling guide provides an expanded presentation of the Joint Contract Tribunal (JCT) standard contracts, the most common forms of building contract. The JCT Contract With Contractor's Design (WCD 98), also known as 'the design and build form', is now covered alongside the other three major forms of contract in the JCT series: JCT 98, IFC 98 and MW 98.

David Chappell has updated the book in line with amendments to the contracts and recent case law. He avoids legal jargon but writes with authority and precision, in a style which won, for the fifth edition of the book, the first prize in the Best Textbook category of the 1999 Chartered Institute of Building Literary Awards.

Architects, quantity surveyors and contractors should find this a straightforward and practical reference tool arranged by topic.

March 2003: 234x156 mm: 160 pp.
PB: 0-415-30631-0: £18.99

To Order: Tel: +44 (0) 1264 343071 Fax: +44 (0) 1264 343005, or
Post: Taylor and Francis Customer Services, Thomson Publishing Services, Cheriton House, Andover, Hants, SP10 5BE, UK Email: book.orders@tandf.co.uk

For a complete listing of all our titles visit:
www.sponpress.com

Taylor & Francis
Taylor & Francis Group plc

Part 2

Cost Information

2.1 LABOUR RATES USED IN THIS EDITION

Based on surveys carried out on a cross section of external works contractors, the following rates for labour have been used in this edition.
These rates include for company overheads such as employee administration, transport insurance, and on costs such as National Insurance. The rates do not include for profit.

The rates for all labour used in this edition is as follows

General contracting	£ 18.75 / hour
Maintenance contracting	£ 15.00 / hour

2.2 RATES OF WAGES

2.2.1 Building Industry

Authorized rates of wages, etc., in the building industry in England and Wales and Scotland agreed by the Construction Industry Joint Council. Effective from 27 June 2005. Basic pay. Weekly rate based on 39 hours.

Craft Rate £351.00	Skill Rate 1	£334.62	General Operative £264.03
	Skill Rate 2	£322.14	
	Skill Rate 3	£301.47	
	Skill Rate 4	£284.31	

The Prices for Measured Works used in this book are based upon the commercial wage rates currently used in the landscaping industry and are typical of the London area.
However, it is recognized that some contractors involved principally in soft landscaping and planting works may base their wages on the rates determined by the Agricultural Wages Board (England and Wales) and therefore the following information is given to assist readers to adjust the prices if necessary.

2.2.2   Agricultural Wages, England and Wales

Minimum payments agreed by the Agricultural Wages Board for standard 39 hour, five day week. Effective 1 October 2004.

Age	Unit	Basic pay	Craft Grade	Appointment Grade	
				Grade 2	Grade 1
Weekly pay:					
19+	£/ week	210.60	248.43	263.25	284.31
18	£/ week	179.01			
17	£/ week	147.42			
16	£/ week	126.36			
15 & under	£/ week	-			
Overtime pay:					
19+	£/ hour	8.10	9.56	10.13	10.94
18	£/ hour	6.89			
17	£/ hour	5.67			
16	£/ hour	4.86			
15 & under	£/ hour	4.05			

Since 1997 there have been three main types of worker:
- Full time workers (standard and flexible)
- Part time workers (standard and flexible)
- Casual workers

Weekly rates apply to Full Time Standard Workers for 39 hours work.

Hourly rates apply to the hours worked in a week by a Part Time Standard worker.

Overtime rates apply to all overtime hours worked by Standard Workers -
- more than 39 hours in any week
- or more than 8 hours on any day
- or Saturday or Sunday
- or Public Holiday

From 1 June 2000 the minimum rate for holiday pay will be the same as the minimum rate for basic pay.

Copies of the Agricultural Wages Order 2004 (Number 1) may be obtained from

The Agricultural Wages Board for England and Wales
Zone 2C Ergon House
Horseferry Road
London SW1P 2AL

Telephone: 020 7238 6523
Fax: 020 7238 6553
Email: m.agriwages@arp.defra.gsi.gov.uk

Grades and responsibilities

Craft Grade
Must have worked in agriculture for at least 3 out of last 4 years and
hold a valid craft certificate issued by the National Proficiency Tests Council in an approved subject or:
hold a valid Level 3 National Vocational Qualification (NVQ) in an approved subject or:
hold a Craft Certificate issued by an Agricultural Wages Committee based on an Employer's declaration or:
be a qualified former Council apprentice or:
hold a valid qualification as specified in sub-section 3.9 of the Agricultural Wages Order

Appointment Grade 2
Day to day responsibility for supervising the work on a farm or part of it, and implementing management decisions, or responsibility for the instruction and supervision of staff.

Appointment Grade 1
A management responsibility for an entire farm or part of it run as a separate operation or business, or responsibility for employing and disciplining staff.

2.2.3   Agricultural Wages, Scotland

New rules relating to agricultural wages and other conditions of service took effect from 1 January 2005. The new rules are contained in the Agricultural Wages (Scotland) Order No. 52 2004. The Order contains the detailed legal requirements for the calculations of minimum pay etc, but the Scottish Office has produced 'A Guide for Workers and Employers' which attempts to explain the new rules in simpler terms.

Both 'The Agricultural Wages (Scotland) Order No. 52 2004' and 'A Guide for Workers and Employers' may be obtained from:

Scottish Agricultural Wages Board
Pentland House
47 Robb's Loan
Edinburgh EH14 1TY
Telephone: 0131 244 6397
Fax: 0131 244 6551
Email: sawb@scotland.gsi.gov.uk

The new Order applies equally to all workers employed in agriculture in Scotland. It makes no distinction between full-time employees, part-time employees and students.

The old classifications of general farm worker, shepherd, stockworker, tractorman and supervisory grade have been removed and replaced by a single category covering all agricultural workers. The one exception to this is that special arrangements have been made for hill-shepherds.

Instead of minimum weekly rates of pay, the new Order works on minimum hourly rates of pay (with the exception hill-shepherds). It is very important to note that although wages are to be calculated on an hourly basis, a worker is still entitled to be paid for the full number of hours for which he is contracted (unless he is unavailable for work). In particular, this means that an employer cannot reduce the number of hours for which he is to be paid simply by e.g. sending him home early.

Minimum rates of wages

There are two main pay scales. One is for workers who have been with their present employer for not more than 10 weeks and the other for workers who have been with their present employer for more than 10 weeks.

Within these two scales, there are varying minimum hourly rates of pay depending on the age of the worker.

The minimum hourly rates of pay are as follows:

	Minimum Hourly Rates (£)	
Age	Up to 10 weeks continuous employment	More than 10 weeks continuous employment
	£	£
Under 16	1.95	2.61
16 and under 17	3.00	3.15
17 and under 18	3.00	3.66
18 and under 19	4.10	4.46
19 and over	4.85	5.25

Workers who have been with the same employer for more than 10 weeks and hold either:

(a) a Scottish, or National, Vocational Qualification in an agricultural subject at Level III or above, or
(b) an apprenticeship certificate approved by Lantra, NTO (formerly ATB Landbase), or a certificate of acquired experience issued by the ATB Landbase.

shall be paid, for each hour worked, an additional sum of not less than £0.79.

Minimum hourly overtime rate

The minimum hourly rate of wages payable to a worker:

(a) for each hour worked in excess of 8 hours on any day, and
(b) for each hour worked in excess of 39 hours in any week.

shall be calculated in accordance with the following formula:

$M \times 1.5$ (where M = the minimum hourly rate of pay to which the worker is entitled)

2.3   DAYWORK AND PRIME COST

The tables below analyse rates for daywork. The calculations used in the measured works and the derived approximate estimates use current commercial rates for the build-up of all items and do not employ the rates used below. Users should refer to Section 2.1 " Labour Rates Used in this Edition" at the beginning of this section for information on the actual hourly rates used.

When work is carried out which cannot be valued in any other way it is customary to assess the value on a cost basis with an allowance to cover overheads and profit. The basis of costing is a matter for agreement between the parties concerned, but definitions of prime cost for the building industry have been prepared and published jointly by The Royal Institution of Chartered Surveyors and the Construction Confederation (formerly the Building Employers Confederation, formerly the National Federation of Building Trades Employers) for the convenience of those who wish to use them. We publish these items with their kind permission. Reference should be made to *Spon's Architects' and Builders' Price Book* where these documents are reproduced by kind permission of the publishers together with the Schedule of Basic Plant Charges (January 1990 revision).

The calculations on page 12 shows an example of the calculation of typical standard hourly base rates for craft and general operatives prepared in accordance with the Definition of Prime Cost of Daywork carried out under a Building Contract. A building contract will normally provide for a contractor to tender a percentage adjustment for the base rates for incidental costs, overheads and profit.

Daywork Rates - Building Operatives

Rates effective from 27 June 2005

		Rate £	Craftsman £	Rate £	Labourer £
Guaranteed minimum weekly earnings					
Standard Basic Rate	46.2 wks	351.00	16216.20	264.03	12198.19
Guaranteed minimum bonus	46.2 wks	0.00	0.00	0.00	0.00
			16216.20		12198.19
Employer's National Insurance Contribution					
(12.8% after the first £91.01 per week)			1500.48		986.18
			17716.68		13184.37
Employer's Contribution to:					
CITB annual levy at 0.50% of payroll			91.26		68.65
Holiday pay	4.2 wks	351.00	1474.20	264.03	1108.93
Public holidays	1.6 wks	351.00	561.60	264.03	422.45
EasyBuild Stakeholder Pension (Death and accident cover is provided free)	52.0 wks	3.50	182.00	3.50	182.00
Annual labour cost as defined in section 3		£	**20025.74**	£	**14966.40**

Hourly rate of labour as defined in section 3, Clause 3.02 (Annual cost/hours per annum)   = £11.11   = £8.31

Note:

1. Calculated following Definition of Prime Cost of Daywork carried out under a Building Contract, published by the Royal Institution of Chartered Surveyors and the Construction Confederation.

2. Standard basic rate effective from 27 June 2005.

3. Standard working hours per annum calculated as follows:

52 weeks @ 39 hours		2028.00
Less		
4.2 weeks holiday @ 39 hours	163.8	
8 days public holidays @ 7.8 hours	62.4	-226.20
		1801.80

4. All labour costs incurred by the Contractor in his capacity as an employer other than those contained in the hourly rate, are to be taken into account under Section 6 of the RICS definitions for labour costs.

5. The above example is for guidance only and does not form part of the Definition; all the basic costs are subject to re-examination according to the time when and in the area where the daywork is executed.

6. All N.I. payments are at non-contracted out rates applicable from April 2005.

7. Basic rate & GMB number of weeks =
    52.0 weeks - 4.2 weeks annual holiday - 1.6 weeks public holiday = 46.2 weeks

2.4   COMPUTATION OF LABOUR RATES, COST OF MATERIALS AND PLANT

Different organizations will have varying views on rates and costs, which will in any event be affected by the type of job, availability of labour and the extent to which mechanical plant can be used. However this information should assist the reader to:

(1)   compare the prices to those used in his own organization;
(2)   calculate the effect of changes in wage rates or prices of materials;
(3)   calculate analogous prices for work similar to but differing in detail from the examples given.

Computation of labour rates

From 27 June 2005 basic weekly rates of pay for craft and general operatives are £351.00 and £264.03 respectively; to these rates have been added allowances for the items below in accordance with the recommended procedure of the Chartered Institute of Building in its Code of Estimating Practice. The resultant hourly rates are £12.44 and £9.31 for craft operatives and general operatives respectively.

The items referred to above for which allowances have been made are:

- Lost time
- Non-productive overtime
- Sick pay
- Construction Industry Training Board Levy
- Employer's contribution towards holidays with pay
- Employer's contribution towards pension and death benefit scheme
- Employer's contribution towards National Insurance
- Severance pay and sundry costs
- Employer's liability and third party insurance

The tables that follow illustrate how the hourly rates referred to above have been calculated. Productive time has been based on a total of 1801.8 hours worked per year for daywork calculations above and for 1953.54 hours worked per year for the all-in labour rates (including 5 hours per week average overtime) for the all in labour rates below.

BUILDING CRAFT AND GENERAL OPERATIVES

Effective from 27 June 2005

			Craft Operatives		General Operatives	
			£	£	£	£
Wages at standard basic rate						
productive time	44.30	wks	351.00	15547.86	264.03	11695.45
lost time allowance	0.904	wks	351.00	317.30	264.03	238.68
overtime	5.80	wks	526.50	3053.70	396.05	2297.06
				18918.86		14231.19
Extra payments under National Working Rules	45.20	wks		-		-
Sick Pay	1	wk		-		-
CITB allowance (0.50% of payroll)	1	year		106.73		80.29
Holiday Pay	4.2	wks	418.56	1757.95	314.85	1322.37
Public Holiday	1.60	wks	418.56	669.70	314.85	503.76
Travelling allowance (say 30 km)	226	dys	2.22		2.22	
Fare allowance (say 30 km)	226	dys	5.20		5.20	
Employers contributions to						
EasyBuild Stakeholder Pension (Death & Accident cover is provided free)	52	wks	3.50	182.00	3.50	182.00
National Insurance (average weekly payment)	48	wks	38.42	1844.16	25.92	1244.16
				23479.41		17563.77
Severance pay and sundry costs	Plus		1.5%	352.19	1.5%	263.46
				23831.60		17827.23
Employers liability and third party insurance	Plus		2.0%	476.63	2.0%	356.54
Total cost per annum				24308.23		18183.77
Total cost per hour				**12.44**		**9.31**

Notes:

1 Absence due to sickness has been assumed to be for periods not exceeding 3 days for which no payment is due (Working Rule 20.7.3).

2 EasyBuild Stakeholder Pension effective from 1 July 2002. Death and accident benefit cover is provided free of charge. Taken as £7.50/week average as range increased for 2005 wage award.

3 All N.I. Payments are at not-contracted out rates applicable from April 2005.

National Insurance is paid for 48 complete weeks (52wks-4.2wks) based on employer making regular monthly payments into the Template holiday pay scheme and by doing so the employer achieves National Insurance savings on holiday wages.

Calculation of Annual Hours Worked

Number of actual hours worked yearly

Normal working hours (52-6.8 =46.2wks x 39 hrs)   1801.80

Less Sick
1 week   -39.00
                                                                                                                                                          1762.80

Less time lost for inclement weather @ 2%   -35.26
                                                                                                                                                          1727.54

Overtime hours
52wksx5hrs (overtime)   260.00
Less

Annual holidays	4.2wks x 5hrs	21.00	
Public holidays	1.6wks x 5hrs	8.00	
Sickness	1 week x 5hrs	5.00	-34.00

                                                                                                                                                            226.00

Number of actual hours worked yearly   1953.54

COMPUTATION OF LABOUR RATES IN PRICES FOR MEASURED WORK

A survey of typical landscape/ external works companies indicates that they are currently paying above average wages for multi skilled operatives regardless of specialist or supervisory capability. In our current overhaul of labour constants and costs we have departed from our previous policy of labour rates based on national awards and used instead a gross rate per hour for all rate calculations. This rate is tabled at the beginning of this section. Estimators can readily adjust this rate if they feel it inappropriate for their work.

The following section illustrates the calculation of the labour rate of £18.75 per hour used in this year's book.

Calculation of Spon's Landscape and External Works Labour Rate

The productive labour of any organisation must return their salary plus the cost of the administration which supports the labour force.

The Labour rate used in this edition is calculated as follows:

Team size
A three man team with annual salaries as shown and allowances for National insurance, uniforms and site tools plus a company vehicle returns a basic labour rate of £13.38 per hour.

Basic Labour Rate Calculation for Spon's 2006

Working Hours per year (2006) = 1953.5							
	Number of	Nett Salary	NI	Nett Cost	Uniforms	Site Tools	Total 3 Man Team
Foreman	1	21000.00	2499.00	23499.00	50.00	100.00	
Craftsman	1	18000.00	2142.00	20142.00	50.00	100.00	
Labourer	1	14000.00	1666.00	15666.00	50.00	100.00	
Overtime (Hrs)	225	17.49	2.08	4403.62			
1.5 times							
normal rate	225	14.99	1.78	3774.53			
	225	11.66	1.39	2935.74			
Total staff in Team	3.00			70420.89	150.00	300.00	£70,870.89
Vehicle Costs Inclusive of Fuel Insurances etc	**Working Days**	**£ /Day**					
	225.13	36.00			>		£ 8,104.50
						TOTAL	**£78,975.39**

The Basic Average labour rate excluding overhead costs (below) is

$$\frac{£78975.39}{\text{(Total staff in team)} \times \text{(Working hours per year)}}$$

$$= £13.48$$

Overhead costs

Add to the above basic rate the company overhead costs for a small company as per the table below. These costs are absorbed by the number of working men multiplied by the number of working hours supplied in the table below. This then generates an hourly overhead rate which is added to the Nett cost rate above.

The table on the next page summarises these costs for a small to medium sized company.

Illustrative Overhead for Small Company Employing 9 - 18 Landscape Operatives

Cost Centre	Number of	Cost	Total
MD salary only; excludes profits	1.0	40000.00	40000.00
Contracts managers salaries and vehicles	1.0	30000.00	30000.00
Secretary	1.0	15000.00	15000.00
Book keeper	1.0	22000.00	22000.00
Rental	12.0	500.00	6000.00
Insurances	1.0	4000.00	4000.00
Telephone and mobiles	12.0	200.00	2400.00
Office equipment	1.0	2000.00	2000.00
Stationary	12.0	50.00	600.00
Advertising	12.0	100.00	1200.00
Other vehicles not allocated to contract teams	3.0	8000.00	24000.00
Other consultants	1.0	5000.00	5000.00
Accountancy	1.0	800.00	800.00
Lights & water	1.0	1000.00	1000.00
Other	1.0	6000.00	6000.00
TOTAL OFFICE OVERHEAD			160000.00

The final labour rate used is then generated as follows:

Nett labour rate   + $\dfrac{\text{Total Office Overhead}}{(\text{Working Hours per year x Labour resources employed})}$

Total nr of site staff	Admin Cost per hour	Total Rate per Man Hour
6	13.65	27.13
9	9.10	22.58
12	6.83	20.31
15	5.46	18.94
18	4.55	18.03

Hourly Labour Rates used for Spon's Landscape and External Works 2006

General contracting	£18.75
Maintenance contracting	£15.00

COMPUTATION OF THE COST OF MATERIALS

Percentages of default waste are placed against material resources within the supplier database. These range from 2.5% (for bricks) to 20% for topsoil.
An allowance for the cost of unloading, stacking etc. should be added to the cost of materials.

The following are typical hours of labour for unloading and stacking some of the more common building materials.

Material	Unit	Labourer hour
Cement	tonne	0.67
Lime	tonne	0.67
Common bricks	1000	1.70
Light engineering bricks	1000	2.00
Heavy engineering bricks	1000	2.40

COMPUTATION OF MECHANICAL PLANT COSTS

Plant used within resource calculations in this book have been used according to the following principles:

1. Plant which is non-pedestrian is fuelled by the Plant Hire company.

2. Operators of excavators above the size of 5 tonnes are provided by the Plant hire company and no labour calculation is allowed for these operators within the rate.

3. Three to five tonne excavators and dumpers are operated by the landscape contractor and additional labour is shown within the calculation of the rate.

4. Small plant is operated by the landscape contractor.

5. No allowance for delivery or collection has been made within the rates shown.

6. Downtime is shown against each plant type either for non-productive time or mechanical down time.

The following tables lists the plant resources used in this years book.

Prices shown reflect 32 working hours per week.

Plant Supplied by Road Equipment Ltd

Description	Unit	Supply Quantity	Rate Used £	Down Time %
Excavator 360 Tracked 21 tonne *fuelled and operated*	hour	1	41.25	20
Excavator 360 Tracked 5 tonne *fuelled only*	hour	1	10.93	20
Excavator 360 Tracked 7 tonne *fuelled only*	hour	1	13.13	20
Mini Excavator JCB 803 Rubber Tracks Self Drive *fuelled only*	hour	1	8.75	20
Mini Excavator JCB 803 Steel Tracks Self Drive *fuelled only*	hour	1	7.50	20
JCB 3Cx 4x4 Sitemaster, *fuelled and operated*	hour	1	30.00	20
JCB 3Cx 4x4 Sitemaster + Breaker *fuelled and operated*	hour	1	31.25	20
Dumper 3 tonne, Thwaites Self Drive *fuelled only*	hour	1	3.44	20
Dumper 5 tonne, Thwaites Self Drive *fuelled only*	hour	1	4.69	20
Dumper 6 tonne, Thwaites Self Drive *fuelled only*	hour	1	5.00	20
Manitou Telehandler *fuelled only*	hour	1	18.75	40

Plant Supplied by HSS Hire

Description	Unit	Supply Quantity	Supply Cost £	Rate Used £	Down Time %
Auger Mechanical 2 Man Weekly Rate	hour	40	80.00	3.20	37.5
Engine Poker Vibrator	hour	40	40	1.60	37.5
Vibrating Plate Compactor	hour	40	60	2.40	37.5
Petrol Masonry Saw Bench	hour	40	110.00	4.40	37.5
Diamond Blade Consumable, 230 mm	mm	1	8.00	8.00	-
Access Platform To 5.2 M	hour	40	139.00	27.80	-
Cutting Torch Oxy Acetylene	hour	40	95.00	5.94	80
Hydraulic Breaker Diesel 5Hrs /Day	hour	40	158.00	6.32	37.5
Rotavator	hour	40	200.00	8.00	37.5
Rotavator Tractor Mounted 1200mm + Operator	hour	1	26.00	26.00	-
Compactor Bomag 136Kg/ 10.1Kn 4 Hrs /Day	week	40	110.00	4.67	25
Compactor Bomag 470 Kg /18.7Kn 4 Hrs /Day	week	40	140.00	5.97	25

2.5 COST INDICES

The purpose of this section is to show changes in the cost of carrying out landscape work (hard surfacing and planting) since 1990. It is important to distinguish between costs and tender prices: the following table reflects the change in cost to contractors but does not necessarily reflect changes in tender prices. In addition to changes in labour and material costs, which are reflected in the indices given below, tender prices are also affected by factors such as the degree of competition at the time of tender and in the particular area where the work is to be carried out, the availability of labour and materials, and the general economic situation. This can mean that in a period when work is scarce tender prices may fall despite the fact that costs are rising, and when there is plenty of work available, tender prices may increase at a faster rate than costs.

The Constructed Cost Index

A Constructed Cost Index based on PSA Price Adjustment Formulae for Construction Contracts (Series 2). Cost indices for the various trades employed in a building contract are published monthly by HMSO and are reproduced in the technical press.

The indices comprise 49 Building Work indices plus seven 'Appendices' and other specialist indices. The Building Work indices are compiled by monitoring the cost of labour and materials for each category and applying a weighting to these to calculate a single index.

Although the PSA indices are prepared for use with price-adjustment formulae for calculating reimbursement of increased costs during the course of a contract, they also present a time series of cost indices for the main components of landscaping projects. They can therefore be used as the basis of an index for landscaping costs.

The method used here is to construct a composite index by allocating weightings to the indices representing the usual work categories found in a landscaping contract, the weightings being established from an analysis of actual projects. These weightings totalled 100 in 1976 and the composite index is calculated by applying the appropriate weightings to the appropriate PSA indices on a monthly basis, which is then compiled into a quarterly index and rebased to 1976 = 100.

Constructed Landscaping (Hard Surfacing and Planting) Cost Index

Based on approximately 50% soft landscaping area and 50% hard external works.
1976 = 100

Year	First Quarter	Second Quarter	Third Quarter	Fourth quarter	Annual average
1990	330	334	353	357	344
1991	356	356	364	364	360
1992	364	365	376	378	371
1993	380	381	383	384	382
1994	387	388	394	397	391
1995	399	404	413	413	407
1996	417	419	428	430	424
1997	430	432	435	442	435
1998	442	445	466	466	455
1999	466	468	488	490	478
2000	493	496	513	514	504
2001	512	514	530	530	522
2002	531	540	573	574	555
2003	577	582	603	602	591
2004	602	610	639	639	623
2005	640	644*			

* Provisional

This index is updated every quarter in Spon's Price Book Update. The updating service is available, free of charge, to all purchasers of Spon's Price Books. (Complete the reply-paid card enclosed)

2.6 REGIONAL VARIATIONS

Prices in Spon's Landscape and External Works Price Book are based upon conditions prevailing for a competitive tender in the outer London area. For the benefit of readers, this edition includes regional variation adjustment factors which can be used for an assessment of price levels in other regions.

Special further adjustment may be necessary when considering city centre or very isolated locations.

Region	Adjustment Factor
Outer London	1.00
Inner London	1.10
South East	0.97
Midlands	0.86
East Anglia	0.93
Northern	0.85
North West	0.92
Scotland	0.85
Wales	0.87
Northern Ireland	0.65
Channel Islands	1.32

The following example illustrates the adjustment of prices for regions other than outer London, by use of regional variation adjustment factors.

		£
A.	Value of items priced using Spon's Landscape and External Works Price Book	100 000
B.	Adjustment to value of A. to reflect Midlands Region price level 100 000 x 0.86	86 000

www.davislangdon.com

DAVIS LANGDON

Maximising value and reducing risk for clients investing in infrastructure, construction and property

managed
solutions

Project Management | Cost Management | Management Consulting | Legal Support | Specification Consulting | Engineering Services | Property Tax & Finance

DAVIS LANGDON

EUROPE & MIDDLE EAST
office locations

ENGLAND

DAVIS LANGDON

LONDON
Mid City Place
71 High Holborn
London WC1V 6QS
Tel: (020) 7061 7000
Fax: (020) 7061 7061
Email: neill.morrison@davislangdon.com

BIRMINGHAM
75-77 Colmore Row
Birmingham
B3 2HD
Tel: (0121) 710 1100
Fax: (0121) 710 1399
Email: david.daly@davislangdon.com

BRISTOL
St Lawrence House
29/31 Broad Street
Bristol BS1 2HF
Tel: (0117) 927 7832
Fax: (0117) 925 1350
Email: alan.francis@davislangdon.com

CAMBRIDGE
36 Storey's Way
Cambridge
CB3 0DT
Tel: (01223) 351 258
Fax: (01223) 321 002
Email: laurence.brett@davislangdon.com

LEEDS
No 4 The Embankment
Victoria Wharf
Sovereign Street
Leeds LS1 4BA
Tel: (0113) 243 2481
Fax: (0113) 242 4601
Email: duncan.sissons@davislangdon.com

LIVERPOOL
Cunard Building
Water Street
Liverpool L3 1JR
Tel: (0151) 236 1992
Fax: (0151) 227 5401
Email: andrew.stevenson@davislangdon.com

MAIDSTONE
11 Tower View
Kings Hill
West Malling
Kent ME19 4UY
Tel: (01732) 840 429
Fax: (01732) 842 305
Email: nick.leggett@davislangdon.com

MANCHESTER
Cloister House
Riverside
New Bailey Street
Manchester M3 5AG
Tel: (0161) 819 7600
Fax: (0161) 819 1818
Email: paul.stanion@davislangdon.com

MILTON KEYNES
Everest House
Rockingham Drive
Linford Wood
Milton Keynes
MK14 6LY
Tel: (01908) 304 700
Fax: (01908) 660 059
Email: kevin.sims@davislangdon.com

NORWICH
63 Thorpe Road
Norwich NR1 1UD
Tel: (01603) 628 194
Fax: (01603) 615 928
Email: michael.ladbrook@davislangdon.com

OXFORD
Avalon House
Marcham Road
Abingdon
Oxford OX14 1TZ
Tel: (01235) 555 025
Fax: (01235) 554 909
Email: paul.coomber@davislangdon.com

PETERBOROUGH
Clarence House
Minerva Business Park
Lynchwood
Peterborough PE2 6FT
Tel: (01733) 362 000
Fax: (01733) 230 875
Email: stuart.bremner@davislangdon.com

PLYMOUTH
1 Ensign House
Parkway Court
Longbridge Road
Plymouth PL6 8LR
Tel: (01752) 827 444
Fax: (01752) 221 219
Email: gareth.steventon@davislangdon.com

SOUTHAMPTON
Brunswick House
Brunswick Place
Southampton SO15 2AP
Tel: (023) 8033 3438
Fax: (023) 8022 6099
Email: chris.tremellen@davislangdon.com/
peter.boote@davislangdon.com

**DAVIS LANGDON
LEGAL SUPPORT**
Mid City Place
71 High Holborn
London WC1V 6QS
Tel: (020) 7061 7000
Fax: (020) 7061 7061
Email: mark.hackett@davislangdon.com

**DAVIS LANGDON
CONSULTANCY**
Mid City Place
71 High Holborn
London WC1V 6QS
Tel: (020) 7061 7007
Fax: (020) 7061 7005
Email: john.connaughton@davislangdon.com

**DAVIS LANGDON
SCHUMANN SMITH**
Southgate House
St Georges Way
Stevenage
Hertfordshire SG1 1HG
Tel: (01438) 742 642
Fax: (01438) 742 632
Email: nick.schumann@schumannsmith.com

**DAVIS LANGDON
MOTT GREEN & WALL**
Mid City Place
71 High Holborn
London WC1V 6QS
Tel: (020) 7061 7777
Fax: (020) 7061 7009
Email: general@mottgreenwall.co.uk

**DAVIS LANGDON
CROSHER & JAMES**
Mid City Place
71 High Holborn
London WC1V 6QS
Tel: (020) 7061 7077
Fax: (020) 7061 7078
Email: tony.llewellyn@crosherjames.com

BIRMINGHAM
102 New Street
Birmingham B2 4HQ
Tel: (0121) 632 3600
Fax: (0121) 632 3601
Email: clive.searle@crosherjames.com

CARDIFF
4 Piershead Street
Capital Waterside
Cardiff
CF10 4QP
Tel: (029) 2049 7497
Fax: (029) 2049 7111
Email: michael.murraym@crosherjames.com

EDINBURGH
39 Melville Street
Edinburgh
EH3 7JF
Tel: (0131) 220 4225
Fax: (0131) 220 4226
Email: ian.mcfarlane@crosherjames.com

GLASGOW
Monteith House
11 George Square
Glasgow
G2 1DY
Tel: (0141) 248 0333
Fax: (0141) 248 0313
Email: fraserk@nbwcrosherjames.com

MANCHESTER
Cloister House
Riverside
New Bailey Street
Manchester M3 5AG
Tel: (0161) 819 7600
Fax: (0161) 819 1818
Email: sharmas@nbwcrosherjames.com

SOUTHAMPTON
Brunswick House
Brunswick Place
Southampton SO15 2AP
Tel: (023) 8068 2800
Fax: (023) 8033 6360
Email: reesd@nbwcrosherjames.com

SCOTLAND

DAVIS LANGDON

GLASGOW
Monteith House
11 George Square
Glasgow G2 1DY
Tel: (0141) 248 0300
Fax: (0141) 248 0303
Email:
sam.mackenzie@davislangdon.com

EDINBURGH
39 Melville Street
Edinburgh
EH3 7JF
Tel: (0131) 240 1350
Fax: (0131) 240 1399
Email: erland.rendall@davislangdon.com

WALES

CARDIFF
4 Pierhead Street
Capital Waterside
Cardiff CF10 4QP
Tel: (029) 2049 7497
Fax: (029) 2049 7111
Email: paul.edwards@davislangdon.com

IRELAND

DAVIS LANGDON PKS

DUBLIN
24 Lower Hatch Street
Dublin 2
Ireland
Tel: (00 353 1) 676 3671
Fax: (00 353 1) 676 3672
Email: mwebb@dlpks.ie

GALWAY
Heritage Hall
Kirwan's Lane
Galway, Ireland
Tel: (00 353 91) 530 199
Fax: (00 353 91) 530 198
Email: joregan@dlpks.ie

LIMERICK
8 The Crescent
Limerick
Ireland
Tel: (00 353 61) 318 870
Fax: (00 353 61) 318 871
Email: cbarry@dlpks.ie

SPAIN

DAVIS LANGDON EDETCO

BARCELONA
C/Muntaner, 479, 12"
Barcelona 08021
Spain
Tel: (00 34 93) 418 6899
Fax: (00 34 93) 211 0003
Email: fmonells@barcelona.edetco.com

GIRONA
C/Salt 10
Girona 17005
Spain
Tel: (00 34 97) 223 8000
Fax: (00 34 97) 224 2661
Email: girona@girona.edetco.com

FRANCE

DAVIS LANGDON
5 Rue St Germain l'Auxerrois
75001 Paris
France
Tel: (00 33 1) 5340 9480
Fax: (00 33 1) 5340 9481
Email: andrew.richardson@dleparis.com

POLAND

DAVIS LANGDON
Warsaw Trade Tower
ul. Chlodna 51, 26th Floor
00-867 Warsaw, Poland
Tel: (00 48 22) 455 39 00
Fax: (00 48 22) 455 39 01
Email: warsaw@davislangdon-polska.pl

RUSSIA

DAVIS LANGDON
Office 5
Myasnitskaya
Moscow, 101000
Russia
Tel: (00 7 095) 933 7810
Fax: (00 7 095) 933 7811
Email: stephen.thomas@davislangdon.com

MIDDLE EAST

DAVIS LANGDON
PO Box 13-5422-Shouran
Beirut
Lebanon
Tel: (00 9611) 780 111
Fax: (00 9611) 809 045
Email: DLL.MI@cyberia.net.lb

ARABIAN GULF

DAVIS LANGDON

BAHRAIN
3rd Floor Building 256
Road No 3605
Area No 336
PO Box 640, Manama
State of Bahrain
Arabian Gulf
Tel: (00 973) 1782 7567
Fax: (00 973) 1772 8257
Email: david.galbraith@davislangdon-bahrain.com

UNITED ARAB EMIRATES
PO Box 7856
Office 410
Oud Metha Office Building
Dubai, UAE
Tel: (00 9714) 32 42 919
Fax: (00 9714) 32 42 838
Email: neil.taylor@davislangdon-dubai.com

QATAR
PO Box 3206, Doha
State of Qatar
Tel: (00 974) 4580 150
Fax: (00 974) 4697 905
Email: david.craig@davislangdon-qatar.com

EGYPT
35 Misr Helwan Road
Maadi 11431
Cairo
Egypt
Tel: (00 20 2) 526 2319
Fax: (00 20 2) 527 1338
Email: dlegypt@link.net

Specialist Service Lines
Project Management | Cost Management | Management Consulting | Legal Support | Specification Consulting | Engineering Services | Property Tax & Finance

Specialist Sectors
Arts | Commercial Offices | Distribution | Education | Food Processing | Health | Heritage | Hotels & Leisure | Industrial | Infrastructure | Public Buildings | Regeneration | Residential | Retail | Sports | Transportation

Davis Langdon LLP is a member firm of Davis Langdon & Seah International, with offices throughout Europe and the Middle East, Asia, Australasia, Africa and the USA

Spon's International Construction Costs Handbook

This practical series of five easy-to-use Handbooks gathers together all the essential overseas price information you need. The Hand-books provide data on a country and regional basis about economic trends and construction indicators, basic data about labour and materials' costs, unit rates (in local currency), approximate estimates for building types and plenty of contact information.

Spon's African Construction Costs Handbook
Countries covered: Algeria, Cameroon, Chad, Cote d'Ivoire, Gabon, The Gambia, Ghana, Kenya, Liberia, Nigeria, Senegal, South Africa, Zambia
2005: 234x156: 368 pp Hb: 0-415-36314-4: £170.00

Spon's Latin American Construction Costs Handbook
Countries covered: Argentina, Brazil, Chile, Colombia, Ecuador, French Guiana, Guyana, Mexico, Paraguay, Peru, Suriname, Uruguay, Venezuela
2000: 234x156: 332 pp Hb: 0-415-23437-9: £85.00

Spon's Middle East Construction Costs Handbook
Countries covered: Bahrain, Eqypt, Iran, Jordan, Kuwait, Lebanon, Oman, Quatar, Saudi Arabia, Syria, Turkey, UAE
2005: 234x156: 384 pp Hb: 0-415-36315-2: £170.00

Spon's European Construction Costs Handbook
Countries covered: Austria, Belgium, Cyprus, Czek Republic, Finland, France, Greece, Germany, Italy, Ireland, Netherlands, Portugal, Poland, Slovak Republic, Spain, Turkey
2000: 234x156: 332 pp Hb: 0-419-25460-9: £170.00

Spon's Asia Pacific Construction Costs Handbook
Countries covered: Australia, Brunei Darassalem, China, Hong Kong, India, Japan, New Zealand, Indonesia, Malaysia, Philippines, Singapore, South Korea, Sri Lanka, Taiwan, Thailand, Vietnam
2000: 234x156: 332 pp Hb: 0-419-25470-6: £170.00

To Order: Tel: +44 (0) 1264 343071 Fax: +44 (0) 1264 343005, or
Post: Taylor and Francis Customer Services, Thomson Publishing Services, Cheriton House, Andover, Hants, SP10 5BE, UK Email: book.orders@tandf.co.uk

For a complete listing of all our titles visit:
www.sponpress.com

Taylor & Francis
Taylor & Francis Group plc

Part 3

Preliminaries

3.1 PRELIMINARIES/GENERAL CONDITIONS

The number of items priced in the Preliminaries section of Tender Documents and the manner in which they are priced vary considerably between Contractors. Some Contractors, by modifying their percentage factor for overheads and profit, attempt to cover the costs of Preliminary items in their Prices for Measured Work. However, the cost of Preliminaries will vary widely according to job size and complexity, site location, accessibility, degree of mechanization practicable, position of the Contractor's head office and relationships with local labour/domestic sub-contractors. It is therefore usually far safer to price Preliminary items separately on their merits according to the job.

Reference should be made to *Spon's Architects' and Builders' Price Book* for the normal clause descriptions from the Preliminaries section of a bill of quantities where the *JCT Standard Form of Building Contract* 1980 Edition is to be used.

For the convenience of readers the clause descriptions which are likely to be found in the Preliminaries section of the tender documents for a project which consists of solely or mainly soft landscaping works and is therefore to be based on the *JCLI Form of Agreement for Landscape Works* 1998 Edition, February 2002 revision, are given below together with further details against those items which are usually priced in tenders.

Note: The term 'Not priced' where used throughout this section means either that the cost implication is negligible or that it is usually included elsewhere in the tender.

1. **Project, parties and consultants.**	*Not priced*
2. **Description of site.**	*Not priced*
3. **Drawings and other documents.**	*Not priced*
4. **Form, type and conditions of contract.**	

Clause No.

1.0 Intentions of the parties	
1.1 Contractor's obligations	*Not priced*
1.2 Contract Administrator's duties	*Not priced*
1.3 Reappointment of Planning Supervisor or Principal Contractor - notification to Contractor	*Not priced*
1.4 Alternative B2 in the 5th recital - notification by Contractor - regulation 7(5) of the	
CDM Regulations	*Not priced*
1.5 Giving or service of notices or other documents	*Not priced*
1.6 Reckoning periods of days	*Not priced*
1.7 Applicable law	*Not priced*
1.8 Bills of Quantities and SMM	*Not priced*
1.9 Contracts (Rights of Third Parties) Act 1999 – contracting out	*Not priced*

2.0 Commencement and completion

2.1 Commencement and completion *Not priced*

2.2 Extension of contract period *Not priced*

2.3 Damages for non-completion *Not priced*

2.4 Practical completion. *Implications of maintenance from completion until Practical completion may need to be considered*

2.5 Defects liability *Inevitably some defects will arise and an allowance will often be made to cover this either here or against* Clause 2.7.

2.6 Partial possession by Employer *Not priced*

2.7 Failures of plants (pre practical completion) *Inevitably some replacements will be required and an allowance should be made here or in the rates. Plant replacements up to and including 10% loss are usually covered in the planting measured work prices.*

2.8A Plants defects liability and post practical completion care by Contractor

2.8B Plants defects liability and post practical completion care by Employer

2.9A Malicious damage or theft

2.9B Malicious damage or theft

3.0 Control of the Works

3.1 Assignment *Not priced*

3.2 Sub-contracting *Not priced*

3.3 Contractor's representative *Any allowance for site supervision/administration which is not included in the rates should be priced here together with the cost of setting out the works.*

3.4 Exclusion from the Works *Not priced*

3.5 Contract Administrator's instructions *Not priced*

3.6 Variations *Not priced*

3.7 P.C. and Provisional sums *An amount should be added to the PC sum items, if required, for profit and a further sum, where applicable, for attendance.*

3.8 Objections to a Nomination *Not priced*

4.0 Payment

4A Payments subject to Supplemental Condition C

4.1 Correction of inconsistencies *Not priced*

4.2 Progress payments and retention *Not priced*

4.3 Penultimate certificate *Not priced*

4.4 Notices of amounts to be paid and deductions

4.5 Final certificate *Not priced*

4.6 Contribution, levy and tax changes *Not priced*

4.7 Fixed price *The additional cost of providing a fixed price can be included here or in the rates.*

4.8 Right of suspension by Contractor

5.0 Statutory obligations

5.1 Statutory obligations, notices, fees and charges   *The cost of complying with this clause should be included here.*

5.2 Value Added Tax   *Not priced*

5.3 Construction Industry Scheme (CIS)   *Not priced*

5.4 [Number not used]   *Not priced*

5.5 Prevention of corruption   *Not priced*

5.6 Employer's obligation - Planning Supervisor - Principal Contractor   *Not priced*

5.7 Contractor is Principal Contractor   *Not priced*

5.8 Contractor is not Principal Contractor   *Not priced*

5.9 Health and safety file

6.0 Injury, damage and insurance

6.1 Injury to or death of persons   *The cost of maintaining insurance should be included here if not allowed for elsewhere.*

6.2 Injury or damage to property As *above in 6.1.*

6.3A Insurance of the Works by Contractor - Fire, etc.   *If at the Contractor's risk the cost of maintaining insurance cover must be sufficient to include the full cost of reinstatement, all increases in cost, professional fees and any consequential costs such as clearing debris.*

6.3B Insurance of the Works and any existing structures by Employer - Fire, etc. *As above in 6.3 A.*

6.4 Evidence of insurance   *Not priced*

7.0 Determination

7.1 Notices   *Not priced*

7.2 Determination by Employer   *Not priced*

7.3 Determination by Contractor   *Not priced*

8.0 Settlement of Disputes - Adjudication - Arbitration

8.1 Adjudication   *Not priced*

8.2 Arbitration

8.3 Legal proceedings

SUPPLEMENTAL CONDITIONS

A   Contribution, levy and tax changes

B   Value added tax

C   Construction Industry Scheme (CIS)

D   Adjudication

E   Arbitration

ADDITIONAL PRELIMINARY COST ITEMS

(Not referenced in the JCLI Agreement)

Note: The following heads of preliminaries although not specifically mentioned in the JCLI Agreement may need to be addressed. The following items should be regarded as a checklist and may be priced if they have cost implications on the project.

1. **Contractor's liability**   *Not priced*

   If the JCLI Form of Agreement is used this item is covered under Item 6. If a different Form of Contract is used it may be necessary to include for Contractor's insurances, etc., under this item.

2. **Obligations and restrictions imposed by the employer**

   These include the following items and costs can only be assessed in the light of circumstances on a particular job.

   (a) Access to and possession of use of the site.

   (b) Limitations of working space.

   (c) Limitations of working hours.

   (d) The use or disposal of any materials found on site.

   (e) Hoardings, fences, and the like, temporary name boards and advertising rights.

   (f) The maintenance of existing live drainage, water, gas and other main or power services on or over the site.

   (g) The execution or completion of the work in any specific order or in sections or phases.

   (h) Temporary accommodation and facilities for the use of the Employer including heating, lighting, furnishing and attendance.

   This will include an office for the Clerk of Works if there is to be one on the site. Against this item the following should be priced:

       i.   Hire of Clerk of Works Office

       ii. Transport to and from the site

       iii. Erecting on a suitable base and later dismantling

       iv. Lighting, heating and attendance on office

       v. Local Authority rates

   (i) The installation of telephones for the use of the Employer (The cost of his telephone calls should be covered by a provisional sum).

   (j) Any other obligation or restriction.

   Additional obligations may include the provision of a performance bond. If the Contractor is required to provide sureties for the fulfilment of the work the usual method of providing this is by a bond provided by one or more insurance companies. The cost of a performance bond depends largely on the financial standing of the applying contractor. Figures tend to range from 0.25 to 0.5% of the contract sum.

Works by nominated sub-contractors, goods and materials from nominated suppliers and works by public bodies

3. **Works by nominated sub-contractors**   *Not priced*

    Work to be carried out by a nominated sub-contractor and given as a prime cost sum to which an amount should be added, if required, for profit and a further sum for attendance.

4. **Goods and materials from nominated suppliers**

    Goods and materials which are required to be obtained from a nominated supplier and given as a prime cost sum to which an amount should be added, if required, for profit.

5. **Works by Public Bodies**

    Works which are to be carried out by a Local Authority or public undertaking given as a provisional sum.

6. **Works by others directly engaged by the Employer**

    A description given for works by others directly engaged by the employer, and any attendance that is required shall be priced in the same way as works by nominated sub-contractors.

General facilities and obligations

7. **Pricing**

    For convenience in pricing items will be listed which may include the following. The contractor should include maintaining any temporary works in connection with the items, adapting, clearing away and making good, and all notices and fees to local authorities and public undertakings.

    a)   Plant, tools and vehicles

    The sixth edition of the Standard Method of Measurement of Building Works requires that items for plant be given at the beginning of each section, whereas the seventh edition provides for these items to be covered under A42 and A43. However, Contractors often include the cost of their own plant within the measured rates e.g. earthmoving.

    b)   Site administration and security

    If the JCLI Form of Agreement is used the cost of administrative staff is often included against Item 3. When required allow for the provision of a watchman or inspection by a security organization.

    c)   Transport for workpeople

    The labour rates per hour on which Prices for Measured Work have been based do not cover travel and lodging allowances which must be assessed according to the particular circumstances.

d)   Protecting the works from inclement weather

In areas likely to suffer particularly inclement weather, some nominal allowance should be included for tarpaulins, polythene sheeting, etc., and the effect of any delays in concreting or brickwork by such weather.

e)   Water for the works

Charges should properly be ascertained from the local Water Authority. If these are not readily available, an allowance of 0.33% of the value of the contract is probably adequate, providing water can be obtained directly from the mains. Failing this, each case must be dealt with on its merits. In all cases an allowance should also be made for temporary plumbing including site storage of water if required.

f)   Lighting and power for the works

The Contractor is usually responsible for providing all temporary lighting and power for the works and all charges involved.

g)   Temporary roads, hardstandings, crossings and similar items

Quite often consolidated bases of eventual site roads are used throughout a contract to facilitate movement of materials around the site. However, during the initial setting up of a site, with drainage works outstanding this is not always possible and occasionally temporary roadways have to be formed and ground levels later reinstated.

h)   Temporary accommodation for the use of the Contractor

This includes all temporary offices and sheds for the Contractor and his domestic sub-contractors' use (temporary office for a Clerk of Works is covered under obligations and restrictions imposed by the employer).

Allowances must also be made for transport, erection, dismantling, etc., as previously shown under the item for Clerk of Works office.

i)   Temporary telephones for the use of the Contractor

Against this item should be included the cost of installation, rental and an assessment of the cost of calls made during the contract.

j)   Traffic regulations

Waiting and unloading restrictions can occasionally add considerably to costs, resulting in forced overtime or additional weekend working. Any such restrictions must be carefully assessed for the job in hand.

k)   Safety, health and welfare of workpeople

The Contractor is required to comply with all relevant codes, regulations, agreements and statutes and in particular should allow for provision of the following:

1. Shelter from inclement weather
2. Accommodation for clothing
3. Accommodation and provision for meals
4. Provision of drinking water
5. Sanitary conveniences
6. Washing facilities
7. First aid
8. Site conditions

A variety of self-contained mobile or jack-type units are available for hire and allowance must be made in addition for transport costs to and from site, setting up costs, connections to mains, fuel supplies and attendance. A general provision to comply with the above code is often 0.50 to 0.75% of the contract value. The cost of safety supervisors (required for firms employing more than 20 people) is usually part of head office overhead costs.

l)   Disbursements arising from the employment of workpeople
Travelling and lodging allowances have been dealt with under Transport for Workpeople and usually all other on-costs and disbursements are included in the all-in hourly rate used in the calculation of Prices for Measured Work. However, it is as well to check that all such disbursements have been included elsewhere.

m)   Maintenance of public and private roads
Some additional insurance or value may be required against this item to insure against damage to entrance gates, kerbs or bridges caused by extraordinary traffic in the execution of the works.

n)   Removing rubbish, protective casings and coverings and cleaning the works on completion
This includes removing surplus materials and final cleaning of the site prior to handover. Allow for sufficient 'bins' for the site throughout contract duration and for some operative time at the end of the contract for final clearing and cleaning ready for handover. A general allowance of 0.20% of contract value is probably sufficient.

o)   Temporary fencing, hoardings and similar items
This item must be considered in some detail as it is dependent on site perimeter, phasing of the work, etc.

p)   Control of noise, pollution and all other statutory obligations
The Local Authority may impose restrictions on the timing of certain operations, particularly noisy or dust-producing operations, and may necessitate the carrying out of these outside normal working hours or using special tools and equipment. The situation is most likely to occur in built up areas such as city centres, etc., where the site is likely to be in close proximity to offices or residential property.

8.   Contingencies
Generally. Provision for contingencies is normally given as a provisional sum.

Contemporary Landscapes of Contemplation

Rebecca Krinke

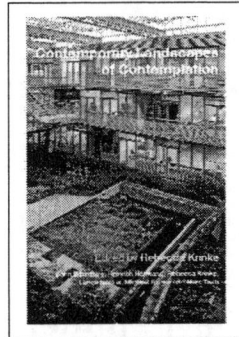

Contemplative landscape and contemplative space are familiar terms in the areas of design, landscape architecture and architecture. Krinke and her highly regarded contributors set out to explore definitions, theories, and case studies of contemplative landscapes and to secure the subject as a scholarly interest. The contributors, Marc Treib, John Beardsley, Michael Singer, Lance Neckar, Heinrich Hermann and Rebecca Krinke have spent their careers researching, critiquing, and making landscapes. Here they investigate the role of contemplative space in a post-modern world and examine the impact of nature and culture on the design or interpretation of contemplative landscapes.

The authors investigate principles and strategies often used as guidance for creating contemplative landscapes, as well as the relationships and differences between contemplative and commemorative space. The essays, drawn from both scholarship and personal experience explore the links between spaces designed to provide health benefits and contemplative space.

July 2005: 234x156 mm: 240 pages
HB: 0-415-70068-X: £70.00
PB: 0-415-70069-8: £27.50

To Order: Tel: +44 (0) 1264 343071 Fax: +44 (0) 1264 343005, or
Post: Taylor and Francis Customer Services, Thomson Publishing Services, Cheriton House, Andover, Hants, SP10 5BE, UK Email: book.orders@tandf.co.uk

For a complete listing of all our titles visit:
www.sponpress.com

Taylor & Francis
Taylor & Francis Group plc

2nd Edition
Urban Drainage

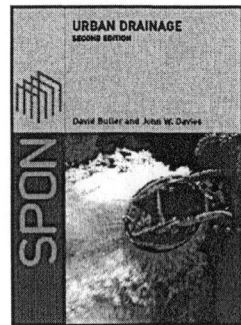

David Butler and John W. Davies

Environmental and engineering aspects are both involved in the drainage of rainwater and wastewater from areas of human development. *Urban Drainage* deals comprehensively not only with the design of new systems, but also the analysis and upgrading of existing infrastructure, and the environmental issues involved. Each chapter contains a descriptive overview of the complex issues involved, the basic engineering principles, and analysis for each topic. Extensive examples are used to support and demonstrate the key issues explained in the text.

Urban Drainage is an essential text for undergraduates and postgraduate students, lecturers and researchers in water engineering, environmental engineering, public health engineering and engineering hydrology. It is a useful reference for drainage design and operation engineers in the water industry and local authorities, and for consulting engineers. It will also be of interest to students, researchers and practitioners in environmental science, technology, policy and planning, geography and health studies.

May 2004: 234x156 mm: 588 pages
159 line figures, 78 tables, 13 b+w photos
HB: 0-415-30606-X: £100.00
PB: 0-415-30607-8: £35.00

To Order: Tel: +44 (0) 1264 343071 Fax: +44 (0) 1264 343005, or
Post: Taylor and Francis Customer Services, Thomson Publishing Services, Cheriton House, Andover, Hants, SP10 5BE, UK Email: book.orders@tandf.co.uk

For a complete listing of all our titles visit:
www.sponpress.com

Taylor & Francis
Taylor & Francis Group plc

Part 4

Fees for Professional Services

LANDSCAPE ARCHITECTS' FEES

The Landscape Institute no longer sets any fee scales for its members. A publication which offers guidance in determining fees for different types of project is available from

The Landscape Institute
33 Great Portland Street London W1W 8QG   Telephone: 020 7299 4500

The publication is entitled
"Engaging a Landscape Consultant – Guidance for Clients on Fees"

The document refers to suggested fee systems on the following basis:

- Time Charged Fee Basis
- Lump Sum Fee Basis
- Percentage Fee Basis
- Retainer Fee Basis

In regard to the 'Percentage Fee Basis' the publication lists various project types and relates them to complexity ratings for works valued at £22,500 and above. The suggested fee scales are dependent on the complexity rating of the project.

The following are samples of the Complexity Categories

Category 1:   Golf Courses, Country Parks and Estates, Planting Schemes

Category 2:   Coastal, River and Agricultural Works, Roads, Rural Recreation Schemes

Category 3:   Hospital Grounds, Sport Stadia, Urban Offices and Commercial Properties, Housing

Category 4:   Domestic/Historic Garden Design, Urban Rehabilitation and Environmental Improvements

Table Showing Suggested Percentage Fees for Various Categories of Complexity of Landscape Design or Consultancy

	Complexity			
Project Value	1	2	3	4
22,500	15.00%	16.50%	18.00%	21.00%
30,000	13.75	15.00	16.50	19.25
50,000	10.50	12.75	14.00	16.25
100,000	9.50	10.50	11.50	13.50
150,000	8.75	9.50	10.50	12.50
200,000	8.00	8.75	9.75	11.25
300,000	7.50	8.25	9.00	10.50
500,000	6.75	7.50	8.25	9.75
750,000	6.50	7.25	7.75	9.25
1,000,000	6.25	7.00	7.50	8.75
10,000,000	6.25	6.75	7.25	8.25

(extrapolated from the graph/curve chart in the aforementioned publication)

Guide to Stage Payments of Fees, Relevant Fee Basis and Proportion of Fee Applicable to Lump Sum and Percentage Fee Basis. *Details of Preliminary, Standard and Other Services are set out in detail in the Landscape Consultant's Appointment.*

Work Stage		Relevant Fee Basis			Proportion of Fee	
		Time	Lump	%age	Proportion of fee	Total
Preliminary Services						
A	Inception	✓	✓	n/a	n/a	n/a
B	Feasibility	✓	✓	n/a	n/a	n/a
Standard Services						
C	Outline Proposals	✓	✓	✓	15%	15%
D	Sketch Scheme Proposals	✓	✓	✓	15%	30%
E	Detailed Proposals	✓	✓	✓	15%	45%
FG	Production Information	✓	✓	✓	20%	65%
HJ	Tender Action & Contract Preparation	✓	✓	✓	5%	70%
K	Operations on Site	✓	✓	✓	25%	95%
L	Completion	✓	✓	✓	5%	100%
Other Services		✓	✓	n/a		

Timing of Fee Payments

Percentage fees are normally paid at the end of each work stage. Time based fees are normally paid at monthly intervals. Lump sum fees are normally paid at intervals by agreement. Retainer or term commission fees are normally paid in advance, for predetermined periods of service.

WORKED EXAMPLES OF PERCENTAGE FEE CALCULATIONS

Worked Example 1

Project Type   Caravan Site
Services Required   To Detailed Proposals - Work Stages C to E
Budget   £120,000

Step 1   Decide on Work Type and therefore Complexity Rating - Complexity Rating 2
Step 2   Decide on Services required and Proportion of Fee - To Detailed Proposals, 45%
Step 3   Read off Graph, Complexity Rating 2, the % fee of £120,000 - Graph Fee 9.9%
Step 4   Multiply the Proportion of Fee (45%) by the Graph Fee (9.9%) - Adjusted Fee - 4.46%
Step 5   Calculate the Guide Fee (4.46% of £120,000) - Guide Fee - £5,352
Step 6   Agree fee with Client, complete Memorandum of Agreement & Schedule of Services & Fees

Worked Example 2

Project Type   New Housing
Services Required   Full Standard Services - Work Stages C to L
Budget   £350,000

Step 1   Decide on Work Type and therefore Complexity Rating - Complexity Rating 3
Step 2   Decide on Services required and Proportion of Fee - To Completion, 100%
Step 3   Read off Graph, Complexity Rating 3 the % fee of £350,000 - Graph Fee 8.8%
Step 4   Multiply the Proportion of Fee (100%) by the Graph Fee (8.8%) - Adjusted Fee - 8.8%
Step 5   Calculate the Guide Fee (8.8% of £350,000) - Guide Fee - £30,800
Step 6   Agree fee with Client, complete Memorandum of Agreement & Schedule of Services & Fees

Worked Example 3

Project Type   Urban Environmental Improvements
Services Required   To Production Information - Work Stages C to G
Budget   £1,250,000

Step 1   Decide on Work Type and therefore Complexity Rating - Complexity Rating 4
Step 2   Decide on Services required and Proportion of Fee - To Production Information, 65%
Step 3   Read off Graph, Complexity Rating 4 the % fee of £1,250,000 - Graph Fee 8.6%
Step 4   Multiply the Proportion of Fee (65%) by the Graph Fee (8.6%) - Adjusted Fee - 5.59%
Step 5   Calculate the Guide Fee (5.59% of £1,250,000) - Guide Fee - £69,875
Step 6   Agree fee with Client, complete Memorandum of Agreement & Schedule of Services & Fees

Fees for Other Professional Services

Extracts from the scales of fees for architects, quantity surveyors and consulting engineers are given together with extracts from the Town and Country Planning Regulations 1993 and Building Regulation Charges. These documents are referenced by kind permission of the bodies concerned, in the case of Building Regulation Charges, by kind permission of the London Borough of Ealing. Attention is drawn to the fact that the full scales are not reproduced here and that the extracts are given for guidance only. The full authority scales should be studied before concluding any agreement and the reader should ensure that the fees quoted here are still current at the time of reference.

ARCHITECTS' FEES

Standard Form of Agreement for the Appointment of an Architect (SFA/99)
Conditions of Engagement for the Appointment of an Architect (CE/99)
Small Works (SW/99)
Employer's Requirements (DB1/99)
Contractor's Proposals (DB2/99)
Form of Appointment as Planning Supervisor (PS/99)
Form of Appointment as Sub-Consultant (SC/99)
Form of Appointment as Project Manager (PM/99)

2nd Edition
Spon's Irish Construction Price Book

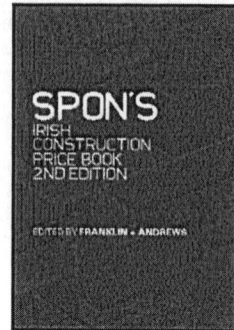

SPON'S
IRISH
CONSTRUCTION
PRICE BOOK
2ND EDITION

EDITED BY FRANKLIN + ANDREWS

Franklin + Andrews

This new edition of *Spon's Irish Construction Price Book*, edited by Franklin + Andrews, is the only complete and up-to-date source of cost data for this important market.

• All the materials costs, labour rates, labour constants and cost per square metre are based on current conditions in Ireland

• Structured according to the new Agreed Rules of Measurement (second edition)

• 30 pages of Approximate Estimating Rates for quick pricing

This price book is an essential aid to profitable contracting for all those operating in Ireland's booming construction industry.

Franklin + Andrews, Construction Economists, have offices in 100 countries and in-depth experience and expertise in all sectors of the construction industry.

April 2004: 246x174 mm: 448 pages
HB: 0-415-34409-3: £125.00

Prices for Measured Works

The Cultured Landscape
Designing the Environment in the 21st Century

Sheila Harvey and Ken Fieldhouse

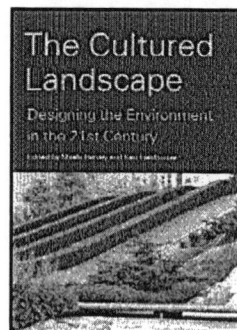

This book poses important philosophical questions about the aims, values and purposes of landscape architecture. The editors, highly regarded in their field, have drawn together a distinguished team of writers who provide unique individual perspectives on contemporary themes from a wide base of knowledge. Altogether, this new international study raises awareness of the landscape and encourages innovative ways of thinking about quality in design.

A Theoretical Framework. Design Context. Testing. The Future. End Matter.

August 2005: 234x156 mm: 208 pages
HB: 0-419-25030-1: £75.00
PB: 0-419-25040-9: £24.99

To Order: Tel: +44 (0) 1264 343071 Fax: +44 (0) 1264 343005, or
Post: Taylor and Francis Customer Services, Thomson Publishing Services, Cheriton House, Andover, Hants, SP10 5BE, UK Email: book.orders@tandf.co.uk

For a complete listing of all our titles visit:
www.sponpress.com

Taylor & Francis
Taylor & Francis Group plc

NEW ITEMS FOR THIS EDITION

Item Excluding site overheads and profit	PC £	Labour hours	Labour £	Plant £	Material £	Unit	Total rate £
D11 SLEEPER WALLS							
Construct retaining wall from railway sleepers; fixed with steel galvanised pins 12 mm driven into the ground; sleepers laid flat							
Grade 1 softwood; 2590 x 250 x 150 mm							
150 mm; 1 sleeper high	5.43	0.50	9.38	-	5.75	m	**15.12**
300 mm; 2 sleepers high	10.86	1.00	18.75	-	11.50	m	**30.25**
450 mm; 3 sleepers high	16.15	0.67	12.50	-	16.99	m	**29.49**
600 mm; 4 sleepers high	21.76	2.00	37.50	-	22.60	m	**60.10**
Grade 1 softwood as above but with 2 nr galvanised angle iron stakes set into concrete internally and screwed to the inside face of the sleepers							
750 mm; 5 sleepers high	26.92	2.50	46.88	-	58.25	m²	**105.13**
900 mm; 6 sleepers high	32.22	2.75	51.56	-	65.16	m²	**116.73**
Grade 1 hardwood; 2590 x 250 x 150 mm							
150 mm; 1 sleeper high	5.74	0.50	9.38	-	6.06	m	**15.43**
300 mm; 2 sleepers high	11.48	1.00	18.75	-	12.12	m	**30.87**
450 mm; 3 sleepers high	17.08	0.67	12.50	-	17.92	m	**30.42**
600 mm; 4 sleepers high	23.01	2.00	37.50	-	23.85	m	**61.35**
Grade 1 hardwood as above but with 2 nr galvanised angle iron stakes set into concrete internally and screwed to the inside face of the sleepers							
750 mm; 5 sleepers high	28.47	2.50	46.88	-	59.80	m²	**106.67**
900 mm; 6 sleepers high	34.07	2.75	51.56	-	67.01	m²	**118.57**
New Pine softwood; 2500 x 230 x 150 mm							
150 mm; 1 sleeper high	6.88	0.50	9.38	-	7.20	m	**16.57**
300 mm; 2 sleepers high	13.76	1.00	18.75	-	14.40	m	**33.15**
450 mm; 3 sleepers high	6.88	0.67	12.50	-	7.72	m	**20.22**
600 mm; 4 sleepers high	27.52	2.00	37.50	-	28.36	m	**65.86**
New pine softwood as above but with 2 nr galvanised angle iron stakes set into concrete internally and screwed to the inside face of the sleepers							
750 mm; 5 sleepers high	34.40	2.50	46.88	-	65.73	m²	**112.60**
900 mm; 6 sleepers high	41.28	2.75	51.56	-	74.22	m²	**125.78**
New Oak Hardwood 2500 x 230 x 130 mm							
150 mm; 1 sleeper high	8.68	0.50	9.38	-	9.00	m	**18.38**
300 mm; 2 sleepers high	17.36	1.00	18.75	-	18.00	m	**36.75**
450 mm; 3 sleepers high	26.04	0.67	12.50	-	26.88	m	**39.38**
600 mm; 4 sleepers high	34.72	2.00	37.50	-	35.56	m	**73.06**
New Oak Hardwood as above but with 2 nr galvanised angle iron stakes set into concrete internally and screwed to the inside face of the sleepers							
750 mm; 5 sleepers high	43.40	2.50	46.88	-	74.73	m²	**121.60**
900 mm; 6 sleepers high	52.08	2.75	51.56	-	85.02	m²	**136.58**
D20 EXCAVATION AND FILLING							
MACHINE SELECTION TABLE Road Equipment Ltd; machine volumes for excavating/filling only and placing excavated material alongside or to a dumper; no bulkages are allowed for in the material volumes; these rates should be increased by user-preferred percentages to suit prevailing site conditions; the figures in the next section for "Excavation mechanical" and filling allow for the use of banksmen within the rates shown below							

NEW ITEMS FOR THIS EDITION

Item Excluding site overheads and profit	PC £	Labour hours	Labour £	Plant £	Material £	Unit	Total rate £
D20 EXCAVATION AND FILLING - cont'd							
MACHINE SELECTION TABLE - cont'd							
1.5 tonne excavators; digging volume							
1 cycle / minute; 0.04 m³	-	0.42	7.81	2.39	-	m³	10.20
2 cycles / minute; 0.08 m³	-	0.21	3.91	1.41	-	m³	5.32
3 cycles / minute; 0.12 m³	-	0.14	2.60	1.09	-	m³	3.70
3 tonne excavators; digging volume							
1 cycle / minute; 0.13 m³	-	0.13	2.40	1.27	-	m³	3.67
2 cycles / minute; 0.26 m³	-	0.06	1.20	1.63	-	m³	2.83
3 cycles / minute; 0.39 m³	-	0.04	0.80	0.75	-	m³	1.55
5 tonne excavators; digging volume							
1 cycle / minute; 0.28 m³	-	0.06	1.12	1.50	-	m³	2.61
2 cycles / minute; 0.56 m³	-	0.03	0.56	1.16	-	m³	1.71
3 cycles / minute; 0.84 m³	-	0.02	0.37	1.17	-	m³	1.55
7 tonne excavators; supplied with operator; digging volume							
1 cycle / minute; 0.28 m³	-	0.06	1.12	2.28	-	m³	3.40
2 cycles / minute; 0.56 m³	-	0.47	8.84	0.41	-	m³	9.24
3 cycles / minute; 0.84 m³	-	0.52	9.72	0.33	-	m³	10.05
21 tonne excavators; supplied with operator; digging volume							
1 cycle / minute; 1.21 m³	-	-	-	0.68	-	m³	0.68
2 cycles / minute; 2.42 m³	-	-	-	0.28	-	m³	0.28
3 cycles / minute; 3.63 m³	-	-	-	0.19	-	m³	0.19
Backhoe loader; excavating; JCB 3CX rear bucket capacity 0.28 m³							
1 cycle / minute; 0.28 m³	-	-	-	1.78	-	m³	1.78
2 cycles / minute; 0.56 m³	-	-	-	0.89	-	m³	0.89
3 cycles / minute; 0.84 m³	-	-	-	0.59	-	m³	0.59
Backhoe loader; loading from stockpile; JCB 3CX front bucket capacity 1.00 m³							
1 cycle / minute; 1.00 m³	-	-	-	0.50	-	m³	0.50
2 cycles / minute; 2.00 m³	-	-	-	0.25	-	m³	0.25
Dumpers; all volumes below are based on excavated "Earth" moist at 1997 /1598 kg/m³ solid/loose; a 25% bulkage factor has been used; the weight capacities below exceed the volume capacities of the machine in most cases; see the memorandum section at the back of this book for further weights of materials							
1 tonne high tip skip loader; volume 0.485 m³ (775 kg)							
5 loads per hour	-	0.41	7.73	1.56	-	m³	9.30
7 loads per hour	-	0.29	5.52	1.15	-	m³	6.67
10 loads per hour	-	0.21	3.87	0.84	-	m³	4.71
3 tonne dumper; max. volume 2.40 m³ (3.38 t); available volume 1.9 m³							
4 loads per hour	-	0.14	2.60	0.50	-	m³	3.10
5 loads per hour	-	0.11	2.08	0.40	-	m³	2.49
7 loads per hour	-	0.08	1.49	0.30	-	m³	1.79
10 loads per hour	-	0.06	1.04	0.22	-	m³	1.26
6 tonne dumper; max. volume 3.40 m³ (5.4 t); available volume 3.77 m³							
4 loads per hour/	-	0.07	1.24	0.35	-	m³	1.59
5 loads per hour	-	0.05	1.00	0.28	-	m³	1.28
7 loads per hour	-	0.04	0.71	0.21	-	m³	0.92
10 loads per hour	-	0.03	0.50	0.16	-	m³	0.66

NEW ITEMS FOR THIS EDITION

Item Excluding site overheads and profit	PC £	Labour hours	Labour £	Plant £	Material £	Unit	Total rate £
Excavate foundation trench; set railway sleepers vertically on end in concrete 1:3:6 continuous foundation to 33.3% of their length to form retaining wall Grade 1 Hardwood; finished height above ground level							
300 mm	10.18	3.00	56.25	1.88	19.71	m	77.83
500 mm	14.75	3.00	56.25	1.88	26.56	m	84.69
600 mm	20.50	3.00	56.25	1.88	39.29	m	97.42
750 mm	22.13	3.50	65.63	1.88	41.73	m	109.22
1.00 m	30.24	3.75	70.31	1.88	47.16	m	119.34
Excavate and place vertical steel universal beams 165 wide in concrete base at 2.590 centres; fix railway sleepers set horizontally between beams to form horizontal fence or retaining wall Grade 1 Hardwood; bay length 2.590 m							
0.50 m high; (2 sleepers)	11.81	2.50	46.88	-	44.29	bay	91.16
750 m high; (3 sleepers)	17.35	2.60	48.75	-	65.74	bay	114.49
1.00 m high; (4 sleepers)	23.26	3.00	56.25	-	84.49	bay	140.74
1.25 m high; (5 sleepers)	29.70	3.50	65.63	-	109.24	bay	174.87
1.50 m high; (6 sleepers)	34.89	3.00	56.25	-	130.01	bay	186.26
1.75 m high; (7 sleepers)	40.61	3.00	56.25	-	135.74	bay	191.99
E30 REINFORCEMENT FOR IN SITU CONCRETE							
Reinforcement bar to concrete formwork							
6 mm straight bar							
100 ccs	0.88	0.33	6.19	-	0.88	m^2	7.07
200 ccs	0.56	0.25	4.69	-	0.56	m^2	5.25
300 ccs	0.28	0.15	2.81	-	0.28	m^2	3.09
8 mm bar							
100 ccs	1.56	0.33	6.19	-	1.56	m^2	7.75
200 ccs	0.76	0.25	4.69	-	0.76	m^2	5.45
300 ccs	0.52	0.15	2.81	-	0.52	m^2	3.33
10 mm bar							
100 ccs	2.44	0.33	6.19	-	2.44	m^2	8.63
200 ccs	1.20	0.25	4.69	-	1.20	m^2	5.89
300 ccs	0.80	0.15	2.81	-	0.80	m^2	3.61
12 mm bar							
100 ccs	3.52	0.33	6.19	-	3.52	m^2	9.71
200 ccs	1.76	0.25	4.69	-	1.76	m^2	6.45
300 ccs	1.16	0.15	2.81	-	1.16	m^2	3.97
16 mm bar							
100 ccs	6.28	0.40	7.50	-	6.28	m^2	13.78
200 ccs	3.16	0.33	6.19	-	3.16	m^2	9.35
300 ccs	2.12	0.25	4.69	-	2.12	m^2	6.81
25 mm bar							
100 ccs	15.40	0.40	7.50	-	15.40	m^2	22.90
200 ccs	7.68	0.33	6.19	-	7.68	m^2	13.87
300 ccs	5.12	0.25	4.69	-	5.12	m^2	9.81
32 mm bar							
100 ccs	25.24	0.40	7.50	-	25.24	m^2	32.74
200 ccs	12.60	0.33	6.19	-	12.60	m^2	18.79
300 ccs	8.40	0.25	4.69	-	8.40	m^2	13.09

NEW ITEMS FOR THIS EDITION

Item Excluding site overheads and profit	PC £	Labour hours	Labour £	Plant £	Material £	Unit	Total rate £
Q10 ALUMINIUM EDGINGS							
Permaloc "AshphaltEdge" extruded aluminium alloy L shaped edging with 5.33 mm exposed upper lip; edging fixed to roadway base and edge profile with 250 mm steel fixing spike; laid to straight or curvilinear road edge; subsequently filled with macadam (not included) Depth of macadam							
38 mm	5.82	0.02	0.31	-	6.66	m	**6.98**
58 mm	6.53	0.02	0.32	-	7.39	m	**7.71**
64 mm	7.19	0.02	0.34	-	8.07	m	**8.41**
76 mm	8.65	0.02	0.38	-	9.56	m	**9.94**
102 mm	10.02	0.02	0.39	-	10.97	m	**11.36**
Permaloc "Cleanline"; heavy duty straight profile edging; for edgings to soft landscape beds or turf areas; 3.2 mm x 102 mm high; 3.2 mm thick with 4.75 mm exposed upper lip; fixed to form straight or curvilinear edge with 305 mm fixing spike Milled aluminium							
100 deep	6.64	0.02	0.34	-	7.50	m	**7.84**
Black							
100 deep	7.31	0.02	0.34	-	8.19	m	**8.53**
Permaloc "Permastrip"; heavy duty L shaped profile maintenance strip; 3.2 mm x 89 mm high with 5.2 mm exposed top lip; for straight or gentle curves on paths or bed turf interfaces; fixed to form straight or curvilinear edge with standard 305 mm stake; other stake lengths available Milled aluminium							
90 deep	6.64	0.02	0.34	-	7.50	m	**7.84**
Black							
90 deep	7.31	0.02	0.34	-	8.19	m	**8.53**
Permaloc "Proline"; medium duty straight profiled maintenance strip; 3.2 mm x 102 mm high with 3.18 mm exposed top lip; for straight or gentle curves on paths or bed turf interfaces; fixed to form straight or curvilinear edge with standard 305 mm stake; other stake lengths available Milled aluminium							
90 deep	5.03	0.02	0.34	-	5.84	m	**6.18**
Black							
90 deep	5.80	0.02	0.34	-	6.63	m	**6.97**
Q23 FOOTPATH GRAVELS							
CED Ltd; footpath gravels; porous self-binding gravel CED Ltd; "Cedec Gravel" rolling wet; on hardcore base (not included); for pavements; to falls and crossfalls and to slopes not exceeding 15 degrees from horizontal; over 300 mm wide							
50 mm thick	7.80	0.10	1.88	0.38	7.80	m²	**10.05**
75 mm thick	11.70	0.10	1.88	0.46	11.70	m²	**14.03**

NEW ITEMS FOR THIS EDITION

Item Excluding site overheads and profit	PC £	Labour hours	Labour £	Plant £	Material £	Unit	Total rate £
Q25 NATURAL STONE PAVING							
CED Ltd; Indian Sandstone, riven							
pavings or edgings; 25-35 mm thick; on							
prepared base measured separately;							
bedding on 25 mm cement:sand (1:3);							
cement:sand (1:3) joints							
Paving							
laid to random rectangular pattern	20.00	2.40	45.01	-	24.24	m²	**69.25**
laid to coursed laying pattern; 3 sizes	20.00	2.00	37.50	-	24.24	m²	**61.74**
Paving; single size							
600 x 600 mm	20.00	1.00	18.75	-	24.24	m²	**42.99**
600 x 400 mm	20.00	1.25	23.44	-	24.24	m²	**47.68**
400 x 400 mm	20.00	1.67	31.25	-	24.24	m²	**55.49**
Natural stone, slate or granite flag							
pavings; CED Ltd; on prepared base							
(not included); bedding on 25 mm							
cement:sand (1:3); cement:sand (1:3)							
joints							
Granite paving; sawn 6 sides; textured top, silver							
grey							
new slabs; 50 mm thick	25.00	1.71	32.14	-	29.66	m²	**61.80**
Q31 CONTAINERISED TREES							
Tree planting; "Airpot" container grown							
trees; advanced nursery stock and							
semi-mature; Deepdale Trees Ltd.							
"Acer platanoides Emerald Queen"; including							
backfilling with excavated material (other							
operations not included)							
16 - 18 cm girth	95.00	1.98	37.13	40.47	95.00	nr	**172.59**
18 - 20 cm girth	130.00	2.18	40.84	43.26	130.00	nr	**214.10**
20 - 25 cm girth	190.00	2.38	44.55	48.56	190.00	nr	**283.11**
25 - 30 cm girth	250.00	2.97	55.69	72.98	250.00	nr	**378.67**
30 - 35 cm girth	350.00	3.96	74.25	80.94	350.00	nr	**505.19**
"Aesculus briotti"; including backfilling with							
excavated material (other operations not							
included)							
16 - 18 cm girth	90.00	1.98	37.13	40.47	90.00	nr	**167.59**
18 - 20 cm girth	130.00	1.60	30.00	43.26	130.00	nr	**203.26**
20 - 25 cm girth	180.00	2.38	44.55	48.56	180.00	nr	**273.11**
25 - 30 cm girth	240.00	2.97	55.69	70.94	240.00	nr	**366.62**
30 - 35 cm girth	330.00	3.96	74.25	80.94	330.00	nr	**485.19**
"Prunus avium Flora Plena"; including backfilling							
with excavated material (other operations not							
included)							
16 - 18 cm girth	95.00	1.98	37.13	40.47	95.00	nr	**172.59**
18 - 20 cm girth	130.00	1.60	30.00	43.26	130.00	nr	**203.26**
20 - 25 cm girth	190.00	2.38	44.55	48.56	190.00	nr	**283.11**
25 - 30 cm girth	250.00	2.97	55.69	72.98	250.00	nr	**378.67**
30 - 35 cm girth	350.00	3.96	74.25	80.94	350.00	nr	**505.19**
"Quercus palustris - Pin Oak"; including							
backfilling with excavated material (other							
operations not included)							
16 - 18 cm girth	95.00	1.98	37.13	40.47	95.00	nr	**172.59**
18 - 20 cm girth	130.00	1.60	30.00	43.26	130.00	nr	**203.26**
20 - 25 cm girth	190.00	2.38	44.55	48.56	190.00	nr	**283.11**
25 - 30 cm girth	250.00	2.97	55.69	72.98	250.00	nr	**378.67**
30 - 35 cm girth	350.00	3.96	74.25	80.94	350.00	nr	**505.19**

NEW ITEMS FOR THIS EDITION

Item Excluding site overheads and profit	PC £	Labour hours	Labour £	Plant £	Material £	Unit	Total rate £
Q31 CONTAINERISED TREES - cont'd							
Tree planting - cont'd							
"Betula pendula multistem"; including backfilling with excavated material (other operations not included)							
3.0 - 3.5 m high	125.00	1.98	37.13	40.47	125.00	nr	202.59
3.5 - 4.0 m high	150.00	1.60	30.00	43.26	150.00	nr	223.26
4.0 - 4.5 m high	185.00	2.38	44.55	48.56	185.00	nr	278.11
4.5 - 5.0 m high	220.00	2.97	55.69	72.98	220.00	nr	348.67
5.0 - 6.0 m high	275.00	3.96	74.25	80.94	275.00	nr	430.19
6.0 - 7.0 m high	320.00	4.50	84.38	96.83	320.00	nr	501.20
"Pinus sylvestris"; including backfilling with excavated material (other operations not included)							
3.0 - 3.5 m high	300.00	1.98	37.13	40.47	300.00	nr	377.59
3.5 - 4.0 m high	350.00	1.60	30.00	43.26	350.00	nr	423.26
4.0 - 4.5 m high	400.00	1.60	30.00	45.76	400.00	nr	475.76
4.5 - 5.0 m high	450.00	2.97	55.69	72.98	450.00	nr	578.67
5.0 - 6.0 m high	550.00	3.96	74.25	80.94	550.00	nr	705.19
6.0 - 7.0 m high	750.00	4.50	84.38	99.65	750.00	nr	934.02
Tree planting; "Airpot" container grown trees; semi-mature and mature trees; Deepdale Trees Ltd; planting and back filling; planted by tele handler or by crane; delivery included; all other operations priced separately							
Semi-mature trees indicative prices							
40 - 45 cm girth	550.00	4.00	75.00	48.43	550.00	nr	673.43
45 - 50 cm girth	750.00	4.00	75.00	48.43	750.00	nr	873.43
55 - 60 cm girth	1000.00	6.00	112.50	48.43	1000.00	nr	1160.93
60 - 70 cm girth	2000.00	7.00	131.25	65.58	2000.00	nr	2196.83
70 - 80 cm girth	3000.00	7.50	140.63	82.73	3000.00	nr	3223.36
80 - 90 cm girth	4000.00	8.00	150.00	96.86	4000.00	nr	4246.86
Q50 SITE FURNITURE (NATURAL STONE)							
Standing Stones CED Ltd; Erect standing stones; vertical height above ground; in concrete base; including excavation setting in concrete to 1/3 depth and crane offload into position							
Purple schist							
1.00 m high	47.50	1.50	28.13	14.13	62.41	nr	104.65
1.25 m high	95.00	1.50	28.13	14.13	104.29	nr	146.54
1.50 m high	152.00	2.00	37.50	16.95	163.85	nr	218.30
2.00 m high	247.00	3.00	56.25	28.25	281.24	nr	365.74
2.50 m high	380.00	1.50	28.13	56.50	446.37	nr	531.00
Q50 STREET FURNITURE							
CED Ltd; Stone bench; "Sinuous bench"; Stone type bench to organic "S" pattern; laid to concrete base; 500 high x 500 wide (not included)							
2.00 m long	1000.00	4.00	75.00	-	1000.00	nr	1075.00
5.00 m long	2500.00	8.00	150.00	-	2500.00	nr	2650.00
10.00 m long	5000.00	11.00	206.25	-	5000.00	nr	5206.25

NEW ITEMS FOR THIS EDITION

Item Excluding site overheads and profit	PC £	Labour hours	Labour £	Plant £	Material £	Unit	Total rate £
S14 IRRIGATION							
Quality Irrigation Ltd; main or ring main supply							
Excavate and lay mains supply pipe 32 mm to supply irrigated area							
32 mm MDPE	-	-	-	-	-	m	4.00
25 mm MDPE	-	-	-	-	-	m	3.00
20 mm LDPE	-	-	-	-	-	m	1.50
Install low voltage electrical cable for station or solenoid control with main supply							
12 core cable	-	-	-	-	-	m	1.50
16 core cable	-	-	-	-	-	m	2.50
20 core cable	-	-	-	-	-	m	3.10
Quality Irrigation Ltd; Supply irrigation infrastructure for irrigation system							
Header tank and submersible pump and pressure stat							
2270 litre (500 gallon 25 mm / 10 days to 1500 m^2)	-	-	-	-	-	nr	2250.00
4540 litre (1000 gallon 25 mm / 10 days to 3500 m^2)	-	-	-	-	-	nr	4050.00
9080 litre (2000 gallon 25 mm / 10 days to 7000 m^2)	-	-	-	-	-	nr	5250.00
Electric multistation controllers with rain cut out							
6 station controller	-	-	-	-	-	nr	560.00
12 station controller	-	-	-	-	-	nr	760.00
24 station controller	-	-	-	-	-	nr	1200.00
Solenoid valve; 25 mm with chamber; extra over for each active station	-	-	-	-	-	nr	80.00
Quality Irrigation Ltd; Station consisting of multiple sprinklers; inclusive of all trenching wiring and connections to ring main or main supply							
Gear drive sprinklers placed to provide head to head (100%) overlap; average 4 sprinklers per station; inclusive of trenching and supply pipework from ring main; price per sprinkler							
300 mm pop-up; max 10.5 m centres; 350 m^2 average cover	-	-	-	-	-	nr	155.00
100 mm pop-up; max 15 m centres; 350 m^2 maximum cover	-	-	-	-	-	nr	125.00
Sprinkler on fixed riser; max 15 m centres; 350 m^2 maximum cover	-	-	-	-	-	nr	115.00
Pop-up Sprays; installed to 80% spray overlap; (average 6 sprays per station); price per spray							
300 mm; 4.6m spray radius; 3.7 m centres; 66 m^2 maximum cover	-	-	-	-	-	nr	91.00
100 mm; 4.6m spray radius; 3.7 m centres; 66 m^2 maximum cover	-	-	-	-	-	nr	71.00
Extra over for supply to each area treated by mini sprays	-	-	-	-	-	nr	55.00
Mini Sprayers: installed to 60% spray overlap; (average 30 sprays per station) price per spray							
1.50 m radius; 7.0 m^2 maximum cover	-	-	-	-	-	nr	8.00
3.00 m radius; 28 m^2 maximum cover	-	-	-	-	-	nr	9.00
Extra over for supply to each area treated by mini sprays	-	-	-	-	-	nr	55.00

NEW ITEMS FOR THIS EDITION

Item Excluding site overheads and profit	PC £	Labour hours	Labour £	Plant £	Material £	Unit	Total rate £
S14 IRRIGATION - cont'd							
Commissioning and testing of irrigation system							
per station	-	-	-	-	-	nr	**30.00**
Annual maintenance costs of irrigation system							
Call out charge per visit	-	-	-	-	-	nr	**250.00**
extra over per station	-	-	-	-	-	nr	**10.00**

A PRELIMINARIES

Item Excluding site overheads and profit	PC £	Labour hours	Labour £	Plant £	Material £	Unit	Total rate £
A11 TENDER AND CONTRACT DOCUMENTS							
Health and Safety							
Produce health and safety file including							
preliminary meeting and subsequent progress							
meetings with external planning officer in							
connection with health and safety. Project							
value:							
£35,000	-	8.00	212.00	-	-	nr	**212.00**
£75000	-	12.00	318.00	-	-	nr	**318.00**
£100000	-	16.00	424.00	-	-	nr	**424.00**
£200,000 to £500,000	-	40.00	1060.00	-	-	nr	**1060.00**
Maintain health and safety file for project duration							
£35,000	-	4.00	106.00	-	-	week	**106.00**
£75000	-	4.00	106.00	-	-	week	**106.00**
£100000	-	8.00	212.00	-	-	week	**212.00**
£200,000 to £500,000	-	8.00	212.00	-	-	week	**212.00**
Produce written risk assessments on all areas of							
operations within the scope of works of the							
contract. Project value:							
£35,000	-	2.00	53.00	-	-	week	**53.00**
£75000	-	3.00	79.50	-	-	week	**79.50**
£100000	-	5.00	132.50	-	-	week	**132.50**
£200,000 to £500,000	-	8.00	212.00	-	-	week	**212.00**
Produce Coshh assessments on all substances							
to be used in connection with the contract.							
Project value:							
£35,000	-	1.50	39.75	-	-	nr	**39.75**
£75000	-	2.00	53.00	-	-	nr	**53.00**
£100000	-	3.00	79.50	-	-	nr	**79.50**
£200,000 to £500,000	-	3.00	79.50	-	-	nr	**79.50**
Method Statements							
Provide detailed method statements on all							
aspects of the works. Project value:							
£ 30,000	-	2.50	66.25	-	-	nr	**66.25**
£ 50,000	-	3.50	92.75	-	-	nr	**92.75**
£ 75,000	-	4.00	106.00	-	-	nr	**106.00**
£ 100,000	-	5.00	132.50	-	-	nr	**132.50**
£ 200,000	-	6.00	159.00	-	-	nr	**159.00**
A32 EMPLOYER'S REQUIREMENTS:							
MANAGEMENT OF THE WORKS							
Programmes							
Allow for production of works programmes prior to							
the start of the works. Project value:							
£ 30,000	-	3.00	79.50	-	-	nr	**79.50**
£ 50,000	-	6.00	159.00	-	-	nr	**159.00**
£ 75,000	-	8.00	212.00	-	-	nr	**212.00**
£ 100,000	-	10.00	265.00	-	-	nr	**265.00**
£ 200,000	-	14.00	371.00	-	-	nr	**371.00**
Allow for updating the works programme during							
the course of the works. Project value							
£ 30,000	-	1.00	26.50	-	-	nr	**26.50**
£ 50,000	-	1.50	39.75	-	-	nr	**39.75**
£ 75,000	-	2.00	53.00	-	-	nr	**53.00**
£ 100,000	-	3.00	79.50	-	-	nr	**79.50**
£ 200,000	-	5.00	132.50	-	-	nr	**132.50**

A PRELIMINARIES

Item Excluding site overheads and profit	PC £	Labour hours	Labour £	Plant £	Material £	Unit	Total rate £
A32 EMPLOYER'S REQUIREMENTS: **MANAGEMENT OF THE WORKS** - cont'd							
Setting out							
Setting out for external works operations comprising hard and soft works elements; placing of pegs and string lines to Landscape Architects drawings; surveying levels and placing level pegs; obtaining approval from the Landscape Architect to commence works; areas of entire site							
1,000 m²	-	5.00	93.75	-	5.60	nr	99.35
2,500 m²	-	8.00	150.00	-	11.20	nr	161.20
5,000 m²	-	8.00	150.00	-	11.20	nr	161.20
10,000 m²	-	32.00	600.00	-	28.00	nr	628.00
A34 EMPLOYER'S REQUIREMENTS: **SECURITY/SAFETY/PROTECTION**							
Jacksons Fencing; Express fence; **framed mesh unclimbable fencing;** **including precast concrete supports and** **couplings** weekly hire							
2.0 m high; weekly hire rate	-	-	-	1.40	-	m	1.40
erection of fencing; labour only	-	0.10	1.88	-	-	m	1.88
delivery charge	-	-	-	0.80	-	m	0.80
return haulage charge	-	-	-	0.60	-	m	0.60
A41 CONTRACTOR'S GENERAL COST ITEMS: **SITE ACCOMMODATION**							
General							
The following items are instances of the commonly found preliminary costs associated with external works contracts. The assumption is made that the external works contractor is sub-contracted to a main contractor.							
Elliotthire; erect temporary **accommodation and storage on concrete** **base measured separately** Prefabricated office hire; jackleg; open plan							
3.6 x 2.4 m	31.76	-	-	-	31.76	week	31.76
4.8 x 2.4 m	33.80	-	-	-	33.80	week	33.80
Armoured store; erect temporary secure storage container for tools and equipment							
3.0 x 2.4 m	-	-	-	-	11.00	week	11.00
3.6 x 2.4 m	-	-	-	-	12.00	week	12.00
Delivery and collection charges on site offices							
Delivery charge	-	-	-	-	130.00	load	130.00
Collection charge	-	-	-	-	130.00	load	130.00
Toilet facilities							
John Anderson Ltd; serviced self-contained toilet delivered to and collected from site; maintained by toilet supply company							
single chemical toilet	-	-	-	-	21.00	week	21.00
single chemical toilet including wash hand basin & water	-	-	-	-	23.00	week	23.00
delivery and collection; each way	-	-	-	-	18.00	nr	18.00

A PRELIMINARIES

Item Excluding site overheads and profit	PC £	Labour hours	Labour £	Plant £	Material £	Unit	Total rate £
A43 CONTRACTOR'S GENERAL COST ITEMS: MECHANICAL PLANT							
Southern Conveyors; moving only of granular material by belt conveyor; conveyors fitted with troughed belts, receiving hopper, front and rear undercarriages and driven by Electrical and Air motors; support work for conveyor installation (scaffolding), delivery, collection and installation all excluded							
Conveyor belt width 400 mm; mechanically loaded and removed at offload point; conveyor length							
Up to 5 m	-	1.50	28.13	6.94	-	m³	35.06
10 m	-	1.50	28.13	7.71	-	m³	35.84
12.5 m	-	1.50	28.13	8.05	-	m³	36.18
15 m	-	1.50	28.13	8.48	-	m³	36.61
20 m	-	1.50	28.13	9.26	-	m³	37.38
25 m	-	1.50	28.13	9.95	-	m³	38.07
30 m	-	1.50	28.13	10.72	-	m³	38.84
Conveyor belt width 600 mm; mechanically loaded and removed at offload point; conveyor length							
Up to 5 m	-	1.00	18.75	5.03	-	m³	23.78
10 m	-	1.00	18.75	5.66	-	m³	24.41
12.5 m	-	1.00	18.75	6.00	-	m³	24.75
15 m	-	1.00	18.75	6.35	-	m³	25.10
20 m	-	1.00	18.75	7.03	-	m³	25.79
25 m	-	1.00	18.75	7.69	-	m³	26.45
30 m	-	1.00	18.75	8.35	-	m³	27.10
Conveyor belt width 400 mm; mechanically loaded and removed by hand at offload point; conveyor length							
Up to 5 m	-	3.19	59.82	5.19	-	m³	65.01
10 m	-	3.19	59.82	6.22	-	m³	66.04
12.5 m	-	3.19	59.82	6.68	-	m³	66.50
15 m	-	3.19	59.82	7.25	-	m³	67.07
20 m	-	3.19	59.82	8.28	-	m³	68.10
25 m	-	3.19	59.82	9.20	-	m³	69.02
30 m	-	3.19	59.82	10.23	-	m³	70.05
Terranova Lifting Ltd Crane Hire; materials handling and lifting; telescopic cranes supply and management; exclusive of roadway management or planning applications; prices below illustrate lifts of 1 tonne at maximum crane reach; lift cycle of 0.25 hours; mechanical filling of material skip if appropriate							
35 tonne mobile crane; lifting capacity of 1 tonne at 26 meters; C.P.A hire (all management by hirer)							
granular materials including concrete	-	0.75	14.06	15.68	-	tonne	29.74
palletised or packed materials	-	0.50	9.38	12.95	-	tonne	22.32
35 tonne mobile crane; lifting capacity of 1 tonne at 26 meters; contract lift							
granular materials or concrete	-	0.25	4.69	42.89	-	tonne	47.58
palletised or packed materials	-	-	-	40.16	-	tonne	40.16
50 tonne mobile crane; lifting capacity of 1 tonne at 34 meters; C.P.A hire (all management by hirer)							
granular materials including concrete	-	0.75	14.06	21.23	-	tonne	35.29
palletised or packed materials	-	0.50	9.38	18.50	-	tonne	27.87

A PRELIMINARIES

Item Excluding site overheads and profit	PC £	Labour hours	Labour £	Plant £	Material £	Unit	Total rate £
A43 CONTRACTOR'S GENERAL COST ITEMS: **MECHANICAL PLANT** - cont'd							
Terranova Lifting Ltd Crane Hire; **materials handling and lifting** - cont'd							
50 tonne mobile crane; lifting capacity of 1 tonne at 34 meters; contract lift							
granular materials or concrete	-	0.25	4.69	49.55	-	tonne	**54.24**
palletised or packed materials	-	-	-	46.82	-	tonne	**46.82**
80 tonne mobile crane; lifting capacity of 1 tonne at 44 meters; C.P.A hire (all management by hirer)							
granular materials including concrete	-	0.75	14.06	30.45	-	tonne	**44.51**
palletised or packed materials	-	0.50	9.38	27.71	-	tonne	**37.09**
80 tonne mobile crane; lifting capacity of 1 tonne at 44 meters; contract lift							
granular materials or concrete	-	0.25	4.69	60.58	-	tonne	**65.27**
palletised or packed materials	-	-	-	57.85	-	tonne	**57.85**
A44 CONTRACTOR'S GENERAL COST ITEMS: **TEMPORARY WORKS**							
Eve Trakway; portable roadway systems Temporary roadway system laid directly onto existing surface or onto PVC matting to protect existing surface. Most systems are based on a weekly hire charge with transportation, installation and recovery charges included.							
Heavy Duty Trakpanel - per panel (3.05m x 2.59m) per week	-	-	-	-	-	m²	**5.06**
Outrigger Mats for use in conjunction with Heavy Duty Trakpanels - per set of 4 mats per week	-	-	-	-	-	set	**82.40**
Medium Duty Trakpanel - per panel (2.44m x 3.00m) per week	-	-	-	-	-	m²	**5.46**
LD20 Eveolution - Light Duty Trakway - roll out system minimum delivery 50 m	-	-	-	-	-	m²	**4.86**
Terraplas Walkways - turf protection system - per section (1m x 1m) per week	-	-	-	-	-	m²	**3.50**

B COMPLETE BUILDINGS/STRUCTURES/UNITS

Item Excluding site overheads and profit	PC £	Labour hours	Labour £	Plant £	Material £	Unit	Total rate £
B10 PREFABRICATED BUILDINGS/STRUCTURES							
Cast stone buildings; Haddonstone Ltd;							
ornamental garden buildings in Portland							
Bath or Terracotta finished cast stone;							
prices for stonework and facades only;							
excavations, foundations, reinforcement,							
concrete infill, roofing and floors all							
priced separately							
Pavilion Venetian Folly L9400; Tuscan columns,							
pedimented arch, quoins and optional							
balustrading							
4184 high x 4728 mm wide x 3147 deep	9148.00	175.00	3281.25	194.60	9254.19	nr	**12730.04**
Pavilion L9300; Tuscan columns							
3496 mm high x 3634 wide	4798.00	144.00	2700.00	166.80	4904.19	nr	**7770.99**
Small Classical Temple L9250; 6 column with							
fibreglass lead effect finish dome roof							
Overall height 3610 mm, diameter 2540 mm	5554.00	130.00	2437.50	139.00	5660.19	nr	**8236.69**
Large Classical Temple L9100; 8 column with							
fibreglass lead effect finish dome roof							
overall height 4664 mm, diameter 3190 mm	9124.00	165.00	3093.75	139.00	9230.19	nr	**12462.94**
Stepped floors to temples							
Single step; Large Classical Temple	1053.00	26.00	487.50	-	1115.25	nr	**1602.75**
Single step; Small Classical Temple	744.00	24.00	450.00	-	795.88	nr	**1245.88**
Stone Structures; Architectural Heritage							
Ltd; hand carved from solid natural							
limestone with wrought iron domed roof,							
decorated frieze and base and integral							
seats; supply and erect only;							
excavations and concrete bases priced							
separately							
'The Park Temple', five columns; 3500 high x							
1650 diameter	9800.00	96.00	1800.00	60.00	9800.00	nr	**11660.00**
'The Estate Temple', six columns; 4000 high x							
2700 diameter	16800.00	120.00	2250.00	120.00	16800.00	nr	**19170.00**
Stone Structures; Architectural Heritage							
Ltd; hand carved from solid natural							
limestone with solid elm trelliage; prices							
for stonework and facades only; supply							
and erect only; excavations and							
concrete bases priced separately							
'The Pergola', 2240 high x 2640 wide x 6990							
long	11000.00	96.00	1800.00	240.00	11031.47	nr	**13071.47**
Ornamental Timber Buildings;							
Architectural Heritage Ltd; hand built in							
English oak with fire retardant wheat							
straw thatch; hand painted, fired lead							
glass windows							
'The Thatched Edwardian Summer House', 3710							
high x 3540 wide x 2910 deep overall, 2540 x							
1910 internal	19500.00	32.00	600.00	-	19500.00	nr	**20100.00**
Ornamental Stone Structures;							
Architectural Heritage Ltd; setting to							
bases or plinths (not included)							
'The Obelisk', classic natural stone obelisk,							
tapering square form on panelled square base,							
1860 high x 360 square	1200.00	2.00	37.50	-	1200.00	1	**1237.50**
'The Narcissus Column', natural limestone column							
on pedestal surmounted by bronze sculpture,							
overall height 3440 with 440 square base	3200.00	2.00	37.50	-	3200.00	1	**3237.50**
'The Armillary Sundial', artificial stone baluster							
pedestal base surmounted by verdigris copper							
armillary sphere calibrated sundial	2200.00	1.00	18.75	-	2200.00	1	**2218.75**

D GROUNDWORK

Item Excluding site overheads and profit	PC £	Labour hours	Labour £	Plant £	Material £	Unit	Total rate £
D11 SOIL STABILIZATION							
Soil Stabilization - General							
Preamble - Earth-retaining and stabilizing materials are often specified as part of the earth-forming work in landscape contracts, and therefore this section lists a number of products specially designed for large-scale earth control. There are two types: rigid units for structural retention of earth on steep slopes; and flexible meshes and sheets for control of soil erosion where structural strength is not required. Prices for these items depend on quantity, difficulty of access to the site and availability of suitable filling material: estimates should be obtained from the manufacturer when the site conditions have been determined.							
Crib Walls; Keller Comtec							
"Timbercrib" timber crib walling system; machine filled with crushed rock inclusive of reinforced concrete footing cribfill stone, rear wall land drain and rear wall drainage/ separation membrane; excluding excavation							
ref 450/38; for retaining walls up to 1.20 m high	-	-	-	-	-	m²	115.00
ref 600/38; for retaining walls up to 2.20 m high	-	-	-	-	-	m²	120.00
ref 750/38; for retaining walls up to 3.10 m high	-	-	-	-	-	m²	125.00
ref 900/48; for retaining walls up to 3.60 m high	-	-	-	-	-	m²	155.00
ref 1050/48; for retaining walls up to 4.40 m high	-	-	-	-	-	m²	165.00
ref 1200/48; for retaining walls up to 5.20 m high	-	-	-	-	-	m²	185.00
ref 1500/48; for retaining walls up to 6.50 m high	-	-	-	-	-	m²	205.00
ref 1800/48; for retaining walls up to 8.20 m high	-	-	-	-	-	m²	230.00
Retaining Walls; Keller Comtec							
"Textomur" reinforced soil system embankments reinforced soil slopes at angles of 60 - 70 degrees to the horizontal; as an alternative to reinforced concrete or gabion solutions; finishing with excavated material/grass or shrubs	-	-	-	-	-	m²	95.00
Retaining Walls; RCC							
Retaining walls of units with plain concrete finish; prices based on 24 tonne loads but other quantities available (excavation, temporary shoring, foundations and backfilling not included)							
1000 wide x 1250 mm high	92.70	1.27	23.82	7.22	95.29	m	126.33
1000 wide x 1750 mm high	113.30	1.27	23.82	7.22	116.77	m	147.81
1000 wide x 2400 mm high	180.25	1.27	23.82	7.22	185.30	m	216.34
1000 wide x 2690 mm high	190.55	1.27	23.82	43.31	197.04	m	264.17
1000 wide x 3000 mm high	202.91	1.47	27.56	54.14	211.26	m	292.96
1000 wide x 3750 mm high	292.52	1.57	29.53	64.97	305.79	m	400.29
Retaining Walls; Maccaferri Ltd							
Wire mesh gabions; galvanized mesh 80 mm x 100 mm; filling with broken stones 125 mm - 200 mm size; wire down securely to manufacturer's instructions; filling front face by hand							
2 x 1 x 0.50 m	33.98	2.00	37.50	4.81	101.81	nr	144.12
2 x 1 x 1.00 m	47.87	4.00	75.00	9.63	183.53	nr	268.16
"Reno" mattress gabions							
6 x 2 x 0.17 m	84.24	3.00	56.25	7.22	222.61	nr	286.08
6 x 2 x 0.23 m	92.17	4.50	84.38	9.63	279.38	nr	373.38
6 x 2 x 0.30 m	102.35	6.00	112.50	10.83	346.54	nr	469.87

D GROUNDWORK

Item Excluding site overheads and profit	PC £	Labour hours	Labour £	Plant £	Material £	Unit	Total rate £
Retaining Walls; Tensar International "Tensar" retaining wall system; modular dry laid concrete blocks; 220 mm x 400 mm long x 150 mm high connected to "Tensar RE" geogrid with proprietary connectors; geogrid laid horizontally within the fill at 300 mm centres; on 150 x 450 concrete foundation; filling with imported granular material							
1.00 m high	75.00	3.00	56.25	8.25	83.60	m^2	**148.10**
2.00 m high	75.00	4.00	75.00	8.25	83.60	m^2	**166.85**
3.00 m high	75.00	4.50	84.38	8.25	83.60	m^2	**176.22**
Retaining Walls; Grass Concrete Ltd "Betoflor" precast concrete landscape retaining walls including soil filling to pockets (excavation, concrete foundations, backfilling stones to rear of walls and planting not included)							
"Betoflor" interlocking units; 250 mm long x 250 mm x 200 mm modular deep; in walls 250 mm wide	-	-	-	-	-	m^2	**70.16**
extra over "Betoflor" interlocking units for colours	-	-	-	-	-	m^2	**1.67**
"Betoatlas" interlocking units; 250 mm long x 500 mm wide x 200 mm modular deep; in walls 500 mm wide	-	-	-	-	-	m^2	**90.65**
extra over "Betoatlas" interlocking units for colours	-	-	-	-	-	m^2	**3.44**
ABG Ltd "Webwall"; honeycomb structured HDPE polymer strips with interlocking joints; laid to prepared terraces and backfilled with excavated material to form raked wall for planting 500 mm high units laid and backfilled horizontally within the face of the slope to produce stepped gradient; measured as bank face area; wall height:							
up to 1.00 m high	-	-	-	-	-	m^2	**80.00**
up to 2.00 m high	-	-	-	-	-	m^2	**100.00**
over 2.00 m high	-	-	-	-	-	m^2	**110.00**
250 mm high units laid and backfilled horizontally within the face of the slope to produce stepped gradient; measured as bank face area; wall height:							
up to 1.00 m high	-	-	-	-	-	m^2	**90.00**
up to 2.00 m high	-	-	-	-	-	m^2	**110.00**
over 2.00 m high	-	-	-	-	-	m^2	**120.00**
Setting out; grading and levelling; compacting bottoms of excavations	-	0.13	2.50	0.29	-	m	**2.79**
Extra over "Betoflor" retaining walls for C7P concrete foundations							
700 mm x 300 mm deep	14.33	0.27	5.00	-	14.33	m	**19.34**
Retaining Walls; Forticrete Ltd "Keystone" pc concrete block retaining wall; geogrid included for walls over 1.00 m high; excavation, concrete foundation, stone backfill to rear of wall all measured separately							
1.0m high	57.76	2.40	45.00	-	59.20	m^2	**104.20**
2.0m high	57.76	2.40	45.00	6.95	72.89	m^2	**124.84**
3.0m high	57.76	2.40	45.00	9.27	82.40	m^2	**136.67**
4.0m high	57.76	2.40	45.00	11.12	89.77	m^2	**145.89**

D GROUNDWORK

Item Excluding site overheads and profit	PC £	Labour hours	Labour £	Plant £	Material £	Unit	Total rate £
D11 SOIL STABILIZATION - cont'd							
Retaining Walls; Forticrete Ltd Stepoc Blocks; interlocking blocks; 10 mm reinforcing laid loose horizontally to preformed notches and vertical reinforcing nominal size 10 mm fixed to starter bars; infilling with concrete; foundations and starter bars measured separately							
Type 256 400 x 225 x 256 mm	42.00	1.20	22.50	-	59.66	m²	**82.16**
Type 190 400 x 225 x 190 mm	42.00	1.00	18.75	-	56.62	m²	**75.37**
Embankments; Tensar International Embankments; reinforced with "Tensar Uniaxial Geogrid"; ref 40 RE; 40 kN/m width; Geogrid laid horizontally within fill to 100% of vertical height of slope at 1.00 m centres; filling with excavated material							
slopes less than 45 degrees	2.30	0.19	3.48	1.75	2.42	m³	**7.64**
slopes exceeding 45 degrees	4.95	0.19	3.48	2.67	5.19	m³	**11.34**
Embankments; reinforced with "Tensar Uniaxial Geogrid"; ref 55 RE; 55 kN/m width; Geogrid laid horizontally within fill to 100% of vertical height of slope at 1.00 m centres; filling with excavated material							
slopes less than 45 degrees	2.90	0.19	3.48	1.75	3.04	m³	**8.28**
slopes exceeding 45 degrees	6.24	0.19	3.48	2.67	6.55	m³	**12.70**
Extra for "Tensar Mat"; erosion control mats; to faces of slopes of 45 degrees or less; filling with 20 mm fine topsoil; seeding	3.10	0.02	0.38	0.17	4.12	m²	**4.66**
Embankments; reinforced with "Tensar Uniaxial Geogrid" ref "55 RE"; 55 kN/m width; Geogrid laid horizontally within fill to 50% of horizontal length of slopes at specified centres; filling with imported fill PC £16.64/m³							
300 mm centres; slopes up to 45 degrees	4.81	0.23	4.38	2.01	21.69	m³	**28.09**
600 mm centres; slopes up to 45 degrees	2.41	0.19	3.48	2.67	22.50	m³	**28.64**
Extra for "Tensar Mat"; erosion control mats; to faces of slopes of 45 degrees or less; filling with 20 mm fine topsoil; seeding	3.10	0.02	0.38	0.17	4.12	m²	**4.66**
Embankments; reinforced with "Tensar Uniaxial Geogrid" ref "55 RE"; 55 kN/m width; Geogrid laid horizontally within fill; filling with excavated material; wrapping around at faces							
300 mm centres; slopes exceeding 45 degrees	9.66	0.19	3.48	1.75	10.14	m³	**15.37**
600 mm centres; slopes exceeding 45 degrees	4.81	0.19	3.48	1.75	5.05	m³	**10.28**
Extra over embankments for bagwork face supports for slopes exceeding 45 degrees	-	-	-	-	12.50	m²	**12.50**
Extra over embankments for seeding of bags	-	-	-	-	1.00	m²	**1.00**
Extra over embankments for "Bodkin" joints	-	-	-	-	1.00	m	**1.00**
Extra over embankments for temporary shuttering to slopes exceeding 45 degrees	-	-	-	-	10.00	m²	**10.00**
Anchoring Systems; surface **stabilization; Platipus Anchors Ltd;** **centres & depths shown** **should be verified with a design** **engineer; they may vary either way** **dependent on circumstances** Concrete revetments; hard anodised stealth anchors with stainless steel accessories installed to a depth of 1 m							
"SO4" anchors to a depth of 1.00 - 1.50 m at 1.00 m ccs	23.00	0.33	6.25	1.58	23.00	m²	**30.83**

D GROUNDWORK

Item Excluding site overheads and profit	PC £	Labour hours	Labour £	Plant £	Material £	Unit	Total rate £
Loadlocking and crimping tool; for stressing and crimping the anchors							
Purchase price	350.00	-	-	-	350.00	nr	**350.00**
Hire rate per week	60.00	-	-	-	60.00	nr	**60.00**
Surface erosion; geotextile anchoring on slopes; Aluminium alloy Stealth anchors							
"SO4", "SO6" to a depth of 1.0 - 2.0 m at 2.0 m ccs	10.50	0.20	3.75	1.58	10.50	m^2	**15.83**
Brick block or in situ concrete, distressed retaining walls; anchors installed in two rows at varying tensions between 18 kN - 50 kN along the length of the retaining wall; core drilling of wall not included							
"SO8", "BO6", "BO8" bronze Stealth and Bat anchors installed in combination with stainless steel accessories to a depth of 6 m; average cost base on 1.50 m ccs; long term solution	330.00	0.50	9.38	1.58	330.00	m^2	**340.95**
"SO8", "BO6", "BO8" Cast SG Iron Stealth and Bat anchors installed in combination with galvanised steel accessories to a depth of 6 m; average cost base on 1.50 m ccs; short term solution	141.00	0.50	9.38	1.58	141.00	m^2	**151.96**
Timber retaining wall support; anchors installed 19 kN along the length of the retaining wall							
"SO6", "SO8", bronze Stealth anchors installed in combination with stainless steel accessories to a depth of 3.0 m. 1.50 centres; average cost	49.50	0.50	9.38	1.26	49.50	m^2	**60.14**
Gabions; anchors for support and stability to moving, overturning or rotating gabion retaining walls							
"SO8", "BO6" anchors including base plate at 1.00 m centres to a depth of 4.00 m	250.00	0.33	6.25	1.58	250.00	m^2	**257.83**
Soil nailing of geofabrics; anchors fixed through surface of fabric or erosion control surface treatment (not included)							
"SO8", "BO6" anchors including base plate at 1.50 m centres to a depth of 3.00 m	100.00	0.17	3.13	1.05	100.00	m^2	**104.18**
Embankments; Grass Concrete Ltd							
"Grasscrete"; in situ reinforced concrete surfacing; to 20 mm thick sand blinding layer (not included); including soiling and seeding							
ref GC1; 100 mm thick	-	-	-	-	-	m^2	**24.07**
ref GC2; 150 mm thick	-	-	-	-	-	m^2	**28.81**
"Grassblock 103"; solid matrix precast concrete blocks; to 20 mm thick sand blinding layer; excluding edge restraint; including soiling and seeding							
406 mm x 406 mm x 103 mm; fully interlocking	-	-	-	-	-	m^2	**25.75**
Grass Reinforcement; Farmura Environmental Ltd							
"Matrix" Grass Paver; recycled polyethylene and polypropylene mixed interlocking erosion control and grass reinforcement system laid to rootzone prepared separately and filled with screened topsoil and seeded with grass seed; green							
640 x 330 x38 mm	12.50	0.13	2.34	-	14.99	m^2	**17.34**
330 x 330 x 38 mm	12.50	0.17	3.13	-	14.99	m^2	**18.12**
extra over for coloured material	4.00	-	-	-	4.00	m^2	**4.00**

Prices for Measured Works

D GROUNDWORK

Item Excluding site overheads and profit	PC £	Labour hours	Labour £	Plant £	Material £	Unit	Total rate £
D11 SOIL STABILIZATION - cont'd							
Embankments; Cooper Clarke Group Ltd							
"Ecoblock" polyethylene; 925 mm x 310 mm x 50 mm; heavy duty for car parking and fire paths							
to firm sub-soil (not included)	14.00	0.04	0.75	-	14.42	m²	15.17
to 100 mm granular fill and "Geotextile"	14.00	0.08	1.50	0.13	19.29	m²	20.92
to 250 mm granular fill and "Geotextile"	14.00	0.20	3.75	0.22	25.50	m²	29.46
Extra for filling "Ecoblock" with topsoil; seeding with rye grass at 50 g/m²	0.88	0.02	0.38	0.18	1.05	m²	1.61
"Geoweb" polyethylene soil-stabilizing panels; to soil surfaces brought to grade (not included); filling with excavated material							
panels; 2.40 x 6.10 m x 100 mm deep	6.75	0.10	1.88	2.53	6.88	m²	11.29
panels; 2.40 x 6.10 m x 200 mm deep	13.55	0.13	2.50	3.03	13.82	m²	19.35
"Geoweb" polyethylene soil-stabilizing panels; to soil surfaces brought to grade (not included); filling with ballast							
panels; 2.40 x 6.10 m x 100 mm deep	6.75	0.11	2.08	2.53	8.79	m²	13.41
panels; 2.40 x 6.10 m x 200 mm deep	13.55	0.15	2.88	3.03	17.63	m²	23.55
"Geoweb" polyethylene soil-stabilizing panels; to soil surfaces brought to grade (not included); filling with ST2 10 N/mm² concrete							
panels; 2.40 x 6.10 m x 100 mm deep	6.75	0.16	3.00	-	13.36	m²	16.36
panels; 2.40 x 6.10 m x 200 mm deep	13.55	0.20	3.75	-	26.76	m²	30.51
Extra over for filling "Geoweb" with imported topsoil; seeding with rye grass at 50 g/m²	1.84	0.07	1.25	0.30	2.19	m²	3.74
Timber log retaining walls; Longlyf Timber Products Ltd							
Machine rounded softwood logs to trenches priced separately; disposal of excavated material priced separately; inclusive of 75 mm hardcore blinding to trench and backfilling trench with site mixed concrete 1:3:6; geofabric pinned to rear of logs; heights of logs above ground							
500 mm (constructed from 1.80 m lengths)	38.50	1.50	28.13	-	47.71	m	75.84
1.20 m (constructed from 1.80 m lengths)	77.00	1.30	24.38	-	96.69	m	121.06
1.60 m (constructed from 2.40 m lengths)	49.40	2.50	46.88	-	74.52	m	121.39
2.00 m (constructed from 3.00 m lengths)	66.60	3.50	65.63	-	97.15	m	162.78
As above but with 150 mm machine rounded timbers							
500 mm	27.48	2.50	46.88	-	36.69	m	83.57
1.20 m	91.51	1.75	32.81	-	111.19	m	144.01
1.60 m	91.60	3.00	56.25	-	116.72	m	172.97
As above but with 200 mm machine rounded timbers							
1.80 m (constructed from 2.40 m lengths)	123.90	4.00	75.00	-	154.45	m	229.45
2.40 m (constructed from 3.60 m lengths)	185.85	4.50	84.38	-	216.58	m	300.95
Railway Sleeper Walls							
Construct retaining wall from railway sleepers; fixed with steel galvanised pins 12 mm driven into the ground; sleepers laid flat							
Grade 1 softwood; 2590 x 250 x 150 mm							
150 mm; 1 sleeper high	5.43	0.50	9.38	-	5.75	m	15.12
300 mm; 2 sleepers high	10.86	1.00	18.75	-	11.50	m	30.25
450 mm; 3 sleepers high	16.15	0.67	12.50	-	16.99	m	29.49
600 mm; 4 sleepers high	21.76	2.00	37.50	-	22.60	m	60.10
Grade 1 softwood as above but with 2 nr galvanised angle iron stakes set into concrete internally and screwed to the inside face of the sleepers							
750 mm; 5 sleepers high	26.92	2.50	46.88	-	58.25	m²	105.13
900 mm; 6 sleepers high	32.22	2.75	51.56	-	65.16	m²	116.73

D GROUNDWORK

Item Excluding site overheads and profit	PC £	Labour hours	Labour £	Plant £	Material £	Unit	Total rate £
Grade 1 hardwood; 2590 x 250 x 150 mm							
150 mm; 1 sleeper high	5.74	0.50	9.38	-	6.06	m	15.43
300 mm; 2 sleepers high	11.48	1.00	18.75	-	12.12	m	30.87
450 mm; 3 sleepers high	17.08	0.67	12.50	-	17.92	m	30.42
600 mm; 4 sleepers high	23.01	2.00	37.50	-	23.85	m	61.35
Grade 1 hardwood as above but with 2 nr galvanised angle iron stakes set into concrete internally and screwed to the inside face of the sleepers							
750 mm; 5 sleepers high	28.47	2.50	46.88	-	59.80	m²	106.67
900 mm; 6 sleepers high	34.07	2.75	51.56	-	67.01	m²	118.57
New Pine softwood; 2500 x 230 x 150 mm							
150 mm; 1 sleeper high	6.88	0.50	9.38	-	7.20	m	16.57
300 mm; 2 sleepers high	13.76	1.00	18.75	-	14.40	m	33.15
450 mm; 3 sleepers high	6.88	0.67	12.50	-	7.72	m	20.22
600 mm; 4 sleepers high	27.52	2.00	37.50	-	28.36	m	65.86
New pine softwood as above but with 2 nr galvanised angle iron stakes set into concrete internally and screwed to the inside face of the sleepers							
750 mm; 5 sleepers high	34.40	2.50	46.88	-	65.73	m²	112.60
900 mm; 6 sleepers high	41.28	2.75	51.56	-	74.22	m²	125.78
New Oak Hardwood 2500 x 230 x 130 mm							
150 mm; 1 sleeper high	8.68	0.50	9.38	-	9.00	m	18.38
300 mm; 2 sleepers high	17.36	1.00	18.75	-	18.00	m	36.75
450 mm; 3 sleepers high	26.04	0.67	12.50	-	26.88	m	39.38
600 mm; 4 sleepers high	34.72	2.00	37.50	-	35.56	m	73.06
New Oak Hardwood as above but with 2 nr galvanised angle iron stakes set into concrete internally and screwed to the inside face of the sleepers							
750 mm; 5 sleepers high	43.40	2.50	46.88	-	74.73	m²	121.60
900 mm; 6 sleepers high	52.08	2.75	51.56	-	85.02	m²	136.58
Excavate foundation trench; set railway sleepers vertically on end in concrete 1:3:6 continuous foundation to 33.3% of their length to form retaining wall							
Grade 1 Hardwood; finished height above ground level							
300 mm	10.18	3.00	56.25	1.88	19.71	m	77.83
500 mm	14.75	3.00	56.25	1.88	26.56	m	84.69
600 mm	20.50	3.00	56.25	1.88	39.29	m	97.42
750 mm	22.13	3.50	65.63	1.88	41.73	m	109.22
1.00 m	30.24	3.75	70.31	1.88	47.16	m	119.34
Excavate and place vertical steel universal beams 165 wide in concrete base at 2.590 centres; fix railway sleepers set horizontally between beams to form horizontal fence or retaining wall							
Grade 1 Hardwood; bay length 2.590 m							
0.50 m high; (2 sleepers)	11.81	2.50	46.88	-	44.29	bay	91.16
750 m high; (3 sleepers)	17.35	2.60	48.75	-	65.74	bay	114.49
1.00 m high; (4 sleepers)	23.26	3.00	56.25	-	84.49	bay	140.74
1.25 m high; (5 sleepers)	29.70	3.50	65.63	-	109.24	bay	174.87
1.50 m high; (6 sleepers)	34.89	3.00	56.25	-	130.01	bay	186.26
1.75 m high; (7 sleepers)	40.61	3.00	56.25	-	135.74	bay	191.99
Willow Walling to Riverbanks; LDC Ltd							
Woven willow walling as retention to riverbanks; driving or concreting posts in at 2 m centres; intermediate posts at 500 mm centres							
1.20 m high	-	-	-	-	-	m	92.40
1.50 m high	-	-	-	-	-	m	117.81

Prices for Measured Works

D GROUNDWORK

Item Excluding site overheads and profit	PC £	Labour hours	Labour £	Plant £	Material £	Unit	Total rate £
D11 SOIL STABILIZATION - cont'd							
Flexible sheet materials; Tensar International							
"Tensar Mat"; erosion mats; 3 m - 4.50 m wide; securing with "Tensar" pegs; lap rolls 100 mm; anchors at top and bottom of slopes; in trenches	3.23	0.02	0.31	-	3.55	m^2	3.86
Topsoil filling to "Tensar Mat"; including brushing and raking	0.43	0.02	0.31	0.10	0.54	m^2	0.95
"Tensar Bi-axial Geogrid"; to graded compacted base; filling with 200 mm granular fill; compacting (turf or paving to surfaces not included); 400 mm laps							
ref SS20; 39 mm x 39 mm mesh	1.50	0.01	0.25	0.14	3.54	m^2	3.94
ref SS30; 39 mm x 39 mm mesh	2.00	0.01	0.25	0.14	4.04	m^2	4.44
Flexible sheet materials; Terram Ltd							
"Terram" synthetic fibre filter fabric; to graded base (not included)							
"Terram 1000", 0.70 mm thick; mean water flow 50 litre/m^2/s	0.39	0.02	0.31	-	0.39	m^2	0.70
"Terram 2000"; 1.00 mm thick; mean water flow 33 litre/m^2/s	0.87	0.02	0.31	-	1.04	m^2	1.35
"Terram Minipack"	0.50	-	0.06	-	0.50	m^2	0.56
Flexible sheet materials; Greenfix Ltd							
"Greenfix"; erosion control mats; 10 mm -15 mm thick; fixing with 4 nr crimped pins in accordance with manufacturer's instructions; to graded surface (not included)							
unseeded "Eromat 1"; 2.40 m wide	155.00	2.00	37.50	-	223.30	100 m^2	260.80
unseeded "Eromat 2"; 2.40 m wide	175.00	2.00	37.50	-	245.30	100 m^2	282.80
unseeded "Eromat 3"; 2.40 m wide	185.00	2.00	37.50	-	256.30	100 m^2	293.80
seeded "Covamat 1"; 2.40 m wide	220.00	2.00	37.50	-	283.80	100 m^2	321.30
seeded "Covamat 2"; 2.40 m wide	250.00	2.00	37.50	-	315.30	100 m^2	352.80
seeded "Covamat 3"; 2.40 m wide	250.00	2.00	37.50	-	315.30	100 m^2	352.80
"Bioroll" 300 mm diameter to river banks and revetments	12.00	0.13	2.34	-	13.39	m	15.74
Extra over "Greenfix" erosion control mats for fertilizer applied at 70 g/m^2	3.11	0.20	3.75	-	3.11	100 m^2	6.86
Extra over "Greenfix" erosion control mats for Geojute; fixing with steel pins	1.33	0.02	0.42	-	1.45	m^2	1.86
Extra over "Greenfix" erosion control mats for laying to slopes exceeding 30 degrees	-	-	-	-	-	-	25%
Extra for the following operations							
Spreading 25 mm approved topsoil							
by machine	0.43	0.01	0.12	0.10	0.52	m^2	0.74
by hand	0.43	0.04	0.66	-	0.52	m^2	1.18
Grass seed; PC £2.80/kg; spreading in two operations; by hand							
35 g/m^2	9.80	0.17	3.13	-	9.80	100 m^2	12.93
50 g/m^2	14.00	0.17	3.13	-	14.00	100 m^2	17.13
70 g/m^2	19.60	0.17	3.13	-	19.60	100 m^2	22.73
100 g/m^2	28.00	0.20	3.75	-	28.00	100 m^2	31.75
125 g/m^2	35.00	0.20	3.75	-	35.00	100 m^2	38.75
Extra over seeding by hand for slopes over 30 degrees (allowing for the actual area but measured in plan)							
35 g/m^2	1.46	-	0.07	-	1.46	100 m^2	1.53
50 g/m^2	2.10	-	0.07	-	2.10	100 m^2	2.17
70 g/m^2	2.94	-	0.07	-	2.94	100 m^2	3.01
100 g/m^2	4.20	-	0.08	-	4.20	100 m^2	4.28
125 g/m^2	5.24	-	0.08	-	5.24	100 m^2	5.32

D GROUNDWORK

Item Excluding site overheads and profit	PC £	Labour hours	Labour £	Plant £	Material £	Unit	Total rate £
Grass seed; PC £2.80/kg; spreading in two operations; by machine							
35 g/m²	9.80	-	-	0.50	9.80	100 m²	10.30
50 g/m²	14.00	-	-	0.50	14.00	100 m²	14.50
70 g/m²	19.60	-	-	0.50	19.60	100 m²	20.10
100 g/m²	28.00	-	-	0.50	28.00	100 m²	28.50
125 kg/ha	350.00	-	-	49.52	350.00	ha	399.52
150 kg/ha	420.00	-	-	49.52	420.00	ha	469.52
200 kg/ha	560.00	-	-	49.52	560.00	ha	609.52
250 kg/ha	700.00	-	-	49.52	700.00	ha	749.52
300 kg/ha	840.00	-	-	49.52	840.00	ha	889.52
350 kg/ha	980.00	-	-	49.52	980.00	ha	1029.52
400 kg/ha	1120.00	-	-	49.52	1120.00	ha	1169.52
500 kg/ha	1400.00	-	-	49.52	1400.00	ha	1449.52
700 kg/ha	1960.00	-	-	49.52	1960.00	ha	2009.52
1400 kg/ha	3920.00	-	-	49.52	3920.00	ha	3969.52
Extra over seeding by machine for slopes over 30 degrees (allowing for the actual area but measured in plan)							
35 g/m²	9.80	-	-	0.07	1.47	100 m²	1.54
50 g/m²	14.00	-	-	0.07	2.10	100 m²	2.17
70 g/m²	19.60	-	-	0.07	2.94	100 m²	3.01
100 g/m²	28.00	-	-	0.07	4.20	100 m²	4.27
125 kg/ha	350.00	-	-	7.43	52.50	ha	59.93
150 kg/ha	420.00	-	-	7.43	63.00	ha	70.43
200 kg/ha	560.00	-	-	7.43	84.00	ha	91.43
250 kg/ha	700.00	-	-	7.43	105.00	ha	112.43
300 kg/ha	840.00	-	-	7.43	126.00	ha	133.43
350 kg/ha	980.00	-	-	7.43	147.00	ha	154.43
400 kg/ha	1120.00	-	-	7.43	168.00	ha	175.43
500 kg/ha	1400.00	-	-	7.43	210.00	ha	217.43
700 kg/ha	1960.00	-	-	7.43	294.00	ha	301.43
1400 kg/ha	3920.00	-	-	7.43	588.00	ha	595.43
D20 EXCAVATION AND FILLING							
MACHINE SELECTION TABLE **Road Equipment Ltd; machine volumes for excavating/filling only and placing excavated material alongside or to a dumper; no bulkages are allowed for in the material volumes; these rates should be increased by user-preferred percentages to suit prevailing site conditions; the figures in the next section for "Excavation mechanical" and filling allow for the use of banksmen within the rates shown below**							
1.5 tonne excavators; digging volume							
1 cycle / minute; 0.04 m³	-	0.42	7.81	2.39	-	m³	10.20
2 cycles / minute; 0.08 m³	-	0.21	3.91	1.41	-	m³	5.32
3 cycles / minute; 0.12 m³	-	0.14	2.60	1.09	-	m³	3.70
3 tonne excavators; digging volume							
1 cycle / minute; 0.13 m³	-	0.13	2.40	1.27	-	m³	3.67
2 cycles / minute; 0.26 m³	-	0.06	1.20	1.63	-	m³	2.83
3 cycles / minute; 0.39 m³	-	0.04	0.80	0.75	-	m³	1.55
5 tonne excavators; digging volume							
1 cycle / minute; 0.28 m³	-	0.06	1.12	1.50	-	m³	2.61
2 cycles / minute; 0.56 m³	-	0.03	0.56	1.16	-	m³	1.71
3 cycles / minute; 0.84 m³	-	0.02	0.37	1.17	-	m³	1.55

D GROUNDWORK

Item Excluding site overheads and profit	PC £	Labour hours	Labour £	Plant £	Material £	Unit	Total rate £
D20 EXCAVATION AND FILLING - cont'd							
MACHINE SELECTION TABLE - cont'd							
7 tonne excavators; supplied with operator; digging volume							
1 cycle / minute; 0.28 m³	-	0.06	1.12	2.28	-	m³	3.40
2 cycles / minute; 0.56 m³	-	0.47	8.84	0.41	-	m³	9.24
3 cycles / minute; 0.84 m³	-	0.52	9.72	0.33	-	m³	10.05
21 tonne excavators; supplied with operator; digging volume							
1 cycle / minute; 1.21 m³	-	-	-	0.68	-	m³	0.68
2 cycles / minute; 2.42 m³	-	-	-	0.28	-	m³	0.28
3 cycles / minute; 3.63 m³	-	-	-	0.19	-	m³	0.19
Backhoe loader; excavating; JCB 3CX **rear bucket capacity 0.28 m³**							
1 cycle / minute; 0.28 m³	-	-	-	1.78	-	m³	1.78
2 cycles / minute; 0.56 m³	-	-	-	0.89	-	m³	0.89
3 cycles / minute; 0.84 m³	-	-	-	0.59	-	m³	0.59
Backhoe loader; loading from stockpile; **JCB 3CX Front bucket capacity 1.00 m³**							
1 cycle / minute; 1.00 m³	-	-	-	0.50	-	m³	0.50
2 cycles / minute; 2.00 m³	-	-	-	0.25	-	m³	0.25
Dumpers; all volumes below are based **on excavated "Earth" moist at 1997** **/1598 kg/m³ solid/loose; a 25% bulkage** **factor has been used; the weight** **capacities below exceed the volume** **capacities of the machine in most cases;** **see the memorandum section at the** **back of this book for further weights of** **materials**							
1 tonne high tip skip loader; volume 0.485 m³ (775 kg)							
5 loads per hour	-	0.41	7.73	1.56	-	m³	9.30
7 loads per hour	-	0.29	5.52	1.15	-	m³	6.67
10 loads per hour	-	0.21	3.87	0.84	-	m³	4.71
3 tonne dumper; max. volume 2.40 m³ (3.38 t); available volume 1.9 m³							
4 loads per hour	-	0.14	2.60	0.50	-	m³	3.10
5 loads per hour	-	0.11	2.08	0.40	-	m³	2.49
7 loads per hour	-	0.08	1.49	0.30	-	m³	1.79
10 loads per hour	-	0.06	1.04	0.22	-	m³	1.26
6 tonne dumper; max. volume 3.40 m³ (5.4 t); available volume 3.77 m³							
4 loads per hour/	-	0.07	1.24	0.35	-	m³	1.59
5 loads per hour	-	0.05	1.00	0.28	-	m³	1.28
7 loads per hour	-	0.04	0.71	0.21	-	m³	0.92
10 loads per hour	-	0.03	0.50	0.16	-	m³	0.66
Market Prices of Topsoil; Charles Morris **(Fertilizers) Ltd**							
Multiple source screened topsoil	-	-	-	-	12.48	m³	12.48
Single source screened topsoil	-	-	-	-	17.32	m³	17.32
"P30" high grade topsoil	-	-	-	-	28.40	m³	28.40
Site preparation							
Removing trees							
girth 600 mm - 1.50 m	-	5.00	93.75	18.48	-	nr	112.22
girth 1.50 m - 3.00 m	-	17.00	318.75	73.90	-	nr	392.65
girth over 3.00 m girth	-	48.53	909.90	118.24	-	nr	1028.14

D GROUNDWORK

Item Excluding site overheads and profit	PC £	Labour hours	Labour £	Plant £	Material £	Unit	Total rate £
Removing tree stumps							
girth 600 mm - 1.50 m	-	2.00	37.50	66.00	-	nr	**103.50**
girth 1.50 m - 3.00 m	-	7.00	131.25	101.06	-	nr	**232.31**
girth over 3.00 m	-	12.00	225.00	173.25	-	nr	**398.25**
Stump grinding; disposing to spoil heaps							
girth 600 mm - 1.50 m	-	2.00	37.50	31.30	-	nr	**68.80**
girth 1.50 m - 3.00 m	-	2.50	46.88	62.60	-	nr	**109.47**
girth over 3.00 m	-	4.00	75.00	54.77	-	nr	**129.78**
Clearing site vegetation							
mechanical clearance	-	0.25	4.69	11.56	-	100 m²	**16.25**
hand clearance	-	2.00	37.50	-	-	100 m²	**37.50**
Lifting turf for preservation							
machine lift and stack	-	0.75	14.06	9.25	-	100 m²	**23.31**
hand lift and stack	-	8.33	156.24	-	-	100 m²	**156.24**
Note: The figures in this section relate to the machine capacities shown earlier in this section. The figures below however allow for dig efficiency based on depth. The figures below also allow for a banksman. The figures below allow for bulkages of 25% on loamy soils; adjustments should be made for different soil types.							
Excavating; mechanical; topsoil for preservation							
3 tonne tracked excavator (bucket volume 0.13 m³)							
average depth 100 mm	-	2.40	45.00	67.80	-	100 m²	**112.80**
average depth 150 mm	-	3.36	63.00	94.92	-	100 m²	**157.92**
average depth 200 mm	-	4.00	75.00	113.00	-	100 m²	**188.00**
average depth 250 mm	-	4.40	82.50	124.30	-	100 m²	**206.80**
average depth 300 mm	-	4.80	90.00	135.60	-	100 m²	**225.60**
JCB Sitemaster 3CX (bucket volume 0.28 m³)							
average depth 100 mm	-	1.00	18.75	33.00	-	100 m²	**51.75**
average depth 150 mm	-	1.25	23.44	41.25	-	100 m²	**64.69**
average depth 200 mm	-	1.78	33.38	58.74	-	100 m²	**92.12**
average depth 250 mm	-	1.90	35.63	62.70	-	100 m²	**98.33**
average depth 300 mm	-	2.00	37.50	66.00	-	100 m²	**103.50**
Excavating; mechanical; to reduce levels							
5 tonne excavator (bucket volume 0.28 m³)							
maximum depth not exceeding 0.25 m	-	0.14	2.63	0.77	-	m³	**3.39**
maximum depth not exceeding 1.00 m	-	0.10	1.78	0.52	-	m³	**2.31**
JCB Sitemaster 3CX (bucket volume 0.28 m³)							
maximum depth not exceeding 0.25 m	-	0.07	1.31	2.31	-	m³	**3.62**
maximum depth not exceeding 1.00 m	-	0.06	1.12	1.96	-	m³	**3.08**
21 tonne 360 Tracked excavator; (bucket volume 1.21 m³)							
maximum depth not exceeding 1.00 m	-	0.01	0.21	0.45	-	m³	**0.66**
maximum depth not exceeding 2.00 m	-	0.02	0.31	0.68	-	m³	**0.99**
Pits; 3 ton tracked excavator							
maximum depth not exceeding 0.25 m	-	0.58	10.88	6.09	-	m³	**16.96**
maximum depth not exceeding 1.00 m	-	0.50	9.38	5.25	-	m³	**14.63**
maximum depth not exceeding 2.00 m	-	0.60	11.25	6.30	-	m³	**17.55**
Trenches; width not exceeding 0.30 m; 3 tonne excavator							
maximum depth not exceeding 0.25 m	-	1.33	25.00	6.00	-	m³	**31.00**
maximum depth not exceeding 1.00 m	-	0.69	12.84	3.08	-	m³	**15.93**
maximum depth not exceeding 2.00 m	-	0.60	11.25	2.70	-	m³	**13.95**
Trenches; width exceeding 0.30 m; 3 tonne excavator							
maximum depth not exceeding 0.25 m	-	0.60	11.25	2.70	-	m³	**13.95**
maximum depth not exceeding 1.00 m	-	0.50	9.38	2.25	-	m³	**11.63**
maximum depth not exceeding 2.00 m	-	0.38	7.21	1.73	-	m³	**8.94**

D GROUNDWORK

Item Excluding site overheads and profit	PC £	Labour hours	Labour £	Plant £	Material £	Unit	Total rate £
D20 EXCAVATION AND FILLING - cont'd							
Excavating; mechanical; to reduce **levels** - cont'd							
Extra over any types of excavating irrespective of depth for breaking out existing materials; JCB with breaker attachment							
hard rock	-	0.50	9.38	62.50	-	m³	71.88
concrete	-	0.50	9.38	23.44	-	m³	32.81
reinforced concrete	-	1.00	18.75	43.13	-	m³	61.88
brickwork, blockwork or stonework	-	0.25	4.69	23.44	-	m³	28.13
Extra over any types of excavating irrespective of depth for breaking out existing hard pavings; JCB with breaker attachment							
concrete; 100 mm thick	-	-	-	1.72	-	m²	1.72
concrete; 150 mm thick	-	-	-	2.86	-	m²	2.86
concrete; 200 mm thick	-	-	-	3.44	-	m²	3.44
concrete; 300 mm thick	-	-	-	5.16	-	m²	5.16
reinforced concrete; 100 mm thick	-	0.08	1.56	2.30	-	m²	3.87
reinforced concrete; 150 mm thick	-	0.08	1.41	3.01	-	m²	4.42
reinforced concrete; 200 mm thick	-	0.10	1.88	4.02	-	m²	5.89
reinforced concrete; 300 mm thick	-	0.15	2.81	6.02	-	m²	8.84
tarmacadam; 75 mm thick	-	-	-	1.72	-	m²	1.72
tarmacadam and hardcore; 150 mm thick	-	-	-	2.75	-	m²	2.75
Extra over any types of excavating irrespective of depth for taking up							
precast concrete paving slabs	-	0.07	1.25	0.63	-	m²	1.88
natural stone paving	-	0.10	1.88	0.95	-	m²	2.82
cobbles	-	0.13	2.34	1.19	-	m²	3.53
brick paviors	-	0.13	2.34	1.19	-	m²	3.53
Filling to make up levels; mechanical							
Arising from the excavations							
average thickness 100 mm maximum thickness	-	-	0.08	0.13	-	m²	0.21
average thickness 150 mm maximum thickness	-	0.01	0.11	0.20	-	m²	0.31
average thickness 200 mm maximum thickness	-	0.01	0.15	0.26	-	m²	0.41
average thickness 250 mm maximum thickness	-	0.01	0.19	0.33	-	m²	0.52
average thickness 300 mm maximum thickness	-	0.01	0.25	0.44	-	m²	0.69
average thickness 400 mm maximum thickness	-	0.02	0.30	0.53	-	m²	0.83
average thickness 500 mm maximum thickness	-	0.02	0.38	0.66	-	m²	1.04
average thickness 750 mm maximum thickness	-	0.03	0.56	0.99	-	m²	1.55
Obtained from on site spoil heaps; average 25 m distance; multiple handling							
average thickness 100 mm	-	0.01	0.19	0.38	-	m²	0.57
average thickness 150 mm	-	0.02	0.28	0.57	-	m²	0.85
average thickness 250 mm	-	0.02	0.38	0.76	-	m²	1.14
average thickness 300 mm	-	0.02	0.45	0.91	-	m²	1.36
average thickness 400 mm	-	0.03	0.60	1.22	-	m²	1.82
average thickness 500 mm	-	0.04	0.78	1.59	-	m²	2.37
average thickness 750 mm	-	0.06	1.17	2.38	-	m²	3.55
Obtained off site; planting quality topsoil PC £17.32 / m³							
average thickness 100 mm	1.73	0.01	0.13	0.22	2.08	m²	2.43
average thickness 150 mm	2.60	0.01	0.19	0.33	3.12	m²	3.64
average thickness 200 mm	3.46	0.01	0.25	0.44	4.16	m²	4.84
average thickness 250 mm	4.33	0.02	0.31	0.55	5.20	m²	6.06
average thickness 300 mm	5.20	0.02	0.28	0.50	6.24	m²	7.01
average thickness 400 mm	6.93	0.02	0.38	0.66	8.31	m²	9.35
average thickness 500 mm	8.66	0.03	0.47	0.82	10.39	m²	11.69
average thickness 750 mm	12.99	0.04	0.70	1.24	15.59	m²	17.53

D GROUNDWORK

Item Excluding site overheads and profit	PC £	Labour hours	Labour £	Plant £	Material £	Unit	Total rate £
Obtained off site; hardcore; PC £12.00 / m³							
average thickness 100 mm	1.20	0.01	0.13	0.22	1.32	m²	**1.67**
average thickness 200 mm	2.40	0.01	0.25	0.44	2.64	m²	**3.33**
average thickness 250 mm	3.00	0.02	0.31	0.55	3.30	m²	**4.17**
average thickness 300 mm	3.60	0.02	0.28	0.50	3.96	m²	**4.74**
average thickness 400 mm	4.80	0.02	0.38	0.66	5.28	m²	**6.32**
average thickness 500 mm	6.00	0.03	0.47	0.82	6.60	m²	**7.89**
average thickness 750 mm	9.00	0.04	0.70	1.24	9.90	m²	**11.84**
Filling to make up levels; mechanical							
Arising from the excavations							
average thickness not exceeding 0.25 m; depositing in layers 150 mm maximum thickness	-	0.04	0.75	1.32	-	m³	**2.07**
average thickness exceeding 0.25 m; depositing in layers 150 mm maximum thickness	-	0.03	0.62	1.10	-	m³	**1.72**
Obtained off site; topsoil; blended grade PC £12.48/m³							
average thickness; not exceeding 0.25 m; depositing in layers 150 maximum thickness (JCB)	12.48	0.07	1.25	2.20	14.98	m³	**18.43**
average thickness; exceeding 0.25 m; depositing in layers 150 maximum thickness. (JCB)	12.48	0.03	0.62	1.10	14.98	m³	**16.70**
average thickness; exceeding 0.25 m; depositing in layers 150 maximum thickness. (360 degree excavator 21 tonne)	12.48	0.02	0.38	1.03	14.98	m³	**16.38**
Filling to make up levels; hand							
Arising from the excavations							
average thickness exceeding 0.25 m; depositing in layers 150 mm maximum thickness	-	0.60	11.25	-	-	m³	**11.25**
Obtained from on site spoil heaps; average 25 m distance; multiple handling							
average thickness exceeding 0.25 m thick; depositing in layers 150 mm maximum thickness	-	1.00	18.75	-	-	m³	**18.75**
Excavating; hand							
Topsoil for preservation; loading to barrows							
average depth 100 mm	-	0.24	4.50	-	-	m²	**4.50**
average depth 150 mm	-	0.36	6.75	-	-	m²	**6.75**
average depth 200 mm	-	0.58	10.80	-	-	m²	**10.80**
average depth 250 mm	-	0.72	13.50	-	-	m²	**13.50**
average depth 300 mm	-	0.86	16.20	-	-	m²	**16.20**
Excavating; hand							
Topsoil to reduce levels							
average depth 100 mm	-	0.33	6.25	-	-	m²	**6.25**
average depth 150 mm	-	0.53	9.84	-	-	m²	**9.84**
average depth 200 mm	-	0.73	13.75	-	-	m²	**13.75**
average depth 250 mm	-	1.00	18.75	-	-	m²	**18.75**
average depth 300 mm	-	1.30	24.38	-	-	m²	**24.38**
average depth 400 mm	-	1.87	35.00	-	-	m²	**35.00**
average depth 600 mm	-	3.00	56.25	-	-	m²	**56.25**
average depth 750 mm	-	4.00	75.00	-	-	m²	**75.00**
average depth 1.00 mm	-	6.67	124.99	-	-	m²	**124.99**
Pits							
maximum depth not exceeding 0.25 m	-	2.67	50.00	-	-	m³	**50.00**
maximum depth not exceeding 1.00 m	-	3.47	65.00	-	-	m³	**65.00**
maximum depth not exceeding 2.00 m	-	6.93	130.00	-	-	m³	**130.00**
Trenches; width not exceeding 0.30 m							
maximum depth not exceeding 0.25 m	-	2.86	53.57	-	-	m³	**53.57**
maximum depth not exceeding 1.00 m	-	3.72	69.77	-	-	m³	**69.77**
Trenches; width exceeding 0.30 m wide							
maximum depth not exceeding 0.25 m	-	2.86	53.57	-	-	m³	**53.57**
maximum depth not exceeding 1.00 m	-	4.00	75.00	-	-	m³	**75.00**
maximum depth not exceeding 2.00 m	-	8.00	150.00	-	-	m³	**150.00**

D GROUNDWORK

Item Excluding site overheads and profit	PC £	Labour hours	Labour £	Plant £	Material £	Unit	Total rate £
D20 EXCAVATION AND FILLING - cont'd							
Excavating; hand - cont'd							
Extra over any types of excavating irrespective of depth for breaking out existing materials; hand held pneumatic breaker							
rock	-	5.00	93.75	31.60	-	m³	125.35
concrete	-	2.50	46.88	15.80	-	m³	62.67
reinforced concrete	-	4.00	75.00	31.22	-	m³	106.22
brickwork, blockwork or stonework	-	1.50	28.13	9.48	-	m³	37.61
Disposal							
Note to users. Most commercial site disposal is carried out by 20 tonne - 8 wheeled vehicles. It has been customary to calculate disposal from construction sites in terms of full 15 m³ loads. Spon's research has found that based on weights of common materials such as clean hardcore and topsoil, that vehicles could not load more than 12 m³ at a time. Most hauliers do not make it apparent that their loads are calculated by weight and not by volume. The rates below reflect these lesser volumes which are limited by the 20 tonne limit.							
The volumes shown below are based on volumes "in the solid". weight and bulking factors have been applied. For further information please see the weights of typical materials in the Earthworks section of the Memoranda at the back of this book.							
Disposal; mechanical; Recycled Materials Ltd							
Light soils and loams (bulking factor - 1.25); 40 tonnes (2 loads per hour)							
Excavated material; off site; to tip; mechanically loaded (JCB)							
inert	-	0.04	0.78	1.25	14.85	m³	16.88
clean concrete	-	0.04	0.78	1.25	4.95	m³	6.98
soil (sandy and loam) dry	-	0.04	0.78	1.25	14.85	m³	16.88
soil (sandy and loam) wet	-	0.04	0.78	1.25	15.75	m³	17.78
broken out compacted materials such as road bases and the like	-	0.04	0.78	1.25	13.50	m³	15.53
soil (clay) dry	-	0.04	0.78	1.25	19.27	m³	21.31
soil (clay) wet	-	0.04	0.78	1.25	20.02	m³	22.06
slightly contaminated	-	0.04	0.78	1.25	21.45	m³	23.48
rubbish	-	0.04	0.78	1.25	25.74	m³	27.77
As above but allowing for 3 loads (36 m³ - 60 tonne) per hour removed							
Inert material	-	0.03	0.52	0.83	14.85	m³	16.21
Excavated material; off site; to tip; mechanically loaded by grab; capacity of load 7.25 m³							
inert material	-	-	-	-	-	m³	36.30
soil (sandy and loam) dry	-	-	-	-	-	m³	36.30
soil (sandy and loam) wet	-	-	-	-	-	m³	38.50
broken out compacted materials such as road bases and the like	-	-	-	-	-	m³	40.33
soil (clay) dry	-	-	-	-	-	m³	47.12
soil (clay) wet	-	-	-	-	-	m³	48.95

D GROUNDWORK

Item Excluding site overheads and profit	PC £	Labour hours	Labour £	Plant £	Material £	Unit	Total rate £
Disposal by skip; 7 yd3 (5.35 m³)							
Excavated material; off site; to tip							
by machine	-	0.13	2.34	1.13	24.17	m³	27.63
by hand	-	3.00	56.25	-	36.25	m³	92.50
Excavated material; on site; in spoil heaps							
average 25 m distance	-	0.02	0.42	0.88	-	m³	1.30
average 50 m distance	-	0.04	0.73	1.52	-	m³	2.25
average 100 m distance	-	0.06	1.16	2.42	-	m³	3.58
average 200 m distance	-	0.12	2.25	4.68	-	m³	6.93
Excavated material; spreading on site							
average 25 m distance	-	0.02	0.42	4.06	-	m³	4.48
average 50 m distance	-	0.04	0.73	4.58	-	m³	5.31
average 100 m distance	-	0.06	1.16	5.61	-	m³	6.77
average 200 m distance	-	0.12	2.25	5.34	-	m³	7.59
Disposal; hand							
Excavated material; on site; in spoil heaps							
average 25 m distance	-	2.40	45.00	-	-	m³	45.00
average 50 m distance	-	2.64	49.50	-	-	m³	49.50
average 100 m distance	-	3.00	56.25	-	-	m³	56.25
average 200 m distance	-	3.60	67.50	-	-	m³	67.50
Excavated material; spreading on site							
average 25 m distance	-	2.64	49.50	-	-	m³	49.50
average 50 m distance	-	3.00	56.25	-	-	m³	56.25
average 100 m distance	-	3.60	67.50	-	-	m³	67.50
average 200 m distance	-	4.20	78.75	-	-	m³	78.75
Cultivating							
Ripping up subsoil; using approved subsoiling machine; minimum depth 250 mm below topsoil; at 1.20 m centres; in							
gravel or sandy clay	-	-	-	2.34	-	100m²	2.34
soil compacted by machines	-	-	-	2.73	-	100m²	2.73
clay	-	-	-	2.92	-	100m²	2.92
chalk or other soft rock	-	-	-	5.84	-	100m²	5.84
Extra for subsoiling at 1 m centres	-	-	-	0.58	-	100m²	0.58
Breaking up existing ground; using pedestrian operated tine cultivator or rotavator; loam or sandy soil							
100 mm deep	-	0.22	4.13	2.13	-	100m²	6.26
150 mm deep	-	0.28	5.16	2.67	-	100m²	7.82
200 mm deep	-	0.37	6.87	3.56	-	100m²	10.43
As above but in heavy clay or wet soils							
100 mm deep	-	0.44	8.25	4.27	-	100m²	12.52
150 mm deep	-	0.66	12.38	6.40	-	100m²	18.77
200 mm deep	-	0.82	15.47	8.00	-	100m²	23.47
Breaking up existing ground; using tractor drawn tine cultivator or rotavator							
100 mm deep	-	-	-	0.57	-	100 m²	0.57
150 mm deep	-	-	-	0.71	-	100 m²	0.71
200 mm deep	-	-	-	0.95	-	100 m²	0.95
400 mm deep	-	-	-	2.86	-	100 m²	2.86
Cultivating ploughed ground; using disc, drag, or chain harrow							
4 passes	-	-	-	3.43	-	100 m²	3.43
Rolling cultivated ground lightly; using self-propelled agricultural roller	-	0.06	1.04	0.58	-	100 m²	1.63
Importing and storing selected and approved topsoil; to BS 3882; 20 tonne load = 11.88 m³ average							
1 - 20 tonne	17.32	-	-	-	51.96	m³	51.96
over 20 tonne	17.32	-	-	-	17.32	m³	17.32

D GROUNDWORK

Item Excluding site overheads and profit	PC £	Labour hours	Labour £	Plant £	Material £	Unit	Total rate £
D20 EXCAVATION AND FILLING - cont'd							
Surface treatments							
Compacting							
bottoms of excavations	-	0.01	0.09	0.02	-	m²	**0.12**
Grading							
Trimming surfaces of cultivated ground to final levels, removing roots stones and debris exceeding 50 mm in any direction to tip offsite; slopes less than 15 degrees							
clean ground with minimal stone content	-	0.25	4.69	-	-	100 m²	**4.69**
slightly stony - 0.5 kg stones per m²	-	0.33	6.24	-	-	100 m²	**6.25**
very stony - 1.0 - 3.00 kg stones per m²	-	0.50	9.38	-	0.01	100 m²	**9.38**
clearing mixed slightly contaminated rubble inclusive of roots and vegetation	-	0.50	9.38	-	0.03	100 m²	**9.40**
clearing brick-bats stones and clean rubble	-	0.60	11.25	-	0.02	100 m²	**11.27**

E IN SITU CONCRETE/LARGE PC CONCRETE

Item Excluding site overheads and profit	PC £	Labour hours	Labour £	Plant £	Material £	Unit	Total rate £
E10 IN SITU CONCRETE							
General							
The concrete mixes used here are referred to as							
"Design" "Standard" and "Designated" mixes.							
The BS references on these are used to denote							
the concrete strength and mix volumes. Please							
refer to the definitions and the tables in the							
Memoranda for each Trade, Concrete Work, at							
the back of this book.							
Designed mix							
User specified performance of the concrete.							
Producer responsible for selecting appropriate							
mix. Strength testing is essential.							
Prescribed mix							
User specified mix constituents and is responsible							
for ensuring that the concrete meets performance							
requirements. The mix proportion is essential.							
Standard mix							
Specified from the list in BS 5328 Pt 2 1991 s.4.							
Made with a restricted range of materials.							
Specification to include the proposed use of the							
material as well as; the standard mix reference,							
the type of cement, type and size of aggregate,							
slump (workability). Quality assurance required.							
Designated mix							
Mix specified in BS 5328 Pt 2: 1991 s.5.							
Producer to hold current product conformity							
certification and quality approval to BS 5750 Pt 1							
(EN 29001). Quality assurance essential. The mix							
may not be modified.							
Concrete mixes; mixed on site; costs for							
producing concrete; prices for commonly							
used mixes for various types of work;							
based on bulk load 20 tonne rates for							
aggregates							
Roughest type mass concrete such as footings,							
road haunchings 300 thick							
1:3:6	62.37	1.00	18.75	-	62.37	m³	**81.12**
1:3:6 sulphate resisting	67.53	1.00	18.75	-	67.53	m³	**86.28**
As above but aggregates delivered in 10 tonne							
loads							
1:3:6	64.07	1.00	18.75	-	64.07	m³	**82.82**
1:3:6 sulphate resisting	69.23	1.00	18.75	-	69.23	m³	**87.98**
As above but aggregates delivered in 900 Kg							
bulk bags							
1:3:6	97.07	1.00	18.75	-	97.07	m³	**115.82**
1:3:6 sulphate resisting	102.23	1.00	18.75	-	102.23	m³	**120.98**
Most ordinary use of concrete such as mass							
walls above ground, road slabs etc. and general							
reinforced concrete work							
1:2:4	74.40	1.00	18.75	-	74.40	m³	**93.15**
1:2:4 sulphate resisting	81.70	1.00	18.75	-	81.70	m³	**100.45**
As above but aggregates delivered in 10 tonne							
loads							
1:2:4	76.10	1.00	18.75	-	76.10	m³	**94.85**
1:2:4 sulphate resisting	83.40	1.00	18.75	-	83.40	m³	**102.15**
As above but aggregates delivered in 900 kg							
bulk bags							
1:2:4	109.10	1.00	18.75	-	109.10	m³	**127.85**
1:2:4 sulphate resisting	116.40	1.00	18.75	-	116.40	m³	**135.15**

Prices for Measured Works

E IN SITU CONCRETE/LARGE PC CONCRETE

Item Excluding site overheads and profit	PC £	Labour hours	Labour £	Plant £	Material £	Unit	Total rate £
E10 IN SITU CONCRETE - cont'd							
Concrete mixes; mixed on site - cont'd							
Watertight floors, pavements and walls, tanks pits steps paths surface of two course roads, reinforced concrete where extra strength is required							
1:1.5:3	85.89	1.00	18.75	-	85.89	m³	104.64
As above but aggregates delivered in 10 tonne loads							
1:1.5:3	87.59	1.00	18.75	-	87.59	m³	106.34
As above but aggregates delivered in 900 kg bulk bags							
1:1.5:3	120.59	1.00	18.75	-	120.59	m³	139.34
Plain in situ concrete; site mixed; 10 N/mm² - 40 aggregate (1:3:6); (aggregate delivery indicated)							
Foundations							
ordinary portland cement; 10 tonne ballast loads	82.82	1.00	18.75	-	84.89	m³	103.64
ordinary portland cement; 900 kg bulk bags	115.82	1.00	18.75	-	118.72	m³	137.47
sulphate resistant cement; 10 tonne ballast loads	83.40	2.00	37.50	-	83.40	m³	120.90
sulphate resistant cement; 900 kg bulk bags	120.98	1.00	18.75	-	124.00	m³	142.75
Foundations; poured on or against earth or unblinded hardcore							
ordinary portland cement; 10 tonne ballast loads	82.82	1.00	18.75	-	86.96	m³	105.71
ordinary portland cement; 900 kg bulk bags	115.82	1.00	18.75	-	121.61	m³	140.36
sulphate resistant cement; 10 tonne ballast loads	83.40	2.00	37.50	-	83.40	m³	120.90
sulphate resistant cement; 900 kg bulk bags	120.98	1.00	18.75	-	127.03	m³	145.78
Isolated foundations							
ordinary portland cement; 10 tonne ballast loads	82.82	1.10	20.63	-	84.89	m³	105.52
ordinary portland cement; 900 kg bulk bags	115.82	1.10	20.63	-	118.72	m³	139.34
sulphate resistant cement; 10 tonne ballast loads	83.40	2.10	39.38	-	83.40	m³	122.77
sulphate resistant cement; 900 kg bulk bags	120.98	1.10	20.63	-	124.00	m³	144.63
Plain in situ concrete; site mixed; 21 N/mm² - 20 aggregate (1:2:4)							
Foundations							
ordinary portland cement; 10 tonne ballast loads	94.85	1.00	18.75	-	97.22	m³	115.97
ordinary portland cement; 900 kg bulk bags	127.85	1.00	18.75	-	131.05	m³	149.80
sulphate resistant cement; 10 tonne ballast loads	83.40	2.00	37.50	-	83.40	m³	120.90
sulphate resistant cement; 900 kg bulk bags	120.98	1.00	18.75	-	124.00	m³	142.75
Foundations; poured on or against earth or unblinded hardcore							
ordinary portland cement; 900 kg bulk bags	127.85	1.65	30.94	-	131.05	m³	161.98
sulphate resistant cement; 900 kg bulk bags	135.14	1.65	30.94	-	138.52	m³	169.46
Isolated foundations							
ordinary portland cement	127.85	2.20	41.25	-	131.05	m³	172.30
sulphate resistant cement	135.14	2.20	41.25	-	138.52	m³	179.77
Reinforced insitu concrete; site mixed; 21 N/mm² - 20 aggregate (1:2:4); aggregates delivered in 10 tonne loads							
Foundations							
ordinary portland cement	94.85	2.20	41.25	-	97.22	m³	138.47
sulphate resistant cement	102.15	2.20	41.25	-	104.70	m³	145.95
Foundations; poured on or against earth or unblinded hardcore							
ordinary portland cement	94.85	2.20	41.25	-	97.22	m³	138.47
sulphate resistant cement	135.14	2.20	41.25	-	138.52	m³	179.77
Isolated foundations							
ordinary portland cement	94.85	2.75	51.56	-	97.22	m³	148.78
sulphate resistant cement	135.14	2.75	51.56	-	138.52	m³	190.08

E IN SITU CONCRETE/LARGE PC CONCRETE

Item Excluding site overheads and profit	PC £	Labour hours	Labour £	Plant £	Material £	Unit	Total rate £
Plain in situ concrete; ready mixed; **Tarmac Southern, 10 N/mm mixes;** **suitable for mass concrete fill and** **blinding**							
Foundations							
GEN1;Designated mix	66.10	1.50	28.13	-	66.10	m³	94.22
ST2; Standard mix	64.71	1.50	28.13	-	67.95	m³	96.07
Foundations; poured on or against earth or unblinded hardcore							
GEN1;Designated mix	66.10	1.57	29.53	-	66.10	m³	95.63
ST2; Standard mix	64.71	1.57	29.53	-	69.56	m³	99.09
Isolated foundations							
GEN1;Designated mix	66.10	2.00	37.50	-	66.10	m³	103.60
ST2; Standard mix	64.71	2.00	37.50	-	67.95	m³	105.45
Plain in situ concrete; ready mixed; **Tarmac Southern, 15 N/mm mixes;** **suitable for oversite below suspended** **slabs and strip footings in non** **aggressive soils**							
Foundations							
GEN 2; Designated mix	68.20	1.50	28.13	-	71.61	m³	99.73
ST3; Standard mix	66.81	1.50	28.13	-	70.15	m³	98.28
Foundations; poured on or against earth or unblinded hardcore							
GEN2; Designated mix	68.20	1.57	29.53	-	68.20	m³	97.73
ST3; Standard mix	66.81	1.57	29.53	-	66.81	m³	96.34
Isolated foundations							
GEN2; Designated mix	68.20	2.00	37.50	-	68.20	m³	105.70
ST3; Standard mix	66.81	2.00	37.50	-	66.81	m³	104.31
Plain in situ concrete; ready mixed; **Tarmac Southern; air entrained** **mixes suitable for paving**							
Beds or slabs; house drives parking and external paving							
PAV 1; 35 N/mm²; Designated mix	77.08	1.50	28.13	-	80.93	m³	109.06
Beds or slabs; heavy duty external paving							
PAV 2; 40 N/mm²; Designated mix	79.12	1.50	28.13	-	83.08	m³	111.20
Reinforced in situ concrete; ready **mixed; Tarmac Southern; 35 N/mm²** **mix; suitable for foundations in class 2** **sulphate conditions**							
Foundations							
RC 35; Designated mix	72.41	2.00	37.50	-	76.03	m³	113.53
Foundations; poured on or against earth or unblinded hardcore							
RC 35; Designated mix	72.41	2.10	39.38	-	77.84	m³	117.22
Isolated foundations							
RC 35; Designated mix	72.41	2.00	37.50	-	72.41	m³	109.91
Ready mix concrete; extra for loads of **less than 6 m³; Ace Minimix**							
Ready mix concrete small load surcharge	-	-	-	-	15.00	m³	15.00

E IN SITU CONCRETE/LARGE PC CONCRETE

Item Excluding site overheads and profit	PC £	Labour hours	Labour £	Plant £	Material £	Unit	Total rate £
E20 FORMWORK FOR IN SITU CONCRETE							
Plain vertical formwork; basic finish							
Sides of foundations							
height exceeding 1.00 m	-	2.00	37.50	-	4.58	m²	42.08
height not exceeding 250 mm	-	1.00	18.75	-	1.28	m	20.03
height 250 - 500 mm	-	1.00	18.75	-	2.29	m	21.04
height 500 mm - 1.00 m	-	1.50	28.13	-	4.58	m	32.70
Sides of foundations; left in							
height over 1.00 m	-	2.00	37.50	-	9.58	m²	47.08
height not exceeding 250 mm	-	1.00	18.75	-	4.97	m	23.72
height 250 - 500 mm	-	1.00	18.75	-	9.67	m	28.42
height 500 mm - 1.00 m	-	1.50	28.13	-	19.33	m	47.46
E30 REINFORCEMENT FOR IN SITU CONCRETE							
The rates for reinforcement shown for steel bar below are based on prices which would be supplied on a typical landscape contract. The steel prices shown have been priced on a selection of steel delivered to site where the total order quantity is in the region of 2 tonnes. The assumption is that should larger quantities be required, the work would fall outside the scope of the typical landscape contract defined in the front of this book. Keener rates can be obtained for larger orders.							
Reinforcement bars; BS 4449; hot rolled plain round mild steel; straight							
Bars							
8 mm nominal size	400.00	27.00	506.25	-	400.00	tonne	906.25
10 mm nominal size	400.00	26.00	487.50	-	400.00	tonne	887.50
12 mm nominal size	400.00	25.00	468.75	-	400.00	tonne	868.75
16 mm nominal size	400.00	24.00	450.00	-	400.00	tonne	850.00
20 mm nominal size	400.00	23.00	431.25	-	400.00	tonne	831.25
Reinforcement bars; BS 4449; hot rolled plain round mild steel; bent							
Bars							
8 mm nominal size	400.00	27.00	506.25	-	400.00	tonne	906.25
10 mm nominal size	400.00	26.00	487.50	-	400.00	tonne	887.50
12 mm nominal size	400.00	25.00	468.75	-	400.00	tonne	868.75
16 mm nominal size	400.00	24.00	450.00	-	400.00	tonne	850.00
20 mm nominal size	400.00	23.00	431.25	-	400.00	tonne	831.25
Reinforcement bar to concrete formwork							
6 mm Straight bar							
100 ccs	0.88	0.33	6.19	-	0.88	m²	7.07
200 ccs	0.56	0.25	4.69	-	0.56	m²	5.25
300 ccs	0.28	0.15	2.81	-	0.28	m²	3.09
8 mm bar							
100 ccs	1.56	0.33	6.19	-	1.56	m²	7.75
200 ccs	0.76	0.25	4.69	-	0.76	m²	5.45
300 ccs	0.52	0.15	2.81	-	0.52	m²	3.33
10 mm bar							
100 ccs	2.44	0.33	6.19	-	2.44	m²	8.63
200 ccs	1.20	0.25	4.69	-	1.20	m²	5.89
300 ccs	0.80	0.15	2.81	-	0.80	m²	3.61
12 mm bar							
100 ccs	3.52	0.33	6.19	-	3.52	m²	9.71
200 ccs	1.76	0.25	4.69	-	1.76	m²	6.45
300 ccs	1.16	0.15	2.81	-	1.16	m²	3.97

E IN SITU CONCRETE/LARGE PC CONCRETE

Item Excluding site overheads and profit	PC £	Labour hours	Labour £	Plant £	Material £	Unit	Total rate £
16 mm bar							
100 ccs	6.28	0.40	7.50	-	6.28	m²	**13.78**
200 ccs	3.16	0.33	6.19	-	3.16	m²	**9.35**
300 ccs	2.12	0.25	4.69	-	2.12	m²	**6.81**
25 mm bar							
100 ccs	15.40	0.40	7.50	-	15.40	m²	**22.90**
200 ccs	7.68	0.33	6.19	-	7.68	m²	**13.87**
300 ccs	5.12	0.25	4.69	-	5.12	m²	**9.81**
32 mm bar							
100 ccs	25.24	0.40	7.50	-	25.24	m²	**32.74**
200 ccs	12.60	0.33	6.19	-	12.60	m²	**18.79**
300 ccs	8.40	0.25	4.69	-	8.40	m²	**13.09**
Reinforcement fabric; BS 4483; lapped;							
in beds or suspended slabs							
Fabric							
ref A98 (1.54 kg/m²)	1.20	0.22	4.13	-	1.32	m²	**5.45**
ref A142 (2.22 kg/m²)	1.50	0.22	4.13	-	1.65	m²	**5.78**
ref A193 (3.02 kg/m²)	2.05	0.22	4.13	-	2.25	m²	**6.38**
ref A252 (3.95 kg/m²)	2.75	0.24	4.50	-	3.02	m²	**7.53**
ref A393 (6.16 kg/m²)	4.20	0.28	5.25	-	4.62	m²	**9.87**

F MASONRY

Item Excluding site overheads and profit	PC £	Labour hours	Labour £	Plant £	Material £	Unit	Total rate £
F MARKET PRICES OF MATERIALS							
Cement; Builder Centre							
Portland cement	-	-	-	-	3.38	25 kg	3.38
Sulphate resistant cement	-	-	-	-	3.98	25 kg	3.98
White cement	-	-	-	-	6.60	25 kg	6.60
Sand; Builder Centre							
Building sand							
loose	-	-	-	-	34.00	m³	34.00
900 kg bulk bags	-	-	-	-	29.95	nr	29.95
Sharp sand							
loose	-	-	-	-	34.00	m³	34.00
900 kg bulk bags	-	-	-	-	34.00	nr	34.00
Sand; Yeoman Aggregates Ltd							
Sharp sand	-	-	-	-	17.60	tonne	17.60
Bricks; E.T. Clay Products							
Ibstock; facing bricks; 215 x 102.5 x 65 mm							
Leicester Red Stock	-	-	-	-	404.25	1000	404.25
Leicester Yellow Stock	-	-	-	-	341.25	1000	341.25
Himley Mixed Russet	-	-	-	-	352.27	1000	352.27
Himley Worcs. Mixture	-	-	-	-	341.25	1000	341.25
Roughdales Red Multi Rustic	-	-	-	-	375.38	1000	375.38
Ashdown Cottage Mixture	-	-	-	-	415.80	1000	415.80
Ashdown Crowborough Multi	-	-	-	-	482.79	1000	482.79
Ashdown Pevensey Multi	-	-	-	-	433.13	1000	433.13
Chailey Stock	-	-	-	-	471.45	1000	471.45
Dorking Multi Coloured	-	-	-	-	358.05	1000	358.05
Holbrook Smooth Red	-	-	-	-	444.68	1000	444.68
Stourbridge Kenilworth Multi	-	-	-	-	340.73	1000	340.73
Stourbridge Pennine Pastone	-	-	-	-	309.75	1000	309.75
Stratford Red Rustic	-	-	-	-	323.40	1000	323.40
Swanage Restoration Red	-	-	-	-	675.67	1000	675.67
Laybrook Sevenoaks Yellow	-	-	-	-	336.00	1000	336.00
Laybrook Arundel Yellow	-	-	-	-	357.00	1000	357.00
Laybrook Thakeham Red	-	-	-	-	304.50	1000	304.50
Funton Second Hard Stock	-	-	-	-	409.50	1000	409.50
Hanson Brick Ltd, London Brand; facing bricks; 215 x 102.5 x 65 mm							
Capel Multi Stock	-	-	-	-	290.00	1000	290.00
Rusper Stock	-	-	-	-	295.00	1000	295.00
Delph Autumn	-	-	-	-	330.00	1000	330.00
Regency	-	-	-	-	330.00	1000	330.00
Sandfaced	-	-	-	-	330.00	1000	330.00
Saxon Gold	-	-	-	-	330.00	1000	330.00
Tudor	-	-	-	-	330.00	1000	330.00
Windsor	-	-	-	-	330.00	1000	330.00
Autumn Leaf	-	-	-	-	330.00	1000	330.00
Claydon Red Multi	-	-	-	-	330.00	1000	330.00
Other facing bricks; 215 x 102.5 x 65 mm							
Soft Reds - Milton Hall	-	-	-	-	375.00	1000	375.00
Staffs Blues - Blue Smooth	-	-	-	-	465.00	1000	465.00
Staffs Blues - Blue Brindle	-	-	-	-	425.00	1000	425.00
Reclaimed (second hand) bricks							
Yellows	-	-	-	-	980.00	1000	980.00
Yellow Multi	-	-	-	-	980.00	1000	980.00
Mixed London Stock	-	-	-	-	700.00	1000	700.00
Red Multi	-	-	-	-	650.00	1000	650.00
Gaults	-	-	-	-	540.00	1000	540.00
Red Rubbers	-	-	-	-	780.00	1000	780.00
Common bricks	-	-	-	-	185.00	1000	185.00
Engineering bricks	-	-	-	-	280.00	1000	280.00
Kempston facing bricks; 215 x 102.5 x 65 mm							
Melford Yellow	-	-	-	-	330.00	1000	330.00

F MASONRY

Item Excluding site overheads and profit	PC £	Labour hours	Labour £	Plant £	Material £	Unit	Total rate £
F10 BRICK/BLOCK WALLING							
Mortar mixes; common mixes for various							
types of work; mortar mixed on site;							
prices based on builders merchant rates							
for cement; aggregates delivered in 900							
kg bulk bags; mechanically mixed;							
batching quantities for these mortar							
mixes may be found in the memorandum							
section of this book							
1:3	-	0.75	14.06	-	138.90	m³	**152.96**
1:4	-	0.75	14.06	-	110.57	m³	**124.63**
1:1:6	-	0.75	14.06	-	111.63	m³	**125.69**
1:1:6 Sulphate resisting	-	0.75	14.06	-	118.11	m³	**132.17**
Mortar mixes; common mixes for various							
types of work; mortar mixed on site;							
prices based on builders merchant rates							
for cement; aggregates delivered in 900							
kg bulk bags; hand mixed							
1:3	-	1.00	18.75	-	138.90	m³	**157.65**
1:4	-	1.00	18.75	-	109.19	m³	**127.94**
1:1:6	-	1.00	18.75	-	111.63	m³	**130.38**
1:1:6 Sulphate resisting	-	1.00	18.75	-	118.11	m³	**136.86**
Mortar mixes; common mixes for various							
types of work; mortar mixed on site;							
prices based on builders merchant rates							
for cement; aggregates delivered in 10							
tonne loads; mechanically mixed;							
batching quantities for these mortar							
mixes may be found in the memorandum							
section of this book							
1:3	-	0.75	14.06	-	115.58	m³	**129.64**
1:4	-	0.75	14.06	-	89.68	m³	**103.75**
1:1:6	-	0.75	14.06	-	92.13	m³	**106.19**
1:1:6 Sulphate resisting	-	0.75	14.06	-	98.61	m³	**112.67**
Variation in brick prices							
Add or subtract the following amounts for every							
£1.00/1000 difference in the PC price of the							
measured items below							
half brick thick	-	-	-	-	0.06	m²	**0.06**
one brick thick	-	-	-	-	0.13	m²	**0.13**
one and a half brick thick	-	-	-	-	0.19	m²	**0.19**
two brick thick	-	-	-	-	0.25	m²	**0.25**
Common bricks; PC £200.00 /1000;							
English garden wall bond; in gauged							
mortar (1:1:6)							
Mechanically offloading; maximum 25 m distance;							
loading to wheel barrows and transporting to							
location; per 215 mm thick walls	-	0.42	7.81	-	-	m²	**7.81**
Walls							
half brick thick (strecher bond)	-	1.80	33.75	-	14.19	m2	**47.94**
half brick thick (using site cut snap headers to							
form bond)	-	2.40	45.06	-	14.19	m2	**59.25**
one brick thick	-	3.60	67.49	-	27.91	m2	**95.40**
one and a half brick thick	-	5.40	101.24	-	41.86	m2	**143.10**
two brick thick	-	7.20	134.99	-	55.81	m2	**190.80**
Walls; curved; mean radius 6 m							
half brick thick	-	1.35	25.38	-	15.32	m2	**40.70**
one brick thick	-	4.27	80.09	-	29.59	m2	**109.68**
Walls; curved; mean radius 1.50 m							
half brick thick	-	2.62	49.10	-	15.92	m2	**65.02**
one brick thick	-	5.24	98.17	-	30.79	m2	**128.95**

F MASONRY

Item Excluding site overheads and profit	PC £	Labour hours	Labour £	Plant £	Material £	Unit	Total rate £
F10 BRICK/BLOCK WALLING - cont'd							
Common bricks; PC £200.00 /1000 - cont'd							
Extra for cement mortar (1:3) in lieu of gauged mortar							
half brick thick	-	-	-	-	0.35	m²	0.35
one brick thick	-	-	-	-	0.70	m²	0.70
one and a half brick thick	-	-	-	-	1.06	m²	1.06
two brick thick	-	-	-	-	1.41	m²	1.41
Walls; stretcher bond; wall ties at 450 centres vertically and horizontally							
one brick thick	0.44	1.71	32.15	-	28.35	m²	60.50
two brick thick	1.31	3.43	64.29	-	57.16	m²	121.45
Facing bricks; PC £300.00 /1000; English garden wall bond; in gauged mortar (1:1:6); facework one side							
Mechanically offloading; maximum 25 m distance; loading to wheel barrows and transporting to location; per 215 mm thick walls	-	0.42	7.81	-	-	m²	7.81
Walls							
half brick thick	-	1.80	33.75	-	20.49	m2	54.24
half brick thick half brick thick (using site cut snap headers to form bond)	-	2.40	45.00	-	20.49	m2	65.49
one brick thick	-	3.60	67.49	-	40.99	m2	108.48
one and a half brick thick	-	5.40	101.24	-	61.48	m2	162.72
two brick thick	-	7.20	134.99	-	81.97	m2	216.96
Walls; curved; mean radius 6 m							
half brick thick	-	1.35	25.38	-	21.92	m2	47.30
one brick thick	-	4.27	80.09	-	42.79	m2	122.88
Walls; curved; mean radius 1.50 m							
half brick thick	18.00	2.62	49.10	-	21.47	m2	70.57
one brick thick	36.00	5.24	98.17	-	41.89	m2	140.05
Walls; tapering; one face battering; average							
one and a half brick thick	-	5.60	105.01	-	64.18	m²	169.19
two brick thick	-	7.47	139.98	-	85.57	m²	225.55
Walls; battering (retaining)							
one and a half brick thick	54.00	5.60	105.01	-	64.18	m²	169.19
two brick thick	72.00	7.47	139.98	-	85.57	m²	225.55
Isolated piers; English bond; facework all round							
one brick thick	36.00	7.00	131.25	-	45.58	m²	176.83
one and a half brick thick	54.00	9.00	168.75	-	68.37	m²	237.12
two brick thick	72.00	10.00	187.50	-	92.45	m²	279.95
three brick thick	108.00	12.40	232.50	-	136.74	m²	369.24
Projections; vertical							
one brick x half brick	4.00	0.70	13.13	-	4.57	m	17.70
one brick x one brick	8.00	1.40	26.25	-	9.14	m	35.39
one and a half brick x one brick	12.00	2.10	39.38	-	13.71	m	53.09
two brick x one brick	16.00	2.30	43.13	-	18.29	m	61.41
Brickwork fair faced both sides; facing bricks in gauged mortar (1:1:6)							
Extra for fair face both sides; flush, struck, weathered, or bucket-handle pointing	-	0.67	12.50	-	-	m²	12.50
Extra for cement mortar (1:3) in lieu of gauged mortar							
half brick thick	-	-	-	-	0.35	m²	0.35
one brick thick	-	-	-	-	0.70	m²	0.70
one and a half brick thick	-	-	-	-	1.06	m²	1.06
two brick thick	-	-	-	-	1.41	m²	1.41

F MASONRY

Item Excluding site overheads and profit	PC £	Labour hours	Labour £	Plant £	Material £	Unit	Total rate £
Class B engineering bricks; PC £260.00 /1000; double Flemish bond in cement mortar (1:3) Mechanically offloading; maximum 25 m distance; loading to wheel barrows; transporting to location; per 215 mm thick walls	-	0.42	7.81	-	-	m²	7.81
Walls							
half brick thick	-	1.80	33.75	-	18.32	m2	52.07
one brick thick	-	3.60	67.49	-	36.65	m2	104.14
one and a half brick thick	-	5.40	101.24	-	62.83	m2	164.07
two brick thick	-	7.20	134.99	-	73.30	m2	208.29
Walls; curved; mean radius 6 m							
half brick thick	-	1.35	25.38	-	18.97	m2	44.35
one brick thick	-	4.27	80.09	-	36.65	m2	116.74
Walls; curved; mean radius 1.50 m							
half brick thick	-	2.62	49.10	-	18.97	m2	68.07
one brick thick	-	5.24	98.17	-	36.65	m2	134.82
Walls; tapering; one face battering; average							
one and a half brick thick	-	5.60	105.01	-	54.97	m2	159.99
two brick thick	-	7.47	139.98	-	73.30	m2	213.27
Walls; tapering; one face battering; average							
one and a half brick thick	-	5.60	105.01	-	54.97	m²	159.99
two brick thick	-	7.47	139.98	-	73.30	m²	213.27
Walls; battering (retaining)							
one and a half brick thick	-	5.60	105.01	-	54.97	m²	159.99
two brick thick	-	7.47	139.98	-	73.30	m²	213.27
Isolated piers							
one brick thick	-	7.00	131.25	-	40.54	m²	171.79
one and a half brick thick	-	9.00	168.75	-	60.81	m²	229.56
two brick thick	-	10.00	187.50	-	82.37	m²	269.87
three brick thick	-	12.40	232.50	-	121.62	m²	354.12
Projections; vertical							
one brick x half brick	-	0.70	13.13	-	4.09	m	17.22
one brick x one brick	-	1.40	26.25	-	8.19	m	34.44
one and a half brick x one brick	-	2.10	39.38	-	12.28	m	51.66
two brick by one brick	-	2.30	43.13	-	16.37	m	59.50
Walls; half brick thick							
in honeycomb bond	-	1.80	33.75	-	13.25	m²	47.00
in quarter bond	-	1.67	31.25	-	17.93	m²	49.18
Brickwork fair faced both sides; facing bricks in gauged mortar (1:1:6)							
Extra for fair face both sides; flush, struck, weathered, or bucket-handle pointing	-	0.67	12.50	-	-	m²	12.50
Brick copings Copings; all brick headers-on-edge; to BS 4729; two angles rounded 53 mm radius; flush pointing top and both sides as work proceeds; one brick wide; horizontal							
machine-made specials	24.66	0.16	2.93	-	26.18	m	29.11
hand-made specials	24.67	0.16	2.93	-	26.18	m	29.12
Extra over copings for two courses machine-made tile creasings, projecting 25 mm each side; 260 mm wide copings; horizontal	5.26	0.50	9.38	-	5.91	m	15.29
Copings; all brick headers-on-edge; flush pointing top and both sides as work proceeds; one brick wide; horizontal							
facing bricks PC £300.00/1000	4.00	0.31	5.86	-	4.30	m	10.16
engineering bricks PC £260.00/1000	3.47	0.31	5.86	-	3.64	m	9.50

F MASONRY

Item Excluding site overheads and profit	PC £	Labour hours	Labour £	Plant £	Material £	Unit	Total rate £
F10 BRICK/BLOCK WALLING - cont'd							
Dense aggregate concrete blocks; **"Tarmac Topblock" or other equal and** **approved; in gauged mortar (1:2:9)**							
Walls							
Solid Blocks 7N/mm^2							
440 x 215 x 100 mm thick	7.85	1.20	22.50	-	8.89	m^2	**31.39**
440 x 215 x 140 mm thick	11.90	1.30	24.38	-	13.07	m^2	**37.44**
Solid Blocks 7N/mm^2 laid flat							
440 x 100 x 215 mm thick	16.01	3.22	60.38	-	18.80	m^2	**79.18**
Hollow Concrete blocks							
440 x 215 x 215 mm thick	14.25	1.30	24.38	-	15.42	m^2	**39.79**
Filling of hollow concrete blocks with concrete as work proceeds; tamping and compacting							
440 x 215 x 215 mm thick	17.34	0.20	3.75	-	17.34	m^2	**21.09**
F20 NATURAL STONE RUBBLE WALLING							
Granite walls							
Granite random rubble walls; laid dry							
200 mm thick; single faced	33.33	6.66	124.88	-	33.33	m^2	**158.20**
Granite walls; one face battering to 50 degrees; pointing faces							
450 mm (average) thick	75.00	5.00	93.75	-	81.44	m^2	**175.19**
Dry stone walling - General							
Preamble: In rural areas where natural stone is a traditional material, it may be possible to use dry stone walling or dyking as an alternative to fences or brick walls. Many local authorities are willing to meet the extra cost of stone walling in areas of high landscape value, and they may hold lists of available craftsmen. DSWA Office, Westmorland County Showground, Lane Farm, Crooklands, Milnthorpe, Cumbria, LA7 7NH; Tel: 01539 567953; information@dswa.org.uk							
Note: Traditional walls are not built on concrete foundations.							
Dry stone wall; wall on concrete foundation (not included) dry-stone coursed wall inclusive of locking stones and filling to wall with broken stone or rubble; walls up to 1.20 m high; battered; 2 sides fair faced							
Yorkstone	-	6.50	121.88	-	42.00	m^2	**163.88**
Cotswold stone	-	6.50	121.88	-	48.00	m^2	**169.88**
Purbeck	-	6.50	121.88	-	66.00	m^2	**187.88**
Rockery stone - General							
Preamble: Rockery stone prices vary considerably with source, carriage, distance and load. Typical garden centre prices for small quantities are in the range of £30 - £40 per tonne collected.							
Rockery stone; Breedon							
Lumping limestone 9 - 12 inches; minimum 15/20 tonne loads							
boulders; maximum diameter 400 mm; to positions maximum 25 m distance from offload	56.50	2.60	48.75	-	56.50	tonne	**105.25**

F MASONRY

Item Excluding site overheads and profit	PC £	Labour hours	Labour £	Plant £	Material £	Unit	Total rate £
Rockery stone; Civil Engineering Developments							
Boulders; maximum distance 25 m; by machine							
750 mm diameter	120.00	0.90	16.88	7.22	120.00	nr	144.09
1 m diameter	111.40	2.00	37.50	14.44	111.40	nr	163.34
1.5 m diameter	111.40	2.00	37.50	45.20	111.40	nr	194.10
2 m diameter	1735.00	2.00	37.50	45.20	1735.00	nr	1817.70
Boulders; maximum distance 25 m; by hand							
not exceeding 750 mm diameter	55.70	0.75	14.06	-	55.70	nr	69.76
not exceeding 1 m diameter	120.00	1.89	35.35	-	120.00	nr	155.35
F22 CAST STONE WALLING							
Brett Landscaping and Building Products; dwarf walls; in Cotswold stone; grey, brown or red							
"Gloucester" stone walls							
uniform size precast units	18.00	2.50	46.88	-	25.78	m²	72.65
random size precast units	18.00	4.00	75.00	-	25.78	m²	100.78
Brett Landscaping and Building Products; cast stone characteristic walls with irregular size stone facing							
"Weathered Cotswold" drystone walls							
600 x 100 x 125 mm	47.48	0.67	12.50	-	51.37	m²	63.87
300 x 100 x 125 mm	71.66	1.50	28.13	-	75.55	m²	103.67
Copings 400 x 150 x 150	4.14	0.40	7.50	-	8.03	m	15.53
Copings							
flat	4.14	0.20	3.75	-	4.23	m	7.98
Haddonstone Ltd; cast stone piers; ornamental masonry in Portland Bath or Terracotta finished cast stone							
Gate Pier S120; to foundations and underground work measured separately; concrete infill							
S120G base unit to pier; 699 x 699 x 172 mm	114.00	0.75	14.06	-	118.63	nr	132.69
S120F/F shaft base unit; 533 x 533 x 280 mm	91.00	1.00	18.75	-	95.76	nr	114.51
S120E/E main shaft unit; 533 x 533 x 280 mm; nr of units required dependent on height of pier	91.00	1.00	18.75	-	95.76	nr	114.51
S120D/D top shaft unit; 33 x 533 x 280 mm	91.00	1.00	18.75	-	95.76	nr	114.51
S120C pier cap unit; 737 x 737 x 114 mm	109.00	0.50	9.38	-	113.76	nr	123.13
S120B pier block unit; base for finial 533 x 533x 64 mm	47.00	0.33	6.19	-	47.13	nr	53.31
Pier blocks; flat to receive gate finial							
S100B; 440 x 440 x 63 mm	30.00	0.33	6.25	-	30.30	nr	36.55
S120B; 546 x 546 x 64 mm	47.00	0.33	6.25	-	47.30	nr	53.55
S150B; 330 x 330 x 51 mm	16.00	0.33	6.25	-	16.30	nr	22.55
Pier caps; part weathered							
S100C; 915 x 915 x 150 mm	258.00	0.50	9.38	-	258.59	nr	267.97
S120C; 737 x 737 x 114 mm	109.00	0.50	9.38	-	109.30	nr	118.67
S150C; 584 x 584 x 120 mm	79.00	0.50	9.38	-	79.30	nr	88.67
Pier caps; weathered							
S230C; 1029 x 1029 x 175 mm	339.00	0.50	9.38	-	339.30	nr	348.67
S215C; 687 x 687 x 175 mm	148.00	0.50	9.38	-	148.30	nr	157.67
S210C; 584 x 584 x 175 mm	93.00	0.50	9.38	-	93.30	nr	102.67
Pier strings							
S100S; 800 x 800 x 55 mm	88.00	0.50	9.38	-	88.18	nr	97.56
S120S; 555 x 555 x 44 mm	40.00	0.50	9.38	-	40.18	nr	49.56
S150S; 457 x 457 x 48 mm	25.00	0.50	9.38	-	25.18	nr	34.56
Balls and bases							
E150A Ball 535 mm and E150C collared base	237.45	0.50	9.38	-	237.63	nr	247.01
E120A Ball 330 mm and E120C collared base	83.40	0.50	9.38	-	83.58	nr	92.96
E110A Ball 230 mm and E110C collared base	61.28	0.50	9.38	-	61.46	nr	70.84
E100A Ball 170 mm and E100B plain base	38.29	0.50	9.38	-	38.47	nr	47.85

F MASONRY

Item Excluding site overheads and profit	PC £	Labour hours	Labour £	Plant £	Material £	Unit	Total rate £
F22 CAST STONE WALLING - cont'd							
Haddonstone Ltd; cast stone copings; **ornamental masonry in Portland Bath or** **Terracotta finished cast stone**							
Copings for walls; bedded, jointed and pointed in approved coloured cement-lime mortar 1:2:9							
T100 weathered coping 102 mm high 178 mm wide x 914 mm	28.34	0.33	6.24	-	28.59	m	**34.84**
T140 weathered coping 102 mm high 337 mm wide x 914 mm	46.87	0.33	6.25	-	47.12	m	**53.37**
T200 weathered coping 127 mm high 508 mm wide x 750 mm	71.94	0.33	6.25	-	72.19	m	**78.44**
T170 weathered coping 108 mm high 483 mm wide x 914	76.30	0.33	6.25	-	76.55	m	**82.80**
T340 raked coping 100-75 mm high 290 wide x 900 mm	43.60	0.33	6.25	-	43.85	m	**50.10**
T310 raked coping 89-76 mm high 381 wide x 914 mm	51.23	0.33	6.25	-	51.48	m	**57.73**
Bordeaux Walling; Forticrete Ltd							
Dry stacked random sized units 150-400 mm long x 150 high cast stone wall mechanically interlocked with fibreglass pins; constructed to levelling pad of coarse compacted granular material back filled behind the elevation with 300 wide granular drainage material; walls to 5.00 m high (Retaining walls over heights shown below require individual design)							
Gravity wall - near vertical wall 250 mm thick	53.00	1.00	18.75	-	53.00	m²	**71.75**
Gravity wall 9.5 deg battered	53.00	2.00	37.50	-	53.00	m²	**90.50**
Copings to Bordeaux wall 70 thick random lengths	8.00	0.25	4.69	-	8.00	m	**12.69**
Retaining Wall; as above but reinforced **with Tensar geogrid 40 RE laid between** **every two courses horizontally into the** **face of the excavation (Excavation not** **included)**							
Near vertical wall 250 mm thick; 1.50 m of geogrid length							
Up to 1.20 m high; 2 layers of geogrid	53.00	1.50	28.13	-	59.90	m²	**88.03**
1.20 m - 1.50 m high; 3 layers of geogrid	53.00	2.00	37.50	-	63.35	m²	**100.85**
1.50 - 1.8 m high; 4 courses of geogrid	53.00	2.25	42.19	-	66.80	m²	**108.99**
Battered walls max 1:3 slope							
Up to 1.20 m high; 3 layers of geogrid	53.00	2.50	46.88	-	63.35	m²	**110.22**
1.20 m - 1.50 m high; 4 layers of geogrid	53.00	2.50	46.88	-	66.80	m²	**113.67**
1.50 - 1.8 m high; 5 layers of geogrid	53.00	3.00	56.25	-	69.67	m²	**125.92**
Blanc de Bierges; cast stone hand **textured characteristic walls with regular** **size stone facing; light cream-buff**							
Textured concrete walls; mortar 1:3							
440 x 125 x 100 mm	61.39	0.67	12.50	-	66.81	m²	**79.31**
600 x 125 x 100	61.39	1.25	23.44	-	66.81	m²	**90.25**
440 x 215 x 100 mm	61.39	1.50	28.13	-	66.81	m²	**94.94**

F MASONRY

Item Excluding site overheads and profit	PC £	Labour hours	Labour £	Plant £	Material £	Unit	Total rate £
F30 ACCESSORIES/SUNDRY ITEMS FOR BRICK/BLOCK STONE WALLING							
Damp proof courses; pitch polymer; 150 mm laps							
Horizontal							
width not exceeding 225 mm	3.90	1.14	21.38	-	5.36	m^2	**26.74**
width exceeding 225 mm	3.90	0.58	10.88	-	5.85	m^2	**16.72**
Vertical							
width not exceeding 225 mm	3.90	1.72	32.25	-	5.36	m^2	**37.61**
Two courses slates in cement mortar (1:3)							
Horizontal							
width exceeding 225 mm	8.43	3.46	64.88	-	12.32	m^2	**77.20**
Vertical							
width exceeding 225 mm	8.43	5.18	97.13	-	12.32	m^2	**109.45**
F31 PRECAST CONCRETE SILLS/LINTELS/COPINGS/FEATURES							
Mix 21.00 N/mm^2 - 20 aggregate (1:2:4)							
Copings; once weathered; twice grooved							
152 x 75 mm	4.68	0.40	7.50	-	5.01	m	**12.51**
178 x 65 mm	6.85	0.40	7.50	-	7.17	m	**14.67**
305 x 75 mm	9.78	0.50	9.38	-	10.30	m	**19.67**
Pier caps; four sides weathered							
305 x 305 mm	5.10	1.00	18.75	-	5.26	nr	**24.01**
381 x 381 mm	5.35	1.00	18.75	-	5.51	nr	**24.26**
533 x 533 mm	5.90	1.20	22.50	-	6.06	nr	**28.56**
Copings; Milner Delvaux; "Blanc de Bierges"							
200 x 50 mm	11.85	0.50	9.38	-	12.59	m	**21.97**
300 x 50 mm	15.14	0.50	9.38	-	15.98	m	**25.36**
400 x 50 mm	16.48	0.50	9.38	-	17.36	m	**26.74**
300 x 80 mm	22.76	0.50	9.38	-	23.83	m	**33.21**
400 x 80 mm	26.57	0.50	9.38	-	27.95	m	**37.33**

Prices for Measured Works

G STRUCTURAL/CARCASSING METAL/TIMBER

Item Excluding site overheads and profit	PC £	Labour hours	Labour £	Plant £	Material £	Unit	Total rate £
G31 PREFABRICATED TIMBER UNIT DECKING							
Timber Decking; Wyckham Blackwell							
Supports for timber decking; treated softwood							
joists to receive decking boards; joists at 400 mm							
centres							
50 x 100 mm	6.90	1.00	18.75	-	9.35	m^2	**28.10**
50 x 150 mm	10.35	1.00	18.75	-	13.14	m^2	**31.90**
50 x 125 mm	8.60	1.00	18.75	-	11.22	m^2	**29.97**
Hardwood Decking Yellow Balau; grooved or							
smooth; 6 mm joints							
Deck boards; 90 mm wide x 19 mm thick	18.11	1.00	18.75	-	20.13	m^2	**38.88**
Deck boards; 145 mm wide x 21 mm thick	20.26	1.00	18.75	-	24.30	m^2	**43.05**
Deck boards; 145 mm wide x 28 mm thick	27.34	1.00	18.75	-	32.09	m^2	**50.84**
Hardwood Decking Tatajuba; grooved or							
smooth; 6 mm joints							
Deck boards; 90 mm wide x 19 mm thick	16.14	1.00	18.75	-	18.15	m^2	**36.90**
Deck boards; 145 mm wide x 21 mm thick	18.27	1.00	18.75	-	20.28	m^2	**39.03**
Deck boards; 145 mm wide x 28 mm thick	24.36	1.00	18.75	-	26.37	m^2	**45.12**
Western Red Cedar; 6 mm joints							
Prime Deck Grade; 90 mm wide x 40 mm thick	41.64	1.00	18.75	-	47.82	m^2	**66.57**
Prime Deck Grade; 142 mm wide x 40 mm thick	43.88	1.00	18.75	-	50.28	m^2	**69.03**
Patio Grade; 90 mm wide x 35 mm thick	36.54	1.00	18.75	-	42.21	m^2	**60.96**
Redwood; kiln dried PAR Tanalised; 6 mm joints							
Deck boards 120 mm wide x 28 mm thick;							
grooved	10.31	0.75	14.06	-	13.35	m^2	**27.41**
Handrails and base rail; fixed to posts at 2.00 m							
centres							
Posts 100 x 100 x 1370 high	5.82	1.00	18.75	-	6.77	m	**25.52**
Posts turned 1220 high	12.94	1.00	18.75	-	13.89	m	**32.64**
Handrails; balusters							
Square balusters at 100 mm centres	34.60	0.50	9.38	-	35.08	m	**44.45**
Square balusters at 300 mm centres	11.52	0.33	6.24	-	11.90	m	**18.15**
Turned balusters at 100 mm centres	54.40	0.50	9.38	-	54.88	m	**64.25**
Turned balusters at 300 mm centres	18.12	0.33	6.19	-	18.50	m	**24.68**

H CLADDING/COVERING

Item Excluding site overheads and profit	PC £	Labour hours	Labour £	Plant £	Material £	Unit	Total rate £
H51 NATURAL STONE SLAB CLADDING/ FEATURES							
Sawn Yorkstone cladding; Johnsons Wellfield Quarries Ltd Six sides sawn stone; rubbed face; sawn and jointed edges; fixed to blockwork (not included) with stainless steel fixings "Ancon Ltd" grade 304 stainless steel frame cramp and dowel 7 mm; cladding units drilled 4 x to receive dowels 440 x 200 x 50 mm thick	86.28	1.87	35.06	-	89.53	m²	**124.59**
H52 CAST STONE SLAB CLADDING/FEATURES							
Cast Stone cladding; Haddonstone Ltd Reconstituted stone in Portland Bath or Terracotta; fixed to blockwork or concrete (not included) with stainless steel fixings "Ancon Ltd" grade 304 stainless steel frame cramp and dowel M6 mm; cladding units drilled 4 x to receive dowels 440 x 200 x 50 mm thick	94.65	1.87	35.06	-	97.90	m²	**132.96**
Cast masonry cladding; 60 N/mm²; Blanc de Bierges; pre-drilled to accommodate dowels; fixing to vertical face with stainless steel cramps 300 x 400 x 50 400 x 400 x 50 300 x 600 x 50 400 x 600 x 50	77.25 77.25 77.25 77.25	2.50 2.00 1.75 1.50	46.88 37.50 32.81 28.13	- - - -	90.69 87.75 86.58 84.25	m² m² m² m²	**137.56** **125.25** **119.39** **112.37**
Cast masonry cladding; 60 N/mm²; Blanc de Bierges; fixing to vertical face and jointing with mortar 1:1:6 300 x 400 x 100 400 x 400 x 100 300 x 600 x 100 400 x 600 x 100	100.94 100.94 100.94 100.94	3.00 2.75 2.50 2.00	56.25 51.56 46.88 37.50	- - - -	103.86 103.86 103.86 103.86	m² m² m² m²	**160.11** **155.42** **150.74** **141.36**

J WATERPROOFING

Item Excluding site overheads and profit	PC £	Labour hours	Labour £	Plant £	Material £	Unit	Total rate £
J10 SPECIALIST WATERPROOF RENDERING							
"Sika" waterproof rendering; steel **trowelled**							
Walls; 20 thick; three coats; to concrete base							
width exceeding 300 mm	-	-	-	-	-	m²	48.84
width not exceeding 300 mm	-	-	-	-	-	m²	76.53
Walls; 25 thick; three coats; to concrete base							
width exceeding 300 mm	-	-	-	-	-	m²	55.37
width not exceeding 300 mm	-	-	-	-	-	m²	87.93
J20 MASTIC ASPHALT TANKING/DAMP **PROOFING**							
Tanking and damp proofing; mastic **asphalt; to BS 6925; type T 1097;** **Coverite Ltd**							
13 mm thick; one coat covering; to concrete base; flat; work subsequently covered							
width exceeding 300 mm	-	-	-	-	-	m²	10.22
20 mm thick; two coat coverings; to concrete base; flat; work subsequently covered							
width exceeding 300 mm	-	-	-	-	-	m²	12.80
30 mm thick; three coat coverings; to concrete base; flat; work subsequently covered							
width exceeding 300 mm	-	-	-	-	-	m²	17.39
13 mm thick; two coat coverings; to brickwork base; vertical; work subsequently covered							
width exceeding 300 mm	-	-	-	-	-	m²	33.95
20 mm thick; three coat coverings; to brickwork base; vertical; work subsequently covered							
width exceeding 300 mm	-	-	-	-	-	m²	46.35
Internal angle fillets; work subsequently covered	-	-	-	-	-	m	3.61
Turning asphalt nibs into grooves; 20 mm deep	-	-	-	-	-	m	2.29
J30 LIQUID APPLIED TANKING/DAMP **PROOFING**							
Tanking and damp proofing; Ruberoid **Building Products, "Synthaprufe" cold** **applied bituminous emulsion waterproof** **coating**							
"Synthaprufe"; to smooth finished concrete or screeded slabs; flat; blinding with sand							
two coats	1.52	0.22	4.17	-	1.82	m²	5.98
three coats	2.29	0.31	5.81	-	2.66	m²	8.47
"Synthaprufe"; to fair faced brickwork with flush joints, rendered brickwork, or smooth finished concrete walls; vertical							
two coats	1.71	0.29	5.36	-	2.03	m²	7.38
three coats	2.51	0.40	7.50	-	2.91	m²	10.41
Tanking and damp proofing; RIW Ltd							
Liquid asphaltic composition; to smooth finished concrete screeded slabs or screeded slabs; flat							
two coats	3.73	0.33	6.25	-	3.73	m²	9.98
Liquid asphaltic composition; fair-faced brickwork with flush joints, rendered brickwork, or smooth finished concrete walls; vertical							
two coats	3.73	0.50	9.38	-	3.73	m²	13.11

J WATERPROOFING

Item Excluding site overheads and profit	PC £	Labour hours	Labour £	Plant £	Material £	Unit	Total rate £
"Heviseal"; to smooth finished concrete or screeded slabs; to surfaces of ponds, tanks, planters; flat							
two coats	5.67	0.33	6.25	-	6.24	m²	**12.49**
"Heviseal"; to fair faced brickwork with flush joints, rendered brickwork, or smooth finished concrete walls; to surfaces of retaining walls, ponds, tanks, planters; vertical							
two coats	5.67	0.50	9.38	-	6.24	m²	**15.61**
J40 FLEXIBLE SHEET TANKING/DAMP PROOFING							
Tanking and damp proofing; Grace Construction Products "Bitu-thene 2000"; 1.00 mm thick; overlapping and bonding; including sealing all edges							
to concrete slabs; flat	2.90	0.25	4.69	-	3.19	m²	**7.88**
to brick/concrete walls; vertical	2.90	0.40	7.50	-	3.44	m²	**10.94**
J50 GREEN ROOF SYSTEMS							
Preamble - A variety of systems are available which address all the varied requirements for a successful Green Roof. For installation by approved contractors only the prices shown are for budgeting purposes only as each installation is site specific and may incorporate some or all of the resources shown. Specifiers should verify that the systems specified include for design liability and inspections by the suppliers. The systems below assume commercial insulation levels are required to the space below the proposed Green Roof. Extensive Green roofs are those of generally lightweight construction with low maintenance planting and shallow soil designed for aesthetics only; Intensive Green roofs are designed to allow use for recreation and trafficking. They require more maintenance and allow a greater variety of surfaces and plant types.							
Intensive Green Roof; Bauder Ltd; soil based systems able to provide a variety of hard and soft landscaping; laid to the surface of an unprepared roof deck Vapour barrier laid to prevent intersticial condensation from spaces below the roof applied by torching to the roof deck							
VB4-Expal aluminium lined	-	-	-	-	-	m²	**10.51**
Insulation laid and hot bitumen bonded to vapour barrier							
PIR Insulation 100 mm	-	-	-	-	-	m²	**23.67**
Underlayer to receive rootbarrier partially bonded to insulation by torching							
G4E	-	-	-	-	-	m²	**10.29**

J WATERPROOFING

Item Excluding site overheads and profit	PC £	Labour hours	Labour £	Plant £	Material £	Unit	Total rate £
J50 GREEN ROOF SYSTEMS - cont'd							
Intensive Green Roof; Bauder Ltd - cont'd							
Root barrier							
"Plant E"; chemically treated root resistant capping sheet fully bonded to G4E underlayer by torching	-	-	-	-	-	m²	14.07
Slip layers to absorb differential movement							
PE Foil 2 layers laid to root barriers	-	-	-	-	-	m²	2.85
Optional protection layer to prevent mechanical damage							
"Protection mat" 6 mm thick rubber matting loose laid	-	-	-	-	-	m²	8.93
Drainage medium laid to root barrier							
"Drainage board" free draining EPS 50 mm thick	-	-	-	-	-	m²	9.90
"Reservoir board" up to 21.5 litre water storage capacity EPS 75 mm thick	-	-	-	-	-	m²	12.47
Filtration to prevent soil migration to drainage system							
"Filter fleece" 3 mm thick polyester geotextile loose laid over drainage/reservoir layer	-	-	-	-	-	m²	2.96
For hard landscaped areas incorporate Rigid Drainage Board laid to the protection mat							
PLT 60 Drainage Board	-	-	-	-	-	m²	16.21
Extensive Green Roof System; Bauder Ltd; low maintenance soil free system incorporating single layer growing and planting medium							
Vapour barrier laid to prevent intersticial condensation applied by torching to the roof deck							
VB4-Expal aluminium lined	-	-	-	-	-	m²	10.51
Insulation laid and hot bitumen bonded to vapour barrier							
PIR Insulation 100 mm	-	-	-	-	-	m²	23.67
Underlayer to receive rootbarrier partially bonded to insulation by torching							
G4E	-	-	-	-	-	m²	10.29
Root barrier							
"Plant E"; chemically treated root resistant capping sheet fully bonded to G4E underlayer by torching	-	-	-	-	-	m²	14.07
Landscape Options							
Hydroplanting system; Bauder Ltd; to Extensive Green Roof as detailed above							
Ecomat 6 mm thick geotextile loose laid Hydroplanting with sedum and succulent coagulant directly onto plant substrate to 60 mm deep	-	-	-	-	-	m²	12.90
Xeroflor vegetation blanket; Bauder Ltd; to Extensive Green Roof as detailed above							
Xeroflor Xf 301 pre cultivated sedum blanket incorporating 800 gram recycled fibre water retention layer laid loose	-	-	-	-	-	m²	33.93
Waterproofing to Upstands; Bauder Ltd							
Bauder Vapour Barrier; Bauder G4E & Bauder Plant E							
up to 200 mm high	-	-	-	-	-	m	16.94
up to 400 mm high	-	-	-	-	-	m	24.58
up to 600 mm high	-	-	-	-	-	m	32.44

J WATERPROOFING

Item Excluding site overheads and profit	PC £	Labour hours	Labour £	Plant £	Material £	Unit	Total rate £
Inverted waterproofing systems; Alumasc Exterior Building Products Ltd; to roof surfaces to receive Green Roof systems "Hydrotech 6125"; monolithic hot melt rubberised bitumen; applied in two 3 mm layers incorporating a polyester reinforcing sheet with 4 mm thick protection sheet and chemically impregnated root barrier; fully bonded into the Hydrotech; applied to plywood or suitably prepared wood float finish and primed concrete deck or screeds							
10 mm thick	-	-	-	-	-	m^2	27.00
"Alumasc Roofmate"; extruded polystyrene insulation; optional system; thickness to suit required U value; calculated at design stage; indicative thicknesses; laid to Hydrotech 6125							
0.25 U value; average requirement 120 mm	-	-	-	-	-	m^2	16.00
Warm Roof waterproofing systems; Alumasc Exterior Building Products Ltd; to roof surfaces to receive Green Roof systems "Derbigum" system							
"Nilperm" aluminium lined vapour barrier; 2 mm thick bonded in hot bitumen to the roof deck	-	-	-	-	-	m^2	7.50
"Korklite" insulation bonded to the vapour barrier in hot bitumen;U value dependent; 80 mm thick	-	-	-	-	-	m^2	17.00
"Hi-Ten Universal" 2 mm thick underlayer; fully bonded to the insulation	-	-	-	-	-	m^2	6.25
"Derbigum Anti-Root" cap sheet impregnated with root resisting chemical bonded to the underlayer	-	-	-	-	-	m^2	15.00
Intensive Green Roof Systems; Alumasc Exterior Building Products Ltd; components laid to the insulation over the Hydrotech or Derbigum waterproofing above Optional inclusion; moisture retention layer;							
SSM-45; moisture mat	-	-	-	-	-	m^2	4.50
Drainage layer; "Floradrain" recycled polypropylene; providing water reservoir, multi-directional drainage and mechanical damage protection							
FD25; 25 mm deep; rolls inclusive of filter sheet SF	-	-	-	-	-	m^2	19.00
FD40; 40 mm deep; rolls inclusive of filter sheet SF	-	-	-	-	-	m^2	23.75
FD25; 25 mm deep; sheets excluding filter sheet	-	-	-	-	-	m^2	16.50
FD40; 40 mm deep; sheets excluding filter sheet	-	-	-	-	-	m^2	18.75
FD60; 60 mm deep; sheets excluding filter sheet	-	-	-	-	-	m^2	33.10
Drainage layer; "Elastodrain" recycled rubber mat; providing multi-directional drainage and mechanical damage protection							
EL200; 20 mm deep	-	-	-	-	-	m^2	26.25
"Zincolit" recycled crushed brick; optional drainage infil to Floradrain layers							
FD40; 17 litres per m^2	-	0.03	0.64	-	4.00	m^2	4.64
FD60; 27 litres per m^2	-	0.05	1.01	-	5.00	m^2	6.01
Filter sheet; rolled out onto drainage layer							
"Filter sheet TG" for Elastodrain range	-	-	-	-	-	m^2	6.25
"Filter sheet SF" for Floradrain range	-	-	-	-	-	m^2	5.00
Intensive substrate; lightweight growing medium laid to filter sheet	-	-	-	-	-	m^2	195.00

J WATERPROOFING

Item Excluding site overheads and profit	PC £	Labour hours	Labour £	Plant £	Material £	Unit	Total rate £
J50 GREEN ROOF SYSTEMS - cont'd							
Extensive Green Roof Systems; Alumasc Exterior Building Products Ltd; components laid to the Hydrotech or insulation layers above							
Moisture retention layer							
SSM-45; moisture mat	-	-	-	-	-	m²	4.50
Drainage layer for flat roofs; "Floradrain"; recycled polypropylene; providing water reservoir, multi-directional drainage and mechanical damage protection; supplied in sheet or roll form							
FD25; 25 mm deep; rolls inclusive of filter sheet SF	-	-	-	-	-	m²	19.00
FD40; 40 mm deep; rolls inclusive of filter sheet SF	-	-	-	-	-	m²	23.75
FD25; 25 mm deep; sheets excluding filter sheet	-	-	-	-	-	m²	16.50
FD40; 40 mm deep; sheets excluding filter sheet	-	-	-	-	-	m²	18.75
FD60; 60 mm deep; sheets excluding filter sheet	-	-	-	-	-	m²	33.10
Drainage layer for pitched roofs; "Floratec"; recycled polystyrene; providing water reservoir and multi-directional drainage; laid to moisture mat or to the waterproofing							
FS50	-	-	-	-	-	m²	18.75
FS75	-	-	-	-	-	m²	20.00
Landscape Options							
Sedum Mat vegetation layer; Alumasc Exterior Building Products Ltd; to Extensive							
Green Roof as detailed above	-	-	-	-	-	m²	34.00
Green Roof Components; Alumasc Exterior Building Products Ltd							
Outlet inspection chambers							
"KS 15" 150 mm deep	-	-	-	-	-	nr	65.00
Height extension piece 100 mm	-	-	-	-	-	nr	21.00
Height extension piece 200 mm	-	-	-	-	-	nr	31.00
Outlet and Irrigation control chambers							
"B32" 300 x 300 x 300 mm high	-	-	-	-	-	nr	285.00
"B52" 400 x 500 x 500 mm high	-	-	-	-	-	nr	345.00
Outlet damming piece for water retention	-	-	-	-	-	nr	60.00
Linear drainage channel; collects surface water from adjacent hard surfaces or down pipes for distribution to the drainage layer	-	-	-	-	-	m	70.00

M SURFACE FINISHES

Item Excluding site overheads and profit	PC £	Labour hours	Labour £	Plant £	Material £	Unit	Total rate £
M10 CEMENT: SAND/CONCRETE SCREEDS/ TOPPINGS							
Granolithic paving; cement and granite chippings 5 mm down (1:1:2); steel trowelled							
Floors; one coat; laid on concrete while green; width exceeding 300 mm							
20 mm thick	-	-	-	-	-	m²	17.31
25 mm thick	-	-	-	-	-	m²	18.68
Floors; two coats; laid on hacked concrete with slurry; width exceeding 300 mm							
38 mm thick	-	-	-	-	-	m²	22.28
50 mm thick	-	-	-	-	-	m²	25.62
75 mm thick	-	-	-	-	-	m²	34.63
M20 PLASTERED/RENDERED/ROUGHCAST COATING							
Cement:lime:sand (1:1:6); 19 mm thick; two coats; wood floated finish							
Walls							
width exceeding 300 mm; to brickwork or blockwork base	-	-	-	-	-	m²	13.29
Extra over cement:sand:lime (1:1:6) coatings for decorative texture finish with water repellent cement							
combed or floated finish	-	-	-	-	-	m²	1.84
M40 STONE/CONCRETE/QUARRY/CERAMIC TILING							
Ceramic tiles; unglazed slip resistant; various colours and textures; jointing							
Floors							
level or to falls only not exceeding 15 degrees from horizontal; 150 x 150 x 8 mm thick	16.44	1.00	18.75	-	19.85	m²	38.60
level or to falls only not exceeding 15 degrees from horizontal; 150 x 150 x 12 mm thick	20.03	1.00	18.75	-	23.62	m²	42.37
Clay tiles - General Preamble: Typical Specification - Clay tiles should be to BS 6431 and shall be reasonably true to shape, flat, free from flaws, frost resistant and true to sample approved by the Landscape Architect prior to laying. Quarry Tiles (or semi-vitrified tiles) shall be of external quality, either heather brown or blue, to size specified, laid on 1:2:4 concrete, with 20 maximum aggregate 100 thick, on 100 hardcore. The hardened concrete should be well wetted and the surplus water taken off. Clay tiles shall be thoroughly wetted immediately before laying and then drained and shall be bedded to 19 thick cement: sand (1:3) screed. Joints should be approximately 4 mm (or 3 mm for vitrified tiles) grouted in cement: sand (1:2) and cleaned off immediately.							

Prices for Measured Works

M SURFACE FINISHES

Item Excluding site overheads and profit	PC £	Labour hours	Labour £	Plant £	Material £	Unit	Total rate £
M40 STONE/CONCRETE/QUARRY/CERAMIC TILING - cont'd							
Quarry tiles; external quality; including bedding; jointing							
Floors							
level or to falls only not exceeding 15 degrees from horizontal; 150 x 150 x 12.5 mm thick; heather brown	20.54	0.80	15.00	-	24.16	m²	**39.16**
level or to falls only not exceeding 15 degrees from horizontal; 225 x 225 x 29 mm thick; heather brown	20.54	0.67	12.50	-	24.16	m²	**36.66**
level or to falls only not exceeding 15 degrees from horizontal; 150 x 150 x 12.5 mm thick; blue/black	13.05	1.00	18.75	-	16.30	m²	**35.05**
level or to falls only not exceeding 15 degrees from horizontal; 194 x 194 x 12.5 mm thick; heather brown	12.02	0.80	15.00	-	15.21	m²	**30.21**
M60 PAINTING/CLEAR FINISHING EXTERNALLY							
Prepare; touch up primer; two undercoats and one finishing coat of gloss oil paint; on metal surfaces							
General surfaces							
girth exceeding 300 mm	0.75	0.33	6.25	-	0.75	m²	**7.00**
isolated surfaces; girth not exceeding 300 mm	0.21	0.13	2.50	-	0.21	m	**2.71**
isolated areas not exceeding 0.50 m²							
irrespective of girth	0.40	0.13	2.50	-	0.40	nr	**2.90**
Ornamental railings; each side measured overall							
girth exceeding 300 mm	0.75	0.75	14.06	-	0.75	m²	**14.81**
Prepare; one coat primer; two undercoats and one finishing coat of gloss oil paint; on wood surfaces							
General Surfaces							
girth exceeding 300 mm	1.01	0.40	7.50	-	1.01	m²	**8.51**
isolated areas not exceeding 0.50 m²							
irrespective of girth	0.39	0.36	6.82	-	0.39	nr	**7.21**
isolated surfaces; girth not exceeding 300 mm	0.33	0.20	3.75	-	0.33	m	**4.08**
Prepare; two coats of creosote; on wood surfaces							
General surfaces							
girth exceeding 300 mm	0.17	0.21	3.89	-	0.17	m²	**4.06**
isolated surfaces; girth not exceeding 300 mm	0.05	0.12	2.16	-	0.05	m	**2.21**
Prepare, proprietary solution primer; two coats of dark stain; on wood surfaces							
General surfaces							
girth exceeding 300 mm	0.12	0.10	1.88	-	0.12	m²	**2.00**
isolated surfaces; girth not exceeding 300 mm	0.02	0.05	0.94	-	0.02	m	**0.95**
Three coats "Dimex Shield"; to clean, dry surfaces; in accordance with manufacturer's instructions							
Brick or block walls							
girth exceeding 300 mm	0.92	0.28	5.25	-	0.92	m²	**6.17**
Cement render or concrete walls							
girth exceeding 300 mm	0.72	0.25	4.69	-	0.72	m²	**5.41**
Two coats resin based paint; "Sandtex Matt"; in accordance with manufacturer's instructions							
Brick or block walls							
girth exceeding 300 mm	6.16	0.20	3.75	-	6.16	m²	**9.91**
Cement render or concrete walls							
girth exceeding 300 mm	4.19	0.17	3.13	-	4.19	m²	**7.32**

P BUILDING FABRIC SUNDRIES

Item Excluding site overheads and profit	PC £	Labour hours	Labour £	Plant £	Material £	Unit	Total rate £
P30 TRENCHES/PIPEWAYS/PITS FOR BURIED ENGINEERING SERVICES							
Excavating trenches; using 3 tonne tracked excavator; to receive pipes; grading bottoms; earthwork support; filling with excavated material to within 150 mm of finished surfaces and compacting; completing fill with topsoil; disposal of surplus soil							
Services not exceeding 200 mm nominal size							
average depth of run not exceeding 0.50 m	0.78	0.12	2.25	0.99	0.94	m	**4.17**
average depth of run not exceeding 0.75 m	0.78	0.16	3.05	1.37	0.94	m	**5.36**
average depth of run not exceeding 1.00 m	0.78	0.28	5.31	2.40	0.94	m	**8.64**
average depth of run not exceeding 1.25 m	0.78	0.38	7.19	3.22	0.78	m	**11.19**
Excavating trenches; using 3 tonne tracked excavator; to receive pipes; grading bottoms; earthwork support; filling with imported granular material and compacting; disposal of surplus soil							
Services not exceeding 200 mm nominal size							
average depth of run not exceeding 0.50 m	4.63	0.09	1.63	0.69	5.98	m	**8.30**
average depth of run not exceeding 0.75 m	6.95	0.11	2.03	0.86	8.97	m	**11.87**
average depth of run not exceeding 1.00 m	9.27	0.14	2.62	1.13	11.97	m	**15.71**
average depth of run not exceeding 1.25 m	11.58	0.23	4.29	1.93	14.96	m	**21.17**
Excavating trenches, using 3 tonne tracked excavator, to receive pipes; grading bottoms; earthwork support; filling with lean mix concrete; disposal of surplus soil							
Services not exceeding 200 mm nominal size							
average depth of run not exceeding 0.50 m	9.71	0.11	2.00	0.36	11.06	m	**13.42**
average depth of run not exceeding 0.75 m	14.56	0.13	2.44	0.45	16.58	m	**19.47**
average depth of run not exceeding 1.00 m	19.41	0.17	3.13	0.60	22.11	m	**25.84**
average depth of run not exceeding 1.25 m	24.27	0.23	4.22	0.90	27.65	m	**32.76**
Earthwork Support; providing support to opposing faces of excavation; moving along as work proceeds; A Plant Acrow							
Maximum depth not exceeding 2.00 m							
distance between opposing faces not exceeding 2.00 m	-	0.80	15.00	16.95	-	m	**31.95**

Q PAVING/PLANTING/FENCING/SITE FURNITURE

Item Excluding site overheads and profit	PC £	Labour hours	Labour £	Plant £	Material £	Unit	Total rate £
Q10 STONE/CONCRETE/BRICK **KERBS/EDGINGS/CHANNELS**							
Foundations to kerbs							
Excavating trenches; width 300 mm; 3 tonne excavator; disposal off site							
depth 300 mm	-	0.10	1.88	0.45	1.11	m	**3.44**
depth 400 mm	-	0.11	2.08	0.50	1.49	m	**4.07**
by hand	-	0.60	11.29	-	1.85	m	**13.14**
Excavating trenches; width 450 mm; 3 tonne excavator; disposal off site							
depth 300 mm	-	0.13	2.34	0.56	1.67	m	**4.57**
depth 400 mm	-	0.14	2.68	0.64	2.23	m	**5.55**
Foundations; to kerbs, edgings, or **channels; in situ concrete; 21 N/mm² -** **20 aggregate (1:2:4) site mixed; one** **side against earth face, other against** **formwork (not included); site mixed** **concrete**							
Site mixed concrete							
150 wide x 100 mm deep	-	0.13	2.50	-	1.42	m	**3.92**
150 wide x 150 mm deep	-	0.17	3.12	-	2.13	m	**5.26**
200 wide x 150 mm deep	-	0.20	3.75	-	2.85	m	**6.60**
300 wide x 150 mm deep	-	0.23	4.38	-	4.27	m	**8.64**
600 wide x 200 mm deep	-	0.29	5.36	-	11.38	m	**16.74**
Ready mixed concrete							
150 wide x 100 mm deep	-	0.13	2.50	-	1.03	m	**3.53**
150 wide x 150 mm deep	-	0.17	3.12	-	1.55	m	**4.67**
200 wide x 150 mm deep	-	0.20	3.75	-	2.07	m	**5.82**
300 wide x 150 mm deep	-	0.23	4.38	-	3.10	m	**7.47**
600 wide x 200 mm deep	-	0.29	5.36	-	8.26	m	**13.62**
Formwork; sides of foundations (this will usually be required to one side of each kerb foundation adjacent to road sub-bases)							
100 mm deep	-	0.04	0.78	-	0.15	m	**0.93**
150 mm deep	-	0.04	0.78	-	0.23	m	**1.01**
Precast concrete kerbs, channels, **edgings etc.; to BS 340; Marshalls** **Mono; bedding, jointing and pointing in** **cement mortar (1:3); including haunching** **with in situ concrete; 11.50 N/mm² - 40** **aggregate one side**							
Kerbs; straight							
150 x 305 mm; ref HB1	6.49	0.50	9.38	-	8.98	m	**18.35**
125 x 255 mm; ref HB2; SP	3.38	0.44	8.33	-	5.86	m	**14.20**
125 x 150 mm; ref BN	2.27	0.40	7.50	-	4.75	m	**12.25**
Dropper kerbs; left and right handed							
125 x 255 - 150 mm; ref DL1 or DR1	4.80	0.50	9.38	-	7.28	m	**16.66**
125 x 255 - 150 mm; ref DL2 or DR2	4.80	0.50	9.38	-	7.28	m	**16.66**
Quadrant kerbs							
305 mm radius	7.32	0.50	9.38	-	8.98	nr	**18.35**
455 mm radius	7.88	0.50	9.38	-	10.36	nr	**19.74**
Internal or external angles							
127 x 254 mm section	13.77	0.50	9.38	-	15.43	m	**24.80**
Straight kerbs or channels; to radius; 125 x 255 mm							
0.90 m radius (2 units per quarter circle)	5.94	0.80	15.00	-	8.50	m	**23.50**
1.80 m radius (4 units per quarter circle)	5.94	0.73	13.63	-	8.50	m	**22.14**
2.40 m radius (5 units per quarter circle)	5.94	0.68	12.71	-	8.50	m	**21.21**
3.00 m radius (5 units per quarter circle)	5.94	0.67	12.50	-	8.50	m	**21.00**
4.50 m radius (2 units per quarter circle)	5.94	0.60	11.26	-	8.50	m	**19.76**
6.10 m radius (11 units per quarter circle)	5.94	0.58	10.87	-	8.50	m	**19.37**
7.60 m radius (14 units per quarter circle)	5.94	0.57	10.71	-	8.50	m	**19.22**
9.15 m radius (17 units per quarter circle)	5.94	0.56	10.42	-	8.50	m	**18.92**
10.70 m radius (20 units per quarter circle)	5.94	0.56	10.42	-	8.50	m	**18.92**
12.20 m radius (22 units per quarter circle)	5.94	0.53	9.87	-	8.50	m	**18.37**

Q PAVING/PLANTING/FENCING/SITE FURNITURE

Item Excluding site overheads and profit	PC £	Labour hours	Labour £	Plant £	Material £	Unit	Total rate £
Kerbs; "Conservation Kerb" units; to simulate natural granite kerbs							
255 x 150 x 914 mm; laid flat	16.02	0.57	10.71	-	20.48	m	**31.19**
150 x 255 x 914 mm; laid vertical	16.02	0.57	10.71	-	19.75	m	**30.46**
145 x 255 mm; radius internal 3.25 m	20.27	0.83	15.63	-	24.50	m	**40.13**
150 x 255 mm; radius external 3.40 m	19.36	0.83	15.63	-	23.57	m	**39.20**
145 x 255 mm; radius internal 6.50 m	18.56	0.67	12.50	-	22.29	m	**34.79**
145 x 255 mm; radius external 6.70 m	18.01	0.67	12.50	-	22.19	m	**34.68**
150 x 255 mm; radius internal 9.80 m	17.91	0.67	12.50	-	22.08	m	**34.58**
150 x 255 mm; radius external 10.00 m	17.40	0.67	12.50	-	21.13	m	**33.63**
305 x 305 x 255 mm; solid quadrants	22.76	0.57	10.71	-	26.49	nr	**37.20**
Channels; square							
125 x 225 x 915 mm long; ref CS1	3.45	0.40	7.50	-	11.96	m	**19.46**
125 x 150 x 915 mm long; ref CS2	2.38	0.40	7.50	-	9.39	m	**16.89**
Channels; dished							
305 x 150 x 915 mm long; ref CD	7.99	0.40	7.50	-	17.21	m	**24.71**
150 x 125 x 915 mm	2.48	0.40	7.50	-	10.77	m	**18.27**
150 x 100 x 915 mm	3.41	0.40	7.50	-	8.63	m	**16.13**
Precast concrete edging units; including haunching with in situ concrete; 11.50 N/mm^2 - 40 aggregate both sides							
Edgings; rectangular, bullnosed, or chamfered							
50 x 150 mm	1.75	0.33	6.25	-	5.90	m	**12.15**
125 x 150 mm bullnosed	2.20	0.33	6.25	-	6.34	m	**12.59**
50 x 200 mm	1.83	0.33	6.25	-	5.97	m	**12.22**
50 x 250 mm	2.12	0.33	6.25	-	6.27	m	**12.52**
50 x 250 mm flat top	2.41	0.33	6.25	-	6.55	m	**12.80**
Precast concrete kerbs; to BS 340; Cemex Ltd; on 150 mm deep concrete foundation; including haunching with in situ concrete; 11.50 N/mm^2 - 40 aggregate one side							
Kerbs; "Kerb-sett" units (prices shown are for grey units, prices for other colours vary)							
standard blocks; 100 wide x 190 x 160 mm (set as high or low rise)	0.75	0.80	15.00	-	11.04	m	**26.04**
external angles; high or low profiles	5.20	0.80	15.00	-	15.49	m	**30.49**
Dropper kerbs (handed pairs); ref H-L	12.40	1.00	18.75	-	15.25	set	**34.00**
Crossovers; low; ref L-X	12.40	0.20	3.75	-	13.04	set	**16.79**
Crossovers; high; ref H-X	12.40	0.20	3.75	-	13.59	set	**17.34**
Curves; internal; low; ref IR/L	20.00	0.80	15.00	-	22.85	m	**37.85**
Curves; external; high; ref ER/H	20.00	0.80	15.00	-	22.85	m	**37.85**
Dressed natural stone kerbs - General Preamble: BS 435 includes the following conditions for dressed natural stone kerbs. The kerbs are to be good, sound, and uniform in texture and free from defects; worked straight or to radius, square and out of wind, with the top front and back edges parallel or concentric to the dimensions specified. All drill and pick holes shall be removed from dressed faces. Standard dressings shall be in accordance with one of three illustrations in BS 435; and designated as either fine picked, single axed or nidged, or rough punched.							

Q PAVING/PLANTING/FENCING/SITE FURNITURE

Item Excluding site overheads and profit	PC £	Labour hours	Labour £	Plant £	Material £	Unit	Total rate £
Q10 STONE/CONCRETE/BRICK **KERBS/EDGINGS/CHANNELS** - cont'd							
Dressed natural stone kerbs; to BS435; **CED Ltd; on concrete foundations (not** **included); including haunching with in** **situ concrete; 11.50 N/mm^2 - 40** **aggregate one side**							
Granite kerbs; 125 x 250 mm							
special quality, straight, random lengths	12.00	0.80	15.00	-	14.97	m	**29.97**
Granite kerbs; 125 x 250 mm; curved to mean radius 3 m							
special quality, random lengths	14.00	0.91	17.06	-	16.97	m	**34.03**
Second-hand granite setts; 100 x 100 **mm; bedding in cement mortar (1:4); on** **150 mm deep concrete foundations;** **including haunching with in situ** **concrete; 11.50 N/mm^2 - 40 aggregate** **one side**							
Edgings							
300 mm wide	13.54	1.33	25.00	-	17.37	m	**42.37**
Concrete setts; Blanc de Bierges; **bedding in cement mortar (1:4); on 150** **mm deep concrete foundations,** **including haunching with in situ** **concrete; 11.50 N/mm^2 - 40 aggregate** **one side**							
Edgings							
70 x 70 x 70 mm; in strips 230 mm wide	5.48	1.33	25.00	-	9.31	m	**34.31**
140 x 140 x 80 mm; in strips 310 mm wide	7.04	1.33	25.00	-	11.23	m	**36.23**
210 x 140 x 80 mm; in strips 290 mm wide	6.59	1.33	25.00	-	10.75	m	**35.75**
Brick or block stretchers; bedding in **cement mortar (1:4); on 150 mm deep** **concrete foundations, including** **haunching with in situ concrete; 11.50** **N/mm^2 - 40 aggregate one side**							
Single course							
concrete paving blocks; PC £8.72 /m^2; 200 x 100 x 60 mm	1.74	0.31	5.77	-	4.22	m	**9.99**
engineering bricks; PC £260.00/1000; 215 x 102.5 x 65 mm	1.16	0.40	7.50	-	3.69	m	**11.19**
paving bricks; PC £450.00/1000; 215 x 102.5 x 65 mm	2.00	0.40	7.50	-	4.48	m	**11.98**
Two courses							
concrete paving blocks; PC £8.72 /m^2; 200 x 100 x 60 mm	3.49	0.40	7.50	-	6.43	m	**13.93**
engineering bricks; PC £260.00/1000; 215 x 102.5 x 65 mm	2.31	0.57	10.71	-	5.37	m	**16.09**
paving bricks; PC £450.00/1000; 215 x 102.5 x 65 mm	4.00	0.57	10.71	-	6.94	m	**17.66**
Three courses							
concrete paving blocks; PC £8.72/m^2; 200 x 100 x 60 mm	5.23	0.44	8.33	-	8.18	m	**16.51**
engineering bricks; PC £260.00/1000; 215 x 102.5 x 65 mm	3.47	0.67	12.50	-	6.58	m	**19.08**
paving bricks; PC £450.00/1000; 215 x 102.5 x 65 mm	6.00	0.67	12.50	-	8.94	m	**21.44**

Q PAVING/PLANTING/FENCING/SITE FURNITURE

Item Excluding site overheads and profit	PC £	Labour hours	Labour £	Plant £	Material £	Unit	Total rate £
Bricks on edge; bedding in cement mortar (1:4); on 150 mm deep concrete foundations; including haunching with in situ concrete; 11.50 N/mm^2 - 40 aggregate one side							
One brick wide							
engineering bricks; 215 x 102.5 x 65 mm	3.47	0.57	10.71	-	6.70	m	**17.41**
paving bricks; 215 x 102.5 x 65 mm	6.00	0.57	10.71	-	9.06	m	**19.77**
Two courses; stretchers laid on edge; 225 mm wide							
engineering bricks; 215 x 102.5 x 65 mm	6.93	1.20	22.50	-	11.38	m	**33.88**
paving bricks; 215 x 102.5 x 65 mm	12.00	1.20	22.50	-	16.09	m	**38.59**
Extra over bricks on edge for standard kerbs to one side; haunching in concrete							
125 x 255 mm; ref HB2; SP	3.38	0.44	8.33	-	5.86	m	**14.20**
Channels; bedding in cement mortar (1:3); joints pointed flush; on concrete foundations (not included)							
Three courses stretchers; 350 mm wide; quarter bond to form dished channels							
engineering bricks; PC £260.00/1000 ; 215 x 102.5 x 65 mm	3.47	1.00	18.75	-	4.69	m	**23.44**
paving bricks; PC £450.00/1000; 215 x 102.5 x 65 mm	6.00	1.00	18.75	-	7.05	m	**25.80**
Three courses granite setts; 340 mm wide; to form dished channels							
340 mm wide	13.54	2.00	37.50	-	15.94	m	**53.44**
Marshalls Plc "Mini Beany" combined kerb and channel drainage system; to trenches (not included)							
Precast concrete drainage channel base; 185 mm - 385 mm deep; bedding, jointing and pointing in cement mortar (1:3); on 150 mm deep concrete (ready mixed) foundation; including haunching with in situ concrete; 11.50 N/mm^2 - 40 aggregate one side; channels 250 mm wide x 1.00 long							
straight; 1.00 m long	20.77	1.00	18.75	-	23.90	m	**42.65**
straight; 500 mm long	25.97	1.05	19.69	-	29.10	m	**48.78**
radial; 30 -10 m or 9-6 m internal or external	31.17	1.33	25.00	-	34.30	m	**59.30**
angles 45 or 90 degree	31.16	1.00	18.75	-	34.29	nr	**53.04**
Mini Beany Top Block; perforated kerb unit to drainage channel above; natural grey							
straight	20.79	0.33	6.25	-	22.03	m	**28.28**
radial; 30 -10 m or 9-6 m internal or external	12.98	0.50	9.38	-	14.22	m	**23.60**
angles 45 or 90 degree	25.96	0.50	9.38	-	27.20	nr	**36.58**
Mini Beany; outfalls; two section concrete trapped outfall with Mini Beany cast iron access cover and frame; to concrete foundation							
High capacity outfalls; silt box 150/225 outlet; two section trapped outfall silt box and cast iron access cover	138.32	1.00	18.75	-	302.17	nr	**320.92**
Inline Side or End outlet Outfall 150 mm; 2 section concrete trapped outfall; cast iron Mini Beany access cover and frame	168.76	1.00	18.75	-	169.41	nr	**188.16**
Ancillaries to Mini Beany							
End cap	9.46	0.25	4.69	-	9.46	nr	**14.15**
End cap outlets	24.56	0.25	4.69	-	24.56	nr	**29.25**

Q PAVING/PLANTING/FENCING/SITE FURNITURE

Item Excluding site overheads and profit	PC £	Labour hours	Labour £	Plant £	Material £	Unit	Total rate £
Q10 STONE/CONCRETE/BRICK **KERBS/EDGINGS/CHANNELS** - cont'd							
Precast concrete channels; Charcon **Hard Landscaping; on 150 mm deep** **concrete foundations; including** **haunching with in situ concrete; 21.00** **N/mm^2 - 20 aggregate; both sides**							
"Charcon Safeticurb"; slotted safety channels; for pedestrians and light vehicles							
ref DBJ; 305 x 305 mm	58.19	0.67	12.50	-	74.31	m	86.81
ref DBA; 250 x 250 mm	25.81	0.67	12.50	-	41.93	m	54.43
"Charcon Safeticurb"; slotted safety channels; for heavy vehicles							
ref DBM; 248 x 248 mm	45.91	0.80	15.00	-	62.03	m	77.03
ref Clearway; 305 x 305 mm x 400 mm long	121.14	0.80	15.00	-	137.26	m	152.26
Inspection units; cast iron lids; including jointing to drainage channels							
248 x 248 x 914 mm	62.14	1.50	28.13	-	63.09	nr	91.21
Silt box tops; concrete frame; cast iron grid lids; type 1; set over gully							
457 x 610 mm	276.60	2.00	37.50	-	277.55	nr	315.05
Manhole covers; type K; cast iron; providing inspection to blocks and back gullies	289.34	2.00	37.50	-	291.54	nr	329.04
Linear Drainage; channels; Ensor **Building Products; on 100 mm deep** **concrete bed; 100 mm concrete fill both** **sides**							
Polymer concrete drain units; tapered to falls							
channel units; Grade 'A', ref Stora-drain 100; 100 mm wide including galvanised grate and locks	20.45	0.80	15.00	-	26.62	m	41.62
channel units; Grade 'B', ref Stora-drain 100; 100 mm wide including galvanised grate and locks	24.75	0.80	15.00	-	30.92	m	45.92
channel units; Grade 'C', ref Stora-drain 100; 100 mm wide including galvanised grate and locks	26.64	0.80	15.00	-	32.81	m	47.81
channel units; Grade 'C', ref Stora-drain 100; 100 mm wide including anti-heel grate and locks	27.75	0.80	15.00	-	33.92	m	48.92
Stora Drain Sump unit, steel bucket, Class A15 galv. slotted grating	41.67	2.00	37.50	-	47.84	nr	85.34
Stora Drain Sump unit, steel bucket, Class B125 galv. mesh grating	43.30	2.00	37.50	-	49.47	nr	86.97
Stora Drain Sump unit, steel bucket, Class C250 galv. slotted grating	44.50	2.00	37.50	-	50.67	nr	88.17
Stora Drain Sump unit, steel bucket, C250 'Heelsafe' slotted grating	41.25	2.00	37.50	-	47.42	nr	84.92
Channels; Hodkin & Jones (Sheffield) Ltd; **on well rammed subsoil base; set to** **gradient; backfilling with compacted** **excavated material**							
"Decathlon"; precast concrete drainage channels							
150 mm wide channels	44.00	1.00	18.75	-	44.00	m	62.75
250 mm wide channels	55.00	1.00	18.75	-	55.00	m	73.75
375 mm wide channels	74.00	1.10	20.63	-	74.00	m	94.63
525 mm wide channels	89.00	1.20	22.50	0.22	89.00	m	111.72
600 mm wide channels	105.00	1.60	30.00	0.28	105.00	m	135.28

Q PAVING/PLANTING/FENCING/SITE FURNITURE

Item Excluding site overheads and profit	PC £	Labour hours	Labour £	Plant £	Material £	Unit	Total rate £
Permaloc "AshphaltEdge"; extruded aluminium alloy L shaped edging with 5.33 mm exposed upper lip; edging fixed to roadway base and edge profile with 250 mm steel fixing spike; laid to straight or curvilinear road edge; subsequently filled with macadam (not included)							
Depth of macadam							
38 mm	5.82	0.02	0.31	-	6.66	m	**6.98**
58 mm	6.53	0.02	0.32	-	7.39	m	**7.71**
64 mm	7.19	0.02	0.34	-	8.07	m	**8.41**
76 mm	8.65	0.02	0.38	-	9.56	m	**9.94**
102 mm	10.02	0.02	0.39	-	10.97	m	**11.36**
Permaloc "Cleanline"; heavy duty straight profile edging; for edgings to soft landscape beds or turf areas; 3.2 mm x 102 high; 3.2 mm thick with 4.75 mm exposed upper lip; fixed to form straight or curvilinear edge with 305 mm fixing spike							
Milled aluminium							
100 deep	6.64	0.02	0.34	-	7.50	m	**7.84**
Black							
100 deep	7.31	0.02	0.34	-	8.19	m	8.53
Permaloc "Permastrip"; heavy duty L shaped profile maintenance strip; 3.2 mm x 89 mm high with 5.2 mm exposed top lip ; for straight or gentle curves on paths or bed turf interfaces; fixed to form straight or curvilinear edge with standard 305 mm stake; other stake lengths available							
Milled aluminium							
90 deep	6.64	0.02	0.34	-	7.50	m	7.84
Black							
90 deep	7.31	0.02	0.34	-	8.19	m	8.53
Permaloc "Proline"; medium duty straight profiled maintenance strip; 3.2 mm x 102 mm high with 3.18 mm exposed top lip; for straight or gentle curves on paths or bed turf interfaces; fixed to form straight or curvilinear edge with standard 305 mm stake; other stake lengths available							
Milled aluminium							
90 deep	5.03	0.02	0.34	-	5.84	m	6.18
Black							
90 deep	5.80	0.02	0.34	-	6.63	m	6.97
Q20 BASES AND SUB-BASES TO ROADS/PAVINGS							
Hardcore bases; obtained off site; PC £17.00 /m³							
by machine							
100 mm thick	1.70	0.05	0.94	0.93	1.70	m²	**3.57**
100 mm thick	17.00	0.50	9.38	9.35	17.00	m³	**35.73**
150 mm thick	2.55	0.07	1.25	1.21	2.55	m²	**5.01**
200 mm thick	3.40	0.08	1.50	1.27	3.40	m²	**6.17**
300 mm thick	5.10	0.07	1.25	1.35	5.10	m²	**7.70**
exceeding 300 mm thick	17.00	0.17	3.12	3.11	20.40	m³	**26.64**

Q PAVING/PLANTING/FENCING/SITE FURNITURE

Item Excluding site overheads and profit	PC £	Labour hours	Labour £	Plant £	Material £	Unit	Total rate £
Q20 BASES AND SUB-BASES TO ROADS/PAVINGS - cont'd							
Hardcore bases; obtained off site; PC **£17.00 /m³** - cont'd							
by hand							
100 mm thick	1.70	0.20	3.75	0.11	1.70	m²	**5.56**
150 mm thick	2.55	0.30	5.63	0.11	3.06	m²	**8.80**
200 mm thick	3.40	0.40	7.50	0.18	4.08	m²	**11.76**
300 mm thick	5.10	0.60	11.25	0.11	6.12	m²	**17.48**
exceeding 300 mm thick	17.00	2.00	37.50	0.37	20.40	m³	**58.27**
Hardcore; difference for each £1.00 increase/decrease in PC price per m³; price will vary with type and source of hardcore							
average 75 mm thick	-	-	-	-	0.08	m²	**0.08**
average 100 mm thick	-	-	-	-	0.10	m²	**0.10**
average 150 mm thick	-	-	-	-	0.15	m²	**0.15**
average 200 mm thick	-	-	-	-	0.20	m²	**0.20**
average 250 mm thick	-	-	-	-	0.25	m²	**0.25**
average 300 mm thick	-	-	-	-	0.30	m²	**0.30**
exceeding 300 mm thick	-	-	-	-	1.10	m³	**1.10**
Type 1 bases; PC £18.50/tonne (£40.70/m³ compacted)							
Type 1 granular fill base; PC £18.50 **/tonne (£40.70/m³ compacted)**							
by machine							
100 mm thick	4.07	0.03	0.53	0.31	4.07	m²	**4.90**
150 mm thick	6.11	0.03	0.47	0.47	6.11	m²	**7.04**
250 mm thick	10.18	0.02	0.39	0.78	10.18	m²	**11.34**
over 250 mm thick	40.70	0.20	3.75	2.84	40.70	m³	**47.29**
by hand; (mechanical compaction)							
100 mm thick	4.07	0.17	3.13	0.04	4.07	m²	**7.23**
150 mm thick	6.11	0.25	4.69	0.06	6.11	m²	**10.85**
250 mm thick	10.18	0.42	7.81	0.09	10.18	m²	**18.08**
Over 250 mm thick	40.70	0.47	8.76	0.09	40.70	m³	**49.55**
Surface treatments							
Sand blinding; to hardcore base (not included); 25 mm thick	0.73	0.03	0.62	-	0.73	m²	**1.35**
Sand blinding; to hardcore base (not included); 50 mm thick	1.46	0.05	0.94	-	1.46	m²	**2.40**
Filter fabrics; to hardcore base (not included)	0.39	0.01	0.19	-	0.41	m²	**0.60**
Herbicides; ICI							
"Casoron G" (residual) herbicide; treating substrate before laying base							
at 1 kg/125 m²	2.82	0.05	0.94	-	3.10	100 m²	**4.04**
Q21 IN SITU CONCRETE ROADS/PAVINGS/BASES							
Unreinforced concrete; on prepared **sub-base (not included)**							
Roads; 21.00 N/mm² - 20 aggregate (1:2:4) mechanically mixed on site							
100 mm thick	9.48	0.13	2.34	-	9.96	m²	**12.30**
150 mm thick	14.23	0.17	3.13	-	14.94	m²	**18.06**

Q PAVING/PLANTING/FENCING/SITE FURNITURE

Item Excluding site overheads and profit	PC £	Labour hours	Labour £	Plant £	Material £	Unit	Total rate £
Reinforced in situ concrete; **mechanically mixed on site; normal** **Portland cement; on hardcore base (not** **included); reinforcement (not included)**							
Roads; 11.50 N/mm^2 - 40 aggregate (1:3:6)							
100 mm thick	8.28	0.40	7.50	-	8.49	m^2	**15.99**
150 mm thick	12.42	0.60	11.25	-	12.74	m^2	**23.99**
200 mm thick	16.56	0.80	15.00	-	17.39	m^2	**32.39**
250 mm thick	20.70	1.00	18.75	0.18	21.22	m^2	**40.15**
300 mm thick	24.85	1.20	22.50	0.18	25.47	m^2	**48.15**
Roads; 21.00 N/mm^2 - 20 aggregate (1:2:4)							
100 mm thick	9.48	0.40	7.50	-	9.72	m^2	**17.22**
150 mm thick	14.23	0.60	11.25	-	14.59	m^2	**25.84**
200 mm thick	18.97	0.80	15.00	-	19.44	m^2	**34.44**
250 mm thick	23.71	1.00	18.75	0.18	24.30	m^2	**43.23**
300 mm thick	28.45	1.20	22.50	0.18	29.17	m^2	**51.85**
Roads; 25.00 N/mm^2 - 20 aggregate GEN 4 ready mixed							
100 mm thick	6.88	0.40	7.50	-	7.05	m^2	**14.55**
150 mm thick	10.32	0.60	11.25	-	10.58	m^2	**21.83**
200 mm thick	13.76	0.80	15.00	-	14.10	m^2	**29.10**
250 mm thick	17.20	1.00	18.75	0.18	17.62	m^2	**36.56**
300 mm thick	20.64	1.20	22.50	0.18	21.15	m^2	**43.83**
Reinforced in situ concrete; ready **mixed; discharged directly into location** **from supply lorry; normal Portland** **cement; on hardcore base (not** **included); reinforcement (not included)**							
Roads; 11.50 N/mm^2 - 40 aggregate (1:3:6)							
100 mm thick	6.47	0.16	3.00	-	6.47	m^2	**9.47**
150 mm thick	9.71	0.24	4.50	-	9.71	m^2	**14.21**
200 mm thick	12.94	0.36	6.75	-	12.94	m^2	**19.69**
250 mm thick	16.18	0.54	10.13	0.18	16.18	m^2	**26.48**
300 mm thick	19.41	0.66	12.38	0.18	19.41	m^2	**31.97**
Roads; 21.00 N/mm^2 - 20 aggregate (1:2:4)							
100 mm thick	6.83	0.16	3.00	-	6.83	m^2	**9.82**
150 mm thick	10.24	0.24	4.50	-	10.24	m^2	**14.74**
200 mm thick	13.65	0.36	6.75	-	13.65	m^2	**20.40**
250 mm thick	17.06	0.54	10.13	0.18	17.06	m^2	**27.37**
300 mm thick	20.48	0.66	12.38	0.18	20.48	m^2	**33.03**
Roads; 26.00 N/mm^2 - 20 aggregate (1:1.5:3)							
100 mm thick	7.15	0.16	3.00	-	7.15	m^2	**10.15**
150 mm thick	10.72	0.24	4.50	-	10.72	m^2	**15.22**
200 mm thick	13.65	0.36	6.75	-	13.65	m^2	**20.40**
250 mm thick	17.87	0.54	10.13	0.18	17.87	m^2	**28.18**
300 mm thick	21.45	0.66	12.38	0.18	21.45	m^2	**34.00**
Roads; PAV1 Concrete - 35 N/mm^2; Designated mix							
100 mm thick	7.71	0.16	3.00	-	7.90	m^2	**10.90**
150 mm thick	11.56	0.24	4.50	-	11.85	m^2	**16.35**
200 mm thick	15.42	0.36	6.75	-	15.80	m^2	**22.55**
250 mm thick	19.27	0.54	10.13	0.18	19.75	m^2	**30.05**
300 mm thick	23.12	0.66	12.38	0.18	23.70	m^2	**36.26**
Concrete sundries Treating surfaces of unset concrete; grading to cambers, tamping with 75 mm thick steel shod tamper or similar	-	0.13	2.50	-	-	m^2	**2.50**
Expansion joints 13 mm thick joint filler; formwork							
width or depth not exceeding 150 mm	1.45	0.20	3.75	-	2.06	m	**5.81**
width or depth 150 - 300 mm	1.05	0.25	4.69	-	2.27	m	**6.96**
width or depth 300 - 450 mm	0.47	0.30	5.63	-	2.30	m	**7.93**

Q PAVING/PLANTING/FENCING/SITE FURNITURE

Item Excluding site overheads and profit	PC £	Labour hours	Labour £	Plant £	Material £	Unit	Total rate £
Q21 IN SITU CONCRETE ROADS/PAVINGS/BASES - cont'd							
Expansion joints - cont'd							
25 mm thick joint filler; formwork							
width or depth not exceeding 150 mm	1.90	0.20	3.75	-	2.51	m	6.26
width or depth 150 - 300 mm	1.90	0.25	4.69	-	3.12	m	7.81
width or depth 300 - 450 mm	1.90	0.30	5.63	-	3.73	m	9.36
Sealants; sealing top 25 mm of joint with rubberized bituminous compound	1.01	0.25	4.69	-	1.01	m	5.70
Formwork for in situ concrete							
Sides of foundations							
height exceeding 1.00 m	1.54	2.00	37.50	-	4.58	m²	42.08
height not exceeding 250 mm	0.38	1.00	18.75	-	1.28	m²	20.03
height 250 - 500 mm	0.77	1.00	18.75	-	2.29	m²	21.04
height 500 mm - 1.00 m	1.54	1.50	28.13	-	4.58	m²	32.70
Extra over formwork for curved work 6 m radius	-	0.25	4.69	-	-	m²	4.69
Steel road forms; to edges of beds or faces of foundations							
150 mm wide	-	0.20	3.75	0.53	-	m	4.28
Reinforcement; fabric; BS 4483; side laps 150 mm; head laps 300 mm; mesh 200 x 200 mm; in roads, footpaths or pavings							
Fabric							
ref A142 (2.22 kg/m²)	1.50	0.08	1.56	-	1.65	m²	3.21
ref A193 (3.02 kg/m²)	2.05	0.08	1.56	-	2.25	m²	3.81
Q22 COATED MACADAM/ASPHALT ROADS/PAVINGS							
Coated macadam/asphalt roads/pavings - General Preamble: The prices for all in situ finishings to roads and footpaths include for work to falls, crossfalls or slopes not exceeding 15 degrees from horizontal; for laying on prepared bases (not included) and for rolling with an appropriate roller.							
Preamble: Users should note the new terminology for the surfaces described below which is to European standard descriptions. The now redundant descriptions for each course are shown in brackets.							
Macadam Surfacing; Spadeoak Construction Co Ltd; Surface (Wearing) Course; 20 mm of 6 mm dense bitumen macadam to BS4987-1 2001 ref 7.5 Machine lay; areas 1000 m² and over							
limestone aggregate	-	-	-	-	-	m²	4.40
granite aggregate	-	-	-	-	-	m²	4.46
red	-	-	-	-	-	m²	7.91
Hand Lay; areas 400 m² and over							
limestone aggregate	-	-	-	-	-	m²	5.96
granite aggregate	-	-	-	-	-	m²	6.03
red	-	-	-	-	-	m²	9.71

Q PAVING/PLANTING/FENCING/SITE FURNITURE

Item Excluding site overheads and profit	PC £	Labour hours	Labour £	Plant £	Material £	Unit	Total rate £
Macadam Surfacing; Spadeoak **Construction Co Ltd; Surface (Wearing)** **Course; 30 mm of 10 mm dense bitumen** **macadam to BS4987-1 2001 ref 7.4**							
Machine lay; areas 1000 m^2 and over							
limestone aggregate	-	-	-	-	-	m^2	5.27
granite aggregate	-	-	-	-	-	m^2	5.35
red	-	-	-	-	-	m^2	11.14
Hand Lay; areas 400 m^2 and over							
limestone aggregate	-	-	-	-	-	m^2	6.89
granite aggregate	-	-	-	-	-	m^2	6.98
red	-	-	-	-	-	m^2	13.16
Macadam Surfacing; Spadeoak **Construction Co Ltd; Surface (Wearing)** **Course; 40 mm of 10 mm dense bitumen** **macadam to BS4987-1 2001 ref 7.4**							
Machine lay; areas 1000 m^2 and over							
limestone aggregate	-	-	-	-	-	m^2	6.68
granite aggregate	-	-	-	-	-	m^2	6.80
red	-	-	-	-	-	m^2	12.63
Hand Lay; areas 400 m^2 and over							
limestone aggregate	-	-	-	-	-	m^2	8.40
granite aggregate	-	-	-	-	-	m^2	8.52
red	-	-	-	-	-	m^2	14.74
Macadam Surfacing; Spadeoak **Construction Co Ltd; Binder (Base)** **Course; 50 mm of 20 mm dense bitumen** **macadam to BS4987-1 2001 ref 6.5**							
Machine lay; areas 1000 m^2 and over							
limestone aggregate	-	-	-	-	-	m^2	7.16
granite aggregate	-	-	-	-	-	m^2	7.29
Hand Lay; areas 400 m^2 and over							
limestone aggregate	-	-	-	-	-	m^2	8.91
granite aggregate	-	-	-	-	-	m^2	9.05
Macadam Surfacing; Spadeoak **Construction Co Ltd; Binder (Base)** **Course; 60 mm of 20 mm dense bitumen** **macadam to BS4987-1 2001 ref 6.5**							
Machine lay; areas 1000 m^2 and over							
limestone aggregate	-	-	-	-	-	m^2	7.88
granite aggregate	-	-	-	-	-	m^2	8.03
Hand Lay; areas 400 m^2 and over							
limestone aggregate	-	-	-	-	-	m^2	9.68
granite aggregate	-	-	-	-	-	m^2	9.84
Macadam Surfacing; Spadeoak **Construction Co Ltd; Base (Roadbase)** **Course; 75 mm of 28 mm dense bitumen** **macadam to BS4987-1 2001 ref 5.2**							
Machine lay; areas 1000 m^2 and over							
limestone aggregate	-	-	-	-	-	m^2	9.49
granite aggregate	-	-	-	-	-	m^2	9.69
Hand Lay; areas 400 m^2 and over							
limestone aggregate	-	-	-	-	-	m^2	11.40
granite aggregate	-	-	-	-	-	m^2	11.60
Macadam Surfacing; Spadeoak **Construction Co Ltd; Base (Roadbase)** **Course; 100 mm of 28 mm dense bitumen** **macadam to BS4987-1 2001 ref 5.2**							
Machine lay; areas 1000 m^2 and over							
limestone aggregate	-	-	-	-	-	m^2	11.98
granite aggregate	-	-	-	-	-	m^2	12.24
Hand Lay; areas 400 m^2 and over							
limestone aggregate	-	-	-	-	-	m^2	14.15
granite aggregate	-	-	-	-	-	m^2	14.33

Q PAVING/PLANTING/FENCING/SITE FURNITURE

Item Excluding site overheads and profit	PC £	Labour hours	Labour £	Plant £	Material £	Unit	Total rate £
Q22 COATED MACADAM/ASPHALT **ROADS/PAVINGS** - cont'd							
Macadam Surfacing; Spadeoak **Construction Co Ltd; Base (Roadbase)** **Course; 150 mm of 28 mm dense bitumen** **macadam in two layers to BS4987-1** **2001 ref 5.2** Machine lay; areas 1000 m² and over							
limestone aggregate	-	-	-	-	-	m²	18.88
granite aggregate	-	-	-	-	-	m²	19.26
Hand Lay; areas 400 m² and over							
limestone aggregate	-	-	-	-	-	m²	22.68
granite aggregate	-	-	-	-	-	m²	23.09
Base (Roadbase) Course; 200 mm of 28 **mm dense bitumen macadam in two** **layers to BS4987-1 2001 ref 5.2** Machine lay; areas 1000 m² and over							
limestone aggregate	-	-	-	-	-	m²	24.01
granite aggregate	-	-	-	-	-	m²	24.53
Hand Lay; areas 400 m² and over							
limestone aggregate	-	-	-	-	-	m²	28.16
granite aggregate	-	-	-	-	-	m²	28.71
Resin Bound Macadam Pavings; **machine laid; Colourpave Ltd** "Naturatex" clear resin bound macadam to pedestrian or vehicular hard landscape areas; laid to base course (not included)							
25 mm thick to pedestrian areas	-	-	-	-	-	m²	25.00
30 mm thick to vehicular areas	-	-	-	-	-	m²	28.00
"Colourtex" coloured resin bound macadam to pedestrian or vehicular hard landscape areas							
25 mm thick	-	-	-	-	-	m²	20.00
Marking car parks Car parking space division strips; in accordance with BS 3262; laid hot at 115 degrees C; on bitumen macadam surfacing							
Minimum daily rate	-	-	-	-	-	m	-
Stainless metal road studs							
100 x 100 mm	4.90	0.25	4.69	-	4.90	nr	9.59
Q23 GRAVEL/HOGGIN ROADS/PAVINGS							
Excavation and path preparation Excavating; 300 mm deep; to width of path; depositing excavated material at sides of excavation							
width 1.00 m	-	-	-	2.20	-	m²	2.20
width 1.50 m	-	-	-	1.84	-	m²	1.84
width 2.00 m	-	-	-	1.57	-	m²	1.57
width 3.00 m	-	-	-	1.32	-	m²	1.32
Excavating trenches; in centre of pathways; 100 flexible drain pipes; filling with clean broken stone or gravel rejects							
300 x 450 mm deep	3.04	0.10	1.88	1.10	3.08	m	6.06
Hand trimming and compacting reduced surface of pathway; by machine							
width 1.00 m	-	0.05	0.94	0.11	-	m	1.05
width 1.50 m	-	0.04	0.83	0.10	-	m	0.93
width 2.00 m	-	0.04	0.75	0.09	-	m	0.84
width 3.00 m	-	0.04	0.75	0.09	-	m	0.84
Permeable membranes; to trimmed and compacted surface of pathway							
"Terram 1000"	0.39	0.02	0.38	-	0.41	m²	0.79

Q PAVING/PLANTING/FENCING/SITE FURNITURE

Item Excluding site overheads and profit	PC £	Labour hours	Labour £	Plant £	Material £	Unit	Total rate £
Permaloc "AshphaltEdge" extruded aluminium alloy L shaped edging with 5.33 mm exposed upper lip; edging fixed to roadway base and edge profile with 250 mm steel fixing spike; laid to straight or curvilinear road edge; subsequently filled with macadam (not included)							
Depth of macadam							
38 mm	5.82	0.02	0.31	-	6.66	m	6.98
58 mm	6.53	0.02	0.32	-	7.39	m	7.71
64 mm	7.19	0.02	0.34	-	8.07	m	8.41
76 mm	8.65	0.02	0.38	-	9.56	m	9.94
102 mm	10.02	0.02	0.39	-	10.97	m	11.36
Permaloc "Cleanline"; heavy duty straight profile edging; for edgings to soft landscape beds or turf areas; 3.2 mm x 102 high; 3.2 mm thick with 4.75 mm exposed upper lip; fixed to form straight or curvilinear edge with 305 mm fixing spike							
Milled aluminium							
100 deep	6.64	0.02	0.34	-	7.50	m	7.84
Black							
100 deep	7.31	0.02	0.34	-	8.19	m	8.53
Permaloc "Permastrip"; heavy duty L shaped profile maintenance strip; 3.2 mm x 89 mm high with 5.2 mm exposed top lip; for straight or gentle curves on paths or bed turf interfaces; fixed to form straight or curvilinear edge with standard 305 mm stake; other stake lengths available							
Milled aluminium							
90 deep	6.64	0.02	0.34	-	7.50	m	7.84
Black							
90 deep	7.31	0.02	0.34	-	8.19	m	8.53
Permaloc "Proline"; medium duty straight profiled maintenance strip; 3.2 mm x 102 mm high with 3.18 mm exposed top lip; for straight or gentle curves on paths or bed turf interfaces; fixed to form straight or curvilinear edge with standard 305 mm stake; other stake lengths available							
Milled aluminium							
90 deep	5.03	0.02	0.34	-	5.84	m	6.18
Black							
90 deep	5.80	0.02	0.34	-	6.63	m	6.97
Edging boards							
Boards; 50 x 50 x 750 mm timber pegs at 1000 mm centres (excavations and hardcore under edgings not included)							
38 x 150 mm treated softwood edge boards	2.28	0.10	1.88	-	2.28	m	4.16
50 x 150 mm treated softwood edge boards	2.49	0.10	1.88	-	2.49	m	4.36
38 x 150 mm hardwood (iroko) edge boards	7.17	0.10	1.88	-	7.17	m	9.05
50 x 150 mm hardwood (iroko) edge boards	9.03	0.10	1.88	-	9.03	m	10.90
Filling to make up levels							
Obtained off site; hardcore; PC £17.00 /m^3							
150 mm thick	2.72	0.04	0.75	0.42	2.72	m^2	3.89
Obtained off site; granular fill type 1; PC £18.50 /tonne (£40.70 /m^3 compacted)							
100 mm thick	4.07	0.03	0.53	0.31	4.07	m^2	4.90
150 mm thick	6.11	0.03	0.47	0.47	6.11	m^2	7.04

Q PAVING/PLANTING/FENCING/SITE FURNITURE

Item Excluding site overheads and profit	PC £	Labour hours	Labour £	Plant £	Material £	Unit	Total rate £
Q23 GRAVEL/HOGGIN ROADS/PAVINGS - cont'd							
Surface treatments							
Sand blinding; to hardcore (not included)							
50 mm thick	1.46	0.04	0.75	-	1.61	m²	**2.36**
Filter fabric; to hardcore (not included)	0.39	0.01	0.19	-	0.41	m²	**0.60**
Herbicides; ICI							
"Casoron G" (residual) herbicide; treating							
substrate before laying base							
at 1 kg/125m²	2.82	0.05	0.94	-	3.10	100 m²	**4.04**
Granular pavings							
Approved washed river or pit gravel; on base							
layer of coarse aggregate; wearing layer 10 mm							
gravel graded; watering; rolling; on hardcore							
base (not included); for pavements; to falls and							
crossfalls and to slopes not exceeding 15							
degrees from horizontal; over 300 mm wide							
50 mm thick	1.85	0.09	1.64	-	2.22	m²	**3.86**
63 mm thick	2.52	0.13	2.36	-	3.02	m²	**5.38**
Ennstone Breedon; "Golden Gravel" or equivalent;							
rolling wet; on hardcore base (not included); for							
pavements; to falls and crossfalls and to slopes							
not exceeding 15 degrees from horizontal; over							
300 mm wide							
50 mm thick	7.80	0.10	1.88	0.38	7.80	m²	**10.05**
75 mm thick	11.70	0.10	1.88	0.46	11.70	m²	**14.03**
Ennstone Breedon; 'Wayfarer' specially formulated							
fine gravel for use on golf course pathways							
50 mm thick	6.25	0.10	1.88	0.38	6.25	m²	**8.50**
75 mm thick	9.37	0.11	2.08	0.46	9.37	m²	**11.91**
Ennstone Breedon; Breedon Buff decorative							
limestone chippings							
50 mm thick	-	0.01	0.19	0.09	5.25	m²	**5.53**
75 mm thick	9.75	0.01	0.23	0.11	10.24	m²	**10.59**
Footpath Gravels; porous self							
binding gravel							
CED Ltd; "Cedec Gravel" rolling wet; on							
hardcore base (not included); for pavements; to							
falls and crossfalls and to slopes not exceeding							
15 degrees from horizontal; over 300 mm wide							
50 mm thick	7.80	0.10	1.88	0.38	7.80	m²	**10.05**
75 mm thick	11.70	0.10	1.88	0.46	11.70	m²	**14.03**
Hoggin (stabilized);PC £14.50/m³ on hardcore							
base (not included); to falls and crossfalls and to							
slopes not exceeding 15 degrees from horizontal;							
over 300 mm wide							
100 mm thick	1.45	0.13	2.50	0.89	2.17	m²	**5.57**
150 mm thick	2.17	0.13	2.36	0.84	3.26	m²	**6.47**
Washed shingle; on prepared base (not included)							
25 - 50 size, 25 mm thick	0.76	0.02	0.33	0.08	0.80	m²	**1.21**
25 - 50 size, 75 mm thick	2.29	0.05	0.99	0.24	2.41	m²	**3.63**
50 - 75 size, 25 mm thick	0.76	0.02	0.38	0.09	0.80	m²	**1.27**
50 - 75 size, 75 mm thick	2.29	0.07	1.25	0.30	2.41	m²	**3.96**
Pea shingle; on prepared base (not included)							
10 - 15 size, 25 mm thick	0.76	0.02	0.33	0.08	0.80	m²	**1.21**
5 - 10 size, 75 mm thick	2.29	0.05	0.99	0.24	2.41	m²	**3.63**
Ballast; as dug; watering; rolling; on hardcore							
base (not included)							
100 mm thick	3.33	0.13	2.50	0.89	4.16	m²	**7.56**
150 mm thick	5.00	0.13	2.36	0.84	6.24	m²	**9.45**
CED Ltd; Cedec golden gravel; self binding; laid							
to base measured separately; compacting							
50 mm thick	6.00	0.03	0.54	0.04	6.00	m²	**6.58**

Q PAVING/PLANTING/FENCING/SITE FURNITURE

Item Excluding site overheads and profit	PC £	Labour hours	Labour £	Plant £	Material £	Unit	Total rate £
Grundon Ltd; Coxwell self-binding path gravels laid and compacted to excavation or base measured separately							
50 mm thick	1.56	0.03	0.54	0.04	1.56	m²	**2.14**
Bark surfaces; ICI "Casoron G" (residual) herbicide; treating substrate before laying base							
at 1 kg / 125m²	2.82	0.05	0.94	-	3.10	100 m²	**4.04**
Bark surfaces; Melcourt Industries Ltd Bark chips; to surface of pathways by machine; material delivered in 80 m³ loads; levelling and spreading by hand (excavation and preparation not included)							
walk chips; 100 mm thick (80 m³ loads)	1.74	0.03	0.63	0.27	1.82	m²	**2.72**
conifer walk chips; 100 mm thick (25 m³ loads)	3.04	0.03	0.63	0.27	3.19	m²	**4.09**
wood fibre; 100 mm thick	1.58	0.03	0.63	0.27	1.66	m²	**2.56**
Bound aggregates; Addagrip Surface Treatments UK Ltd; natural decorative resin bonded surface dressing laid to concrete, macadam or to plywood panels priced separately							
Primer coat to macadam or concrete base	-	-	-	-	-	m²	**4.00**
Golden pea Gravel 1-3 mm							
Buff adhesive	-	-	-	-	-	m²	**20.00**
Red adhesive	-	-	-	-	-	m²	**20.00**
Green adhesive	-	-	-	-	-	m²	**20.00**
Golden pea Gravel 2-5 mm							
Buff adhesive	-	-	-	-	-	m²	**23.00**
Chinese bauxite 1-3 mm							
Buff adhesive	-	-	-	-	-	m²	**20.00**
Q24 INTERLOCKING BRICK/BLOCK ROADS/PAVINGS							
Precast concrete block edgings; PC £7.83 /m²; 200 x 100 x 60 mm; on prepared base (not included); haunching one side							
Edgings; butt joints							
stretcher course	1.57	0.17	3.12	-	3.87	m	**6.99**
header course	0.78	0.27	5.00	-	3.23	m	**8.23**
Precast concrete vehicular paving blocks; Marshalls Plc; on prepared base (not included); on 50 mm compacted sharp sand bed; blocks laid in 7 mm loose sand and vibrated; joints filled with sharp sand and vibrated; level and to falls only							
"Trafica" paving blocks; 450 x 450 x 70 mm							
"Perfecta" finish; colour natural	20.51	0.50	9.38	0.09	22.63	m²	**32.09**
"Perfecta" finish; colour buff	23.76	0.50	9.38	0.09	25.96	m²	**35.43**
"Saxon" finish; colour natural	18.73	0.50	9.38	0.09	20.81	m²	**30.27**
"Saxon" finish; colour buff	22.23	0.50	9.38	0.09	24.40	m²	**33.86**

Q PAVING/PLANTING/FENCING/SITE FURNITURE

Item Excluding site overheads and profit	PC £	Labour hours	Labour £	Plant £	Material £	Unit	Total rate £
Q24 INTERLOCKING BRICK/BLOCK **ROADS/PAVINGS** - cont'd							
Precast concrete vehicular paving **blocks; "Keyblock" Marshalls Plc; on prepared** **base (not included); on 50 mm** **compacted sharp sand bed; blocks laid** **in 7 mm loose sand and vibrated; joints** **filled with sharp sand and vibrated; level** **and to falls only**							
Herringbone bond							
200 x 100 x 60 mm; natural grey	7.83	1.50	28.12	0.09	10.01	m²	**38.22**
200 x 100 x 60 mm; colours	8.72	1.50	28.12	0.09	10.94	m²	**39.15**
200 x 100 x 80 mm; natural grey	8.74	1.50	28.12	0.09	10.96	m²	**39.17**
200 x 100 x 80 mm; colours	10.08	1.50	28.12	0.09	12.37	m²	**40.58**
Basketweave bond							
200 x 100 x 60 mm; natural grey	7.83	1.20	22.50	0.09	10.01	m²	**32.60**
200 x 100 x 60 mm; colours	8.72	1.20	22.50	0.09	10.94	m²	**33.53**
200 x 100 x 80 mm; natural grey	8.74	1.20	22.50	0.09	10.96	m²	**33.55**
200 x 100 x 80 mm; colours	10.08	1.20	22.50	0.09	12.37	m²	**34.96**
Precast concrete vehicular paving **blocks; Charcon Hard Landscaping; on** **prepared base (not included); on 50 mm** **compacted sharp sand bed; blocks laid** **in 7 mm loose sand and vibrated; joints** **filled with sharp sand and vibrated; level** **and to falls only**							
"Europa" concrete blocks							
200 x 100 x 60 mm; natural grey	8.02	1.50	28.12	0.09	10.21	m²	**38.42**
200 x 100 x 60 mm; colours	8.59	1.50	28.12	0.09	10.80	m²	**39.02**
"Parliament" concrete blocks							
200 x 100 x 65 mm; natural grey	19.42	1.50	28.12	0.09	21.69	m²	**49.90**
200 x 100 x 65 mm; colours	19.42	1.50	28.12	0.09	21.69	m²	**49.90**
Recycled polyethylene grassblocks; **Netlon Turf Systems; interlocking units laid to** **prepared base or rootzone (not included)**							
"Netpave 50"; load bearing 150 tonnes per m²; 500 x 500 x 50 mm deep							
minimum area 40 m²	-	0.20	3.75	0.18	13.20	m²	**17.13**
160 - 1039 m²	-	0.20	3.75	0.18	12.90	m²	**16.83**
over 1039 m²	-	0.20	3.75	0.18	11.80	m²	**15.73**
"Netpave 25"; load bearing: light vehicles and pedestrians; 500 x 500 x 25 mm deep; laid onto established grass surface							
minimum area 80 m²	9.90	0.10	1.88	-	9.90	m²	**11.78**
320 - 1039 m²	9.65	0.10	1.88	-	9.65	m²	**11.53**
1040 - 2079 m²	9.25	0.10	1.88	-	9.25	m²	**11.13**
over 2079 m²	8.60	0.10	1.88	-	8.60	m²	**10.47**
Turfguard; extruded polyethethylene flexible mesh laid to existing grass surface or newly seeded areas to provide surface protection from traffic including vehicle or animal wear and tear							
Turfguard 30m x 2m; up to 660 m²	2.66	0.01	0.16	-	2.66	m²	**2.81**
Turfguard 30m x 2m; up to 1500 m²	2.41	0.01	0.16	-	2.41	m²	**2.56**
Turfguard 30m x 2m; over 1500 m²	2.15	0.01	0.16	-	2.15	m²	**2.31**

Q PAVING/PLANTING/FENCING/SITE FURNITURE

Item Excluding site overheads and profit	PC £	Labour hours	Labour £	Plant £	Material £	Unit	Total rate £
"Grassroad"; Cooper Clarke Group; **heavy duty for car parking and fire paths** **verge hardening and shallow** **embankments** Honeycomb cellular polyproylene interconnecting paviors with integral downstead anti-shear cleats including topsoil but excluding edge restraints; to granular sub-base (not included)							
635 x 330 x 42 overall laid to a module of 622 x 311 x 32	-	-	-	-	-	m²	21.23
Extra over for green colour	-	-	-	-	-	m²	0.26
Q25 SLAB/BRICK/SETT/COBBLE PAVINGS							
Bricks - General Preamble: BS 3921 includes the following specification for bricks for paving: bricks shall be hard, well burnt, non-dusting, resistant to frost and sulphate attack and true to shape, size and sample.							
Movement of materials Mechanically offloading bricks; loading wheel barrows; transporting maximum 25 m distance	-	0.20	3.75	-	-	m²	3.75
Edge restraints; to brick paving; on **prepared base (not included); 65 mm** **thick bricks; PC £300.00/1000;** **haunching one side** Header Course							
200 x 100 mm; butt joints	3.00	0.27	5.00	-	5.45	m	10.45
210 x 105 mm; mortar joints	2.67	0.50	9.38	-	5.38	m	14.76
Stretcher course							
200 x 100 mm; butt joints	1.50	0.17	3.12	-	3.80	m	6.92
210 x 105 mm; mortar joints	1.36	0.33	6.25	-	3.79	m	10.04
Variation in brick prices; add or subtract **the following amounts for every** **£1.00/1000 difference in the PC price** Edgings							
100 wide	-	-	-	-	0.05	10 m	0.05
200 wide	-	-	-	-	0.11	10 m	0.11
102.5 wide	-	-	-	-	0.05	10 m	0.05
215 wide	-	-	-	-	0.09	10 m	0.09
Clay brick pavings; on prepared base **(not included); bedding on 50 mm sharp** **sand; kiln dried sand joints** Pavings; Hanson Brick; "Nori"; 200 x 100 x 65 mm wirecut chamfered paviors							
"Sherbourne Red"; PC £17.05/m²	17.05	1.44	27.07	0.23	19.42	m²	46.72
"Allandale Gold"; PC £20.64/m²	20.64	1.44	27.07	0.23	22.59	m²	49.88
Clay brick pavings; 200 x 100 x 50; laid **to running stretcher, or stack bond only;** **on prepared base (not included);** **bedding on cement:sand (1:4) pointing** **mortar as work proceeds** PC £600.00 /1000							
laid on edge	47.62	4.76	89.29	-	55.77	m²	145.05
laid on edge but pavior 65 mm thick	40.00	3.81	71.43	-	47.96	m²	119.39
laid flat	25.97	2.20	41.25	-	30.77	m²	72.02

Q PAVING/PLANTING/FENCING/SITE FURNITURE

Item Excluding site overheads and profit	PC £	Labour hours	Labour £	Plant £	Material £	Unit	Total rate £
Q25 SLAB/BRICK/SETT/COBBLE PAVINGS - cont'd							
Clay brick pavings; 200 x 100 x 50 - cont'd							
PC 500.00/1000							
laid on edge	39.68	4.76	89.29	-	47.63	m²	136.92
laid on edge but pavior 65 mm thick	33.33	3.81	71.43	-	41.12	m²	112.55
laid flat	21.64	2.20	41.25	-	26.33	m²	67.58
PC 400.00/1000							
laid on edge	31.74	4.76	89.29	-	39.50	m²	128.78
laid on edge but pavior 65 mm thick	26.66	3.81	71.43	-	34.29	m²	105.72
laid flat	17.32	2.20	41.25	-	21.90	m²	63.15
PC 300.00/1000							
laid on edge	23.81	4.76	89.29	-	31.36	m²	120.65
laid on edge but pavior 65 mm thick	20.00	3.81	71.43	-	27.46	m²	98.89
laid flat	12.99	2.20	41.25	-	17.46	m²	58.71
Clay brick pavings; 200 x 100 x 50; butt **jointed laid herringbone or basketweave** **pattern only; on prepared base (not** **included); bedding on 50 mm sharp** **sand**							
PC £600.00 /1000							
laid flat	30.00	1.44	27.07	0.29	32.70	m²	60.05
PC 500.00/1000							
laid flat	25.00	1.44	27.07	0.29	27.57	m²	54.93
PC 400.00/1000							
laid flat	20.00	1.44	27.07	0.29	22.45	m²	49.80
PC 300.00/1000							
laid flat	15.00	1.44	27.07	0.29	17.32	m²	44.68
Clay brick pavings; 215 x 102.5 x 65 **mm; on prepared base (not included);** **bedding on cement:sand (1:4) pointing** **mortar as work proceeds**							
Paving bricks; PC £600.00/1000; Herringbone bond							
laid on edge	35.55	3.55	66.65	-	41.97	m²	108.62
laid flat	23.70	2.37	44.44	-	29.23	m²	73.67
Paving bricks; PC £600.00/1000; Basket weave bond							
laid on edge	35.55	2.37	44.44	-	41.97	m²	86.41
laid flat	23.70	1.58	29.63	-	29.23	m²	58.86
Paving bricks; PC £600.00/1000; Running or Stack bond							
laid on edge	35.55	1.90	35.56	-	41.97	m²	77.53
laid flat	23.70	1.26	23.70	-	29.23	m²	52.93
Paving bricks; PC £500.00/1000; Herringbone bond							
laid on edge	29.63	3.55	66.65	-	33.54	m²	100.19
laid flat	19.75	2.37	44.44	-	25.28	m²	69.72
Paving bricks; PC £500.00/1000; Basket weave bond							
laid on edge	29.63	2.37	44.44	-	33.54	m²	77.98
laid flat	19.75	1.58	29.63	-	25.28	m²	54.91
Paving bricks; PC £500.00/1000; Running or Stack bond							
laid on edge	29.63	1.90	35.56	-	33.54	m²	69.10
laid flat	19.75	1.26	23.70	-	25.28	m²	48.98
Paving bricks; PC £400.00/1000; Herringbone bond							
laid flat	15.80	2.37	44.44	-	21.23	m²	65.67
laid on edge	23.70	3.55	66.65	-	27.47	m²	94.12

Q PAVING/PLANTING/FENCING/SITE FURNITURE

Item Excluding site overheads and profit	PC £	Labour hours	Labour £	Plant £	Material £	Unit	Total rate £
Paving bricks; PC £400.00/1000; Basket weave bond							
laid on edge	23.70	2.37	44.44	-	27.47	m²	71.91
laid flat	15.80	1.58	29.63	-	21.23	m²	50.86
Paving bricks; PC £400.00/1000; Running or Stack bond							
laid on edge	23.70	1.90	35.56	-	27.47	m²	63.03
laid flat	15.80	1.26	23.70	-	21.23	m²	44.93
Paving bricks; PC £300.00/1000; Herringbone bond							
laid on edge	17.77	3.55	66.65	-	20.95	m²	87.60
laid flat	11.85	2.37	44.44	-	17.38	m²	61.82
Paving bricks; PC £300.00/1000; Basket weave bond							
laid on edge	17.77	2.37	44.44	-	20.95	m²	65.39
laid flat	11.85	1.58	29.63	-	17.38	m²	47.01
Paving bricks; PC £300.00/1000; Running or Stack bond							
laid on edge	17.77	1.90	35.55	-	20.95	m²	56.50
laid flat	11.85	1.26	23.70	-	17.38	m²	41.08
Cutting							
curved cutting	-	0.44	8.33	4.23	-	m	12.56
raking cutting	-	0.33	6.25	3.25	-	m	9.50
Add or subtract the following amounts for every £10.00 /1000 difference in the prime cost of bricks							
Butt joints							
200 x 100	-	-	-	-	0.50	m²	0.50
215 x 102.5	-	-	-	-	0.45	m²	0.45
10 mm mortar joints							
200 x 100	-	-	-	-	0.43	m²	0.43
215 x 102.5	-	-	-	-	0.40	m²	0.40
Precast concrete pavings; Charcon Hard Landscaping; to BS 7263; on prepared sub-base (not included); bedding on 25 mm thick cement:sand mortar (1:4); butt joints; straight both ways; jointing in cement:sand (1:3) brushed in; on 50 mm thick sharp sand base							
Pavings; natural grey							
450 x 450 x 70 mm chamfered	11.26	0.44	8.33	-	15.31	m²	23.65
450 x 450 x 50 mm chamfered	9.73	0.44	8.33	-	13.78	m²	22.12
600 x 300 x 50 mm	8.78	0.44	8.33	-	12.83	m²	21.16
400 x 400 x 65 mm chamfered	18.13	0.40	7.50	-	22.18	m²	29.68
450 x 600 x 50 mm	7.81	0.44	8.33	-	11.87	m²	20.20
600 x 600 x 50 mm	6.28	0.40	7.50	-	10.33	m²	17.83
750 x 600 x 50 mm	5.93	0.40	7.50	-	9.99	m²	17.49
900 x 600 x 50 mm	5.43	0.40	7.50	-	9.48	m²	16.98
Pavings; coloured							
450 x 450 x 70 mm chamfered	13.19	0.44	8.33	-	17.24	m²	25.57
450 x 600 x 50 mm	11.78	0.44	8.33	-	15.83	m²	24.16
400 x 400 x 65 mm chamfered	18.13	0.40	7.50	-	22.18	m²	29.68
600 x 600 x 50 mm	9.83	0.40	7.50	-	13.89	m²	21.39
750 x 600 x 50 mm	9.07	0.40	7.50	-	13.12	m²	20.62
900 x 600 x 50 mm	8.05	0.40	7.50	-	12.10	m²	19.60
Precast concrete pavings; Charcon Hard Landscaping; to BS 7263; on prepared sub-base (not included); bedding on 25 mm thick cement:sand mortar (1:4); butt joints; straight both ways; jointing in cement:sand (1:3) brushed in; on 50 mm thick sharp sand base							
"Appalacian" rough textured exposed aggregate pebble paving							
600 mm x 600 mm x 65 mm	23.37	0.50	9.38	-	26.55	m²	35.92

Q PAVING/PLANTING/FENCING/SITE FURNITURE

Item Excluding site overheads and profit	PC £	Labour hours	Labour £	Plant £	Material £	Unit	Total rate £
Q25 SLAB/BRICK/SETT/COBBLE PAVINGS - cont'd							
Pavings; Brett Landscaping and Building Products; on prepared base (not included) "Broadway paving" in Cotswold; spot bedded on 5 pads of sand:cement mortar (1:4) on sharp sand							
600 x 600 mm	8.78	0.40	7.50	-	10.51	m^2	18.01
450 x 450 mm	7.61	0.50	9.38	-	9.31	m^2	18.68
Pavings; Marshalls Plc; spot bedding on 5 nr pads of cement:sand mortar (1:4); on sharp sand "Blister Tactile" pavings; specially textured slabs for guidance of blind pedestrians; red or buff							
400 x 400 x 50 mm	21.69	0.50	9.38	-	23.74	m^2	33.12
450 x 450 x 50 mm	19.06	0.50	9.38	-	21.05	m^2	30.43
"Metric Four Square" pavings							
496 x 496 x 50 mm; exposed river gravel aggregate	55.62	0.50	9.38	-	57.13	m^2	66.51
"Metric Four Square" cycle blocks							
496 x 496 x 50 mm; exposed aggregate	55.62	0.25	4.69	-	57.74	m^2	62.42
Precast concrete pavings; Marshalls Plc; "Heritage" imitation riven yorkstone paving; on prepared sub-base measured separately; bedding on 25 mm thick cement:sand mortar (1:4); pointed straight both ways cement:sand (1:3) Square and rectangular paving							
450 x 300 x 38 mm	28.91	1.00	18.75	-	32.59	m^2	51.34
450 x 450 x 38 mm	17.39	0.75	14.06	-	21.08	m^2	35.14
600 x 300 x 38 mm	19.08	0.80	15.00	-	22.76	m^2	37.76
600 x 450 x 38 mm	19.31	0.75	14.06	-	23.00	m^2	37.06
600 x 600 x 38 mm	18.70	0.50	9.38	-	22.45	m^2	31.83
Extra labours for laying the a selection of the above sizes to random rectangular pattern	-	0.33	6.25	-	-	m^2	6.25
Radial paving for circles							
circle with centre stone and first ring (8 slabs), 450 x 230/560 x 38 mm; diameter 1.54 m (total area 1.86 m^2)	50.39	1.50	28.13	-	55.18	nr	83.31
circle with second ring (16 slabs), 450 x 300/460 x 38 mm; diameter 2.48 m (total area 4.83 m^2)	50.39	4.00	75.00	-	142.07	nr	217.07
circle with third ring (16 slabs), 450 x 470/625 x 38 mm; diameter 3.42 m (total area 9.18 m^2)	50.39	8.00	150.00	-	259.67	nr	409.67
Stepping stones							
380 dia x 38 mm	3.60	0.20	3.75	-	8.79	nr	12.54
Asymmetrical 560 x 420 x 38 mm	5.36	0.20	3.75	-	10.55	nr	14.30
Precast concrete pavings; Marshalls Plc; "Chancery" imitation reclaimed riven yorkstone paving; on prepared sub-base measured separately; bedding on 25 mm thick cement:sand mortar (1:4); pointed straight both ways cement:sand (1:3) Square and rectangular paving							
300 x 300 x 45 mm	23.92	1.00	18.75	-	27.60	m^2	46.35
450 x 300 x 45 mm	22.46	0.90	16.88	-	26.15	m^2	43.02
600 x 300 x 45 mm	21.94	0.80	15.00	-	25.62	m^2	40.62
600 x 450 x 45 mm	21.94	0.75	14.06	-	25.62	m^2	39.69
450 x 450 x 45 mm	19.80	0.75	14.06	-	23.48	m^2	37.54
600 x 600 x 45 mm	22.43	0.50	9.38	-	26.11	m^2	35.48

Q PAVING/PLANTING/FENCING/SITE FURNITURE

Item Excluding site overheads and profit	PC £	Labour hours	Labour £	Plant £	Material £	Unit	Total rate £
Extra labours for laying the a selection of the above sizes to random rectangular pattern	-	0.33	6.25	-	-	m²	6.25
Radial paving for circles							
circle with centre stone and first ring (8 slabs), 450 x 230/560 x 38 mm; diameter 1.54 m (total area 1.86 m²)	60.46	1.50	28.13	-	65.25	nr	93.38
circle with second ring (16 slabs), 450 x 300/460 x 38 mm; diameter 2.48 m (total area 4.83 m²)	155.98	4.00	75.00	-	168.30	nr	243.30
circle with third ring (16 slabs), 450 x 470/625 x 38 mm; diameter 3.42 m (total area 9.18 m²)	283.82	8.00	150.00	-	307.02	nr	457.02
Squaring off set for 2 ring circle							
16 slabs; 2.72 m²	87.21	1.00	18.75	-	87.21	nr	105.96
Pavings; Milner Delvaux Ltd; bedding on 50 mm cement:sand mortar (1:4)							
"Blanc de Bierges" precast concrete pavings							
200 x 200 mm	21.89	0.80	15.00	-	27.62	m²	42.62
400 x 200 mm	19.90	0.74	13.89	-	25.58	m²	39.47
400 x 300 mm	20.07	0.71	13.39	-	25.76	m²	39.15
400 x 400 mm	19.90	0.71	13.39	-	25.58	m²	38.97
600 x 300 mm	18.80	0.61	11.36	-	24.46	m²	35.82
400 x 600 mm	18.80	0.50	9.42	-	24.46	m²	33.88
600 x 600 mm	18.80	0.44	8.29	-	24.46	m²	32.75
300 mm diameter	3.03	0.40	7.50	-	3.40	nr	10.90
Marshalls Plc; La Linia Pavings; fine textured exposed aggregate 80 mm thick pavings in various sizes to designed laying patterns; laid to 50 mm sharp sand bed on Type 1 base all priced separately							
Bonded laying patterns 300 x 300							
Light Granite / Anthracite basalt	24.72	0.75	14.06	-	25.34	m²	39.40
Indian granite / Yellow	24.72	0.75	14.06	-	25.34	m²	39.40
Random Scatter pattern incorporating 100 x 200, 200 x 200 and 300 x 200 units							
Light Granite / Anthracite basalt	24.72	1.00	18.75	-	24.72	m²	43.47
Indian granite / Yellow	24.72	1.00	18.75	-	24.72	m²	43.47
Marshalls Plc; "La Linia Grande Paving"; as above but 140 mm thick							
Bonded laying patterns 300 x 300							
All colours	45.46	0.75	14.06	-	45.46	m²	59.52
Random Scatter pattern incorporating 100 x 200, 200 x 200 and 300 x 200 units							
Light Granite / Anthracite basalt	45.46	1.00	18.75	-	45.46	m²	64.21
Extra over to the above for incorporating inlay stones to patterns; blue, light granite or anthracite basalt							
to prescribed patterns	286.74	1.50	28.13	-	286.74	m²	314.87
as individual units; triangular 200 x 200 x 282 mm	1.62	0.05	0.94	-	1.62	nr	2.56
Pedestrian deterrent pavings; Marshalls Plc; on prepared base (not included); bedding on 25 mm cement:sand (1:3); cement:sand (1:3) joints							
"Lambeth" pyramidal pavings							
600 x 600 x 75 mm	37.60	0.25	4.69	-	41.28	m²	45.97
"Thaxted" pavings; granite sett appearance							
600 x 600 x 75 mm	43.59	0.33	6.25	-	47.27	m²	53.52

Q PAVING/PLANTING/FENCING/SITE FURNITURE

Item Excluding site overheads and profit	PC £	Labour hours	Labour £	Plant £	Material £	Unit	Total rate £
Q25 SLAB/BRICK/SETT/COBBLE PAVINGS - cont'd							
Pedestrian deterrent pavings; **Townscape Products Ltd; on prepared base** **(not included); bedding on 25 mm** **cement:sand (1:3); cement:sand (1:3)** **joints**							
"Strata" striated textured slab pavings; giving bonded appearance; grey							
600 x 600 x 60 mm	25.86	0.50	9.38	-	29.54	m^2	**38.92**
"Geoset" raised chamfered studs pavings; grey							
600 x 600 x 60 mm	25.86	0.50	9.38	-	29.54	m^2	**38.92**
"Abbey" square cobble pattern pavings; reinforced							
600 x 600 x 65 mm	24.44	0.80	15.00	-	28.12	m^2	**43.12**
Surface treatments; ICI							
Treating substrate before laying base							
"Casoron G" (residual); at 1 kg/125m^2	2.82	0.05	0.94	-	3.10	100 m^2	**4.04**
Edge restraints; to block paving; on **prepared base (not included); 200 x 100** **x 80 mm; PC £8.74/m^2; haunching one** **side**							
Header course							
200 x 100 mm; butt joints	87.40	0.27	5.00	-	89.85	m	**94.85**
Stretcher course							
200 x 100 mm; butt joints	43.70	0.17	3.12	-	46.00	m	**49.12**
Concrete paviors; Marshalls Plc; on **prepared base (not included); bedding** **on 50 mm sand; kiln dried sand joints** **swept in**							
"Keyblock" paviors							
200 x 100 x 60 mm; grey	7.83	0.40	7.50	0.09	9.61	m^2	**17.20**
200 x 100 x 60 mm; colours	8.72	0.40	7.50	0.09	10.50	m^2	**18.09**
200 x 100 x 80 mm; grey	8.74	0.44	8.33	0.09	10.52	m^2	**18.95**
200 x 100 x 80 mm; colours	10.08	0.44	8.33	0.09	11.86	m^2	**20.29**
Concrete cobble paviors; Charcon Hard **Landscaping; Concrete Products; on** **prepared base (not included); bedding** **on 50 mm sand; kiln dried sand joints** **swept in**							
Paviors							
"Woburn" blocks; 100 - 201 x 134 x 80 mm; random sizes	20.41	0.67	12.50	0.09	22.70	m^2	**35.30**
"Woburn" blocks; 100 - 201 x 134 x 80 mm; single size	20.41	0.50	9.38	0.09	22.70	m^2	**32.17**
"Woburn" blocks; 100 - 201 x 134 x 60 mm; random sizes	16.52	0.67	12.50	0.09	18.72	m^2	**31.31**
"Woburn" blocks; 100 - 201 x 134 x 60 mm; single size	16.52	0.50	9.38	0.09	18.70	m^2	**28.17**
Concrete setts; on 25 mm sand; **compacted; vibrated; joints filled with** **sand; natural or coloured; well rammed** **hardcore base (not included)**							
Marshalls Plc; Tegula Cobble Paving							
60 mm thick; random sizes	19.07	0.57	10.71	0.09	21.95	m^2	**32.76**
60 mm thick; single size	19.07	0.45	8.52	0.09	21.95	m^2	**30.57**
80 mm thick; random sizes	21.49	0.57	10.71	0.09	24.50	m^2	**35.30**
80 mm thick; single size	21.49	0.45	8.52	0.09	24.50	m^2	**33.11**
Cobbles 80 x 80 x 60 mm thick; traditional	28.35	0.56	10.42	0.09	30.99	m^2	**41.50**

Q PAVING/PLANTING/FENCING/SITE FURNITURE

Item Excluding site overheads and profit	PC £	Labour hours	Labour £	Plant £	Material £	Unit	Total rate £
Cobbles							
Charcon Hard Landscaping; Country setts							
100 mm thick; random sizes	25.20	1.00	18.75	0.09	27.13	m²	**45.97**
100 mm thick; single size	25.20	0.67	12.50	0.09	27.76	m²	**40.35**
Blanc de Bierges; cast setts on							
prepared base (not included); bedding							
on 50 mm sand; kiln dried sand joints							
swept in							
"Blanc de Bierges" setts							
70 x 70 x 70 mm	23.83	0.45	8.52	0.09	26.95	m²	**35.56**
140 x 140 x 80 mm	22.72	0.45	8.52	0.09	25.79	m²	**34.40**
210 x 140 x 80 mm	22.72	0.40	7.50	0.09	25.79	m²	**33.38**
Blanc de Bierges; cast setts on							
prepared base (not included); bedding							
on 30 mm cement:sand (1:3);							
cement:sand (1:3) joints							
70 x 70 x 70 mm	23.83	2.50	46.88	-	32.80	m²	**79.67**
140 x 140 x 80 mm	22.72	2.20	41.25	-	31.63	m²	**72.88**
210 x 140 x 80 mm	22.72	2.00	37.50	-	31.63	m²	**69.13**
Natural stone, slab or granite paving -							
General							
Preamble: provide paving slabs of the specified							
thickness in random sizes but not less than 25							
slabs per 10 m² of surface area, to be laid in							
parallel courses with joints alternately broken and							
laid to falls.							
Reconstituted Yorkstone aggregate							
pavings; Marshalls Plc; "Saxon" on prepared							
sub-base measured separately; bedding							
on 25 mm thick cement:sand mortar (1:4);							
on 50 mm thick sharp sand base							
square and rectangular paving in various colours;							
butt joints straight both ways							
300 x 300 x 35 mm	26.53	0.88	16.50	-	31.42	m²	**47.92**
600 x 300 x 35 mm	17.47	0.71	13.41	-	22.37	m²	**35.77**
450 x 450 x 35 mm	9.58	0.71	13.41	-	14.47	m²	**27.88**
450 x 450 x 50 mm	20.05	0.82	15.47	-	24.94	m²	**40.41**
600 x 600 x 35 mm	13.05	0.55	10.31	-	17.94	m²	**28.25**
600 x 600 x 50 mm	17.44	0.60	11.34	-	22.34	m²	**33.68**
square and rectangular paving in natural; butt							
joints straight both ways							
300 x 300 x 35 mm	22.09	0.88	16.50	-	26.98	m²	**43.48**
450 x 450 x 35 mm	9.58	0.77	14.44	-	14.47	m²	**28.91**
450 x 450 x 50 mm	17.35	0.77	14.44	-	22.24	m²	**36.68**
600 x 300 x 35 mm	15.40	0.82	15.47	-	20.29	m²	**35.76**
600 x 600 x 35 mm	11.26	0.55	10.31	-	16.15	m²	**26.46**
600 x 600 x 50 mm	14.78	0.66	12.38	-	19.68	m²	**32.05**
Radial paving for circles; 20 mm joints							
circle with centre stone and first ring (8 slabs),							
450 x 230/560 x 35 mm; diameter 1.54 m (total							
area 1.86 m²)	52.01	1.50	28.13	-	57.07	nr	**85.20**
circle with second ring (16 slabs), 450 x 300/460							
x 35 mm; diameter 2.48 m (total area 4.83 m²)	126.57	4.00	75.00	-	139.64	nr	**214.64**
circle with third ring (24 slabs), 450 x 310/430 x							
35 mm; diameter 3.42 m (total area 9.18 m²)	238.41	8.00	150.00	-	263.41	nr	**413.41**
Granite setts; bedding on 25 mm							
cement:sand (1:3)							
Natural granite setts; 100 x 100 mm to 125 x 150							
mm; x 150 to 250 mm length; riven surface; silver							
grey							
new; standard grade	18.89	2.00	37.50	-	26.67	m²	**64.17**
new; high grade	24.44	2.00	37.50	-	32.22	m²	**69.72**
reclaimed; cleaned	40.00	2.00	37.50	-	48.57	m²	**86.07**

Prices for Measured Works

Q PAVING/PLANTING/FENCING/SITE FURNITURE

Item Excluding site overheads and profit	PC £	Labour hours	Labour £	Plant £	Material £	Unit	Total rate £
Q25 SLAB/BRICK/SETT/COBBLE PAVINGS - cont'd							
Natural stone, slate or granite flag **pavings; CED Ltd; on prepared base** **(not included); bedding on 25 mm** **cement:sand (1:3); cement:sand (1:3)** **joints**							
Yorkstone; riven laid random rectangular							
new slabs; 40 - 60 mm thick	45.00	1.71	32.14	-	52.91	m²	85.05
reclaimed slabs, Cathedral grade; 50 - 75 mm							
thick	60.00	2.80	52.50	-	69.41	m²	121.91
Donegal quartzite slabs; standard tiles							
595 x 595 x 7 - 13 mm	52.00	2.33	43.75	-	58.01	m²	101.76
Natural Yorkstone, pavings or edgings; **Johnsons Wellfield Quarries; sawn 6** **sides; 50 mm thick; on prepared base** **measured separately; bedding on 25 mm** **cement:sand (1:3); cement:sand (1:3)** **joints**							
Paving							
laid to random rectangular pattern	50.50	1.71	32.14	-	56.77	m²	88.91
laid to coursed laying pattern; 3 sizes	54.76	1.72	32.33	-	60.74	m²	93.07
Paving; single size							
600 x 600 mm	60.63	0.85	15.94	-	66.90	m²	82.84
600 x 400 mm	60.63	1.00	18.75	-	66.90	m²	85.65
300 x 200 mm	69.15	2.00	37.50	-	75.85	m²	113.35
215 x 102.5 mm	72.35	2.50	46.88	-	79.21	m²	126.08
Paving; cut to template off site; 600 x 600;							
radius							
1.00 m	173.06	3.33	62.50	-	176.30	m²	238.80
2.50 m	167.73	2.00	37.50	-	170.97	m²	208.47
5.00 m	162.40	2.00	37.50	-	165.64	m²	203.14
Edgings							
100 mm wide x random lengths	6.66	0.50	9.38	-	7.32	m	16.69
100 mm x 100 mm	6.92	0.50	9.38	-	10.50	m	19.88
100 mm x 200 mm	13.83	0.50	9.38	-	17.76	m	27.14
250 mm wide x random lengths	13.05	0.40	7.50	-	16.94	m	24.44
500 mm wide x random lengths	26.12	0.33	6.25	-	30.67	m	36.92
Yorkstone edgings; 600 mm long x 250 mm							
wide; cut to radius							
1.00 m to 3.00	45.26	0.50	9.38	-	47.85	m	57.22
3.00 m to 5.00 m	43.93	0.44	8.33	-	46.45	m	54.78
exceeding 5.00 m	42.60	0.40	7.50	-	45.05	m	52.55
Natural Yorkstone, pavings or edgings; **Johnsons Wellfield Quarries; sawn 6** **sides; 75 mm thick; on prepared base** **measured separately; bedding on 25 mm** **cement:sand (1:3); cement:sand (1:3)** **joints**							
Paving							
laid to random rectangular pattern	55.96	0.95	17.81	-	62.00	m²	79.81
laid to coursed laying pattern; 3 sizes	62.88	0.95	17.81	-	69.27	m²	87.08
Paving; single size							
600 x 600 mm	70.08	0.95	17.81	-	76.83	m²	94.64
600 x 400 mm	70.08	0.95	17.81	-	76.83	m²	94.64
300 x 200 mm	80.47	0.75	14.06	-	87.73	m²	101.80
215 x 102.5 mm	84.46	2.50	46.88	-	91.92	m²	138.80
Paving; cut to template off site; 600 x 600;							
radius							
1.00 m	202.49	4.00	75.00	-	205.73	m²	280.73
2.50 m	197.16	2.50	46.88	-	200.40	m²	247.28
5.00 m	191.84	2.50	46.88	-	195.08	m²	241.96

Q PAVING/PLANTING/FENCING/SITE FURNITURE

Item Excluding site overheads and profit	PC £	Labour hours	Labour £	Plant £	Material £	Unit	Total rate £
Edgings							
100 mm wide x random lengths	8.53	0.60	11.25	-	9.28	m	**20.53**
100 mm x 100 mm	8.45	0.60	11.25	-	12.11	m	**23.36**
100 mm x 200 mm	16.89	0.50	9.38	-	20.98	m	**30.35**
250 mm wide x random lengths	15.72	0.50	9.38	-	19.75	m	**29.12**
500 mm wide x random lengths	31.44	0.40	7.50	-	36.25	m	**43.75**
Edgings; 600 mm long x 250 mm wide; cut to radius							
1.00 m to 3.00	52.98	0.60	11.25	-	55.95	m	**67.20**
3.00 m to 5.00 m	51.65	0.50	9.38	-	54.55	m	**63.93**
exceeding 5.00 m	50.82	0.44	8.33	-	53.69	m	**62.02**
CED Ltd; Indian Sandstone, riven pavings or edgings; 25-35 mm thick; on prepared base measured separately; bedding on 25 mm cement:sand (1:3); cement:sand (1:3) joints							
Paving							
laid to random rectangular pattern	20.00	2.40	45.01	-	24.24	m²	**69.25**
laid to coursed laying pattern; 3 sizes	20.00	2.00	37.50	-	24.24	m²	**61.74**
Paving; single size							
600 x 600 mm	20.00	1.00	18.75	-	24.24	m²	**42.99**
600 x 400 mm	20.00	1.25	23.44	-	24.24	m²	**47.68**
400 x 400 mm	20.00	1.67	31.25	-	24.24	m²	**55.49**
Natural stone, slate or granite flag pavings; CED Ltd; on prepared base (not included); bedding on 25 mm cement:sand (1:3); cement:sand (1:3) joints							
Granite paving; sawn 6 sides; textured top, silver grey							
new slabs; 50 mm thick	25.00	1.71	32.14	-	29.66	m²	**61.80**

Cobble pavings - General

Preamble: Excavate for and supply and lay hard coarse clinker, consolidated to a thickness of 75 mm. Lay a 50 mm bed of concrete (1:2:4) with small aggregate. Cobbles should be embedded by hand, tight-butted, endwise to a depth of 60% of their length. The tops of cobbles generally shall be about 25 mm above the level of the adjoining paving, except where cobbled areas abut flag paving when the last three rows of cobbles shall be graded down so that the tops of the cobbles shall be level with the flags. Each flint shall be consolidated, using a wooden mallet. Preamble: Upon completion of each area, a dry grout of rapid-hardening cement:sand (1:2) shall be brushed over the cobbles until the interstices are filled to the level of the adjoining paving. Surplus grout shall then be brushed off and a light, fine spray of water applied over the area. The area shall again be brushed after 24 hours to remove any free grout adhering to exposed faces of cobbles. Cement for grouting shall be rapid-hardening Portland Cement conforming to BS12:1978, and shall be stored on site in a proper manner to avoid deterioration (manufacturer's instructions to be carefully followed).

Q PAVING/PLANTING/FENCING/SITE FURNITURE

Item Excluding site overheads and profit	PC £	Labour hours	Labour £	Plant £	Material £	Unit	Total rate £
Q25 SLAB/BRICK/SETT/COBBLE PAVINGS - cont'd							
Cobble pavings Cobbles; to present a uniform colour in panels; or varied in colour as required							
Scottish Beach Cobbles; 200 - 100 mm	21.25	2.00	37.50	-	26.90	m^2	**64.40**
Scottish Beach Cobbles; 100 - 75 mm	20.00	2.50	46.88	-	25.65	m^2	**72.52**
Scottish Beach Cobbles; 75 - 50 mm	16.25	3.33	62.50	-	21.90	m^2	**84.39**
Concrete cycle blocks; Marshalls Plc; **bedding in cement:sand (1:4)** "Metric 4 Square" cycle stand blocks, smooth grey concrete							
496 x 496 x 100 mm	25.04	0.25	4.69	-	25.53	nr	**30.22**
Concrete cycle blocks; Townscape **Products Ltd; on 100 mm concrete (1:2:4);** **on 150 mm hardcore; bedding in** **cement:sand (1:4)** Cycle Blocks							
"Cycle Bloc"; in white concrete	18.89	0.25	4.69	-	19.54	nr	**24.23**
"Mountain Cycle Bloc"; in white concrete	26.14	0.25	4.69	-	26.79	nr	**31.48**
Grass concrete - General Preamble: Grass seed should be a perennial ryegrass mixture, with the proportion depending on expected traffic. Hardwearing winter sportsground mixtures are suitable for public areas. Loose gravel, shingle or sand is liable to be kicked out of the blocks; rammed hoggin or other stabilized material should be specified.							
Grass concrete; Grass Concrete Ltd; on **blinded Granular Type 1 sub-base (not** **included)** "Grasscrete" in situ concrete continuously reinforced surfacing; including expansion joints at 10 m centres; soiling; seeding							
ref GC2; 150 mm thick; traffic up to 40.00 tonnes	-	-	-	-	-	m^2	**28.81**
ref GC1; 100 mm thick; traffic up to 13.30 tonnes	-	-	-	-	-	m^2	**24.07**
ref GC3; 76 mm thick; traffic up to 4.30 tonnes	-	-	-	-	-	m^2	**20.63**
Grass concrete; Grass Concrete Ltd; **406 x 406 mm blocks; on 20 mm sharp** **sand; on blinded MOT type 1 sub-base** **(not included); level and to falls only;** **including filling with topsoil; seeding with** **dwarf rye grass at £2.40/kg** Pavings							
ref GB103; 103 mm thick	12.56	0.40	7.50	0.18	14.37	m^2	**22.05**
ref GB83; 83 mm thick	11.27	0.36	6.82	0.18	13.08	m^2	**20.08**
Grass concrete; Charcon Hard **Landscaping; on 25 mm sharp sand;** **including filling with topsoil; seeding with** **dwarf rye grass at £2.40/kg** "Grassgrid" grass/concrete paving blocks							
366 x 274 x 100 mm thick	16.71	0.38	7.03	0.18	18.52	m^2	**25.73**
Grass concrete; Marshalls Plc; on 25 **mm sharp sand; including filling with** **topsoil; seeding with dwarf rye grass at** **£2.40/kg** Concrete grass pavings							
"Grassguard 130"; for light duty applications; (80 mm prepared base not included)	13.69	0.38	7.03	0.18	16.00	m^2	**23.22**

Q PAVING/PLANTING/FENCING/SITE FURNITURE

Item Excluding site overheads and profit	PC £	Labour hours	Labour £	Plant £	Material £	Unit	Total rate £
"Grassguard 160"; for medium duty applications; (80 - 150 mm prepared base not included)	16.12	0.46	8.65	0.18	18.49	m²	27.33
"Grassguard 180"; for heavy duty applications; (150 mm prepared base not included)	18.06	0.60	11.25	0.18	20.48	m²	31.91
Full mortar bedding Extra over pavings for bedding on 25 mm cement:sand (1:4); in lieu of spot bedding on sharp sand	-	0.03	0.47	-	1.24	m²	1.71
Precast Steps; Blanc de Bierges; "L" **shaped units with tapered edges in light** **cream-buff; individually hand textured** **precast concrete; laid to formed** **concrete or brick treads and risers (not** **included); jointed on 1:3 mortar; options** **on all dimensions available; a selection** **is shown below**							
Riser height 150 mm; tread							
450 - 600 deep	56.65	0.56	10.42	-	62.08	m	72.50
200 - 350 deep	56.65	0.67	12.50	-	59.11	m	71.61
140 deep	56.65	0.84	15.76	-	57.42	m	73.17
Riser height 150 mm; tread							
450 - 600 deep	72.10	0.56	10.42	-	77.53	m	87.95
200 - 350 deep	72.10	0.67	12.50	-	74.56	m	87.07
140 deep	72.10	0.84	15.76	-	72.87	m	88.62
Q26 SPECIAL SURFACINGS FOR SPORT							
Market Prices of Surfacing Materials							
Surfacings; Melcourt Industries Ltd							
Playbark 10/50®; per 25m³ load	-	-	-	-	50.10	m³	50.10
Playbark 10/50®; per 80m³ load	-	-	-	-	37.10	m³	37.10
Playbark 8/25®; per 25m³ load	-	-	-	-	48.60	m³	48.60
Playbark 8/25®; per 80m³ load	-	-	-	-	35.60	m³	35.60
Adventure Bark 30; per 25m³ load	-	-	-	-	46.90	m³	46.90
Adventure Bark 30; per 80m³ load	-	-	-	-	33.90	m³	33.90
Playchips®; per 25m³ load	-	-	-	-	33.40	m³	33.40
Playchips®; per 80m³ load	-	-	-	-	20.40	m³	20.40
Kushyfall; per 25m³ load	-	-	-	-	31.60	m³	31.60
Kushyfall; per 80m³ load	-	-	-	-	18.60	m³	18.60
Softfall; per 25m³ load	-	-	-	-	25.30	m³	25.30
Softfall; per 80m³ load	-	-	-	-	12.30	m³	12.30
Playsand; per 10t load	-	-	-	-	78.30	m³	78.30
Playsand; per 20t load	-	-	-	-	66.06	m³	66.06
Walk chips; per 25m³ load	-	-	-	-	30.35	m³	30.35
Walk chips; per 80m³ load	-	-	-	-	17.35	m³	17.35
Woodfibre; per 25m³ load	-	-	-	-	28.85	m³	28.85
Woodfibre; per 80m³ load	-	-	-	-	15.85	m³	15.85

Q PAVING/PLANTING/FENCING/SITE FURNITURE

Item Excluding site overheads and profit	PC £	Labour hours	Labour £	Plant £	Material £	Unit	Total rate £
Q26 SPECIAL SURFACINGS FOR SPORT - cont'd							
Artificial surfaces and finishes - General Preamble: Manufacturer's should be consulted before specifying artificial playing surfaces as usage, local site conditions and maintenance will affect the choice of material. Advice should also be sought from the Technical Unit for Sport, the appropriate regional office of the Sports Council or the National Playing Fields Association as many surfaces have not yet been fully tested in use so that final opinions on durability, value for money or permanence cannot always be formed. Note that finishes which are inexpensive to lay are often expensive to maintain. Prices for the construction of average facilities are given in the Approximate Estimates section (book only). Some of the following prices include base work whereas others are for a specialist surface only on to a base prepared and costed separately.							
Sports areas; Baylis Landscape Contractors Sports tracks; polyurethane rubber surfacing; on bitumen-macadam (not included); (prices for 5500 m^2 minimum)							
"International"	-	-	-	-	-	m^2	45.67
"Club Grade"	-	-	-	-	-	m^2	27.19
Sports areas; multi-component polyurethane-rubber surfacing; on bitumen-macadam; finished with polyurethane or acrylic coat; green or red							
"Permaprene"	-	-	-	-	-	m^2	63.94
Sports areas; Exclusive Leisure Ltd "Tom Graveney Cricketweave" artificial wicket system; including proprietary sub-base (all by specialist sub-contractor)							
28 x 2.74 m	-	-	-	-	-	nr	5500.00
"Tom Graveney Cricketweave" practice batting end							
14 x 2 m	-	-	-	-	-	each	2800.00
"Tom Graveney Cricketweave" practice bowling end							
7 x 2 m	-	-	-	-	-	each	1700.00
Supply and erection of single bay Cricket cage							
18.3 x 3.65 x 3.2 m high	-	-	-	-	-	each	1700.00
Tennis courts; Duracourt (Spadeoak) Ltd Hard playing surfaces to SAPCA Code of Practice minimum requirements; laid on 65 mm thick macadam base on 150 mm thick stone foundation, to include lines, nets, posts, brick edging, 2.75 m high fence and perimeter drainage (excluding excavation, levelling or additional foundation). Based on court size 36.6 m x 18.3 m - 670 m^2							
"Durapore"; 'All Weather' porous macadam and acrylic colour coating	-	-	-	-	-	m^2	33.52
"Tiger Turf TT"; sand filled artificial grass	-	-	-	-	-	m^2	46.23
"Tiger Turf Grand Prix"; short pile sand filled artificial grass	-	-	-	-	-	m^2	48.69
"DecoColour"; impervious acrylic hardcourt (200 mm foundation)	-	-	-	-	-	m^2	44.69
"DecoTurf"; cushioned impervious acrylic tournament surface (200 mm foundation)	-	-	-	-	-	m^2	52.38

Q PAVING/PLANTING/FENCING/SITE FURNITURE

Item Excluding site overheads and profit	PC £	Labour hours	Labour £	Plant £	Material £	Unit	Total rate £
"Porous Kushion Kourt"; porous cushioned acrylic surface	-	-	-	-	-	m²	52.89
"EasiClay"; synthetic clay system	-	-	-	-	-	m²	54.73
"Canada Tenn"; American green clay, fast dry surface (no macadam but including irrigation)	-	-	-	-	-	m²	54.73
Playgrounds; Baylis Landscape Contractors; "Rubaflex" in situ playground surfacing, porous; on prepared stone/granular base Type 1 (not included) and macadam base course (not included)							
Black							
15 mm thick (0.50 m critical fall height)	-	-	-	-	-	m²	26.10
35 mm thick (1.00 m critical fall height)	-	-	-	-	-	m²	40.23
60 mm thick (1.50 m critical fall height)	-	-	-	-	-	m²	53.29
Playgrounds; Baylis Landscape Contractors; "Rubaflex" in situ playground surfacing, porous; on prepared stone/granular base Type 1 (not included) and macadam base course (not included)							
Coloured							
15 mm thick (0.50 m critical fall height)	-	-	-	-	-	m²	61.98
35 mm thick (1.00 m critical fall height)	-	-	-	-	-	m²	67.42
60 mm thick (1.50 m critical fall height)	-	-	-	-	-	m²	76.12
Playgrounds; Wicksteed Leisure Ltd							
Safety tiles; on prepared base (not included)							
1000 x 1000 x 60 mm; red or green	-	-	-	-	43.00	m²	43.00
1000 x 1000 x 60 mm; black	-	-	-	-	40.00	m²	40.00
1000 x 1000 x 43 mm; red or green	-	-	-	-	40.00	m²	40.00
1000 x 1000 x 43 mm; black	-	-	-	-	37.00	m²	37.00
Playgrounds; SMP Playgrounds Ltd							
Tiles; on prepared base (not included)							
Premier 25; 1000 x 1000 x 25 mm; black; for general use	22.00	0.20	3.75	-	27.33	m²	31.08
Premier 70; 1000 x 1000 x 70 mm; black; for higher equipment	42.40	0.20	3.75	-	47.73	m²	51.48
Playgrounds; Melcourt Industries Ltd							
"Playbark"; on drainage layer (not included); to BS5696; minimum 300 mm settled depth							
"Playbark 8/25", 8 - 25 mm particles; red/brown	10.68	0.35	6.58	-	11.75	m²	18.33
"Playbark 10/50"; 10 - 50 mm particles; red/brown	12.37	0.35	6.58	-	13.60	m²	20.18
Playgrounds; timber edgings							
Timber edging boards; 50 x 50 x 750 mm timber pegs at 1000 mm centres; excavations and hardcore under edgings (not included)							
50 x 150 mm; hardwood (iroko) edge boards	9.03	0.10	1.88	-	9.03	m	10.90
38 x 150 mm; hardwood (iroko) edge boards	7.17	0.10	1.88	-	7.17	m	9.05
50 x 150 mm; treated softwood edge boards	2.49	0.10	1.88	-	2.49	m	4.36
38 x 150 mm; treated softwood edge boards	2.28	0.10	1.88	-	2.28	m	4.16

Q PAVING/PLANTING/FENCING/SITE FURNITURE

Item Excluding site overheads and profit	PC £	Labour hours	Labour £	Plant £	Material £	Unit	Total rate £
Q30 SEEDING/TURFING							
Seeding/turfing - General							
Preamble: The following market prices generally reflect the manufacturer's recommended retail prices. Trade and bulk discounts are often available on the prices shown. The manufacturer's of these products generally recommend application rates. Note the following rates reflect the average rate for each product.							
Market Prices of pre-seeding materials							
Scotts UK							
herbicide; "Casoron G"; minimum application rate	-	-	-	-	1.97	100 m²	1.97
"Casoron G" as a selective herbicide	-	-	-	-	4.40	100 m²	4.40
herbicide; "Casoron G"; maximum application rate	-	-	-	-	7.93	100 m²	7.93
"Dextrone X"	-	-	-	-	0.52	100 m²	0.52
"Intrepid"	-	-	-	-	0.54	100 m²	0.54
"Speedway"	-	-	-	-	1.04	100 m²	1.04
turf herbicide; "Tritox", 5 litre	-	-	-	-	2.55	100 m²	2.55
turf herbicide; "Tritox", 20 litre	-	-	-	-	2.38	100 m²	2.38
Rigby Taylor Ltd							
outfield fertilizer; "Taylor's Pre-Seeding"	-	-	-	-	1.28	100 m²	1.28
Boughton Loam							
Screened topsoil; 100 mm	-	-	-	-	52.47	m³	52.47
Screened Kettering loam; 3 mm	-	-	-	-	73.89	m³	73.89
Screened Kettering loam - sterilized; 3 mm	-	-	-	-	85.23	m³	85.23
Top dressing; sand soil mixtures; 90/10 to 50/50	-	-	-	-	58.88	m³	58.88
Market Prices of grass maintenance materials; recommended application rates for most products vary; the rates below reflect an application rate for "normal" application							
Scotts UK; Fungicides							
fungicide; "Daconil"; turf	-	-	-	-	5.54	100 m²	5.54
turf fungicide and wormcast control; "Turfclear"; 5 litre	-	-	-	-	1.05	100 m²	1.05
Scotts UK; Turf chemicals							
growth regulator; "Shortcut"	-	-	-	-	1.64	100 m²	1.64
spray indicator; "Turf Mark"; per litre of water	-	-	-	-	0.38	litre	0.38
wetting and penetrating agent; "Aquamaster"; applied quarterly	-	-	-	-	2.45	100 m²	2.45
Scotts UK; Moss killers							
mosskiller; "Enforcer"	-	-	-	-	5.57	100 m²	5.57
Scotts UK; Turf fertilizers							
grass fertilizer; "Longlife Fine Turf - Spring & Summer"	-	-	-	-	4.47	100 m²	4.47
grass fertilizer; "Longlife Fine Turf - Autumn Feed"	-	-	-	-	2.67	100 m²	2.67
grass fertilizer; "Longlife Fine Turf - Finegreen NK"	-	-	-	-	4.47	100 m²	4.47
grass fertilizer; "Greenmaster - Invigorator"	-	-	-	-	2.46	100 m²	2.46
grass fertilizer; "Longlife - Renovator"	-	-	-	-	5.43	100 m²	5.43
grass fertilizer; slow release; "Sierraform", 18+24+5	-	-	-	-	-	100 m²	-
grass fertilizer; slow release; "Sierraform", 18+9+18+Fe+Mn	-	-	-	-	4.05	100 m²	4.05
grass fertilizer; slow release; "Sierraform", 15+0+26	-	-	-	-	4.86	100 m²	4.86
grass fertilizer; slow release; "Sierraform", 22+5+10	-	-	-	-	3.24	100 m²	3.24
grass fertilizer; slow release; "Sierraform", 16+0+15+Fe+Mn	-	-	-	-	4.05	100 m²	4.05

Q PAVING/PLANTING/FENCING/SITE FURNITURE

Item Excluding site overheads and profit	PC £	Labour hours	Labour £	Plant £	Material £	Unit	Total rate £
grass fertilizer; water soluble; "Sierrasol", 28+5+18+TE	-	-	-	-	6.94	100 m²	6.94
grass fertilizer; water soluble; "Sierrasol", 20+5+30+TE	-	-	-	-	5.40	100 m²	5.40
grass fertilizer; controlled release; "Sierrablen", 28+5+5+Fe	-	-	-	-	4.50	100 m²	4.50
grass fertilizer; controlled release; "Sierrablen", 27+5+5+Fe	-	-	-	-	4.71	100 m²	4.71
grass fertilizer; controlled release; "Sierrablen", 15+0+22+Fe	-	-	-	-	4.50	100 m²	4.50
grass fertilizer; controlled release; "Sierrablen Fine", 38+0+0	-	-	-	-	1.68	100 m²	1.68
grass fertilizer; controlled release; "Sierrablen Fine", 25+5+12	-	-	-	-	2.80	100 m²	2.80
grass fertilizer; controlled release; "Sierrablen Fine", 21+0+20	-	-	-	-	2.80	100 m²	2.80
grass fertilizer; controlled release; "Sierrablen Fine", 15+0+29	-	-	-	-	2.80	100 m²	2.80
grass fertilizer; controlled release; "Sierrablen Mini", 22+5+10	-	-	-	-	8.69	100 m²	8.69
grass fertilizer; controlled release; "Sierrablen Mini", 0+0+37	-	-	-	-	4.01	100 m²	4.01
grass fertilizer; "Greenmaster Turf Tonic"	-	-	-	-	2.28	100 m²	2.28
grass fertilizer; "Greenmaster Spring & Summer"	-	-	-	-	3.26	100 m²	3.26
grass fertilizer; "Greenmaster Zero Phosphate"	-	-	-	-	3.26	100 m²	3.26
grass fertilizer; "Greenmaster Mosskiller"	-	-	-	-	3.63	100 m²	3.63
grass fertilizer; "Greenmaster Extra"	-	-	-	-	4.13	100 m²	4.13
grass fertilizer; "Greenmaster Autumn"	-	-	-	-	3.51	100 m²	3.51
grass fertilizer; "Greenmaster Double K"	-	-	-	-	3.51	100 m²	3.51
grass fertilizer; "Greenmaster NK"	-	-	-	-	3.51	100 m²	3.51
grass & soil fertilizer; "Greenmaster Liquid Seafeed"	-	-	-	-	0.71	100 m²	0.71
outfield turf fertilizer; "Sportsmaster PS3"	-	-	-	-	1.49	100 m²	1.49
outfield turf fertilizer; "Sportsmaster PS4"	-	-	-	-	1.49	100 m²	1.49
outfield turf fertilizer; "Sportsmaster PS5"	-	-	-	-	1.93	100 m²	1.93
outfield turf fertilizer; "Sportsmaster Fairway"	-	-	-	-	3.26	100 m²	3.26
outfield turf fertilizer; "Sportsmaster Municipality"	-	-	-	-	4.47	100 m²	4.47
"TPMC"	-	-	-	-	4.56	80 lt	4.56
Rigby Taylor Ltd; 35 g/m²							
grass fertilizer; "Mascot Microfine 20-0-15" + 2% Mg	-	-	-	-	6.30	100 m²	6.30
grass fertilizer; "Mascot Microfine 15-5-15" +2% Mg	-	-	-	-	5.65	100 m²	5.65
grass fertilizer; "Mascot Microfine 18-0-0" + 6% Fe	-	-	-	-	6.88	100 m²	6.88
grass fertilizer; "Mascot Microfine 14-4-7" + 2% Mg	-	-	-	-	4.02	100 m²	4.02
grass fertilizer; "Mascot Microfine 8-0-0" + 4% Fe + 2% Mg	-	-	-	-	2.89	100 m²	2.89
grass fertilizer; "Mascot Microfine 4-0-8" + 6% Fe + 2% Mg	-	-	-	-	4.47	100 m²	4.47
grass fertilizer; "Taylor's Organic SS2"	-	-	-	-	3.24	100 m²	3.24
grass fertilizer; "Taylor's Mini-Granular 11-2-7"	-	-	-	-	2.74	100 m²	2.74
grass fertilizer/weedkiller; "Taylor's Weed & Feed"	-	-	-	-	2.37	100 m²	2.37
grass fertilizer/mosskiller; "Taylor's Lawn Sand"	-	-	-	-	1.29	100 m²	1.29
outfield fertilizer; "Taylor's Spring/Summer Outfield 9-7-7"	-	-	-	-	1.88	100 m²	1.88
outfield fertilizer; "Taylor's Spring/Summer Outfield 7-7-7"	-	-	-	-	1.56	100 m²	1.56
outfield fertilizer; "Taylor's Spring/Summer Outfield 20-10-10"	-	-	-	-	2.20	100 m²	2.20
outfield fertilizer; "Taylor's Autumn/Winter Outfield"	-	-	-	-	1.82	100 m²	1.82
outfield fertilizer; "Taylor's Spring/Summer Sportsfield"	-	-	-	-	2.20	100 m²	2.20

Q PAVING/PLANTING/FENCING/SITE FURNITURE

Item Excluding site overheads and profit	PC £	Labour hours	Labour £	Plant £	Material £	Unit	Total rate £
Q30 SEEDING/TURFING - cont'd							
Market Prices of grass maintenance materials - cont'd							
liquid fertilizer; "Vitax 50/50 Fine Turf"; 56 ml/100m²	-	-	-	-	13.22	100 m²	13.22
liquid fertilizer; "Vitax 50/50 Fine Turf Special"; 56 ml/100m²	-	-	-	-	11.31	100 m²	11.31
liquid fertilizer; "Vitax 50/50 Autumn & Winter"; 56 ml/100m²	-	-	-	-	12.13	100 m²	12.13
Cultivation							
Treating soil with "Paraquat-Diquat" weedkiller at rate of 5 litre/ha; PC £22.20 per litre; in accordance with manufacturer's instructions; including all safety precautions							
by machine	1.11	-	-	0.24	1.11	100 m²	1.35
by hand	1.11	0.13	2.50	-	1.11	100 m²	3.61
Ripping up subsoil; using approved subsoiling machine; minimum depth 250 mm below topsoil; at 1.20 m centres; in							
gravel or sandy clay	-	-	-	2.34	-	100m²	2.34
soil compacted by machines	-	-	-	2.73	-	100m²	2.73
clay	-	-	-	2.92	-	100m²	2.92
chalk or other soft rock	-	-	-	5.84	-	100m²	5.84
Extra for subsoiling at 1 m centres	-	-	-	0.58	-	100m²	0.58
Breaking up existing ground; using pedestrian operated tine cultivator or rotavator							
100 mm deep	-	0.22	4.13	2.13	-	100m²	6.26
150 mm deep	-	0.28	5.16	2.67	-	100m²	7.82
200 mm deep	-	0.37	6.87	3.56	-	100m²	10.43
As above but in heavy clay or wet soils							
100 mm deep	-	0.44	8.25	4.27	-	100m²	12.52
150 mm deep	-	0.66	12.38	6.40	-	100m²	18.77
200 mm deep	-	0.82	15.47	8.00	-	100m²	23.47
Breaking up existing ground; using tractor drawn tine cultivator or rotavator							
single pass							
100 mm deep	-	-	-	0.57	-	100 m²	0.57
150 mm deep	-	-	-	0.71	-	100 m²	0.71
200 mm deep	-	-	-	0.95	-	100 m²	0.95
600 mm deep	-	-	-	2.86	-	100 m²	2.86
Cultivating ploughed ground; using disc, drag, or chain harrow							
4 passes	-	-	-	3.43	-	100 m²	3.43
Rolling cultivated ground lightly; using self-propelled agricultural roller	-	0.06	1.04	0.58	-	100 m²	1.63
Importing and storing selected and approved topsoil; to BS 3882; from source not exceeding 13 km from site; inclusive of settlement							
small quantities, less than 15 m³	27.00	-	-	-	32.40	m³	32.40
over 15 m³	17.32	-	-	-	20.78	m³	20.78
Spreading and lightly consolidating approved topsoil (imported or from spoil heaps); in layers not exceeding 150 mm; travel distance from spoil heaps not exceeding 100 m; by machine (imported topsoil not included)							
minimum depth 100 mm	-	1.55	29.06	37.20	-	100 m²	66.26
minimum depth 150 mm	-	2.33	43.75	55.98	-	100 m²	99.73
minimum depth 300 mm	-	4.67	87.50	111.96	-	100 m²	199.46
minimum depth 450 mm	-	6.99	131.06	167.76	-	100 m²	298.82

Q PAVING/PLANTING/FENCING/SITE FURNITURE

Item Excluding site overheads and profit	PC £	Labour hours	Labour £	Plant £	Material £	Unit	Total rate £
Spreading and lightly consolidating approved topsoil (imported or from spoil heaps); in layers not exceeding 150 mm; travel distance from spoil heaps not exceeding 100 m; by hand (imported topsoil not included)							
minimum depth 100 mm	-	20.00	375.07	-	-	100 m²	375.07
minimum depth 150 mm	-	30.01	562.61	-	-	100 m²	562.61
minimum depth 300 mm	-	60.01	1125.22	-	-	100 m²	1125.22
minimum depth 450 mm	-	90.02	1687.84	-	-	100 m²	1687.84
Extra over for spreading topsoil to slopes 15 - 30 degrees by machine or hand	-	-	-	-	-	-	10%
Extra over for spreading topsoil to slopes over 30 degrees by machine or hand	-	-	-	-	-	-	25%
Extra over for spreading topsoil from spoil heaps travel exceeding 100 m; by machine							
100 - 150 m	-	0.01	0.23	0.07	-	m³	0.30
150 - 200 m	-	0.02	0.35	0.11	-	m³	0.46
200 - 300 m	-	0.03	0.52	0.17	-	m³	0.69
Extra over spreading topsoil for travel exceeding 100 m; by hand							
100 m	-	0.83	15.63	-	-	m³	15.63
200 m	-	1.67	31.25	-	-	m³	31.25
300 m	-	2.50	46.88	-	-	m³	46.88
Evenly grading; to general surfaces to bring to finished levels							
by machine (tractor mounted rotavator)	-	-	-	0.03	-	m²	0.03
by pedestrian operated rotavator	-	-	0.08	0.04	-	m²	0.12
by hand	-	0.01	0.19	-	-	m²	0.19
Extra over grading for slopes 15 - 30 degrees by machine or hand	-	-	-	-	-	-	10%
Extra over grading for slopes over 30 degrees by machine or hand	-	-	-	-	-	-	25%
Apply screened topdressing to grass surfaces. Spread using Tru-Lute							
Sand soil mixes 90/10 to 50/50	0.09	-	0.04	0.03	0.09	m²	0.16
Spread only existing cultivated soil to final levels using Tru-Lute							
Cultivated soil	-	-	0.04	0.03	-	m²	0.07
Clearing stones; disposing off site; to distance not exceeding 13 km							
by hand; stones not exceeding 50 mm in any direction; loading to skip 5.35 m³	-	0.01	0.19	0.03	-	m²	0.21
by mechanical stone rake; stones not exceeding 50 mm in any direction; loading to 15 m³ truck by mechanical loader	-	-	0.04	0.07	-	m²	0.11
Lightly cultivating; weeding; to fallow areas; disposing debris off site; to distance not exceeding 13 km							
by hand	-	0.01	0.27	-	0.06	m²	0.33
Surface applications; soil additives; pre-seeding; material delivered to a maximum of 25 m from area of application; applied; by machine							
Soil conditioners; to cultivated ground; ground limestone; PC £9.30/tonne; including turning in							
0.25 kg/m² = 2.50 tonnes/ha	0.23	-	-	4.32	0.23	100 m²	4.55
0.50 kg/m² = 5.00 tonnes/ha	0.47	-	-	4.32	0.47	100 m²	4.78
0.75 kg/m² = 7.50 tonnes/ha	0.70	-	-	4.32	0.70	100 m²	5.02
1.00 kg/m² = 10.00 tonnes/ha	0.93	-	-	4.32	0.93	100 m²	5.25
Soil conditioners; to cultivated ground; medium bark; based on deliveries of 15 m³ loads; PC £31.55/m³; including turning in							
1 m³ per 40 m² = 25 mm thick	0.79	-	-	0.10	0.79	m²	0.89
1 m³ per 20 m² = 50 mm thick	-	-	-	0.15	1.58	m²	1.73
1 m³ per 13.33 m² = 75 mm thick	-	-	-	0.21	2.37	m²	2.57
1 m³ per 10 m² = 100 mm thick	3.15	-	-	0.27	3.15	m²	3.42

Q PAVING/PLANTING/FENCING/SITE FURNITURE

Item Excluding site overheads and profit	PC £	Labour hours	Labour £	Plant £	Material £	Unit	Total rate £
Q30 SEEDING/TURFING - cont'd							
Surface applications; soil additives; **pre-seeding** - cont'd							
Soil conditioners; to cultivated ground; peat loose 55 m³ loads; PC £30.00/m³; including turning in							
1 m³ per 40 m² = 25 mm thick	0.75	-	-	0.10	0.75	m²	**0.85**
1 m³ per 20 m² = 50 mm thick	1.50	-	-	0.15	1.50	m²	**1.65**
1 m³ per 13.33 m² = 75 mm thick	2.25	-	-	0.21	2.25	m²	**2.46**
1 m³ per 10 m² = 100 mm thick	3.00	-	-	0.27	3.00	m²	**3.27**
Soil conditioners; to cultivated ground; mushroom compost; delivered in 25 m3 loads; PC £18.10/m3; including turning in							
1 m3 per 40 m2 = 25 mm thick	0.45	0.02	0.31	0.10	0.45	m2	**0.86**
1 m3 per 20 m2 = 50 mm thick	0.91	0.03	0.58	0.15	0.91	m2	**1.63**
1 m3 per 13.33 m2 = 75 mm thick	1.36	0.04	0.75	0.21	1.36	m2	**2.32**
1 m3 per 10 m2 = 100 mm thick	1.81	0.05	0.94	0.27	1.81	m2	**3.01**
Soil conditioners; to cultivated ground; mushroom compost; delivered in 35 m3 loads; PC £8.57/m3; including turning in							
1 m3 per 40 m2 = 25 mm thick	0.21	0.02	0.31	0.10	0.21	m2	**0.63**
1 m3 per 20 m2 = 50 mm thick	0.43	0.03	0.58	0.15	0.43	m2	**1.16**
1 m3 per 13.33 m2 = 75 mm thick	0.64	0.04	0.75	0.21	0.64	m2	**1.60**
1 m3 per 10 m2 = 100 mm thick	0.86	0.05	0.94	0.27	0.86	m2	**2.06**
Surface applications and soil additives; **pre-seeding; material delivered to a** **maximum of 25 m from area of** **application; applied; by hand**							
Soil conditioners; to cultivated ground; ground limestone; PC £9.30/tonne; including turning in							
0.25 kg/m² = 2.50 tonnes/ha	0.23	1.20	22.50	-	0.23	100 m²	**22.73**
0.50 kg/m² = 5.00 tonnes/ha	0.47	1.33	25.00	-	0.47	100 m²	**25.46**
0.75 kg/m² = 7.50 tonnes/ha	0.70	1.50	28.13	-	0.70	100 m²	**28.82**
1.00 kg/m² = 10.00 tonnes/ha	0.93	1.71	32.14	-	0.93	100 m²	**33.07**
Soil conditioners; to cultivated ground; medium bark; based on deliveries of 25 m³ loads; PC £29.35 / m³; including turning in							
1 m³ per 40 m² = 25 mm thick	0.79	0.02	0.42	-	0.79	m²	**1.21**
1 m³ per 20 m² = 50 mm thick	1.58	0.04	0.83	-	1.58	m²	**2.41**
1 m³ per 13.33 m² = 75 mm thick	2.37	0.07	1.25	-	2.37	m²	**3.62**
1 m³ per 10 m² = 100 mm thick	3.15	0.08	1.50	-	3.15	m²	**4.65**
Soil conditioners; to cultivated ground; peat loose 55 m³ loads; PC £30.00 /m³; including turning in							
1 m³ per 40 m² = 25 mm thick	0.75	0.02	0.42	-	0.75	m²	**1.17**
1 m³ per 20 m² = 50 mm thick	1.50	0.04	0.83	-	1.50	m²	**2.33**
1 m³ per 13.33 m² = 75 mm thick	2.25	0.07	1.25	-	2.25	m²	**3.50**
1 m³ per 10 m² = 100 mm thick	3.00	0.08	1.50	-	3.00	m²	**4.50**
Soil conditioners; to cultivated ground; mushroom compost; delivered in 25 m³ loads; PC £18.10/m³; including turning in							
1 m³ per 40 m² = 25 mm thick	0.45	0.02	0.42	-	0.45	m²	**0.87**
1 m³ per 20 m² = 50 mm thick	0.91	0.04	0.83	-	0.91	m²	**1.74**
1 m³ per 13.33 m² = 75 mm thick	1.36	0.07	1.25	-	1.36	m²	**2.61**
1 m³ per 10 m² = 100 mm thick	1.81	0.08	1.50	-	1.81	m²	**3.31**
Soil conditioners; to cultivated ground; mushroom compost; delivered in 55 m³ loads; PC £6.40/m³; including turning in							
1 m³ per 40 m² = 25 mm thick	0.16	0.02	0.42	-	0.16	m²	**0.58**
1 m³ per 20 m² = 50 mm thick	0.28	0.04	0.83	-	0.28	m²	**1.12**
1 m³ per 13.33 m² = 75 mm thick	0.48	0.07	1.25	-	0.48	m²	**1.73**
1 m³ per 10 m² = 100 mm thick	0.64	0.08	1.50	-	0.64	m²	**2.14**

Q PAVING/PLANTING/FENCING/SITE FURNITURE

Item Excluding site overheads and profit	PC £	Labour hours	Labour £	Plant £	Material £	Unit	Total rate £
Preparation of seedbeds - General Preamble: For preliminary operations see "Cultivation" section.							
Preparation of seedbeds; soil **preparation** Lifting selected and approved topsoil from spoil heaps; passing through 6 mm screen; removing debris	-	0.08	1.56	4.08	0.01	m^3	**5.65**
Topsoil; supply only; PC £17.32/m^3; allowing for 20% settlement							
25 mm	0.43	-	-	-	0.52	m^2	**0.52**
50 mm	0.87	-	-	-	1.04	m^2	**1.04**
100 mm	1.73	-	-	-	2.08	m^2	**2.08**
150 mm	2.60	-	-	-	3.12	m^2	**3.12**
200 mm	3.46	-	-	-	4.16	m^2	**4.16**
250 mm	4.33	-	-	-	5.20	m^2	**5.20**
300 mm	5.20	-	-	-	6.24	m^2	**6.24**
400 mm	6.93	-	-	-	8.31	m^2	**8.31**
450 mm	7.79	-	-	-	9.35	m^2	**9.35**
Spreading topsoil to form seedbeds (topsoil not included); by machine							
25 mm deep	-	-	0.05	0.10	-	m^2	**0.15**
50 mm deep	-	-	0.06	0.13	-	m^2	**0.19**
75 mm deep	-	-	0.07	0.15	-	m^2	**0.22**
100 mm deep	-	0.01	0.09	0.20	-	m^2	**0.29**
150 mm deep	-	0.01	0.14	0.29	-	m^2	**0.43**
Spreading only topsoil to form seedbeds (topsoil not included); by hand							
25 mm deep	-	0.03	0.47	-	-	m^2	**0.47**
50 mm deep	-	0.03	0.63	-	-	m^2	**0.63**
75 mm deep	-	0.04	0.80	-	-	m^2	**0.80**
100 mm deep	-	0.05	0.94	-	-	m^2	**0.94**
150 mm deep	-	0.08	1.41	-	-	m^2	**1.41**
Bringing existing topsoil to a fine tilth for seeding; by raking or harrowing; stones not to exceed 6 mm; by machine	-	-	0.08	0.04	-	m^2	**0.12**
Bringing existing topsoil to a fine tilth for seeding; by raking or harrowing; stones not to exceed 6 mm; by hand	-	0.01	0.17	-	-	m^2	**0.17**
Preparation of seedbeds; soil treatments For the following topsoil improvement and seeding operations add or subtract the following amounts for every £0.10 difference in the material cost price							
35 g/m^2	-	-	-	-	0.35	100 m^2	**0.35**
50 g/m^2	-	-	-	-	0.50	100 m^2	**0.50**
70 g/m^2	-	-	-	-	0.70	100 m^2	**0.70**
100 g/m^2	-	-	-	-	1.00	100 m^2	**1.00**
125 kg/ha	-	-	-	-	12.50	ha	**12.50**
150 kg/ha	-	-	-	-	15.00	ha	**15.00**
175 kg/ha	-	-	-	-	17.50	ha	**17.50**
200 kg/ha	-	-	-	-	20.00	ha	**20.00**
225 kg/ha	-	-	-	-	22.50	ha	**22.50**
250 kg/ha	-	-	-	-	25.00	ha	**25.00**
300 kg/ha	-	-	-	-	30.00	ha	**30.00**
350 kg/ha	-	-	-	-	35.00	ha	**35.00**
400 kg/ha	-	-	-	-	40.00	ha	**40.00**
500 kg/ha	-	-	-	-	50.00	ha	**50.00**
700 kg/ha	-	-	-	-	70.00	ha	**70.00**
1000 kg/ha	-	-	-	-	100.00	ha	**100.00**
1250 kg/ha	-	-	-	-	125.00	ha	**125.00**

Q PAVING/PLANTING/FENCING/SITE FURNITURE

Item Excluding site overheads and profit	PC £	Labour hours	Labour £	Plant £	Material £	Unit	Total rate £
Q30 SEEDING/TURFING - cont'd							
Preparation of seedbeds; soil treatments - cont'd							
Pre-seeding fertilizers (6:9:6); PC £0.37/kg; to seedbeds; by machine							
35 g/m^2	1.28	-	-	0.18	1.28	100 m^2	1.47
50 g/m^2	1.83	-	-	0.18	1.83	100 m^2	2.02
70 g/m^2	2.57	-	-	0.18	2.57	100 m^2	2.75
100 g/m^2	3.67	-	-	0.18	3.67	100 m^2	3.85
125 g/m^2	4.58	-	-	0.18	4.58	100 m^2	4.77
125 kg/ha	45.85	-	-	18.23	45.85	ha	64.08
250 kg/ha	91.70	-	-	18.23	91.70	ha	109.93
300 kg/ha	110.04	-	-	18.23	110.04	ha	128.27
350 kg/ha	128.38	-	-	18.23	128.38	ha	146.61
400 kg/ha	146.72	-	-	18.23	146.72	ha	164.95
500 kg/ha	183.40	-	-	29.17	183.40	ha	212.57
700 kg/ha	256.76	-	-	29.17	256.76	ha	285.93
1250 kg/ha	458.50	-	-	48.62	458.50	ha	507.12
Pre-seeding fertilizers (6:9:6); PC £0.37/kg; to seedbeds; by hand							
35 g/m^2	1.28	0.17	3.13	-	1.28	100 m^2	4.41
50 g/m^2	1.83	0.17	3.13	-	1.83	100 m^2	4.96
70 g/m^2	2.57	0.17	3.13	-	2.57	100 m^2	5.69
100 g/m^2	3.67	0.20	3.75	-	3.67	100 m^2	7.42
125 g/m^2	4.58	0.20	3.75	-	4.58	100 m^2	8.34
Pre-seeding pesticides; to seedbeds; in accordance with manufacturer's instructions; by machine							
35 g/m^2	16.45	-	-	0.18	16.45	100 m^2	16.63
50 g/m^2	23.50	-	-	0.18	23.50	100 m^2	23.68
70 g/m^2	32.90	-	-	0.18	32.90	100 m^2	33.08
100 g/m^2	47.00	-	-	0.18	47.00	100 m^2	47.18
125 g/m^2	58.75	-	-	0.18	58.75	100 m^2	58.93
125 kg/ha	587.50	-	-	18.23	587.50	ha	605.73
250 kg/ha	1175.00	-	-	18.23	1175.00	ha	1193.23
300 kg/ha	1410.00	-	-	18.23	1410.00	ha	1428.23
350 kg/ha	1645.00	-	-	18.23	1645.00	ha	1663.23
400 kg/ha	1880.00	-	-	18.23	1880.00	ha	1898.23
500 kg/ha	2350.00	-	-	29.17	2350.00	ha	2379.17
700 kg/ha	3290.00	-	-	29.17	3290.00	ha	3319.17
1250 kg/ha	5875.00	-	-	18.23	5875.00	ha	5893.23
Pre-seeding pesticides; to seedbeds; in accordance with manufacturer's instructions; by hand							
35 g/m^2	16.45	0.17	3.13	-	16.45	100 m^2	19.58
50 g/m^2	23.50	0.17	3.13	-	23.50	100 m^2	26.63
70 g/m^2	32.90	0.17	3.13	-	32.90	100 m^2	36.03
100 g/m^2	47.00	0.20	3.75	-	47.00	100 m^2	50.75
125 g/m^2	58.75	0.20	3.75	-	58.75	100 m^2	62.50
Pre-emergent weedkillers; in accordance with manufacturer's instructions; including all safety precautions; by machine (coverage 0.80 ha/hr)							
15 ml	1.17	-	-	0.18	1.17	100 m^2	1.36
20 ml	1.56	-	-	0.18	1.56	100 m^2	1.75
25 ml	1.95	-	-	0.18	1.95	100 m^2	2.14
50 ml	3.91	-	-	0.18	3.91	100 m^2	4.09

Q PAVING/PLANTING/FENCING/SITE FURNITURE

Item Excluding site overheads and profit	PC £	Labour hours	Labour £	Plant £	Material £	Unit	Total rate £
Pre-emergent liquid applied weedkillers; in accordance with manufacturer's instructions; including all safety precautions; by hand							
10 ml	0.04	0.50	9.38	-	0.04	100 m²	9.42
12 ml	0.05	0.50	9.38	-	0.05	100 m²	9.42
15 ml	0.06	0.50	9.38	-	0.06	100 m²	9.44
20 ml	0.08	0.50	9.38	-	0.08	100 m²	9.46
25 ml	0.10	0.50	9.38	-	0.10	100 m²	9.48
30 ml	0.12	0.50	9.38	-	0.12	100 m²	9.50
35 ml	0.14	0.50	9.38	-	0.14	100 m²	9.52
40 ml	0.16	0.50	9.38	-	0.16	100 m²	9.54
45 ml	0.18	0.50	9.38	-	0.18	100 m²	9.56
50 ml	0.20	0.50	9.38	-	0.20	100 m²	9.58
Lime granules; on seedbeds; by machine							
35 g/m²	0.73	-	-	0.18	0.73	100 m²	0.92
50 g/m²	1.05	-	-	0.18	1.05	100 m²	1.23
70 g/m²	1.47	-	-	0.18	1.47	100 m²	1.65
100 g/m²	2.10	-	-	0.18	2.10	100 m²	2.28
125 g/m²	2.63	-	-	0.18	2.63	100 m²	2.81
125 kg/ha	26.25	-	-	18.23	26.25	ha	44.48
250 kg/ha	52.50	-	-	18.23	52.50	ha	70.73
300 kg/ha	63.00	-	-	18.23	63.00	ha	81.23
350 kg/ha	73.50	-	-	18.23	73.50	ha	91.73
400 kg/ha	84.00	-	-	18.23	84.00	ha	102.23
500 kg/ha	105.00	-	-	29.17	105.00	ha	134.17
700 kg/ha	147.00	-	-	72.93	147.00	ha	219.93
1250 kg/ha	262.50	-	-	72.93	262.50	ha	335.43
Lime granules; on seedbeds; by hand							
35 g/m²	0.73	0.17	3.13	-	0.73	100 m²	3.86
50 g/m²	1.05	0.17	3.13	-	1.05	100 m²	4.18
70 g/m²	1.47	0.17	3.13	-	1.47	100 m²	4.60
100 g/m²	2.10	0.17	3.13	-	2.10	100 m²	5.23
125 g/m²	2.63	0.17	3.13	-	2.63	100 m²	5.75

Seeding grass areas - General
Preamble: The British Standard recommendations for seed and seeding of grass areas are contained in BS 4428: 1989. The prices given in this section are based on compliance with the standard.

Market Prices of Grass Seed
Preamble: The prices shown are for supply only at one number 20 kg or 25 kg bag purchase price unless otherwise stated. Rates shown are based on the manufacturer's maximum recommendation for each seed type. Trade and bulk discounts are often available on the prices shown for quantities of more than one bag.

Item	PC £	Labour hours	Labour £	Plant £	Material £	Unit	Total rate £
Bowling greens; fine lawns; ornamental turf; croquet lawns							
"British Seed Houses"; ref A1 Greens; 35g/m² (New Formulation)	-	-	-	-	20.68	100m²	20.68
"Johnsons Seeds"; ref J1 Golf & Bowling Greens; 34 - 50g/m²	-	-	-	-	29.40	100m²	29.40
"Johnsons Seeds"; ref J2 Greens High Performance; 34 - 50g/m²	-	-	-	-	34.30	100m²	34.30
"Perryfields"; ref Pro 20 Fineturf; 35 - 50g/m²	-	-	-	-	14.40	100m²	14.40
"Perryfields"; ref Pro 10 Traditional Green; 35 - 50g/m²	-	-	-	-	20.15	100m²	20.15

Q PAVING/PLANTING/FENCING/SITE FURNITURE

Item Excluding site overheads and profit	PC £	Labour hours	Labour £	Plant £	Material £	Unit	Total rate £
Q30 SEEDING/TURFING - cont'd							
Market Prices of Grass Seed - cont'd							
Tennis courts; cricket squares							
"British Seed Houses"; ref A2 Lawns & Tennis; 35g/m^2	-	-	-	-	13.18	100m^2	**13.18**
"British Seed Houses"; ref A5 Cricket Square; 35g/m^2	-	-	-	-	17.85	100m^2	**17.85**
"Johnsons Seeds"; ref J2 Greens High Performance; 34 - 50g/m^2	-	-	-	-	34.30	100m^2	**34.30**
"Johnsons Seeds"; ref Taskmaster; 18 - 25g/m^2	-	-	-	-	9.30	100m^2	**9.30**
"Perryfields"; ref Pro 35 Universal; 35g/m^2	-	-	-	-	10.22	100m^2	**10.22**
Amenity grassed areas; general purpose lawns							
"British Seed Houses"; ref A3 Landscape; 25 - 50g/m^2	-	-	-	-	15.65	100m^2	**15.65**
"Johnsons Seeds"; ref J5 Tees Fairways & Cricket; 18 - 25g/m^2	-	-	-	-	11.88	100m^2	**11.88**
"Johnsons Seeds"; ref Taskmaster; 18 - 25g/m^2	-	-	-	-	9.30	100m^2	**9.30**
"Perryfields"; ref Pro 50 Quality Lawn; 25 - 35g/m^2	-	-	-	-	11.06	100m^2	**11.06**
"Perryfields"; ref Pro 120 Slowgrowth; 25 - 35g/m^2	-	-	-	-	11.48	100m^2	**11.48**
Conservation; country parks; slopes and banks							
"British Seed Houses"; ref A4 Parkland; 17 - 35g/m^2	-	-	-	-	11.30	100m^2	**11.30**
"British Seed Houses"; ref A16 Country Park; 8 - 19g/m^2	-	-	-	-	7.26	100m^2	**7.26**
"British Seed Houses"; ref A17 Legume and Clover; 2g/m^2	-	-	-	-	12.54	100m^2	**12.54**
Shaded areas							
"British Seed Houses"; ref A6 Supra Shade; 50g/m^2	-	-	-	-	20.95	100m^2	**20.95**
"Johnsons Seeds"; ref J1 Golf & Bowling Greens; 34 - 50g/m^2	-	-	-	-	29.40	100m^2	**29.40**
"Perryfields"; ref Pro 60 Greenshade; 35 - 50g/m^2	-	-	-	-	17.25	100m^2	**17.25**
Sports pitches; rugby; soccer pitches							
"British Seed Houses"; ref A7 Sportsground; 20g/m^2	-	-	-	-	7.14	100m^2	**7.14**
"Johnson Seeds"; ref Sportsmaster; 18 - 30g/m^2	-	-	-	-	9.18	100m^2	**9.18**
"Perryfields"; ref Pro 70 Recreation; 15 - 35g/m^2	-	-	-	-	10.26	100m^2	**10.26**
"Perryfields"; ref Pro 75 Stadia; 15 - 35g/m^2	-	-	-	-	12.01	100m^2	**12.01**
"Perryfields"; ref Pro 80 Renovator; 17 - 35g/m^2	-	-	-	-	9.55	100m^2	**9.55**
Outfields							
"British Seed Houses"; ref A7 Sportsground; 20g/m^2	-	-	-	-	7.14	100m^2	**7.14**
"British Seed Houses"; ref A9 Outfield; 17 - 35g/m^2	-	-	-	-	8.99	100m^2	**8.99**
"Johnsons Seeds"; ref J4 Fairways & Cricket Outfields; 12 - 25g/m^2	-	-	-	-	8.68	100m^2	**8.68**
"Perryfields"; ref Pro 40 Tee & Fairway; 35g/m^2	-	-	-	-	10.81	100m^2	**10.81**
"Perryfields"; ref Pro 70 Recreation; 15 - 35g/m^2	-	-	-	-	10.26	100m^2	**10.26**
Hockey pitches							
"Johnsons Seeds"; ref J4 Fairways & Cricket Outfields; 12 - 25g/m^2	-	-	-	-	8.68	100m^2	**8.68**
"Perryfields"; ref Pro 70 Recreation; 15 - 35g/m^2	-	-	-	-	10.26	100m^2	**10.26**
Parks							
"British Seed Houses"; ref A7 Sportsground 20g/m^2	-	-	-	-	7.14	100m^2	**7.14**
"British Seed Houses"; ref A9 Outfield; 17 - 35g/m^2	-	-	-	-	8.99	100m^2	**8.99**
"Perryfields"; ref Pro 120 Slowgrowth; 25 - 35g/m^2	-	-	-	-	11.48	100m^2	**11.48**

Q PAVING/PLANTING/FENCING/SITE FURNITURE

Item Excluding site overheads and profit	PC £	Labour hours	Labour £	Plant £	Material £	Unit	Total rate £
Informal playing fields							
"Johnsons seeds"; ref J5 Tees Fairways & Cricket; 18 - 25g/m^2	-	-	-	-	11.88	100m^2	**11.88**
"Perryfields"; ref Pro 45 Tee & Fairway Plus; 35g/m^2	-	-	-	-	10.68	100m^2	**10.68**
Caravan sites							
"British Seed Houses"; ref A9 Outfield; 17 - 35g/m^2	-	-	-	-	8.99	100m^2	**8.99**
Sports pitch re-seeding and repair							
"British Seed Houses"; ref A8 Pitch Renovator; 20 - 35g/m^2	-	-	-	-	10.39	100m^2	**10.39**
"British Seed Houses"; ref A20 Ryesport; 20 - 35g/m^2	-	-	-	-	11.45	100m^2	**11.45**
"Johnson Seeds"; ref Sportsmaster; 18 - 30g/m^2	-	-	-	-	9.18	100m^2	**9.18**
"Perryfields"; ref Pro 80 Renovator; 17 - 35g/m^2	-	-	-	-	9.55	100m^2	**9.55**
"Perryfields"; ref Pro 81 Premier Renovation; 17 - 35g/m^2	-	-	-	-	13.76	100m^2	**13.76**
Racecourses; gallops; polo grounds; horse rides							
"British Seed Houses"; ref A14 Racecourse; 25 - 30g/m^2	-	-	-	-	8.79	100m^2	**8.79**
"Johnsons Seeds"; ref Taskmaster; 18 - 25g/m^2	-	-	-	-	9.30	100m^2	**9.30**
"Perryfields"; ref Pro 65 Gallop; 17 - 35g/m^2	-	-	-	-	13.37	100m^2	**13.37**
Motorway and road verges							
"British Seed Houses"; ref A18 Road Verge; 6 - 15g/m^2	-	-	-	-	4.75	100m^2	**4.75**
"Perryfields"; ref Pro 120 Slowgrowth; 25 - 35g/m^2	-	-	-	-	11.48	100m^2	**11.48**
"Perryfields"; ref Pro 85 DOT; 10g/m^2	-	-	-	-	2.98	100m^2	**2.98**
Golf courses; tees							
"British Seed Houses"; ref A10 Golf Tee, 35 - 50g/m^2	-	-	-	-	18.45	100m^2	**18.45**
"Johnsons Seeds"; ref J3 Golf Tees & Fairways; 18 - 30g/m^2	-	-	-	-	14.61	100m^2	**14.61**
"Johnsons Seeds"; ref Taskmaster; 18 - 25g/m^2	-	-	-	-	9.30	100m^2	**9.30**
"Perryfields"; ref Pro 40 Tee & Fairway; 35g/m^2	-	-	-	-	10.81	100m^2	**10.81**
"Perryfields"; ref Pro 45 Tee & Fairway Plus; 35g/m^2	-	-	-	-	10.68	100m^2	**10.68**
Golf courses; greens							
"British Seed Houses"; ref A11 Golf Green; 35g/m^2	-	-	-	-	22.09	100m^2	**22.09**
"British Seed Houses"; ref A12R Golf Roughs; 8g/m^2	-	-	-	-	2.60	100m^2	**2.60**
"Johnsons Seeds"; ref J1 Golf & Bowling Greens; 34 - 50g/m^2	-	-	-	-	29.40	100m^2	**29.40**
"Johnsons Seeds"; ref Greenmaster; 8g/m^2	-	-	-	-	10.66	100m^2	**10.66**
"Perryfields"; ref Pro 5 Economy Green; 35 - 50g/m^2	-	-	-	-	15.90	100m^2	**15.90**
Golf courses; fairways							
"British Seed Houses"; ref A12 Golf Fairway; 15 - 25g/m^2	-	-	-	-	8.75	100m^2	**8.75**
"Johnsons Seeds"; ref J3 Golf Tees & Fairways; 18 - 30g/m^2	-	-	-	-	14.61	100m^2	**14.61**
"Johnsons Seeds"; ref J5 Tees Fairways & Cricket; 18 - 30g/m^2	-	-	-	-	14.25	100m^2	**14.25**
"Perryfields"; ref Pro 40 Tee & Fairway; 35g/m^2	-	-	-	-	10.81	100m^2	**10.81**
"Perryfields"; ref Pro 45 Tee & Fairway Plus; 35g/m^2	-	-	-	-	10.68	100m^2	**10.68**
Golf courses; roughs							
"Perryfields"; ref Pro 25 Grow-Slow; 17 - 35g/m^2	-	-	-	-	12.32	100m^2	**12.32**
Waste land; spoil heaps; quarries							
"British Seed Houses"; ref A15 Reclamation; 15 - 20g/m^2	-	-	-	-	7.02	100m^2	**7.02**
"Perryfields"; ref Pro 95 Land Reclamation; 12 - 35g/m^2	-	-	-	-	13.47	100m^2	**13.47**
"Perryfields"; ref Pro 105 Fertility; 5g/m^2	-	-	-	-	2.69	100m^2	**2.69**

Q PAVING/PLANTING/FENCING/SITE FURNITURE

Item Excluding site overheads and profit	PC £	Labour hours	Labour £	Plant £	Material £	Unit	Total rate £
Q30 SEEDING/TURFING - cont'd							
Market Prices of Grass Seed - cont'd							
Low maintenance; housing estates; amenity grassed areas							
"British Seed Houses"; ref A19 Housing Estate; 25 - 35g/m^2	-	-	-	-	9.35	100m^2	**9.35**
"British Seed Houses"; ref A22 Low Maintenance; 25 - 35g/m^2	-	-	-	-	12.49	100m^2	**12.49**
"Perryfields"; ref Pro 120 Slowgrowth; 25 - 35g/m^2	-	-	-	-	11.48	100m^2	**11.48**
Saline coastal; roadside areas							
"British Seed Houses"; ref A21 Coastal/Saline Restoration; 15 - 20g/m^2	-	-	-	-	7.46	100m^2	**7.46**
"Perryfields"; ref Pro 90 Coastal; 15 - 35g/m^2	-	-	-	-	12.21	100m^2	**12.21**
Turf production							
"British Seed Houses"; ref A23 Green Velvet; 160 kg/ha	-	-	-	-	537.12	ha	**537.12**
"British Seed Houses"; ref A24 Wear & Tear; 185 kg/ha	-	-	-	-	587.75	ha	**587.75**
Market Prices of wild flora seed mixtures							
Acid soils							
"British Seed Houses"; ref WF1 (Annual Flowering); 1.00 - 2.00 g/m^2	-	-	-	-	19.50	100 m^2	**19.50**
Neutral soils							
"British Seed Houses"; ref WF3 (Neutral Soils); 0.50 - 1.00 g/m^2	-	-	-	-	9.75	100 m^2	**9.75**
Market Prices of wild flora and grass seed mixtures							
General purpose							
"Perryfields"; ref Pro Flora 8 Old English Country Meadow Mix; 5 g/m^2	-	-	-	-	15.00	100 m^2	**15.00**
"Perryfields"; ref Pro Flora 9 General purpose; 5 g/m^2	-	-	-	-	10.00	100 m^2	**10.00**
Acid soils							
"British Seed Houses"; ref WFG2 (Annual Meadow); 5 g/m^2	-	-	-	-	22.25	100 m^2	**22.25**
"Perryfields"; ref Pro Flora 2 Acidic soils; 5 g/m^2	-	-	-	-	11.25	100 m^2	**11.25**
Neutral soils							
"British Seed Houses"; ref WFG4 (Neutral Meadow); 5 g/m^2	-	-	-	-	24.45	100 m^2	**24.45**
"British Seed Houses"; ref WFG13 (Scotland); 5 g/m^2	-	-	-	-	21.55	100 m^2	**21.55**
"Perryfields"; ref Pro Flora 3 Damp loamy soils; 5 g/m^2	-	-	-	-	15.75	100 m^2	**15.75**
Calcareous soils							
"British Seed Houses"; ref WFG5 (Calcareous Soils); 5 g/m^2	-	-	-	-	22.80	100 m^2	**22.80**
"Perryfields"; ref Pro Flora 4 Calcareous soils; 5 g/m^2	-	-	-	-	13.50	100 m^2	**13.50**
Heavy clay soils							
"British Seed Houses"; ref WFG6 (Clay Soils); 5 g/m^2	-	-	-	-	27.50	100 m^2	**27.50**
"British Seed Houses"; ref WFG12 (Ireland); 5 g/m^2	-	-	-	-	21.63	100 m^2	**21.63**
"Perryfields"; ref Pro Flora 5 Wet loamy soils; 5 g/m^2	-	-	-	-	33.00	100 m^2	**33.00**

Q PAVING/PLANTING/FENCING/SITE FURNITURE

Item Excluding site overheads and profit	PC £	Labour hours	Labour £	Plant £	Material £	Unit	Total rate £
Sandy soils							
"British Seed Houses"; ref WFG7 (Free Draining Soils); 5 g/m^2	-	-	-	-	32.25	100 m^2	**32.25**
"British Seed Houses"; ref WFG11 (Ireland); 5 g/m^2	-	-	-	-	22.75	100 m^2	**22.75**
"British Seed Houses"; ref WFG14 (Scotland); 5 g/m^2	-	-	-	-	25.00	100 m^2	**25.00**
"Perryfields"; ref Pro Flora 6 Dry free draining loamy soils; 5 g/m^2	-	-	-	-	29.25	100 m^2	**29.25**
Shaded areas							
"British Seed Houses"; ref WFG8 (Woodland and Hedgerow); 5 g/m^2	-	-	-	-	24.00	100 m^2	**24.00**
"Perryfields"; ref Pro Flora 7 Hedgerow and light shade; 5 g/m^2	-	-	-	-	33.75	100 m^2	**33.75**
Educational							
"British Seed Houses"; ref WFG15 (Schools and Colleges); 5 g/m^2	-	-	-	-	33.75	100 m^2	**33.75**
Wetlands							
"British Seed Houses"; ref WFG9 (Wetlands and Ponds); 5 g/m^2	-	-	-	-	29.75	100 m^2	**29.75**
"Perryfields"; ref Pro Flora 5 Wet loamy soils; 5 g/m^2	-	-	-	-	33.00	100 m^2	**33.00**
Scrub and moorland							
"British Seed Houses"; ref WFG10 (Cornfield Annuals); 5 g/m^2	-	-	-	-	32.80	100 m^2	**32.80**
Hedgerow							
"Perryfields"; ref Pro Flora 7 Hedgerow and light shade; 5 g/m^2	-	-	-	-	33.75	100 m^2	**33.75**
Vacant sites							
"Perryfields"; ref Pro Flora 1 Cornfield annuals; 5 g/m^2	-	-	-	-	30.00	100 m^2	**30.00**
Regional Environmental mixes							
"RE1"; 5g / m^2	-	-	-	-	25.80	100 m^2	**25.80**
"RE6"; 5g / m^2	-	-	-	-	47.45	100 m^2	**47.45**
Seeding							
Seeding labours only in two operations; by machine (for seed prices see above)							
35 g/m^2	-	-	-	0.50	-	100 m^2	**0.50**
Grass seed; spreading in two operations; PC £2.80/kg; (for changes in material prices please refer to table above); by machine							
35 g/m^2	-	-	-	0.50	9.80	100 m^2	**10.30**
50 g/m^2	-	-	-	0.50	14.00	100 m^2	**14.50**
70 g/m^2	-	-	-	0.50	19.60	100 m^2	**20.10**
100 g/m^2	-	-	-	0.50	28.00	100 m^2	**28.50**
125 kg/ha	-	-	-	49.52	350.00	ha	**399.52**
150 kg/ha	-	-	-	49.52	420.00	ha	**469.52**
200 kg/ha	-	-	-	49.52	560.00	ha	**609.52**
250 kg/ha	-	-	-	49.52	700.00	ha	**749.52**
300 kg/ha	-	-	-	49.52	840.00	ha	**889.52**
350 kg/ha	-	-	-	49.52	980.00	ha	**1029.52**
400 kg/ha	-	-	-	49.52	1120.00	ha	**1169.52**
500 kg/ha	-	-	-	49.52	1400.00	ha	**1449.52**
700 kg/ha	-	-	-	49.52	1960.00	ha	**2009.52**
1400 kg/ha	-	-	-	49.52	3920.00	ha	**3969.52**

Q PAVING/PLANTING/FENCING/SITE FURNITURE

Item Excluding site overheads and profit	PC £	Labour hours	Labour £	Plant £	Material £	Unit	Total rate £
Q30 SEEDING/TURFING - cont'd							
Seeding - cont'd							
Extra over seeding by machine for slopes over 30 degrees (allowing for the actual area but measured in plan)							
35 g/m^2	-	-	-	0.07	1.47	100 m^2	**1.54**
50 g/m^2	-	-	-	0.07	2.10	100 m^2	**2.17**
70 g/m^2	-	-	-	0.07	2.94	100 m^2	**3.01**
100 g/m^2	-	-	-	0.07	4.20	100 m^2	**4.27**
125 kg/ha	-	-	-	7.43	52.50	ha	**59.93**
150 kg/ha	-	-	-	7.43	63.00	ha	**70.43**
200 kg/ha	-	-	-	7.43	84.00	ha	**91.43**
250 kg/ha	-	-	-	7.43	105.00	ha	**112.43**
300 kg/ha	-	-	-	7.43	126.00	ha	**133.43**
350 kg/ha	-	-	-	7.43	147.00	ha	**154.43**
400 kg/ha	-	-	-	7.43	168.00	ha	**175.43**
500 kg/ha	-	-	-	7.43	210.00	ha	**217.43**
700 kg/ha	-	-	-	7.43	294.00	ha	**301.43**
1400 kg/ha	-	-	-	7.43	588.00	ha	**595.43**
Seeding labours only in two operations; **by machine (for seed prices see above)**							
35 g/m^2	-	0.17	3.13	-	-	100 m^2	**3.13**
Grass seed; spreading in two operations; PC £2.80/kg; (for changes in material prices please refer to table above); by hand							
35 g/m^2	-	0.17	3.13	-	9.80	100 m^2	**12.93**
50 g/m^2	-	0.17	3.13	-	14.00	100 m^2	**17.13**
70 g/m^2	-	0.17	3.13	-	19.60	100 m^2	**22.73**
100 g/m^2	-	0.20	3.75	-	28.00	100 m^2	**31.75**
125 g/m^2	-	0.20	3.75	-	35.00	100 m^2	**38.75**
Extra over seeding by hand for slopes over 30 degrees (allowing for the actual area but measured in plan)							
35 g/m^2	1.46	-	0.07	-	1.46	100 m^2	**1.53**
50 g/m^2	2.10	-	0.07	-	2.10	100 m^2	**2.17**
70 g/m^2	2.94	-	0.07	-	2.94	100 m^2	**3.01**
100 g/m^2	4.20	-	0.08	-	4.20	100 m^2	**4.28**
125 g/m^2	5.24	-	0.08	-	· 5.24	100 m^2	**5.32**
Harrowing seeded areas; light chain harrow	-	-	-	0.09	-	100 m^2	**0.09**
Raking over seeded areas							
by mechanical stone rake	-	-	-	2.12	-	100 m^2	**2.12**
by hand	-	0.80	15.00	-	-	100 m^2	**15.00**
Rolling seeded areas; light roller							
by tractor drawn roller	-	-	-	0.52	-	100 m^2	**0.52**
by pedestrian operated mechanical roller	-	0.08	1.56	0.37	-	100 m^2	**1.93**
by hand drawn roller	-	0.17	3.13	-		100 m^2	**3.13**
Extra over harrowing, raking or rolling seeded areas for slopes over 30 degrees; by machine or hand	-	-	-	-	-	-	**25%**
Turf edging; to seeded areas; 300 mm wide	1.80	0.05	0.89	-	1.80	m^2	**2.69**
Liquid Sod; Turf Management Spray on grass system of grass plantlets fertilizer, bio-degradable mulch carrier, root enhancer and water							
to prepared ground	-	-	-	-	-	m^2	**2.50**
Preparation of turf beds Rolling turf to be lifted; lifting by hand or mechanical turf stripper; stacks to be not more than 1 m high							
cutting only preparing to lift; pedestrian turf cutter	-	0.75	14.06	9.25	-	100 m^2	**23.31**
lifting and stacking; by hand	-	8.33	156.25	-	-	100 m^2	**156.25**

Q PAVING/PLANTING/FENCING/SITE FURNITURE

Item Excluding site overheads and profit	PC £	Labour hours	Labour £	Plant £	Material £	Unit	Total rate £
Rolling up; moving to stacks							
distance not exceeding 100 m	-	2.50	46.88	-	-	100 m²	**46.88**
extra over rolling and moving turf to stacks to							
transport per additional 100 m	-	0.83	15.62	-	-	100 m²	**15.62**
Lifting selected and approved topsoil from spoil							
heaps							
passing through 6 mm screen; removing debris	-	0.17	3.13	8.17	-	m³	**11.29**
Extra over lifting topsoil and passing through							
screen for imported topsoil; plus 20% allowance							
for settlement	17.32	-	-	-	20.78	m³	**20.78**
Topsoil; PC £17.32/m³; plus 20% allowance for							
settlement							
25 mm deep	-	-	-	-	0.52	m²	**0.52**
50 mm deep	-	-	-	-	1.04	m²	**1.04**
100 mm deep	-	-	-	-	2.08	m²	**2.08**
150 mm deep	-	-	-	-	3.12	m²	**3.12**
200 mm deep	-	-	-	-	4.16	m²	**4.16**
250 mm deep	-	-	-	-	5.20	m²	**5.20**
300 mm deep	-	-	-	-	6.24	m²	**6.24**
400 mm deep	-	-	-	-	8.31	m²	**8.31**
450 mm deep	-	-	-	-	9.35	m²	**9.35**
Spreading topsoil to form turf beds (topsoil not							
included); by machine							
25 mm deep	-	-	0.05	0.10	-	m²	**0.15**
50 mm deep	-	-	0.06	0.13	-	m²	**0.19**
75 mm deep	-	-	0.07	0.15	-	m²	**0.22**
100 mm deep	-	0.01	0.09	0.20	-	m²	**0.29**
150 mm deep	-	0.01	0.14	0.29	-	m²	**0.43**
Spreading topsoil to form turf beds (topsoil not							
included); by hand							
25 mm deep	-	0.03	0.47	-	-	m²	**0.47**
50 mm deep	-	0.03	0.63	-	-	m²	**0.63**
75 mm deep	-	0.04	0.80	-	-	m²	**0.80**
100 mm deep	-	0.05	0.94	-	-	m²	**0.94**
150 mm deep	-	0.08	1.41	-	-	m²	**1.41**
Bringing existing topsoil to a fine tilth for turfing							
by raking or harrowing; stones not to exceed 6							
mm; by machine	-	-	0.08	0.04	-	m²	**0.12**
Bringing existing topsoil to a fine tilth for turfing by							
raking or harrowing; stones not to exceed 6 mm;							
by hand	-	0.01	0.17	-	-	m²	**0.17**
Preparation of cricket squares							
Excavating cricket square size 20 x 22 m to a							
depth of 150 mm; screening and returning topsoil;							
mixing with imported marl or clay loam at the rate							
of 1 kg/m²; bringing to accurate levels; hand	-	-	-	-	-	nr	**4514.37**
Turfing							
Turfing; laying only; to stretcher bond; butt joints;							
including providing and working from barrow							
plank runs where necessary to surfaces not							
exceeding 30 degrees from horizontal; average							
size of turves 0.50 yd2 (0.418 m²); 2.39 turves							
per m²							
specially selected lawn turves from previously							
lifted stockpile	-	0.08	1.41	-	-	m²	**1.41**
cultivated lawn turves; to large open areas	-	0.06	1.09	-	-	m²	**1.09**
cultivated lawn turves; to domestic or garden							
areas	-	0.08	1.46	-	-	m²	**1.46**
road verge quality turf	-	0.04	0.75	-	-	m²	**0.75**

Q PAVING/PLANTING/FENCING/SITE FURNITURE

Item Excluding site overheads and profit	PC £	Labour hours	Labour £	Plant £	Material £	Unit	Total rate £
Q30 SEEDING/TURFING - cont'd							
Industrially grown turf; PC prices listed represent the general range of industrial turf prices for sportsfields and amenity purposes; prices will vary with quantity and site location							
"Rolawn"							
ref RB Medallion; sports fields, domestic lawns, general landscape	2.00	0.07	1.31	-	2.00	m²	3.31
Tensar Ltd							
"Tensar Turf Mat" reinforced turf for embankments; laid to embankments 3.3 m² turves	-	0.11	2.06	-	3.10	m²	5.16
"Inturf"							
ref Inturf 5; fine texture, special golf and bowling greens	2.04	0.10	1.79	-	2.04	m²	3.83
ref Inturf 1; fine lawns, golf greens, bowling greens	2.54	0.10	1.79	-	2.54	m²	4.33
ref Inturf 3; hockey grounds, polo, medium wearing areas	1.79	0.05	1.02	-	1.79	m²	2.81
ref Inturf 4; hardwearing fine turf, low maintenance areas	2.04	0.04	0.79	-	2.04	m²	2.83
ref Inturf 2; football grounds, parks, hardwearing areas	1.54	0.05	0.98	-	1.54	m²	2.53
ref Inturf 2 Bargold; fine turf ; football grounds, parks, hardwearing areas	1.79	0.05	0.98	-	1.79	m²	2.78
Custom Grown Turf; specific seed mixtures to suit soil or site conditions	8.29	0.08	1.50	-	8.29	m²	9.79
Turf Tiles; instant repairs to goal mouths and playing fields providing instant stable surface	-	-	-	-	-	m²	28.62
Turf Modules; for trade shows exhibitions and events	-	-	-	-	-	m²	129.60
Reinforced turf; Netlon Advanced Turf; NetlonTurf Systems; blended mesh fibre elements incorporated into root zone; Rootzone spread and levelled over cultivated, prepared and reduced and levelled ground (not included); compacted with light roller							
"ATS 400" with "R400" topping, seed and fertilizer; 100 thick							
100 -199 m²	16.15	0.03	0.47	0.61	16.15	m²	17.23
200 - 299 m²	14.65	0.03	0.47	0.61	14.65	m²	15.73
300 - 499 m²	14.10	0.03	0.47	0.61	14.10	m²	15.18
over 500 m²	13.70	0.03	0.47	0.61	13.70	m²	14.78
"ATS 400" with "R400" topping, seed and fertilizer; 150 thick							
100 -199 m²	21.10	0.03	0.62	0.81	21.10	m²	22.53
200 - 299 m²	19.60	0.03	0.62	0.81	19.60	m²	21.03
300 - 499 m²	18.95	0.03	0.62	0.81	18.95	m²	20.38
over 500 m²	18.50	0.03	0.62	0.81	18.50	m²	19.93
"ATS 400" with washed turf and fertilizer ("R400" topping not included) 100 thick							
100 -199 m²	19.70	0.03	0.62	0.81	19.70	m²	21.13
200 - 499 m²	18.10	0.03	0.62	0.81	18.10	m²	19.53
over 500 m²	17.15	0.03	0.62	0.81	17.15	m²	18.58
"ATS 400" with washed turf and fertilizer ("R400" topping not included) 150 thick							
100 -199 m²	-	0.04	0.75	0.91	24.55	m²	26.21
200 - 499 m²	22.95	0.04	0.75	0.91	22.95	m²	24.61
over 500 m²	21.85	0.04	0.75	0.91	21.85	m²	23.51

Q PAVING/PLANTING/FENCING/SITE FURNITURE

Item Excluding site overheads and profit	PC £	Labour hours	Labour £	Plant £	Material £	Unit	Total rate £
Firming turves with wooden beater	-	0.01	0.19	-	-	m²	**0.19**
Rolling turfed areas; light roller							
by tractor with turf tyres and roller	-	-	-	0.52	-	100 m²	**0.52**
by pedestrian operated mechanical roller	-	0.08	1.56	0.37	-	100 m²	**1.93**
by hand drawn roller	-	0.17	3.13	-	-	100 m²	**3.13**
Dressing with finely sifted topsoil; brushing into joints	0.01	0.05	0.94	-	0.02	m²	**0.95**
Turfing; laying only							
to slopes over 30 degrees; to diagonal bond (measured as plan area - add 15% to these rates for the incline area of 30 degree slopes)	-	0.12	2.25	-	-	m²	**2.25**
Extra over laying turfing for pegging down turves wooden or galvanized wire pegs; 200 mm long; 2 pegs per 0.50 yd2	1.39	0.01	0.25	-	1.39	m²	**1.64**
Artificial grass; Artificial Lawn Company; laid to sharp sand bed priced separately 15 kg kiln sand brushed in /m²							
"Wimbledon" 20 mm thick artificial sports turf; sand filled	-	-	-	-	-	m²	**24.00**
"Greenfield"; for general use; budget surface; sand filled	-	-	-	-	-	m²	**19.21**
"Garden"; patios conservatories and pool surrounds; sand filled	-	-	-	-	-	m²	**20.09**
"Premier"; lawns and patios	-	-	-	-	-	m²	**25.75**
"Elite"; grass/sand and rubber filled	-	-	-	-	-	m²	**28.84**
"Grassflex"; safety surfacing for play areas	-	-	-	-	-	m²	**41.20**
Maintenance operations (Note: the following rates apply to aftercare maintenance executed as part of a landscaping contract only)							
Initial cutting; to turfed areas							
20 mm high; using pedestrian guided power driven cylinder mower; including boxing off cuttings (stone picking and rolling not included)	-	0.18	3.38	0.25	-	100 m²	**3.63**
Repairing damaged grass areas							
scraping out; removing slurry; from ruts and holes; average 100 mm deep	-	0.13	2.50	-	-	m²	**2.50**
100 mm topsoil	-	0.13	2.50	-	2.08	m²	**4.58**
Repairing damaged grass areas; sowing grass seed to match existing or as specified; to individually prepared worn patches							
35 g/m²	0.10	0.01	0.19	-	0.11	m²	**0.30**
50 g/m²	0.14	0.01	0.19	-	0.16	m²	**0.35**
Sweeping leaves; disposing off site; motorized vacuum sweeper or rotary brush sweeper							
areas of maximum 2500 m² with occasional large tree and established boundary planting; 5.35 m³ (1 skip of material to be removed)	-	0.40	7.50	2.75	-	100 m²	**10.25**
Leaf Clearance; clearing grassed area of leaves and other extraneous debris							
Using equipment towed by tractor							
large grassed areas with perimeters of mature trees such as sports fields and amenity areas	-	0.01	0.23	0.27	-	100 m²	**0.51**
large grassed areas containing ornamental trees and shrub beds	-	0.03	0.47	0.38	-	100 m²	**0.85**
Using pedestrian operated mechanical equipment and blowers							
grassed areas with perimeters of mature trees such as sports fields and amenity areas	-	0.04	0.75	0.09	-	100 m²	**0.84**
grassed areas containing ornamental trees and shrub beds	-	0.10	1.88	0.24	-	100 m²	**2.12**
verges	-	0.07	1.25	0.16	-	100 m²	**1.41**

Q PAVING/PLANTING/FENCING/SITE FURNITURE

Item Excluding site overheads and profit	PC £	Labour hours	Labour £	Plant £	Material £	Unit	Total rate £
Q30 SEEDING/TURFING - cont'd							
Maintenance operations; Leaf Clearance - cont'd							
By hand							
grassed areas with perimeters of mature trees such as sports fields and amenity areas	-	0.05	0.94	0.09	-	100 m^2	1.03
grassed areas containing ornamental trees and shrub beds	-	0.08	1.56	0.15	-	100 m^2	1.72
verges	-	1.00	18.75	1.86	-	100 m^2	20.61
Removal of arisings							
areas with perimeters of mature trees	-	0.01	0.10	0.08	0.72	100 m^2	0.90
areas containing ornamental trees and shrub beds	-	0.02	0.30	0.32	1.80	100 m^2	2.42
Cutting grass to specified height; per cut							
multi unit gang mower	-	0.59	11.03	15.89	-	ha	26.92
ride-on triple cylinder mower	-	0.01	0.26	0.12	-	100 m^2	0.38
ride-on triple rotary mower	-	0.01	0.26	-	-	100 m^2	0.26
pedestrian mower	-	0.18	3.38	0.60	-	100 m^2	3.98
Cutting grass to banks; per cut							
side arm cutter bar mower	-	0.02	0.44	0.29	-	100 m^2	0.73
Cutting rough grass; per cut							
power flail or scythe cutter	-	0.04	0.66	-	-	100 m^2	0.66
Extra over cutting grass for slopes not exceeding 30 degrees	-	-	-	-	-	-	10%
Extra over cutting grass for slopes exceeding 30 degrees	-	-	-	-	-	-	40%
Cutting fine sward							
pedestrian operated seven-blade cylinder lawn mower	-	0.14	2.63	0.19	-	100 m^2	2.82
Extra over cutting fine sward for boxing off cuttings							
pedestrian mower	-	0.03	0.53	0.04	-	100 m^2	0.56
Cutting areas of rough grass							
scythe	-	1.00	18.75	-	-	100 m^2	18.75
sickle	-	2.00	37.50	-	-	100 m^2	37.50
petrol operated strimmer	-	0.30	5.63	0.41	-	100 m^2	6.04
Cutting areas of rough grass which contain trees or whips							
petrol operated strimmer	-	0.40	7.50	0.54	-	100 m^2	8.04
Extra over cutting rough grass for on site raking up and dumping	-	0.33	6.25	-	-	100 m^2	6.25
Trimming edge of grass areas; edging tool							
with petrol powered strimmer	-	0.13	2.50	0.18	-	100 m	2.68
by hand	-	0.67	12.50	-	-	100 m	12.50
Marking out pitches using approved line marking compound; including initial setting out and marking							
discus, hammer, javelin or shot putt area	2.61	2.00	37.50	-	2.61	nr	40.11
cricket square	1.74	2.00	37.50	-	1.74	nr	39.24
cricket boundary	6.09	8.00	150.00	-	6.09	nr	156.09
grass tennis court	2.61	4.00	75.00	-	2.61	nr	77.61
hockey pitch	8.70	8.00	150.00	-	8.70	nr	158.70
football pitch	8.70	8.00	150.00	-	8.70	nr	158.70
rugby pitch	8.70	8.00	150.00	-	8.70	nr	158.70
eight lane running track; 400 m	17.40	16.00	300.00	-	17.40	nr	317.40
Re-marking out pitches using approved line marking compound							
discus, hammer, javelin or shot putt area	1.74	0.50	9.38	-	1.74	nr	11.12
cricket square	1.74	0.50	9.38	-	1.74	nr	11.12
grass tennis court	6.09	1.00	18.75	-	6.09	nr	24.84
hockey pitch	6.09	1.00	18.75	-	6.09	nr	24.84
football pitch	6.09	1.00	18.75	-	6.09	nr	24.84
rugby pitch	6.09	1.00	18.75	-	6.09	nr	24.84
eight lane running track; 400 m	17.40	2.50	46.88	-	17.40	nr	64.28

Q PAVING/PLANTING/FENCING/SITE FURNITURE

Item Excluding site overheads and profit	PC £	Labour hours	Labour £	Plant £	Material £	Unit	Total rate £
Rolling grass areas; light roller							
by tractor drawn roller	-	-	-	0.52	-	100 m²	**0.52**
by pedestrian operated mechanical roller	-	0.08	1.56	0.37	-	100 m²	**1.93**
by hand drawn roller	-	0.17	3.13	-	-	100 m²	**3.13**
Aerating grass areas; to a depth of 100 mm							
using tractor-drawn aerator	-	0.06	1.09	2.08	-	100 m²	**3.17**
using pedestrian-guided motor powered solid or slitting tine turf aerator	-	0.18	3.28	2.75	-	100 m²	**6.03**
using hollow tine aerator; including sweeping up and dumping corings	-	0.50	9.38	5.50	-	100 m²	**14.88**
using hand aerator or fork	-	1.67	31.25	-	-	100 m²	**31.25**
Extra over aerating grass areas for on site sweeping up and dumping corings	-	0.17	3.13	-	-	100 m²	**3.13**
Switching off dew; from fine turf areas	-	0.20	3.75	-	-	100 m²	**3.75**
Scarifying grass areas to break up thatch; removing dead grass							
using tractor-drawn scarifier	-	0.07	1.31	1.45	-	100 m²	**2.76**
using self-propelled scarifier; including removing and disposing of grass on site	-	0.33	6.25	0.44	-	100 m²	**6.68**
Harrowing grass areas							
using drag mat	-	0.03	0.53	0.29	-	100 m²	**0.82**
using chain harrow	-	0.04	0.66	0.37	-	100 m²	**1.02**
using drag mat	-	2.80	52.51	29.39	-	ha	**81.89**
using chain harrow	-	3.50	65.63	36.73	-	ha	**102.36**
Extra for scarifying and harrowing grass areas for disposing excavated material off site; to tip not exceeding 13 km; loading by machine							
slightly contaminated	-	-	-	1.65	13.00	m³	**14.65**
rubbish	-	-	-	1.65	13.00	m³	**14.65**
inert material	-	-	-	1.10	7.50	m³	**8.60**
For the following topsoil improvement and seeding operations add or subtract the following amounts for every £0.10 difference in the material cost price							
35 g/m²	-	-	-	-	0.35	100 m²	**0.35**
50 g/m²	-	-	-	-	0.50	100 m²	**0.50**
70 g/m²	-	-	-	-	0.70	100 m²	**0.70**
100 g/m²	-	-	-	-	1.00	100 m²	**1.00**
125 kg/ ha	-	-	-	-	12.50	ha	**12.50**
150 kg/ ha	-	-	-	-	15.00	ha	**15.00**
175 kg/ ha	-	-	-	-	17.50	ha	**17.50**
200 kg/ ha	-	-	-	-	20.00	ha	**20.00**
225 kg/ ha	-	-	-	-	22.50	ha	**22.50**
250 kg/ ha	-	-	-	-	25.00	ha	**25.00**
300 kg/ ha	-	-	-	-	30.00	ha	**30.00**
350 kg/ ha	-	-	-	-	35.00	ha	**35.00**
400 kg/ ha	-	-	-	-	40.00	ha	**40.00**
500 kg/ ha	-	-	-	-	50.00	ha	**50.00**
700 kg/ ha	-	-	-	-	70.00	ha	**70.00**
1000 kg/ ha	-	-	-	-	100.00	ha	**100.00**
1250 kg/ ha	-	-	-	-	125.00	ha	**125.00**
Selective residual pre-emergent weedkiller; Rigby Taylor Ltd; "Flexidor 125"; in accordance with manufacturer's instructions; PC £78.15 per litre; application rate 1.20 - 2 litre/hectare							
1.20 l /ha	-	0.28	5.25	-	0.94	100 m²	**6.19**
2.00 l /ha	-	0.28	5.25	-	1.56	100 m²	**6.81**
Top dressing fertilizers (7:7:7); PC £0.44/kg; to seedbeds; by machine							
35 g/m²	1.56	-	-	0.18	1.56	100 m²	**1.74**
50 g/m²	2.22	-	-	0.18	2.22	100 m²	**2.41**
300 kg/ha	133.44	-	-	18.23	133.44	ha	**151.67**
350 kg/ha	155.68	-	-	18.23	155.68	ha	**173.91**
400 kg/ha	177.92	-	-	18.23	177.92	ha	**196.15**
500 kg/ha	222.40	-	-	29.17	222.40	ha	**251.57**

Q PAVING/PLANTING/FENCING/SITE FURNITURE

Item Excluding site overheads and profit	PC £	Labour hours	Labour £	Plant £	Material £	Unit	Total rate £
Q30 SEEDING/TURFING - cont'd							
Maintenance operations - cont'd							
Top dressing fertilizers (7:7:7); PC £0.44/kg; to seedbeds; by hand							
35 g/m^2	1.56	0.17	3.13	-	1.56	100 m^2	4.68
50 g/m^2	2.22	0.17	3.13	-	2.22	100 m^2	5.35
70 g/m^2	3.11	0.17	3.13	-	3.11	100 m^2	6.24
Watering turf; evenly; at a rate of 5 litre/m^2							
using movable spray lines powering 3 nr sprinkler heads with a radius of 15 m and allowing for 60% overlap (irrigation machinery costs not included)	-	0.02	0.29	-	-	100 m^2	0.29
using sprinkler equipment and with sufficient water pressure to run 1 nr 15 m radius sprinkler	-	0.02	0.37	-	-	100 m^2	0.37
using hand-held watering equipment	-	0.25	4.69	-	-	100 m^2	4.69
Q31 PLANTING							
Market prices of mulching materials							
Melcourt Industries Ltd; 25 m³ loads							
"Ornamental Bark Mulch"	-	-	-	-	46.00	m³	46.00
"Bark Nuggets®"	-	-	-	-	44.30	m³	44.30
"Graded Bark Flakes"	-	-	-	-	45.90	m³	45.90
"Amenity Bark Mulch"	-	-	-	-	31.55	m³	31.55
"Contract Bark Mulch"	-	-	-	-	29.30	m³	29.30
"Spruce Ornamental"	-	-	-	-	32.05	m³	32.05
"Decorative Biomulch®"	-	-	-	-	31.00	m³	31.00
"Rustic Biomulch®"	-	-	-	-	34.75	m³	34.75
"Mulch 2000"	-	-	-	-	25.00	m³	25.00
"Forest BioMulch®"	-	-	-	-	28.50	m³	28.50
Melcourt Industries Ltd; 50 m³ loads							
"Mulch 2000"	-	-	-	-	14.50	m³	14.50
Melcourt Industries Ltd; 65 m³ loads							
"Contract Bark Mulch"	-	-	-	-	19.20	m³	19.20
Melcourt Industries Ltd; 80 m³ loads							
"Ornamental Bark Mulch"	-	-	-	-	33.00	m³	33.00
"Bark Nuggets®"	-	-	-	-	31.30	m³	31.30
"Graded Bark Flakes"	-	-	-	-	32.90	m³	32.90
"Amenity Bark Mulch"	-	-	-	-	18.55	m³	18.55
"Spruce Ornamental"	-	-	-	-	19.05	m³	19.05
"Decorative Biomulch®"	-	-	-	-	18.00	m³	18.00
"Rustic Biomulch®"	-	-	-	-	21.75	m³	21.75
"Forest BioMulch®"	-	-	-	-	15.50	m³	15.50
Market prices of planting materials (Note: the rates shown generally reflect the manufacturer's recommended retail prices; trade and bulk discounts are often available on the prices shown)							
Topsoil; Boughton Loam Ltd							
50/50; as dug / 10 mm screened	-	-	-	-	41.76	m³	41.76
Mulch; Melcourt Industries Ltd; 25 m³ loads							
"Composted Fine Bark"	-	-	-	-	28.30	m³	28.30
"Humus 2000"	-	-	-	-	23.25	m³	23.25
"Spent Mushroom Compost"	-	-	-	-	18.10	m³	18.10
"Topgrow"	-	-	-	-	27.20	m³	27.20
Mulch; Melcourt Industries Ltd; 50 m³ loads							
"Humus 2000"	-	-	-	-	12.75	m³	12.75
Mulch; Melcourt Industries Ltd; 60 m³ loads							
"Spent Mushroom Compost"	-	-	-	-	6.40	m³	6.40
"Topgrow"	-	-	-	-	16.70	m³	16.70
Mulch; Melcourt Industries Ltd; 65 m³ loads							
"Super Humus"	-	-	-	-	14.50	m³	14.50
"Composted Fine Bark"	-	-	-	-	16.30	m³	16.30

Q PAVING/PLANTING/FENCING/SITE FURNITURE

Item Excluding site overheads and profit	PC £	Labour hours	Labour £	Plant £	Material £	Unit	Total rate £
Fertilizers; Scotts UK							
"Enmag"; 70 g/m^2	-	-	-	-	11.76	100 m^2	11.76
fertilizer; controlled release; "Osmocote Exact",							
15+10+12+2MgO+TE, tablet; 5 gr each	-	-	-	-	3.03	100 nr	3.03
Fertilizers; Scotts UK; granular							
"Sierrablen Flora", 15+9+9+3 MgO; controlled							
release fertilizer; costs for recommended							
application rates							
transplant	-	-	-	-	0.09	nr	0.09
whip	-	-	-	-	0.14	nr	0.14
feathered	-	-	-	-	0.19	nr	0.19
light standard	-	-	-	-	0.19	nr	0.19
standard	-	-	-	-	0.23	nr	0.23
selected standard	-	-	-	-	0.33	nr	0.33
heavy standard	-	-	-	-	0.37	nr	0.37
extra heavy standard	-	-	-	-	0.47	nr	0.47
16 -18 cm girth	-	-	-	-	0.52	nr	0.52
18 - 20 cm girth	-	-	-	-	0.56	nr	0.56
20 - 22 cm girth	-	-	-	-	0.66	nr	0.66
22 - 24 cm girth	-	-	-	-	0.70	nr	0.70
24 - 26 cm girth	-	-	-	-	0.75	nr	0.75
Fertilizers; Farmura Environmental Ltd; Seanure							
Root Dip							
to transplants	-	-	-	-	0.21	10 nr	0.21
medium whips	-	-	-	-	0.05	each	0.05
standard trees	-	-	-	-	0.34	each	0.34
Fertilizers; Farmura Environmental Ltd; Seanure							
Soilbuilder							
soil amelioration; 70 g/m^2	-	-	-	-	7.80	100 m^2	7.80
to plant pits; 300 x 300 x 300	-	-	-	-	0.03	each	0.03
to plant pits; 600 x 600 x 600	-	-	-	-	0.36	each	0.36
to tree pits; 1.00 x 1.00 x 1.00	-	-	-	-	1.67	each	1.67
Fertilizers; Scotts UK							
grass and soil fertilizer; "Greenmaster Liquid							
Seafeed"	-	-	-	-	3.54	litre	3.54
"TPMC" tree and shrub planting compost	-	-	-	-	4.56	bag	4.56
Fertilizers and anti-desiccants; Rigby Taylor Ltd;							
fertilizer application rates 35 g/m^2 unless							
otherwise shown							
straight fertilizer; "Bone Meal"; at 70 g/m^2	-	-	-	-	5.58	100 m^2	5.58
straight fertilizer; "Sulphate of Ammonia"	-	-	-	-	1.60	100 m^2	1.60
straight fertilizer; "Sulphate of Iron"	-	-	-	-	1.49	100 m^2	1.49
straight fertilizer; "Sulphate of Potash"	-	-	-	-	1.98	100 m^2	1.98
straight fertilizer; "Super Phosphate Powder"	-	-	-	-	0.42	100 m^2	0.42
liquid fertilizer; "Vitax 50/50 Soluble Iron"; 56							
ml/100 m^2	-	-	-	-	6.42	100 m^2	6.42
liquid fertilizer; "Vitax 50/50 Standard"; 56							
ml/100 m^2	-	-	-	-	10.49	100 m^2	10.49
liquid fertilizer; "Vitax 50/50 Extra"; 56 ml/100 m^2	-	-	-	-	12.68	100 m^2	12.68
wetting agent; "Breaker liquid"; 10 lt	-	-	-	-	1.12	100 m^2	1.12
wetting agent; "Breaker granules"; 25 kg	-	-	-	-	7.02	100 m^2	7.02
Herbicides; Scotts UK; application							
rates used to produce these rates are the							
maximum rate recommended by the manufacturer							
in each case (Note: lower rates may often be							
appropriate in many cases)							
"Dextrone X"	-	-	-	-	0.52	100 m^2	0.52
"Intrepid"	-	-	-	-	0.54	100 m^2	0.54
"Speedway"	-	-	-	-	1.04	100 m^2	1.04

Q PAVING/PLANTING/FENCING/SITE FURNITURE

Item Excluding site overheads and profit	PC £	Labour hours	Labour £	Plant £	Material £	Unit	Total rate £
Q31 PLANTING - cont'd							
Market prices of planting materials - cont'd							
Herbicides; Rigby Taylor Ltd; application rates used to produce these are the maximum rate recommended by the manufacturer in each case (Note: these application rates will vary dependent on season)							
"Roundup Pro Biactive"; 5 lt/ha	-	-	-	-	0.95	100 m²	0.95
"Flexidor 125 (Isoxaben)"; 2 lt/ha	-	-	-	-	1.56	100 m²	1.56
"Kerb Flowable (Propyzamide)"	-	-	-	-	2.00	100 m²	2.00
"Kerb Granules" (15 x 120 tree pack)	-	-	-	-	0.28	100 m²	0.28
"Timbrel (Triclopyr)"	-	-	-	-	4.38	100 m²	4.38
"Casoron G" as a selective herbicide	-	-	-	-	4.40	100 m²	4.40
"Casoron G" as a residual herbicide	-	-	-	-	7.93	100 m²	7.93
Market prices of trees, shrubs and plants: Notcutts Nurseries Ltd Shrubs							
Acer palmatum Dissectum Atropurpureum; 30-45 3 L	-	-	-	-	16.60	nr	16.60
Arbutus unedo rubra; 30-45 3 L	-	-	-	-	7.50	nr	7.50
Aucuba japonica Variegata; 30-45 3 L	-	-	-	-	3.16	nr	3.16
Berberis candidula; 20-30 2 L	-	-	-	-	1.98	nr	1.98
Berberis thunbergii; 30-45 3 L	-	-	-	-	1.50	nr	1.50
Berberis thunbergii Rose Glow; 30-45 2 L	-	-	-	-	3.00	nr	3.00
Buddleja davidii Black Knight; 45-60 3 L	-	-	-	-	1.93	nr	1.93
Buddleja globosa; 45-60 3 L	-	-	-	-	3.11	nr	3.11
Buxus sempervirens; 30-45 3 L	-	-	-	-	3.11	nr	3.11
Caryopteris clandonensis; 30-45 3 L	-	-	-	-	2.63	nr	2.63
Ceanothus dentatus; 60-90 3 L	-	-	-	-	3.43	nr	3.43
Ceanothus thyrsiflorus Repens; 30-45 2 L	-	-	-	-	2.36	nr	2.36
Chaenomeles superba Jet Trail; 45-60 3 L	-	-	-	-	2.36	nr	2.36
Choisya ternata; 30-45 3 L	-	-	-	-	2.63	nr	2.63
Cornus alba; 60-90 3 L	-	-	-	-	1.50	nr	1.50
Cornus alba Elegantissima; 60-90 3 L	-	-	-	-	2.58	nr	2.58
Cornus stolonifera Flayiramea; 60-90 3 L	-	-	-	-	1.93	nr	1.93
Corylus avellana; 45-60 3 L	-	-	-	-	1.50	nr	1.50
Cotinus coggygria Royal Purple; 30-45 3 L	-	-	-	-	3.53	nr	3.53
Cotoneaster salicifolius Exburyensis; 45-50 3 L	-	-	-	-	2.58	nr	2.58
Cotoneaster suecicus watereri; 60-90 3 L	-	-	-	-	1.93	nr	1.93
Elaeagnus pungens maculata; 45-60 3 L	-	-	-	-	3.43	nr	3.43
Elaeagnus x ebbingei; 45-60 3 L	-	-	-	-	2.36	nr	2.36
Elaeagnus x ebbingei Limelight; 45-60 3 L	-	-	-	-	3.43	nr	3.43
Escallonia Edinensis; 45-60 3 L	-	-	-	-	2.25	nr	2.25
Euonymus fortunei Blondy; 20-30 2 L	-	-	-	-	3.00	nr	3.00
Euonymus fortunei Darts Blanket; 30-45 2 L	-	-	-	-	1.50	nr	1.50
Forsythia intermedia Lynwood; 60-90 3 L	-	-	-	-	1.93	nr	1.93
Genista hispanica; 20-30 2 L	-	-	-	-	1.93	nr	1.93
Griselinia littoralis; 45-60 3 L	-	-	-	-	3.16	nr	3.16
Hebe albicans; 20-30 2 L	-	-	-	-	1.93	nr	1.93
Hebe Great Orme; 20-30 3 L	-	-	-	-	2.78	nr	2.78
Hebe pinguifolia Pagei; 15-20 2 L	-	-	-	-	1.61	nr	1.61
Hebe rakaiensis; 20-30 2 L	-	-	-	-	1.61	nr	1.61
Hedera helix Glacier; 45-60 1.5 L	-	-	-	-	2.25	nr	2.25
Hippophae rhamnoides; 45-60 3 L	-	-	-	-	1.71	nr	1.71
Hydrangea macrophylla Blue Wave; 45-60 4 L	-	-	-	-	4.50	nr	4.50
Hypericum Hidcote; 30-45 3 L	-	-	-	-	1.93	nr	1.93
Ilex aquifolium Pyramidalis; 60-90 3 L	-	-	-	-	4.18	nr	4.18
Lavandula angustifolia Hidcote; 15-20 1.5 L	-	-	-	-	1.50	nr	1.50
Lavandula angustifolia Hidcote; 20-30 3 L	-	-	-	-	2.36	nr	2.36
Lavatera Barnsley; 45-60 3 L	-	-	-	-	2.63	nr	2.63
Leucothoe walteri Rainbow; 30-45 3 L	-	-	-	-	4.82	nr	4.82
Leycesteria formosa; 30-45 3 L	-	-	-	-	3.00	nr	3.00

Q PAVING/PLANTING/FENCING/SITE FURNITURE

Item Excluding site overheads and profit	PC £	Labour hours	Labour £	Plant £	Material £	Unit	Total rate £
Ligustrum ovalifolium; 45-60 2 L	-	-	-	-	1.50	nr	1.50
Ligustrum ovalifolium Aureum; 30-45 2 L	-	-	-	-	1.93	nr	1.93
Ligustrum nitida Baggesens Gold; 30-45 3 L	-	-	-	-	2.04	nr	2.04
Mahonia japonica; 45-60 3 L	-	-	-	-	2.78	nr	2.78
Olearia haastii; 30-45 3 L	-	-	-	-	2.36	nr	2.36
Osmanthus delavayi; 30-45 3 L	-	-	-	-	3.43	nr	3.43
Pachysandra terminalis; 15-20 1.5 L	-	-	-	-	1.71	nr	1.71
Pernettya mucronata Male; 20-30 3 L	-	-	-	-	2.78	nr	2.78
Philadelphus Belle Etoile; 45-60 3 L	-	-	-	-	1.93	nr	1.93
Philadelphus coronarius Aureus; 30-45 3 L	-	-	-	-	3.00	nr	3.00
Philadelphus Virginal; 45-60 3 L	-	-	-	-	2.58	nr	2.58
Phormium tenax; 30-45 3 L	-	-	-	-	3.75	nr	3.75
Photinia x fraseri Red Robin; 30-45 3 L	-	-	-	-	3.85	nr	3.85
Potentilla arbuscula; 20-30 3 L	-	-	-	-	1.71	nr	1.71
Potentilla fruticosa Pretty Polly; 20-30 3 L	-	-	-	-	2.63	nr	2.63
Prunus lusitanica; 45-60 3 L	-	-	-	-	1.93	nr	1.93
Pyracantha Orange Glow; 60-90 3 L	-	-	-	-	1.93	nr	1.93
Pyracantha Teton; 30-45 3 L	-	-	-	-	2.36	nr	2.36
Ribes sanguineum King Edward VII; 45-60 3 L	-	-	-	-	2.25	nr	2.25
Rosmarinus officinalis; 20-30 3 L	-	-	-	-	2.36	nr	2.36
Rubus Betty Ashburner; 45-60 2 L	-	-	-	-	1.39	nr	1.39
Rubus Tridel Benenden; 45-60 3 L	-	-	-	-	2.58	nr	2.58
Salix lanata; 45-60 3 L	-	-	-	-	2.78	nr	2.78
Salvia officinalis Icterina; 20-30 3 L	-	-	-	-	2.78	nr	2.78
Sambucus nigra Aurea; 45-60 3 L	-	-	-	-	1.50	nr	1.50
Sambucus racemosa Plumosa Aurea; 45-60 3 L	-	-	-	-	4.23	nr	4.23
Santolina chamaecyparissus; 20-30 3 L	-	-	-	-	1.71	nr	1.71
Sarcococca confusa; 20-30 2 L	-	-	-	-	3.00	nr	3.00
Senecio monroi; 20-25 2 L	-	-	-	-	2.78	nr	2.78
Senecio Sunshine; 30-45 3 L	-	-	-	-	1.71	nr	1.71
Skimmia japonica; 30-45 3 L	-	-	-	-	2.78	nr	2.78
Spiraea x arguta; 45-60 3 L	-	-	-	-	1.93	nr	1.93
Spiraea bumalda x Goldflame; 20-30 2 L	-	-	-	-	1.71	nr	1.71
Stephanandra incisa Crispa; 30-45 2 L	-	-	-	-	1.71	nr	1.71
Symphoricarpos albus; 45-60 3 L	-	-	-	-	1.71	nr	1.71
Symphoricarpos x chenaultii Hancock; 45-60 2 L	-	-	-	-	1.29	nr	1.29
Tamarix tetrandra; 45-60 3 L	-	-	-	-	2.58	nr	2.58
Thuya plicata; "Dan Flynn" 3 L	-	-	-	-	3.65	nr	3.65
Viburnum x bodnantense Dawn; 45-60 3 L	-	-	-	-	3.65	nr	3.65
Viburnum opulus; 60-90 3 L	-	-	-	-	1.71	nr	1.71
Viburnum opulus Sterile; 45-60 3 L	-	-	-	-	2.58	nr	2.58
Viburnum plicatum Pink Beauty; 45-60 3 L	-	-	-	-	4.60	nr	4.60
Viburnum tinus; 30-45 3 L	-	-	-	-	2.09	nr	2.09
Viburnum tinus Variegatum; 20-30 3 L	-	-	-	-	3.11	nr	3.11
Container grown climbers							
Clematis Jackmanii; 60-90 2 L	-	-	-	-	3.45	nr	3.45
Clematis montana Elizabeth; 60-90 2 L	-	-	-	-	3.45	nr	3.45
Hydrangea petiolaris; 45-60 3 L	-	-	-	-	3.77	nr	3.77
Jasminum nudiflorum; 45-60 2 L	-	-	-	-	2.76	nr	2.76
Lonicera japonica Halliana; 60-90 3 L	-	-	-	-	2.76	nr	2.76
Parthenocissus quinquefolia; 60-90 2 L	-	-	-	-	2.54	nr	2.54
Passiflora caerulea; 60-90 2 L	-	-	-	-	3.71	nr	3.71
Vitis coignetiae; 60-90 2 L	-	-	-	-	5.84	nr	5.84
Wisteria sinensis; 60-90 3 L	-	-	-	-	6.90	nr	6.90
Container grown specimen shrubs							
Amelanchier lamarckii; 90-120 10 L	-	-	-	-	9.11	nr	9.11
Aralia elata; 120-150 10 L	-	-	-	-	21.42	nr	21.42
Aucuba japonica; 45-60 10 L	-	-	-	-	10.18	nr	10.18
Berberis darwinii; 45-60 10 L	-	-	-	-	10.18	nr	10.18
Choisya ternata; 45-60 10 L	-	-	-	-	10.18	nr	10.18
Cotoneaster cornubia; 90-120 10 L	-	-	-	-	9.11	nr	9.11
Cotoneaster Exburiensis; 90-120 10 L	-	-	-	-	10.18	nr	10.18
Cotoneaster watereri; 90-120 10 L	-	-	-	-	9.11	nr	9.11
Elaeagnus pungens Maculata; 45-60 10 L	-	-	-	-	13.39	nr	13.39
Elaeagnus ebbingei; 60-90 10 L	-	-	-	-	9.11	nr	9.11

Q PAVING/PLANTING/FENCING/SITE FURNITURE

Item Excluding site overheads and profit	PC £	Labour hours	Labour £	Plant £	Material £	Unit	Total rate £
Q31 PLANTING - cont'd							
Market prices of trees, shrubs and plants: Notcutts Nurseries Ltd - cont'd							
Elaeagnus ebbingei Limelight; 60-90 10 L	-	-	-	-	13.39	nr	**13.39**
Escallonia rubra Crimson Spire; 60-90 10 L	-	-	-	-	9.11	nr	**9.11**
Euonymus fortunei Emerald N Gold; 30-45 7.3 L	-	-	-	-	8.57	nr	**8.57**
Fatsia japonica; 90-120 15 L	-	-	-	-	13.39	nr	**13.39**
Ilex aquifolium; 90-120 10 L	-	-	-	-	10.18	nr	**10.18**
Ilex aquifolium Golden King; 90-120 10 L	-	-	-	-	16.07	nr	**16.07**
Ilex aquifolium Argentea Marginata; 90-120 10 L	-	-	-	-	16.07	nr	**16.07**
Mahonia japonica; 60-90 10 L	-	-	-	-	13.39	nr	**13.39**
Phormium tenax; 60-90 10 L	-	-	-	-	15.00	nr	**15.00**
Phormium tenax Purpureum; 60-90 10 L	-	-	-	-	15.00	nr	**15.00**
Phormium Yellow Wave; 60-90 10 L	-	-	-	-	19.28	nr	**19.28**
Photinia x Fraseri Red Robin; 60-90 10 L	-	-	-	-	13.39	nr	**13.39**
Prunus x cistena Crimson Dwarf; 45-60 10 L	-	-	-	-	13.39	nr	**13.39**
Prunus laurocerasus; 90-120 10 L	-	-	-	-	9.11	nr	**9.11**
Prunus laurocerasus Otto Luyken; 45-60 10 L	-	-	-	-	9.11	nr	**9.11**
Prunus laurocerasus Zabeliana; 45-60 10 L	-	-	-	-	9.11	nr	**9.11**
Prunus lusitanica; 45-60 10 L	-	-	-	-	10.18	nr	**10.18**
Pyracantha Mohave; 45-60 10 L	-	-	-	-	9.11	nr	**9.11**
Pyracantha Orange Glow; 60-90 10 L	-	-	-	-	9.11	nr	**9.11**
Pyracantha Red Column; 60-90 10 L	-	-	-	-	9.11	nr	**9.11**
Pyracantha Soleil d'Or; 60-90 10 L	-	-	-	-	9.11	nr	**9.11**
Pyracantha Teton; 45-60 10 L	-	-	-	-	9.11	nr	**9.11**
Rhus typhina; 90-120 10 L	-	-	-	-	9.11	nr	**9.11**
Rosmarinus officinalis; 30-45 10 L	-	-	-	-	10.71	nr	**10.71**
Rubus cockburnianus Golden Vale; 90-120 10 L	-	-	-	-	11.25	nr	**11.25**
Skimmia Rubella; 30-45 7.5 L	-	-	-	-	10.18	nr	**10.18**
Spiraea japonica Little Princess; 30-45 7.5 L	-	-	-	-	9.11	nr	**9.11**
Spiraea Snowmound; 45-60 10 L	-	-	-	-	9.11	nr	**9.11**
Spiraea thunbergii; 45-60 10 L	-	-	-	-	9.11	nr	**9.11**
Bamboos							
Arundinaria auricoma; 40-50 7.5 L	-	-	-	-	19.05	nr	**19.05**
Container grown conifers							
X Cupressocyparis leylandii; 120-150 10 L	-	-	-	-	7.95	nr	**7.95**
X Cupressocyparis leylandii Castlewellan; 120-150 10 L	-	-	-	-	7.95	nr	**7.95**
Juniperus communis Repanda; 30-45 10 L	-	-	-	-	9.02	nr	**9.02**
Juniperus x media pfitzeriana Aurea; 45-60 10 L	-	-	-	-	9.02	nr	**9.02**
Juniperus sabina Tamariscifolia; 45-60 10 L	-	-	-	-	9.02	nr	**9.02**
Pinus mugo; 45-60 10 L	-	-	-	-	13.27	nr	**13.27**
Pinus sylvestris; 90-120 10 L	-	-	-	-	9.02	nr	**9.02**
Pinus sylvestris; 150-180 50 L	-	-	-	-	63.65	nr	**63.65**
Pinus sylvestris; 180-210 80 L	-	-	-	-	111.39	nr	**111.39**
Chamaecyparis lawsoniana Ellwoodii; 30-45 2 L	-	-	-	-	2.34	nr	**2.34**
Chamaecyparis lawsoniana Ellwoods Gold; 30-45 2 L	-	-	-	-	3.08	nr	**3.08**
Chamaecyparis lawsoniana Ellwoods Pillar; 30-45 2 L	-	-	-	-	3.08	nr	**3.08**
X Cupressocyparis leylandii; 60-90 3 L	-	-	-	-	1.91	nr	**1.91**
X Cupressocyparis leylandii Castlewellan; 60-90 3 L	-	-	-	-	1.91	nr	**1.91**
Juniperus communis Compressa; 30-45 2 L	-	-	-	-	2.76	nr	**2.76**
Juniperus communis Repanda; 30-45 3 L	-	-	-	-	2.76	nr	**2.76**
Juniperus conferta; 30-45 3 L	-	-	-	-	2.97	nr	**2.97**
Juniperus media Mint Julep; 30-45 3 L	-	-	-	-	3.08	nr	**3.08**
Juniperus media pfitzeriana; 30-45 3 L	-	-	-	-	2.86	nr	**2.86**
Juniperus media pfitzeriana Aurea; 30-45 3 L	-	-	-	-	3.08	nr	**3.08**
Juniperus sabina Tamariscifolia; 30-45 3 L	-	-	-	-	2.86	nr	**2.86**
Juniperus squamata Blue Star; 30-45 2 L	-	-	-	-	3.40	nr	**3.40**
Larix decidua; 60-90 3 L	-	-	-	-	1.70	nr	**1.70**
Picea albertiana Conica; 20-30 3 L	-	-	-	-	2.97	nr	**2.97**
Pinus mugo; 20-30 2 L	-	-	-	-	2.34	nr	**2.34**
Pinus nigra; 45-60 3 L	-	-	-	-	2.02	nr	**2.02**
Pinus nigra austriaca; 45-60 3 L	-	-	-	-	1.70	nr	**1.70**
Pinus nigra maritima; 45-60 3 L	-	-	-	-	1.70	nr	**1.70**
Pinus sylvestris; 45-60 3 L	-	-	-	-	1.70	nr	**1.70**

Q PAVING/PLANTING/FENCING/SITE FURNITURE

Item Excluding site overheads and profit	PC £	Labour hours	Labour £	Plant £	Material £	Unit	Total rate £
Transplants 45-60							
Acer campestre	-	-	-	-	0.20	nr	0.20
Acer platanoides	-	-	-	-	0.20	nr	0.20
Acer pseudoplatanus	-	-	-	-	0.22	nr	0.22
Alnus cordata	-	-	-	-	0.22	nr	0.22
Alnus glutinosa	-	-	-	-	0.22	nr	0.22
Alnus incana	-	-	-	-	0.22	nr	0.22
Betula pendula	-	-	-	-	0.22	nr	0.22
Carpinus betulus	-	-	-	-	0.25	nr	0.25
Cornus sanguinea	-	-	-	-	0.25	nr	0.25
Corylus avellana	-	-	-	-	0.25	nr	0.25
Crataegus monogyna	-	-	-	-	0.16	nr	0.16
Fagus sylvatica	-	-	-	-	0.27	nr	0.27
Fraxinus excelsior	-	-	-	-	0.25	nr	0.25
Hippophae rhamnoides	-	-	-	-	0.29	nr	0.29
Prunus avium	-	-	-	-	0.25	nr	0.25
Quercus robur	-	-	-	-	0.29	nr	0.29
Rosa rugosa	-	-	-	-	0.27	nr	0.27
Rosa rugosa Alba	-	-	-	-	0.27	nr	0.27
Sambucus nigra	-	-	-	-	0.25	nr	0.25
Sorbus aria	-	-	-	-	0.31	nr	0.31
Sorbus intermedia	-	-	-	-	0.33	nr	0.33
Viburnum opulus	-	-	-	-	0.33	nr	0.33
Transplants 60-90							
Acer campestre	-	-	-	-	0.22	nr	0.22
Acer platanoides	-	-	-	-	0.25	nr	0.25
Alnus cordata	-	-	-	-	0.25	nr	0.25
Amelanchier canadensis	-	-	-	-	0.29	nr	0.29
Cornus alba	-	-	-	-	0.25	nr	0.25
Crataegus monogyna	-	-	-	-	0.25	nr	0.25
Fagus sylvatica	-	-	-	-	0.33	nr	0.33
Fraxinus excelsior	-	-	-	-	0.27	nr	0.27
Populus alba	-	-	-	-	0.27	nr	0.27
Populus nigra italica	-	-	-	-	0.29	nr	0.29
Populus trichocarpa	-	-	-	-	0.29	nr	0.29
Prunus avium	-	-	-	-	0.29	nr	0.29
Salix alba	-	-	-	-	0.22	nr	0.22
Salix caprea	-	-	-	-	0.27	nr	0.27
Salix fragilis	-	-	-	-	0.25	nr	0.25
Salix rosmarinifolia	-	-	-	-	0.25	nr	0.25
Salix viminalis	-	-	-	-	0.25	nr	0.25
Salix vitellina	-	-	-	-	0.25	nr	0.25
Sorbus aucuparia	-	-	-	-	0.27	nr	0.27
Herbaceous							
Achillea fil. Cloth of Gold; 10-15 2 L	-	-	-	-	1.92	nr	1.92
Alchemilla mollis; 10-15 2 L	-	-	-	-	1.92	nr	1.92
Anemone hybrida Honorine Jobert; 10-15 3 L	-	-	-	-	2.35	nr	2.35
Artemisia absin. Lambrook Silver; 10-15 3 L	-	-	-	-	2.62	nr	2.62
Artemisia Powis Castle; 10-15 3 L	-	-	-	-	2.62	nr	2.62
Bergenia Bressingham Salmon; 20-30 2 L	-	-	-	-	2.40	nr	2.40
Convallaria majalis; 10-15 2 L	-	-	-	-	1.92	nr	1.92
Euphorbia amygdaloides Purpurea; 15-20 2 L	-	-	-	-	2.35	nr	2.35
Euphorbia wulfenii; 10-15 2 L	-	-	-	-	2.77	nr	2.77
Geranium oxonianum Wargrave Pink; 10-15 3 L	-	-	-	-	1.92	nr	1.92
Geranium Johnson Blue; 10-15 3 L	-	-	-	-	1.92	nr	1.92
Hosta August Moon; 10-15 2 L	-	-	-	-	2.67	nr	2.67
Hosta sieboldii; 10-15 2 L	-	-	-	-	2.67	nr	2.67
Iris foetidissima; 10-15 2 L	-	-	-	-	2.35	nr	2.35
Iris foetidissima Eloise Garrison; 10-15 2 L	-	-	-	-	2.35	nr	2.35
Papaver orientale; 10-15 3 L	-	-	-	-	1.92	nr	1.92
Penstemon Garnet; 10-15 3 L	-	-	-	-	2.40	nr	2.40
Pulmonaria angustifolia; 10-15 2 L	-	-	-	-	1.92	nr	1.92
Salvia East Friesland; 10-15 3 L	-	-	-	-	1.92	nr	1.92
Sedum spectabilis Autumn Joy; 10-15 3 L	-	-	-	-	1.92	nr	1.92
Tiarella cordifolia; 10-15 2 L	-	-	-	-	1.92	nr	1.92
Waldsteinia ternata; 10-15 2 L	-	-	-	-	1.92	nr	1.92

Q PAVING/PLANTING/FENCING/SITE FURNITURE

Item Excluding site overheads and profit	PC £	Labour hours	Labour £	Plant £	Material £	Unit	Total rate £
Q31 PLANTING - cont'd							
Market prices of trees, shrubs and							
plants: Notcutts Nurseries Ltd - cont'd							
Bare root trees 1.50 - 1.80 feathered							
Acer campestre	-	-	-	-	2.14	nr	2.14
Acer platanoides	-	-	-	-	1.61	nr	1.61
Acer pseudoplatanus	-	-	-	-	1.61	nr	1.61
Alnus cordata	-	-	-	-	1.61	nr	1.61
Alnus glutinosa	-	-	-	-	1.61	nr	1.61
Alnus incana	-	-	-	-	3.21	nr	3.21
Amelanchier canadensis	-	-	-	-	6.43	nr	6.43
Betula pendula	-	-	-	-	1.61	nr	1.61
Carpinus betulus	-	-	-	-	2.68	nr	2.68
Fagus sylvatica	-	-	-	-	3.75	nr	3.75
Fagus sylvatica purpurea	-	-	-	-	5.89	nr	5.89
Fraxinus excelsior	-	-	-	-	2.68	nr	2.68
Populus alba	-	-	-	-	1.61	nr	1.61
Populus nigra italica	-	-	-	-	1.61	nr	1.61
Populus tremula	-	-	-	-	5.89	nr	5.89
Prunus avium	-	-	-	-	2.68	nr	2.68
Prunus avium 'Plena'	-	-	-	-	6.43	nr	6.43
Quercus robur	-	-	-	-	10.18	nr	10.18
Quercus rubra	-	-	-	-	4.82	nr	4.82
Sorbus lutescens	-	-	-	-	3.21	nr	3.21
Sorbus aucuparia	-	-	-	-	2.14	nr	2.14
Bare root trees 1.80 - 2.10 feathered							
Acer campestre	-	-	-	-	2.14	nr	2.14
Alnus cordata	-	-	-	-	3.21	nr	3.21
Acer platanoides	-	-	-	-	3.21	nr	3.21
Acer pseudoplatanus	-	-	-	-	3.21	nr	3.21
Alnus glutinosa	-	-	-	-	3.21	nr	3.21
Alnus incana	-	-	-	-	2.68	nr	2.68
Amelanchier canadensis	-	-	-	-	12.85	nr	12.85
Betula pendula	-	-	-	-	3.21	nr	3.21
Carpinus betulus	-	-	-	-	4.82	nr	4.82
Fagus sylvatica	-	-	-	-	5.89	nr	5.89
Fagus sylvatica purpurea	-	-	-	-	10.18	nr	10.18
Fraxinus excelsior	-	-	-	-	4.28	nr	4.28
Populus alba	-	-	-	-	3.75	nr	3.75
Populus nigra italica	-	-	-	-	3.75	nr	3.75
Populus tremula	-	-	-	-	3.75	nr	3.75
Prunus avium	-	-	-	-	4.28	nr	4.28
Prunus avium 'Plena'	-	-	-	-	8.57	nr	8.57
Quercus robur	-	-	-	-	10.18	nr	10.18
Quercus rubra	-	-	-	-	11.25	nr	11.25
Sorbus aucuparia	-	-	-	-	4.82	nr	4.82
Sorbus lutescens	-	-	-	-	9.64	nr	9.64
Bare root trees 8-10 cm							
Acer campestre	-	-	-	-	11.78	nr	11.78
Acer platanoides	-	-	-	-	9.64	nr	9.64
Acer pseudoplatanus	-	-	-	-	9.64	nr	9.64
Alnus cordata	-	-	-	-	9.64	nr	9.64
Alnus glutinosa	-	-	-	-	9.64	nr	9.64
Alnus incana	-	-	-	-	9.64	nr	9.64
Amelanchier canadensis	-	-	-	-	26.78	nr	26.78
Betula pendula	-	-	-	-	8.57	nr	8.57
Carpinus betulus	-	-	-	-	12.85	nr	12.85
Fagus sylvatica	-	-	-	-	12.85	nr	12.85
Fraxinus excelsior	-	-	-	-	13.50	nr	13.50
Fagus sylvatica Purpurea	-	-	-	-	21.42	nr	21.42
Populus tremula	-	-	-	-	11.78	nr	11.78
Populus alba	-	-	-	-	10.18	nr	10.18
Populus nigra Italica	-	-	-	-	10.18	nr	10.18

Q PAVING/PLANTING/FENCING/SITE FURNITURE

Item Excluding site overheads and profit	PC £	Labour hours	Labour £	Plant £	Material £	Unit	Total rate £
Prunus avium	-	-	-	-	10.71	nr	10.71
Prunus avium Plena	-	-	-	-	13.93	nr	13.93
Quercus robor	-	-	-	-	12.34	nr	12.34
Quercus rubra	-	-	-	-	18.21	nr	18.21
Sorbus aucuparia	-	-	-	-	11.78	nr	11.78
Bare root trees 10-12 cm sel. standard							
Acer campestre	-	-	-	-	15.00	nr	15.00
Alnus cordata	-	-	-	-	10.71	nr	10.71
Acer platanoides	-	-	-	-	10.71	nr	10.71
Acer pseudoplatanus	-	-	-	-	10.71	nr	10.71
Amelanchier canadensis	-	-	-	-	42.85	nr	42.85
Betula pendula	-	-	-	-	10.71	nr	10.71
Carpinus betulus	-	-	-	-	27.50	nr	27.50
Fagus sylvatica	-	-	-	-	19.28	nr	19.28
Fagus sylvatica purpurea	-	-	-	-	32.14	nr	32.14
Fraxinus excelsior	-	-	-	-	12.85	nr	12.85
Populus alba	-	-	-	-	13.39	nr	13.39
Populus nigra italica	-	-	-	-	13.39	nr	13.39
Populus tremula	-	-	-	-	17.14	nr	17.14
Prunus avium	-	-	-	-	12.85	nr	12.85
Prunus avium 'Plena'	-	-	-	-	17.14	nr	17.14
Quercus robur	-	-	-	-	22.50	nr	22.50
Quercus rubra	-	-	-	-	22.50	nr	22.50
Sorbus aucuparia	-	-	-	-	16.07	nr	16.07
Sorbus lutescens	-	-	-	-	19.28	nr	19.28
Bare root trees 12-14 cm heavy standard							
Acer campestre	-	-	-	-	25.71	nr	25.71
Alnus cordata	-	-	-	-	21.42	nr	21.42
Acer platanoides	-	-	-	-	21.42	nr	21.42
Acer pseudoplatanus	-	-	-	-	21.42	nr	21.42
Alnus glutinosa	-	-	-	-	21.42	nr	21.42
Alnus incana	-	-	-	-	21.42	nr	21.42
Amelanchier canadensis	-	-	-	-	74.98	nr	74.98
Betula pendula	-	-	-	-	21.42	nr	21.42
Carpinus betulus	-	-	-	-	29.99	nr	29.99
Fagus sylvatica	-	-	-	-	29.99	nr	29.99
Fagus sylvatica Purpurea	-	-	-	-	48.20	nr	48.20
Fraxinus excelsior	-	-	-	-	25.71	nr	25.71
Populus alba	-	-	-	-	23.57	nr	23.57
Populus nigra italica	-	-	-	-	13.39	nr	13.39
Populus tremula	-	-	-	-	17.14	nr	17.14
Prunus avium	-	-	-	-	25.71	nr	25.71
Prunus avium 'Plena'	-	-	-	-	29.99	nr	29.99
Quercus robur	-	-	-	-	34.28	nr	34.28
Quercus rubra	-	-	-	-	34.28	nr	34.28
Sorbus aucuparia	-	-	-	-	23.57	nr	23.57
Sorbus lutescens	-	-	-	-	32.14	nr	32.14
Market prices of trees; Coblands **Nurseries Ltd**							
Bare root trees 14-16 cm extra heavy standard							
Acer campestre	-	-	-	-	55.00	nr	55.00
Acer platanoides	-	-	-	-	40.00	nr	40.00
Acer pseudoplatanus	-	-	-	-	40.00	nr	40.00
Alnus glutinosa	-	-	-	-	40.00	nr	40.00
Alnus incana	-	-	-	-	115.00	nr	115.00
Amelanchier lamarckii	-	-	-	-	55.00	nr	55.00
Betula pendula	-	-	-	-	40.00	nr	40.00
Carpinus betulus	-	-	-	-	55.00	nr	55.00
Fagus sylvatica	-	-	-	-	55.00	nr	55.00
Fagus sylvatica Dawyck purple	-	-	-	-	115.00	nr	115.00
Fraxinus excelsior	-	-	-	-	50.00	nr	50.00
Populus tremula	-	-	-	-	50.00	nr	50.00
Prunus avium	-	-	-	-	40.00	nr	40.00
Prunus avium 'Plena'	-	-	-	-	55.00	nr	55.00

Q PAVING/PLANTING/FENCING/SITE FURNITURE

Item Excluding site overheads and profit	PC £	Labour hours	Labour £	Plant £	Material £	Unit	Total rate £
Q31 PLANTING - cont'd							
Market prices of trees; Coblands							
Nurseries Ltd - cont'd							
Quercus robur	-	-	-	-	50.00	nr	**50.00**
Quercus rubra	-	-	-	-	55.00	nr	**55.00**
Robinia pseudoacasia Frisia	-	-	-	-	115.00	nr	**115.00**
Sorbus aucuparia	-	-	-	-	40.00	nr	**40.00**
Sorbus aria	-	-	-	-	50.00	nr	**50.00**
Root ball trees; extra over for root balling and							
wrapping in hessian	-	-	-	-	30.00	nr	**30.00**
Market prices of aquatic plants in							
prepared growing medium of approved							
hessian bag filled with soil and fertilizer;							
tied, slit and weighted and placed in							
position in pool to receive three							
submerged floating leafed marginal or							
swamp plants; Anglo Aquarium Plant Co Ltd							
Water lilies							
Nymphaea in variety - white	-	-	-	-	7.95	nr	**7.95**
Nymphaea in variety - deep red	-	-	-	-	7.95	nr	**7.95**
Deep water aquatics							
Aponogeton distachyos	-	-	-	-	2.75	nr	**2.75**
Nymphoides peltata	-	-	-	-	2.75	nr	**2.75**
Marginal plants							
Acorus calamus	-	-	-	-	1.90	nr	**1.90**
Butomus umbellatus	-	-	-	-	1.90	nr	**1.90**
Caltha palustris	-	-	-	-	1.90	nr	**1.90**
Carex acuta	-	-	-	-	1.90	nr	**1.90**
Iris pseudacorus	-	-	-	-	1.90	nr	**1.90**
Juncus effusus	-	-	-	-	1.90	nr	**1.90**
Mentha aquatica	-	-	-	-	1.90	nr	**1.90**
Mimulus luteus	-	-	-	-	1.90	nr	**1.90**
Phragmites (australis)	-	-	-	-	1.90	nr	**1.90**
Ranunculus lingua	-	-	-	-	1.90	nr	**1.90**
Scirpus lacustris	-	-	-	-	1.90	nr	**1.90**
Sagittaria sagittifolia var. leucopetala	-	-	-	-	1.90	nr	**1.90**
Typha latifolia	-	-	-	-	1.90	nr	**1.90**
Submerged oxygenating plants							
Ceratophyllum demersum; containerised	-	-	-	-	36.00	100	**36.00**
Lagarosiphon major (Elodea crispa);							
containerised	-	-	-	-	36.00	100	**36.00**
Ranunculus aquatilis; containerised	-	-	-	-	36.00	100	**36.00**
Market Prices of bulbs; prices per each;							
purchased in the quantities shown;							
Autumn Bulbs							
Dutch Hyacinths (prepared for early flowering)							
"Aiolos", white	-	-	-	-	0.30	100	**0.30**
"Anna Marie", clear pink	-	-	-	-	0.30	100	**0.30**
"Carnegie", white, dense spike	-	-	-	-	0.30	100	**0.30**
"Ostara", deep blue	-	-	-	-	0.32	100	**0.32**
"City of Haarlem", primrose yellow	-	-	-	-	0.30	100	**0.30**
Dutch Hyacinths (not prepared); 15 to 16 cm							
"Amsterdam", cerise red, Medium	-	-	-	-	0.22	100	**0.22**
"Anna Marie", clear pink, Early	-	-	-	-	0.20	100	**0.20**
"Carnegie", white, dense spike, Late	-	-	-	-	0.20	100	**0.20**
"Marie", deep blue, Early	-	-	-	-	0.20	100	**0.20**
"City of Haarlem", primrose yellow, Late	-	-	-	-	0.20	100	**0.20**
Tulips, Single Early,							
"Apricot Beauty", soft salmon rose	-	-	-	-	0.10	1000	**0.10**
"Best Seller", bright coppery orange	-	-	-	-	0.11	1000	**0.11**
Mixture, blended colours	-	-	-	-	0.09	1000	**0.09**

Q PAVING/PLANTING/FENCING/SITE FURNITURE

Item Excluding site overheads and profit	PC £	Labour hours	Labour £	Plant £	Material £	Unit	Total rate £
Tulips, Double Early							
"Abba", glowing tomato red	-	-	-	-	0.09	1000	0.09
"Electra", deep cherry red	-	-	-	-	0.12	1000	0.12
"Murillo Sports Mixed", selected species	-	-	-	-	0.11	1000	0.11
Tulips, Triumph							
"Abu Hassan", deep mahogany, broad yellow edge	-	-	-	-	0.09	1000	0.09
"Arabian Mystery", deep purple violet, edged white	-	-	-	-	0.11	1000	0.11
Mixed, blended colours	-	-	-	-	0.08	1000	0.08
Tulips, Darwin Hybrid							
"Ad Rem", scarlet	-	-	-	-	0.09	1000	0.09
"Golden Parade", deep yellow	-	-	-	-	0.09	1000	0.09
Mixed, blended colours	-	-	-	-	0.08	1000	0.08
Tulips, Single Late or Cottage							
"Alabaster", pure white	-	-	-	-	0.10	1000	0.10
"Blushing Lady", red, feathered rosy white	-	-	-	-	0.12	1000	0.12
Mixed, blended colours	-	-	-	-	0.09	1000	0.09
Tulips, Lily Flowered							
"Aladdin", scarlet, edged yellow	-	-	-	-	0.09	1000	0.09
Mixed, blended colours	-	-	-	-	0.10	1000	0.10
Tulips, Parrot							
"Apricot Parrot", pale apricot yellow, tinged creamy white	-	-	-	-	0.12	1000	0.12
"Rococo", carmine edged glowing red	-	-	-	-	0.09	1000	0.09
Mixture, blended colours	-	-	-	-	0.12	1000	0.12
Tulips, Double Late							
"Lilac Perfection", lilac	-	-	-	-	0.14	1000	0.14
"Upstar", pale pink and white	-	-	-	-	0.09	1000	0.09
"Superb Mixed", blended colours	-	-	-	-	0.10	1000	0.10
Tulips, Greigii							
"Ali Baba", rose	-	-	-	-	0.08	1000	0.08
"Mary Ann", carmine red, edged white	-	-	-	-	0.09	1000	0.09
Mixed Hybrids	-	-	-	-	0.08	1000	0.08
Daffodils, Yellow Trumpets; Top Size DN2							
"Dutch Master", large golden yellow	-	-	-	-	0.06	1000	0.06
"King Alfred", medium golden yellow	-	-	-	-	0.06	1000	0.06
Daffodils, Bi-Coloured Trumpets							
"Magnet", perianth creamy white, lemon yellow trumpet	-	-	-	-	0.11	1000	0.11
Daffodils, White Trumpets							
"Mount Hood", pure white perianth, pure white frilled trumpet	-	-	-	-	0.11	1000	0.11
Narcissi, Large Cupped; Top Size DN2							
"Carbineer", rich yellow perianth, bright orange cup	-	-	-	-	0.07	1000	0.07
"St. Patricks Day", lemon yellow	-	-	-	-	0.08	1000	0.08
"Pink Pride", white perianth, large pink rimmed cup	-	-	-	-	0.10	1000	0.10
Narcissi, Small Cupped							
"Aflame", creamy white perianth, orange scarlet crown	-	-	-	-	0.09	1000	0.09
"Verger", white perianth, orange scarlet cup	-	-	-	-	0.09	1000	0.09
Narcissi, Double							
"Flower Drift", white perianth, orange yellow cup	-	-	-	-	0.10	1000	0.10
"Texas", yellow outer petals, inter-mixed with orange scarlet	-	-	-	-	0.11	1000	0.11
Narcissi, Triandrus							
"Ice Wings", multi-headed white	-	-	-	-	0.16	1000	0.16
"Thalia", 2 or 4 white flowers on a stem	-	-	-	-	0.10	1000	0.10
Narcissi, Cyclamineus							
"Jack Snipe", long creamy white reflexing petals, short orange yellow cup	-	-	-	-	0.10	1000	0.10
Narcissi, Jonquilla							
"Jonquilla Single", buttercup yellow flowers, sweet scented	-	-	-	-	0.06	1000	0.06
"Martinette", yellow perianth, orange cup	-	-	-	-	0.07	1000	0.07

Q PAVING/PLANTING/FENCING/SITE FURNITURE

Item Excluding site overheads and profit	PC £	Labour hours	Labour £	Plant £	Material £	Unit	Total rate £
Q31 PLANTING - cont'd							
Market Prices of bulbs - cont'd							
Narcissi, Tazetta or Poetaz							
"Cheerfulness", creamy white double flowers, scented	-	-	-	-	0.06	1000	0.06
"Scarlett Gem", deep lemon yellow perianth, brilliant bright red eye, four to six flowers on stem	-	-	-	-	0.11	1000	0.11
Daffodils, Miniature							
"Little Gem", yellow perianth and trumpet, smallest of all trumpet daffodils	-	-	-	-	0.07	1000	0.07
"Minnow", miniature tazetta hybrid, 2-3 soft flowers on each stem	-	-	-	-	0.06	1000	0.06
"Nanus", a delightful yellow miniature	-	-	-	-	0.08	1000	0.08
Daffodils & Narcissi, Special Mixture							
suitable for planting in parkland, river banks etc.	-	-	-	-	0.07	1000	0.07
Crocus, Spring Flowering Dutch; 8/9cm							
Blue	-	-	-	-	0.04	1000	0.04
Yellow	-	-	-	-	0.04	1000	0.04
Striped	-	-	-	-	0.05	1000	0.05
Mixed	-	-	-	-	0.04	1000	0.04
Crocus, Winter Flowering Species;							
"Ancyrensis" (Golden Bunch), maize yellow with orange interior	-	-	-	-	0.03	1000	0.03
"Chrysanthus Advance", yellow and violet	-	-	-	-	0.04	1000	0.04
"Sieberi Violet Queen", dark blue	-	-	-	-	0.04	1000	0.04
Species Mixed, blended colours	-	-	-	-	0.03	1000	0.03
Iris, Dutch; 7/8cm							
"Golden Harvest", golden yellow	-	-	-	-	0.04	1000	0.04
"Frans Hals", standard violet, falls violet with bronze, blotch aureolin	-	-	-	-	0.06	1000	0.06
"Superb Mixed", blended colours	-	-	-	-	0.04	1000	0.04
Iris, Species							
"Danfordiae", sweet scented, deep lemon yellow with brown markings; dwarf variety	-	-	-	-	0.04	1000	0.04
"Reticulata", deep purple violet with yellow blotch; dwarf variety	-	-	-	-	0.03	1000	0.03
"Reticulata Harmony", royal blue with yellow rimmed white blotch	-	-	-	-	0.03	1000	0.03
"Reticulata Seedlings Mixed", blended colours	-	-	-	-	0.04	1000	0.04
Market Prices of bulbs; prices per each; purchased in the quantities shown; Spring Bulbs							
Allium							
"Giganteum", deep violet, star shaped flowers, height 80 cm	-	-	-	-	0.17	100	0.17
"Neapolitanum", pure white, height 30 cm	-	-	-	-	0 02	1000	0.02
Amaryllis Hippeastrum; 32+cm							
"Apple Blossom", blushing pink	-	-	-	-	2.20	10	2.20
"Minerva" striped, red and white	-	-	-	-	2.20	10	2.20
Anemones; 5/6cm							
"Single De Caen", mixed shades	-	-	-	-	0.03	1000	0.03
"Blanda Pink Star", phlox purple	-	-	-	-	0.08	1000	0.08
Freesia							
Single Flowering Mixture	-	-	-	-	0.05	1000	0.05
Double Flowering Mixture	-	-	-	-	0.05	1000	0.05
Galanthus (Snowdrop)							
"Nivalis Double"	-	-	-	-	0.14	1000	0.14
"Nivalis Single"	-	-	-	-	0.06	1000	0.06
"Nivalis Single", supplied 'in the green'	-	-	-	-	0.06	1000	0.06
Muscari (Grape Hyacinth)							
"Armeniacum", bright blue	-	-	-	-	0.03	1000	0.03
"Botryoides Album", white	-	-	-	-	0.05	1000	0.05

Q PAVING/PLANTING/FENCING/SITE FURNITURE

Item Excluding site overheads and profit	PC £	Labour hours	Labour £	Plant £	Material £	Unit	Total rate £
Ranunculus							
Paeony Flowering Mixed	-	-	-	-	0.05	1000	0.05
Scilla							
"Campanulata Blue"	-	-	-	-	0.68	1000	0.68
Sparaxis							
"Tri Colored Mixed"	-	-	-	-	0.02	1000	0.02
Planting - General							
Preamble: The British Standard recommendations for Nursery Stock are in BS 3936: pts 1-5. Planting of trees and shrubs as well as forestry is covered by the appropriate sections of BS 4428. Transplanting of semi-mature trees is covered by BS 4043. Prices for all planting work are deemed to include carrying out planting in accordance with BS 4428 and good horticultural practice.							
Site protection; temporary protective fencing							
Cleft chestnut rolled fencing; to 100 mm diameter chestnut posts; driving into firm ground at 3 m centres; pales at 50 mm centres							
900 mm high	3.11	0.11	2.00	-	5.77	m	7.77
1100 mm high	3.65	0.11	2.00	-	6.16	m	8.16
1500 mm high	5.80	0.11	2.00	-	8.30	m	10.30
Extra over temporary protective fencing for removing and making good (no allowance for re-use of material)	-	0.07	1.25	0.19	-	m	1.44
Cultivation							
Treating soil with "Paraquat-Diquat" weedkiller at rate of 5 litre/ha; PC £22.20 / litre; in accordance with manufacturer's instructions; including all safety precautions							
by machine	-	-	-	0.24	1.11	100 m²	1.35
by hand	1.11	0.13	2.50	-	1.11	100 m²	3.61
Ripping up subsoil; using approved subsoiling machine; minimum depth 250 mm below topsoil; at 1.20 m centres; in							
gravel or sandy clay	-	-	-	2.34	-	100m²	2.34
soil compacted by machines	-	-	-	2.73	-	100m²	2.73
clay	-	-	-	2.92	-	100m²	2.92
chalk or other soft rock	-	-	-	5.84	-	100m²	5.84
Extra for subsoiling at 1 m centres	-	-	-	0.58	-	100m²	0.58
Breaking up existing ground; using pedestrian operated tine cultivator or rotavator							
100 mm deep	-	0.22	4.13	2.13	-	100m²	6.26
150 mm deep	-	0.28	5.16	2.67	-	100m²	7.82
200 mm deep	-	0.37	6.87	3.56	-	100m²	10.43
As above but in heavy clay or wet soils							
100 mm deep	-	0.44	8.25	4.27	-	100m²	12.52
150 mm deep	-	0.66	12.38	6.40	-	100m²	18.77
200 mm deep	-	0.82	15.47	8.00	-	100m²	23.47
Breaking up existing ground; using tractor drawn tine cultivator or rotavator							
100 mm deep	-	-	-	0.57	-	100 m²	0.57
150 mm deep	-	-	-	0.71	-	100 m²	0.71
200 mm deep	-	-	-	0.95	-	100 m²	0.95
600 mm deep	-	-	-	2.86	-	100 m²	2.86
Cultivating ploughed ground; using disc, drag, or chain harrow							
4 passes	-	-	-	3.43	-	100 m²	3.43
Rolling cultivated ground lightly; using self-propelled agricultural roller	-	0.06	1.04	0.58	-	100 m²	1.63

Q PAVING/PLANTING/FENCING/SITE FURNITURE

Item Excluding site overheads and profit	PC £	Labour hours	Labour £	Plant £	Material £	Unit	Total rate £
Q31 PLANTING - cont'd							
Cultivation - cont'd							
Importing only selected and approved topsoil; to BS 3882; from source not exceeding 13 km from site							
1 - 14 m³	17.32	-	-	-	51.96	m³	**51.96**
over 15 m³	17.32	-	-	-	17.32	m³	**17.32**
Spreading and lightly consolidating approved topsoil (imported or from spoil heaps); in layers not exceeding 150 mm; travel distance from spoil heaps not exceeding 100 m; by machine (imported topsoil not included)							
minimum depth 100 mm	-	1.55	29.06	37.20	-	100 m²	**66.26**
minimum depth 150 mm	-	2.33	43.75	55.98	-	100 m²	**99.73**
minimum depth 300 mm	-	4.67	87.50	111.96	-	100 m²	**199.46**
minimum depth 450 mm	-	6.99	131.06	167.76	-	100 m²	**298.82**
Spreading and lightly consolidating approved topsoil (imported or from spoil heaps); in layers not exceeding 150 mm; travel distance from spoil heaps not exceeding 100 m; by hand (imported topsoil not included)							
minimum depth 100 mm	-	20.00	375.07	-	-	100 m²	**375.07**
minimum depth 150 mm	-	30.01	562.61	-	-	100 m²	**562.61**
minimum depth 300 mm	-	60.01	1125.22	-	-	100 m²	**1125.22**
minimum depth 450 mm	-	90.02	1687.84	-	-	100 m²	**1687.84**
Extra over for spreading topsoil to slopes 15 - 30 degrees by machine or hand	-	-	-	-	-	-	**10%**
Extra over for spreading topsoil to slopes over 30 degrees by machine or hand	-	-	-	-	-	-	**25%**
Extra over spreading topsoil for travel exceeding 100 m; by machine							
100 - 150 m	-	0.02	0.38	0.12	-	m³	**0.50**
150 - 200 m	-	0.03	0.50	0.16	-	m³	**0.66**
200 - 300 m	-	0.04	0.67	0.21	-	m³	**0.88**
Extra over spreading topsoil for travel exceeding 100 m; by hand							
100 m	-	2.50	46.88	-	-	m³	**46.88**
200 m	-	3.50	65.63	-	-	m³	**65.63**
300 m	-	4.50	84.38	-	-	m³	**84.38**
Evenly grading; to general surfaces to bring to finished levels							
by machine (tractor mounted rotavator)	-	-	-	0.03	-	m²	**0.03**
by pedestrian operated rotavator	-	-	0.08	0.04	-	m²	**0.12**
by hand	-	0.01	0.19	-	-	m²	**0.19**
Extra over grading for slopes 15 - 30 degrees by machine or hand	-	-	-	-	-	-	**10%**
Extra over grading for slopes over 30 degrees by machine or hand	-	-	-	-	-	-	**25%**
Clearing stones; disposing off site; to distance not exceeding 13 km							
by hand; stones not exceeding 50 mm in any direction; loading to skip 5.35 m³	-	0.01	0.19	0.03	-	m²	**0.21**
by mechanical stone rake; stones not exceeding 50 mm in any direction; loading to 15 m³ truck by mechanical loader	-	-	0.04	0.07	-	m²	**0.11**
Lightly cultivating; weeding; to fallow areas; disposing debris off site; to distance not exceeding 13 km							
by hand	-	0.01	0.27	-	0.06	m²	**0.33**

Q PAVING/PLANTING/FENCING/SITE FURNITURE

Item Excluding site overheads and profit	PC £	Labour hours	Labour £	Plant £	Material £	Unit	Total rate £
Preparation of planting operations							
For the following topsoil improvement and							
planting operations add or subtract the following							
amounts for every £0.10 difference in the material							
cost price							
35 g/m^2	-	-	-	-	0.35	100 m^2	0.35
50 g/m^2	-	-	-	-	0.50	100 m^2	0.50
70 g/m^2	-	-	-	-	0.70	100 m^2	0.70
100 g/m^2	-	-	-	-	1.00	100 m^2	1.00
150 kg/ ha	-	-	-	-	15.00	ha	15.00
200 kg/ ha	-	-	-	-	20.00	ha	20.00
250 kg/ ha	-	-	-	-	25.00	ha	25.00
300 kg/ ha	-	-	-	-	30.00	ha	30.00
400 kg/ ha	-	-	-	-	40.00	ha	40.00
500 kg/ ha	-	-	-	-	50.00	ha	50.00
700 kg/ ha	-	-	-	-	70.00	ha	70.00
1000 kg/ ha	-	-	-	-	100.00	ha	100.00
1250 kg/ ha	-	-	-	-	125.00	ha	125.00
Selective herbicides; in accordance with							
manufacturer's instructions; PC £24.54 per litre;							
by machine; application rate							
30 ml/100m^2	-	-	-	0.29	0.74	100 m^2	1.03
35 ml/100m^2	-	-	-	0.29	0.86	100 m^2	1.15
40 ml/100m^2	-	-	-	0.29	0.98	100 m^2	1.27
50 ml/100m^2	-	-	-	0.29	1.23	100 m^2	1.52
3.00 l/ha	-	-	-	29.17	73.64	ha	102.81
3.50 l/ ha	-	-	-	29.17	85.91	ha	115.08
4.00 l/ha	-	-	-	29.17	98.18	ha	127.35
Selective herbicides; in accordance with							
manufacturer's instructions; PC £24.54 per litre;							
by hand; application rate							
30 ml/m^2	-	0.17	3.13	-	0.74	100 m^2	3.86
35 ml/m^2	-	0.17	3.13	-	0.86	100 m^2	3.98
40 ml/m^2	-	0.17	3.13	-	0.98	100 m^2	4.11
50 ml/m^2	-	0.17	3.13	-	1.23	100 m^2	4.35
3.00 l/ha	-	16.67	312.50	-	73.64	ha	386.14
3.50 l/ha	-	16.67	312.50	-	85.91	ha	398.41
4.00 l/ha	-	16.67	312.51	-	98.18	ha	410.69
General herbicides; in accordance with							
manufacturer's instructions; "Knapsack" spray							
application							
"Super Verdone" at 100 ml/100 m^2	0.60	0.05	0.94	-	0.60	100 m^2	1.53
"Roundup Pro" at 40 ml/100 m^2	0.37	0.05	0.94	-	0.37	100 m^2	1.31
"Spasor" at 50 ml/100 m^2	0.70	0.05	0.94	-	0.70	100 m^2	1.64
"Casoron G (residual)" at 1 kg/125 m^2	2.82	0.05	0.94	-	3.10	100 m^2	4.04
Fertilizers; in top 150 mm of topsoil; at 35g/m^2							
fertilizer (18:0:0+Mg +Fe)	6.88	0.12	2.30	-	6.88	100 m^2	9.17
"Enmag"	5.88	0.12	2.30	-	6.17	100 m^2	8.47
fertilizer (7:7:7)	1.56	0.12	2.30	-	1.56	100 m^2	3.85
"Superphosphate"	1.46	0.12	2.30	-	1.53	100 m^2	3.83
fertilizer (20:10:10)	2.20	0.12	2.30	-	2.20	100 m^2	4.50
"Hoof and Horn"	2.01	0.12	2.30	-	2.11	100 m^2	4.41
"Bone meal"	2.79	0.12	2.30	-	2.93	100 m^2	5.23
Fertilizers; in top 150 mm of topsoil at 70 g/m^2							
fertilizer (18:0:0+Mg +Fe)	13.76	0.12	2.30	-	13.76	100 m^2	16.05
"Enmag"	11.76	0.12	2.30	-	12.35	100 m^2	14.64
fertilizer (7:7:7)	3.11	0.12	2.30	-	3.11	100 m^2	5.41
"Superphosphate"	2.92	0.12	2.30	-	3.07	100 m^2	5.36
fertilizer (20:10:10)	4.40	0.12	2.30	-	4.40	100 m^2	6.70
"Hoof and Horn"	4.02	0.12	2.30	-	4.22	100 m^2	6.52
"Bone meal"	5.58	0.12	2.30	-	5.86	100 m^2	8.16

Q PAVING/PLANTING/FENCING/SITE FURNITURE

Item Excluding site overheads and profit	PC £	Labour hours	Labour £	Plant £	Material £	Unit	Total rate £
Q31 PLANTING - cont'd							
Preparation of planting operations - cont'd							
Spreading topsoil; from dump not exceeding 100 m distance; by machine (topsoil not included)							
at 1 m³ per 13 m²; 75 mm thick	-	0.56	10.42	28.09	-	100 m²	**38.50**
at 1 m³ per 10 m²; 100 mm thick	-	0.89	16.66	28.89	-	100 m²	**45.55**
at 1 m³ per 6.50 m²; 150 mm thick	-	1.11	20.81	50.62	-	100 m²	**71.43**
at 1 m³ per 5 m²; 200 mm thick	-	1.48	27.75	67.29	-	100 m²	**95.04**
Spreading topsoil; from dump not exceeding 100 m distance; by hand (topsoil not included)							
at 1 m³ per 13 m²; 75 mm thick	-	15.00	281.31	-	-	100 m²	**281.31**
at 1 m³ per 10 m²; 100 mm thick	-	20.00	375.07	-	-	100 m²	**375.07**
at 1 m³ per 6.50 m²; 150 mm thick	-	30.01	562.61	-	-	100 m²	**562.61**
at 1 m³ per 5 m²; 200 mm thick	-	36.67	687.64	-	-	100 m²	**687.64**
Imported topsoil; tipped 100 m from area of application; by machine							
at 1 m³ per 13 m²; 75 mm thick	129.90	0.56	10.42	28.09	155.88	100 m²	**194.38**
at 1 m³ per 10 m²; 100 mm thick	173.20	0.89	16.66	28.89	207.84	100 m²	**253.39**
at 1 m³ per 6.50 m²; 150 mm thick	259.80	1.11	20.81	50.62	311.76	100 m²	**383.19**
at 1 m³ per 5 m²; 200 mm thick	346.40	1.48	27.75	67.29	415.68	100 m²	**510.72**
Imported topsoil; tipped 100 m from area of application; by hand							
at 1 m³ per 13 m²; 75 mm thick	129.90	15.00	281.31	-	155.88	100 m²	**437.19**
at 1 m³ per 10 m²; 100 mm thick	173.20	20.00	375.07	-	207.84	100 m²	**582.91**
at 1 m³ per 6.50 m²; 150 mm thick	259.80	30.01	562.61	-	311.76	100 m²	**874.37**
at 1 m³ per 5 m²; 200 mm thick	346.40	36.67	687.64	-	415.68	100 m²	**1103.32**
Composted bark and manure soil conditioner (20 m³ loads); from not further than 25m from location; cultivating into topsoil by machine							
50 mm thick	131.00	2.86	53.57	1.78	137.55	100 m²	**192.90**
100 mm thick	262.00	6.05	113.42	1.78	275.10	100 m²	**390.30**
150 mm thick	393.00	8.90	166.97	1.78	412.65	100 m²	**581.39**
200 mm thick	524.00	12.90	241.97	1.78	550.20	100 m²	**793.94**
Composted bark soil conditioner (20 m³ loads); placing on beds by mechanical loader; spreading and rotavating into topsoil by machine							
50 mm thick	131.00	-	-	8.09	131.00	100 m²	**139.09**
100 mm thick	262.00	-	-	12.33	275.10	100 m²	**287.43**
150 mm thick	393.00	-	-	16.75	412.65	100 m²	**429.40**
200 mm thick	524.00	-	-	21.09	550.20	100 m²	**571.29**
Mushroom compost (20 m³ loads); from not further than 25m from location; cultivating into topsoil by machine							
50 mm thick	90.50	2.86	53.57	1.78	90.50	100 m²	**145.85**
100 mm thick	181.00	6.05	113.42	1.78	181.00	100 m²	**296.20**
150 mm thick	271.50	8.90	166.97	1.78	271.50	100 m²	**440.24**
200 mm thick	362.00	12.90	241.97	1.78	362.00	100 m²	**605.74**
Mushroom compost (20 m³ loads); placing on beds by mechanical loader; spreading and rotavating into topsoil by machine							
50 mm thick	90.50	-	-	8.09	90.50	100 m²	**98.59**
100 mm thick	181.00	-	-	12.33	181.00	100 m²	**193.33**
150 mm thick	271.50	-	-	16.75	271.50	100 m²	**288.25**
200 mm thick	362.00	-	-	21.09	362.00	100 m²	**383.09**
Manure (60 m³ loads); from not further than 25m from location; cultivating into topsoil by machine							
50 mm thick	125.00	2.86	53.57	1.78	125.00	100 m²	**180.35**
100 mm thick	250.00	6.05	113.42	1.78	262.50	100 m²	**377.70**
150 mm thick	375.00	8.90	166.97	1.78	393.75	100 m²	**562.49**
200 mm thick	500.00	12.90	241.97	1.78	525.00	100 m²	**768.74**

Q PAVING/PLANTING/FENCING/SITE FURNITURE

Item Excluding site overheads and profit	PC £	Labour hours	Labour £	Plant £	Material £	Unit	Total rate £
Surface applications and soil additives; **pre-planting; from not further than 25 m** **from location; by machine**							
Ground limestone soil conditioner; including							
turning in to cultivated ground							
0.25 kg/m^2 = 2.50 tonnes/ha	0.23	-	-	4.32	0.23	100 m^2	**4.55**
0.50 kg/m^2 = 5.00 tonnes/ha	0.47	-	-	4.32	0.47	100 m^2	**4.78**
0.75 kg/m^2 = 7.50 tonnes/ha	0.70	-	-	4.32	0.70	100 m^2	**5.02**
1.00 kg/m^2 = 10.00 tonnes/ha	0.93	-	-	4.32	0.93	100 m^2	**5.25**
Medium bark soil conditioner; A.H.S. Ltd;							
including turning in to cultivated ground;							
delivered in 15 m^3 loads							
1 m^3 per 40 m^2 = 25 mm thick	0.79	-	-	0.10	0.79	m^2	**0.89**
1 m^3 per 20 m^2 = 50 mm thick	1.58	-	-	0.15	1.58	m^2	**1.73**
1 m^3 per 13.33 m^2 = 75 mm thick	2.37	-	-	0.21	2.37	m^2	**2.57**
1 m^3 per 10 m^2 = 100 mm thick	3.15	-	-	0.27	3.15	m^2	**3.42**
Peat soil conditioner; A.H.S. Ltd; including							
turning in to cultivated ground; delivered in 55 m^3							
loads							
1 m^3 per 40 m^2 = 25 mm thick	0.75	-	-	0.10	0.75	m^2	**0.85**
1 m^3 per 20 m^2 = 50 mm thick	1.50	-	-	0.15	1.50	m^2	**1.65**
1 m^3 per 13.33 m^2 = 75 mm thick	2.25	-	-	0.21	2.25	m^2	**2.46**
1 m^3 per 10 m^2 = 100 mm thick	3.00	-	-	0.27	3.00	m^2	**3.27**
Mushroom compost soil conditioner; A.H.S. Ltd;							
including turning in to cultivated ground;							
delivered in 20 m^3 loads							
1 m^3 per 40 m^2 = 25 mm thick	0.45	0.02	0.31	-	0.45	m^2	**0.76**
1 m^3 per 20 m^2 = 50 mm thick	0.91	0.03	0.58	-	0.91	m^2	**1.48**
1 m^3 per 13.33 m^2 = 75 mm thick	1.36	0.04	0.75	-	1.36	m^2	**2.11**
1 m^3 per 10 m^2 = 100 mm thick	1.81	0.05	0.94	-	1.81	m^2	**2.75**
Mushroom compost soil conditioner; A.H.S. Ltd;							
including turning in to cultivated ground;							
delivered in 35 m^3 loads							
1 m^3 per 40 m^2 = 25 mm thick	0.21	0.02	0.31	-	0.21	m^2	**0.53**
1 m^3 per 20 m^2 = 50 mm thick	0.43	0.03	0.58	-	0.43	m^2	**1.01**
1 m^3 per 13.33 m^2 = 75 mm thick	0.64	0.04	0.75	-	0.64	m^2	**1.39**
1 m^3 per 10 m^2 = 100 mm thick	0.86	0.05	0.94	-	0.86	m^2	**1.79**
Surface applications and soil additives; **pre-planting; from not further than 25 m** **from location; by hand**							
Ground limestone soil conditioner; including							
turning in to cultivated ground							
0.25 kg/m^2 = 2.50 tonnes/ha	0.23	1.20	22.50	-	0.23	100 m^2	**22.73**
0.50 kg/m^2 = 5.00 tonnes/ha	0.47	1.33	25.00	-	0.47	100 m^2	**25.46**
0.75 kg/m^2 = 7.50 tonnes/ha	0.70	1.50	28.13	-	0.70	100 m^2	**28.82**
1.00 kg/m^2 = 10.00 tonnes/ha	0.93	1.71	32.14	-	0.93	100 m^2	**33.07**
Medium bark soil conditioner; A.H.S. Limited;							
including turning in to cultivated ground;							
delivered in 15 m^3 loads							
1 m^3 per 40 m^2 = 25 mm thick	0.79	0.02	0.42	-	0.79	m^2	**1.21**
1 m^3 per 20 m^2 = 50 mm thick	1.58	0.04	0.83	-	1.58	m^2	**2.41**
1 m^3 per 13.33 m^2 = 75 mm thick	2.37	0.07	1.25	-	2.37	m^2	**3.62**
1 m^3 per 10 m^2 = 100 mm thick	3.15	0.08	1.50	-	3.15	m^2	**4.65**
Peat soil conditioner; A.H.S. Limited; including							
turning in to cultivated ground; delivered in 55 m^3							
loads							
1 m^3 per 40 m^2 = 25 mm thick	0.75	0.02	0.42	-	0.75	m^2	**1.17**
1 m^3 per 20 m^2 = 50 mm thick	1.50	0.04	0.83	-	1.50	m^2	**2.33**
1 m^3 per 13.33 m^2 = 75 mm thick	2.25	0.07	1.25	-	2.25	m^2	**3.50**
1 m^3 per 10 m^2 = 100 mm thick	3.00	0.08	1.50	-	3.00	m^2	**4.50**

Q PAVING/PLANTING/FENCING/SITE FURNITURE

Item Excluding site overheads and profit	PC £	Labour hours	Labour £	Plant £	Material £	Unit	Total rate £
Q31 PLANTING - cont'd							
Surface applications and soil additives - cont'd							
Mushroom compost soil conditioner; Melcourt Industries Ltd; including turning in to cultivated ground; delivered in 25 m³ loads							
1 m³ per 40 m² = 25 mm thick	0.45	0.02	0.42	-	0.45	m²	0.87
1 m³ per 20 m² = 50 mm thick	0.91	0.04	0.83	-	0.91	m²	1.74
1 m³ per 13.33 m² = 75 mm thick	1.36	0.07	1.25	-	1.36	m²	2.61
1 m³ per 10 m² = 100 mm thick	1.81	0.08	1.50	-	1.81	m²	3.31
Mushroom compost soil conditioner; A.H.S. Limited; including turning in to cultivated ground; delivered in 35 m³ loads							
1 m³ per 40 m² = 25 mm thick	0.21	0.02	0.42	-	0.21	m²	0.63
1 m³ per 20 m² = 50 mm thick	0.43	0.04	0.83	-	0.43	m²	1.26
1 m³ per 13.33 m² = 75 mm thick	0.64	0.07	1.25	-	0.64	m²	1.89
1 m³ per 10 m² = 100 mm thick	0.86	0.08	1.50	-	0.86	m²	2.36
"Super Humus" mixed bark and manure conditioner; Melcourt Industries Ltd; delivered in 25 m³ loads							
1 m³ per 40 m² = 25 mm thick	0.66	0.02	0.42	-	0.69	m²	1.11
1 m³ per 20 m² = 50 mm thick	1.31	0.04	0.83	-	1.38	m²	2.21
1 m³ per 13.33 m² = 75 mm thick	1.97	0.07	1.25	-	2.06	m²	3.31
1 m³ per 10 m² = 100 mm thick	2.62	0.08	1.50	-	2.75	m²	4.25
Tree planting; pre-planting operations							
Excavating tree pits; depositing soil alongside pits; by machine							
600 mm x 600 mm x 600 mm deep	-	0.15	2.76	0.66	-	nr	3.42
900 mm x 900 mm x 600 mm deep	-	0.33	6.19	1.49	-	nr	7.67
1.00 m x 1.00 m x 600 mm deep	-	0.61	11.51	1.84	-	nr	13.35
1.25 m x 1.25 m x 600 mm deep	-	0.96	18.01	2.88	-	nr	20.89
1.00 m x 1.00 m x 1.00 m deep	-	1.02	19.18	3.07	-	nr	22.24
1.50 m x 1.50 m x 750 mm deep	-	1.73	32.36	5.18	-	nr	37.54
1.50 m x 1.50 m x 1.00 m deep	-	2.30	43.04	6.89	-	nr	49.92
1.75 m x 1.75 m x 1.00 mm deep	-	3.13	58.73	9.40	-	nr	68.13
2.00 m x 2.00 m x 1.00 mm deep	-	4.09	76.70	12.27	-	nr	88.97
Excavating tree pits; depositing soil alongside pits; by hand							
600 mm x 600 mm x 600 mm deep	-	0.44	8.25	-	-	nr	8.25
900 mm x 900 mm x 600 mm deep	-	1.00	18.75	-	-	nr	18.75
1.00 m x 1.00 m x 600 mm deep	-	1.13	21.09	-	-	nr	21.09
1.25 m x 1.25 m x 600 mm deep	-	1.93	36.19	-	-	nr	36.19
1.00 m x 1.00 m x 1.00 m deep	-	2.06	38.63	-	-	nr	38.63
1.50 m x 1.50 m x 750 mm deep	-	3.47	65.06	-	-	nr	65.06
1.75 m x 1.50 m x 750 mm deep	-	4.05	75.94	-	-	nr	75.94
1.50 m x 1.50 m x 1.00 m deep	-	4.63	86.81	-	-	nr	86.81
2.00 m x 2.00 m x 750 mm deep	-	6.17	115.69	-	-	nr	115.69
2.00 m x 2.00 m x 1.00 m deep	-	8.23	154.32	-	-	nr	154.32
Breaking up subsoil in tree pits; to a depth of 200 mm	-	0.03	0.62	-	-	m²	0.62
Spreading and lightly consolidating approved topsoil (imported or from spoil heaps); in layers not exceeding 150 mm; distance from spoil heaps not exceeding 100 m (imported topsoil not included); by machine							
minimum depth 100 mm	-	1.55	29.06	37.20	-	100 m²	66.26
minimum depth 150 mm	-	2.33	43.75	55.98	-	100 m²	99.73
minimum depth 300 mm	-	4.67	87.50	111.96	-	100 m²	199.46
minimum depth 450 mm	-	6.99	131.06	167.76	-	100 m²	298.82

Q PAVING/PLANTING/FENCING/SITE FURNITURE

Item Excluding site overheads and profit	PC £	Labour hours	Labour £	Plant £	Material £	Unit	Total rate £
Spreading and lightly consolidating approved topsoil (imported or from spoil heaps); in layers not exceeding 150 mm; distance from spoil heaps not exceeding 100 m (imported topsoil not included); by hand							
minimum depth 100 mm	-	20.00	375.07	-	-	100 m²	**375.07**
minimum depth 150 mm	-	30.01	562.61	-	-	100 m²	**562.61**
minimum depth 300 mm	-	60.01	1125.22	-	-	100 m²	**1125.22**
minimum depth 450 mm	-	90.02	1687.84	-	-	100 m²	**1687.84**
Extra for filling tree pits with imported topsoil; PC £17.32 /m³; plus allowance for 20% settlement							
depth 100 mm	-	-	-	-	2.08	m²	**2.08**
depth 150 mm	-	-	-	-	3.12	m²	**3.12**
depth 200 mm	-	-	-	-	4.16	m²	**4.16**
depth 300 mm	-	-	-	-	6.24	m²	**6.24**
depth 400 mm	-	-	-	-	8.31	m²	**8.31**
depth 450 mm	-	-	-	-	9.35	m²	**9.35**
depth 500 mm	-	-	-	-	10.39	m²	**10.39**
depth 600 mm	-	-	-	-	12.47	m²	**12.47**
Add or deduct the following amounts for every £0.50 change in the material price of topsoil							
depth 100 mm	-	-	-	-	0.06	m²	**0.06**
depth 150 mm	-	-	-	-	0.09	m²	**0.09**
depth 200 mm	-	-	-	-	0.12	m²	**0.12**
depth 300 mm	-	-	-	-	0.18	m²	**0.18**
depth 400 mm	-	-	-	-	0.24	m²	**0.24**
depth 450 mm	-	-	-	-	0.27	m²	**0.27**
depth 500 mm	-	-	-	-	0.30	m²	**0.30**
depth 600 mm	-	-	-	-	0.36	m²	**0.36**
Structural Soils; BSI Metro Sand (Amsterdam Tree Sand); non compressive soil mixture for tree planting in areas to receive compressive surface treatments							
backfilling and lightly compacting in layers; excavation, disposal, moving of material from delivery position and surface treatments not included; by machine							
individual tree pits	49.50	0.50	9.38	2.25	49.50	m³	**61.13**
in trenches	49.50	0.42	7.81	1.88	49.50	m³	**59.19**
backfilling and lightly compacting in layers; excavation, disposal, moving of material from delivery position and surface treatments not included; by hand							
individual tree pits	49.50	1.60	30.00	-	49.50	m³	**79.50**
in trenches	49.50	1.33	24.94	-	49.50	m³	**74.44**
Tree staking							
J Toms Ltd; Extra over trees for tree stake(s); driving 500 mm into firm ground; trimming to approved height; including two tree ties to approved pattern							
one stake; 75 mm diameter x 2.40 m long	3.84	0.20	3.75	-	3.84	nr	**7.59**
two stakes; 60 mm diameter x 1.65 m long	5.29	0.30	5.63	-	5.29	nr	**10.91**
two stakes; 75 mm diameter x 2.40 m long	6.99	0.30	5.63	-	6.99	nr	**12.62**
three stakes; 75 mm diameter x 2.40 m long	10.48	0.36	6.75	-	10.48	nr	**17.23**

Q PAVING/PLANTING/FENCING/SITE FURNITURE

Item Excluding site overheads and profit	PC £	Labour hours	Labour £	Plant £	Material £	Unit	Total rate £
Q31 PLANTING - cont'd							
Tree anchors							
Platipus Anchors Ltd; Extra over trees for							
tree anchors							
ref RF1 rootball kit; for 75 to 220 mm girth, 2 to							
4.5 m high inclusive of "Plati-Mat" PM1	26.58	1.00	18.75	-	26.58	nr	45.33
ref RF2; rootball kit; for 220 to 450 mm girth, 4.5							
to 7.5 m high inclusive of "Plati-Mat" Pm^2	43.54	1.33	24.94	-	43.54	nr	68.48
ref RF3; rootball kit; for 450 to 750 mm girth, 7.5							
to 12 m high "Plati-Mat" Pm^3	92.01	1.50	28.13	-	92.01	nr	120.13
ref CG1; guy fixing kit; 75 to 220 mm girth, 2 to 4.5							
m high	17.30	1.67	31.25	-	17.30	nr	48.55
ref CG2; guy fixing kit; 220 to 450 mm girth, 4.5 to							
7.5 m high	30.44	2.00	37.50	-	30.44	nr	67.94
installation tools; Drive Rod for RF1/CG1 Kits	-	-	-	-	55.83	nr	55.83
installation tools; Drive Rod for RF2/CG2 Kits	-	-	-	-	84.14	nr	84.14
Extra over trees for land drain to tree pits; 100							
mm diameter perforated flexible agricultural drain;							
including excavating drain trench; laying pipe;							
backfilling	1.45	1.00	18.75	-	1.60	m	20.35
Tree planting; tree pit additives							
Melcourt Industries Ltd; "Topgrow"; incorporating							
into topsoil at 1 part "Topgrow" to 3 parts							
excavated topsoil; supplied in 75 l bags; pit size							
600 mm x 600 mm x 600 mm	1.00	0.02	0.38	-	1.00	nr	1.38
900 mm x 900 mm x 900 mm	3.38	0.06	1.13	-	3.38	nr	4.50
1.00 m x 1.00 m x 1.00 m	4.63	0.24	4.50	-	4.63	nr	9.13
1.25 m x 1.25 m x 1.25 m	9.05	0.40	7.50	-	9.05	nr	16.55
1.50 m x 1.50 m x 1.50 m	15.64	0.90	16.88	-	15.64	nr	32.51
Melcourt Industries Ltd; "Topgrow"; incorporating							
into topsoil at 1 part "Topgrow" to 3 parts							
excavated topsoil; supplied in 60 m^3 loose loads;							
pit size							
600 mm x 600 mm x 600 mm	0.90	0.02	0.31	-	0.90	nr	1.21
900 mm x 900 mm x 900 mm	3.04	0.05	0.94	-	3.04	nr	3.98
1.00 m x 1.00 m x 1.00 m	4.17	0.20	3.75	-	4.17	nr	7.92
1.25 m x 1.25 m x 1.25 m	8.15	0.33	6.25	-	8.15	nr	14.40
1.50 m x 1.50 m x 1.50 m	14.09	0.75	14.06	-	14.09	nr	28.15
Alginure Products; "Alginure Root Dip"; to bare							
rooted plants at 1 part "Alginure Root Dip" to 3							
parts water							
transplants; at 3000/15 kg bucket	1.03	0.07	1.25	-	1.03	100 nr	2.28
standard trees; at 112/15 kg bucket of dip	0.28	0.03	0.62	-	0.28	nr	0.90
medium shrubs; at 600/15 kg bucket	0.05	0.02	0.31	-	0.05	nr	0.36
Tree Planting - Root Barriers							
English Woodlands; "Root Director", one-piece							
root control planters for installation at time of							
planting, to divert root growth down away from							
pavements and out for anchorage; excavation							
measured separately							
RD 1050; 1050 x 1050 mm to 1300 x 1300 mm							
at base	70.35	0.25	4.69	-	70.35	nr	75.04
RD 640; 640 x 640 mm to 870 x 870 mm at base	45.05	0.25	4.69	-	45.05	nr	49.74
Greenleaf Horticulture; linear root deflection							
barriers; installed to trench measured separately							
Re-Root 2000, 1.5 mm thick	6.09	0.05	0.94	-	6.09	m	7.03
Re-Root 2000, 1.0 mm thick	5.12	0.05	0.94	-	5.12	m	6.06
Re-Root 600, 1.0 mm thick.	5.78	0.05	0.94	-	5.78	m	6.72

Q PAVING/PLANTING/FENCING/SITE FURNITURE

Item Excluding site overheads and profit	PC £	Labour hours	Labour £	Plant £	Material £	Unit	Total rate £
Greenleaf Horticulture - Irrigation **Systems; "Root Rain" tree pit irrigation** **systems**							
"Metro" - "Small"; 35 mm pipe diameter 1.25 m long for specimen shrubs and standard trees							
Plastic	5.34	0.25	4.69	-	5.34	nr	**10.03**
Plastic with chain	6.98	0.25	4.69	-	6.98	nr	**11.67**
Metal with chain	8.65	0.25	4.69	-	8.65	nr	**13.34**
"Medium" 35 mm pipe diameter 1.75 m long for s standard and selected standard trees							
Plastic	5.67	0.29	5.36	-	5.67	nr	**11.03**
Plastic with chain	7.33	0.29	5.36	-	7.33	nr	**12.69**
Metal with chain	8.98	0.29	5.36	-	8.98	nr	**14.34**
"Large" 35 mm pipe diameter 2.50 m long for selected standards and extra heavy standards							
Plastic	6.55	0.33	6.25	-	6.55	nr	**12.80**
Plastic with chain	8.21	0.33	6.25	-	8.21	nr	**14.46**
Metal with chain	10.96	0.33	6.25	-	10.96	nr	**17.21**
"Large" 35 mm pipe diameter 2.50 m long for selected standards and extra heavy standards							
Plastic	6.55	0.33	6.25	-	6.55	nr	**12.80**
Plastic with chain	8.21	0.33	6.25	-	8.21	nr	**14.46**
Metal with chain	10.96	0.33	6.25	-	10.96	nr	**17.21**
"Urban" Irrigation and aeration systems; for large capacity general purpose irrigation to parkland and street verge planting							
RRUrb1; 3.0 m pipe	-	0.29	5.36	-	13.23	nr	**18.59**
RRUrb2; 5.0 m pipe	15.33	0.33	6.25	-	15.33	nr	**21.58**
RRUrb3; 8.0 m pipe	16.70	0.40	7.50	-	16.70	nr	**24.20**
"Precinct" Irrigation and aeration systems; heavy cast aluminium inlet for heavily trafficked locations							
"5 m pipe"	30.40	0.33	6.25	-	30.40	nr	**36.65**
"8 m pipe"	31.50	0.40	7.50	-	31.50	nr	**39.00**
Mulching of tree pits; Melcourt Industries **Ltd**							
Spreading mulch; to individual trees; maximum distance 25 m (mulch not included)							
50 mm thick	-	0.05	0.91	-	-	m²	**0.91**
75 mm thick	-	0.07	1.37	-	-	m²	**1.37**
100 mm thick	-	0.10	1.82	-	-	m²	**1.82**
Mulch; "Bark Nuggets®"; to individual trees; delivered in 80 m³ loads; maximum distance 25 m							
50 mm thick	1.56	0.05	0.91	-	1.64	m²	**2.55**
75 mm thick	2.35	0.07	1.37	-	2.47	m²	**3.83**
100 mm thick	3.13	0.10	1.82	-	3.29	m²	**5.11**
Mulch; "Bark Nuggets®"; to individual trees; delivered in 25 m³ loads; maximum distance 25 m							
50 mm thick	2.21	0.05	0.91	-	2.33	m²	**3.24**
75 mm thick	3.32	0.05	0.94	-	3.49	m²	**4.43**
100 mm thick	4.43	0.07	1.25	-	4.65	m²	**5.90**
Mulch; "Amenity Bark"; to individual trees; delivered in 80 m³ loads; maximum distance 25 m							
50 mm thick	0.93	0.05	0.91	-	0.97	m²	**1.89**
75 mm thick	1.39	0.07	1.37	-	1.46	m²	**2.83**
100 mm thick	1.85	0.10	1.82	-	1.95	m²	**3.77**
Mulch; "Amenity Bark"; to individual trees; delivered in 25 m³ loads; maximum distance 25 m							
50 mm thick	1.58	0.05	0.91	-	1.66	m²	**2.57**
75 mm thick	2.37	0.07	1.37	-	2.49	m²	**3.85**
100 mm thick	3.15	0.07	1.25	-	3.31	m²	**4.56**

Q PAVING/PLANTING/FENCING/SITE FURNITURE

Item Excluding site overheads and profit	PC £	Labour hours	Labour £	Plant £	Material £	Unit	Total rate £
Q31 PLANTING - cont'd							
Trees; planting labours only							
Bare root trees; including backfilling with							
previously excavated material (all other							
operations and materials not included)							
light standard; 6 - 8 cm girth	-	0.35	6.56	-	-	nr	**6.56**
standard; 8 - 10 cm girth	-	0.40	7.50	-	-	nr	**7.50**
selected standard; 10 - 12 cm girth	-	0.58	10.88	-	-	nr	**10.88**
heavy standard; 12 - 14 cm girth	-	0.83	15.62	-	-	nr	**15.62**
extra heavy standard; 14 - 16 cm girth	-	1.00	18.75	-	-	nr	**18.75**
Root balled trees; including backfilling with							
previously excavated material (all other							
operations and materials not included)							
standard; 8 - 10 cm girth	-	0.50	9.38	-	-	nr	**9.38**
selected standard; 10 - 12 cm girth	-	0.60	11.25	-	-	nr	**11.25**
heavy standard; 12 - 14 cm girth	-	0.80	15.00	-	-	nr	**15.00**
extra heavy standard; 14 - 16 cm girth	-	1.50	28.13	-	-	nr	**28.13**
16 - 18 cm girth	-	1.30	24.30	21.38	-	nr	**45.68**
18 - 20 cm girth	-	1.60	30.00	26.40	-	nr	**56.40**
20 - 25 cm girth	-	4.50	84.38	74.25	-	nr	**158.63**
25 - 30 cm girth	-	6.00	112.50	99.00	-	nr	**211.50**
30 - 35 cm girth	-	11.00	206.25	181.50	-	nr	**387.75**
Tree planting; root balled trees;							
advanced nursery stock and semi-mature							
- General							
Preamble: The cost of planting semi-mature trees							
will depend on the size and species, and on the							
access to the site for tree handling machines.							
Prices should be obtained for individual trees and							
planting.							
Tree planting; bare root trees; nursery							
stock; Coblands Nurseries Ltd							
"Acer platanoides"; including backfilling with							
excavated material (other operations not							
included)							
light standard; 6 - 8 cm girth	9.25	0.35	6.56	-	9.25	nr	**15.81**
standard; 8 - 10 cm girth	11.25	0.40	7.50	-	11.25	nr	**18.75**
selected standard; 10 - 12 cm girth	16.25	0.58	10.88	-	16.25	nr	**27.13**
heavy standard; 12 - 14 cm girth	24.00	0.83	15.62	-	24.00	nr	**39.62**
extra heavy standard; 14 - 16 cm girth	40.00	1.00	18.75	-	40.00	nr	**58.75**
"Carpinus betulus"; including backfilling with							
excavated material (other operations not							
included)							
light standard; 6 - 8 cm girth	14.00	0.35	6.56	-	14.00	nr	**20.56**
standard; 8 - 10 cm girth	16.00	0.40	7.50	-	16.00	nr	**23.50**
selected standard; 10 - 12 cm girth	27.50	0.58	10.88	-	27.50	nr	**38.38**
heavy standard; 12 - 14 cm girth	40.00	0.83	15.63	-	40.00	nr	**55.63**
extra heavy standard; 14 - 16 cm girth	40.00	1.00	18.75	-	40.00	nr	**58.75**
"Fraxinus excelsior"; including backfilling with							
excavated material (other operations not							
included)							
light standard; 6 - 8 cm girth	11.75	0.35	6.56	-	11.75	nr	**18.31**
standard; 8 - 10 cm girth	13.50	0.40	7.50	-	13.50	nr	**21.00**
selected standard; 10 - 12 cm girth	22.00	0.58	10.88	-	22.00	nr	**32.88**
heavy standard; 12 - 14 cm girth	40.00	0.83	15.56	-	40.00	nr	**55.56**
extra heavy standard; 14 - 16 cm girth	50.00	1.00	18.75	-	50.00	nr	**68.75**
"Prunus avium Plena"; including backfilling with							
excavated material (other operations not							
included)							
light standard; 6 - 8 cm girth	11.75	0.36	6.83	-	11.75	nr	**18.57**
standard; 8 - 10 cm girth	13.50	0.40	7.50	-	13.50	nr	**21.00**
selected standard; 10 - 12 cm girth	27.50	0.58	10.88	-	27.50	nr	**38.38**

Q PAVING/PLANTING/FENCING/SITE FURNITURE

Item Excluding site overheads and profit	PC £	Labour hours	Labour £	Plant £	Material £	Unit	Total rate £
heavy standard; 12 - 14 cm girth	45.00	0.83	15.62	-	45.00	nr	60.62
extra heavy standard; 14 - 16 cm girth	55.00	1.00	18.75	-	55.00	nr	73.75
"Quercus robur"; including backfilling with excavated material (other operations not included)							
light standard; 6 - 8 cm girth	11.75	0.35	6.56	-	11.75	nr	18.31
standard; 8 - 10 cm girth	13.50	0.40	7.50	-	13.50	nr	21.00
selected standard; 10 - 12 cm girth	22.00	0.58	10.88	-	22.00	nr	32.88
heavy standard; 12 - 14 cm girth	40.00	0.83	15.62	-	40.00	nr	55.62
extra heavy standard; 14 - 16 cm girth	50.00	1.00	18.75	-	50.00	nr	68.75
"Robinia pseudoacacia Frisia"; including backfilling with excavated material (other operations not included)							
light standard; 6 - 8 cm girth	30.00	0.35	6.56	-	30.00	nr	36.56
standard; 8 - 10 cm girth	40.00	0.40	7.50	-	40.00	nr	47.50
selected standard; 10 - 12 cm girth	55.00	0.58	10.88	-	55.00	nr	65.88
heavy standard; 12 - 14 cm girth	80.00	0.83	15.62	-	80.00	nr	95.62
extra heavy standard; 14 - 16 cm girth	115.00	1.00	18.75	-	115.00	nr	133.75
Tree planting; root balled trees; nursery stock; Coblands Nurseries Ltd							
"Acer platanoides"; including backfilling with excavated material (other operations not included)							
standard; 8 - 10 cm girth	-	0.48	9.00	-	21.25	nr	30.25
selected standard; 10 - 12 cm girth	-	0.56	10.51	-	36.25	nr	46.76
heavy standard; 12 - 14 cm girth	-	0.76	14.33	-	49.00	nr	63.33
extra heavy standard; 14 - 16 cm girth	-	1.20	22.50	-	70.00	nr	92.50
"Carpinus betulus"; including backfilling with excavated material (other operations not included)							
standard; 8 - 10 cm girth	-	0.48	9.00	-	26.00	nr	35.00
selected standard; 10 - 12 cm girth	-	0.56	10.51	-	47.50	nr	58.01
heavy standard; 12 - 14 cm girth	-	0.76	14.33	-	69.00	nr	83.33
extra heavy standard; 14 - 16 cm girth	-	1.20	22.50	-	85.00	nr	107.50
"Fraxinus excelsior"; including backfilling with excavated material (other operations not included)							
standard; 8 - 10 cm girth	-	0.48	9.00	-	23.50	nr	32.50
selected standard; 10 - 12 cm girth	-	0.56	10.51	-	42.00	nr	52.51
heavy standard; 12 - 14 cm girth	-	0.76	14.33	-	65.00	nr	79.33
extra heavy standard; 14 - 16 cm girth	-	1.20	22.50	-	80.00	nr	102.50
"Prunus avium Plena"; including backfilling with excavated material (other operations not included)							
standard; 8 - 10 cm girth	-	0.40	7.50	-	26.00	nr	33.50
selected standard; 10 - 12 cm girth	-	0.56	10.51	-	26.00	nr	36.51
heavy standard; 12 - 14 cm girth	-	0.76	14.33	-	70.00	nr	84.33
extra heavy standard; 14 - 16 cm girth	-	1.20	22.50	-	85.00	nr	107.50
"Quercus robur"; including backfilling with excavated material (other operations not included)							
standard; 8 - 10 cm girth	-	0.48	9.00	-	23.50	nr	32.50
selected standard; 10 - 12 cm girth	-	0.56	10.51	-	42.00	nr	52.51
heavy standard; 12 - 14 cm girth	-	0.76	14.33	-	65.00	nr	79.33
extra heavy standard; 14 - 16 cm girth	-	1.20	22.50	-	80.00	nr	102.50
"Robinia pseudoacacia Frisia"; including backfilling with excavated material (other operations not included)							
standard; 8 - 10 cm girth	-	0.48	9.00	-	40.00	nr	49.00
selected standard; 10 - 12 cm girth	-	0.56	10.51	-	55.00	nr	65.51
heavy standard; 12 - 14 cm girth	-	0.76	14.33	-	80.00	nr	94.33
extra heavy standard; 14 - 16 cm girth	-	1.20	22.50	-	115.00	nr	137.50

Q PAVING/PLANTING/FENCING/SITE FURNITURE

Item Excluding site overheads and profit	PC £	Labour hours	Labour £	Plant £	Material £	Unit	Total rate £
Q31 PLANTING - cont'd							
Tree planting; "Airpot" Container grown trees; advanced nursery stock and semi-mature; Deepdale Trees Ltd.							
"Acer platanoides Emerald Queen"; including backfilling with excavated material (other operations not included)							
16 - 18 cm girth	95.00	1.98	37.13	40.47	95.00	nr	172.59
18 - 20 cm girth	130.00	2.18	40.84	43.26	130.00	nr	214.10
20 - 25 cm girth	190.00	2.38	44.55	48.56	190.00	nr	283.11
25 - 30 cm girth	250.00	2.97	55.69	72.98	250.00	nr	378.67
30 - 35 cm girth	350.00	3.96	74.25	80.94	350.00	nr	505.19
"Aesculus briotti"; including backfilling with excavated material (other operations not included)							
16 - 18 cm girth	90.00	1.98	37.13	40.47	90.00	nr	167.59
18 - 20 cm girth	130.00	1.60	30.00	43.26	130.00	nr	203.26
20 - 25 cm girth	180.00	2.38	44.55	48.56	180.00	nr	273.11
25 - 30 cm girth	240.00	2.97	55.69	72.98	240.00	nr	368.67
30 - 35 cm girth	330.00	3.96	74.25	80.94	330.00	nr	485.19
"Prunus avium Flora Plena"; including backfilling with excavated material (other operations not included)							
16 - 18 cm girth	95.00	1.98	37.13	40.47	95.00	nr	172.59
18 - 20 cm girth	130.00	1.60	30.00	43.26	130.00	nr	203.26
20 - 25 cm girth	190.00	2.38	44.55	48.56	190.00	nr	283.11
25 - 30 cm girth	250.00	2.97	55.69	72.98	250.00	nr	378.67
30 - 35 cm girth	350.00	3.96	74.25	80.94	350.00	nr	505.19
"Quercus palustris - Pin Oak"; including backfilling with excavated material (other operations not included)							
16 - 18 cm girth	95.00	1.98	37.13	40.47	95.00	nr	172.59
18 - 20 cm girth	130.00	1.60	30.00	43.26	130.00	nr	203.26
20 - 25 cm girth	190.00	2.38	44.55	48.56	190.00	nr	283.11
25 - 30 cm girth	250.00	2.97	55.69	72.98	250.00	nr	378.67
30 - 35 cm girth	350.00	3.96	74.25	80.94	350.00	nr	505.19
"Betula pendula multistem"; including backfilling with excavated material (other operations not included)							
3.0 - 3.5 m high	125.00	1.98	37.13	40.47	125.00	nr	202.59
3.5 - 4.0 m high	150.00	1.60	30.00	43.26	150.00	nr	223.26
4.0 - 4.5 m high	185.00	2.38	44.55	48.56	185.00	nr	278.11
4.5 - 5.0 m high	220.00	2.97	55.69	72.98	220.00	nr	348.67
5.0 - 6.0 m high	275.00	3.96	74.25	80.94	275.00	nr	430.19
6.0 - 7.0 m high	320.00	4.50	84.38	96.83	320.00	nr	501.20
"Pinus sylvestris"; including backfilling with excavated material (other operations not included)							
3.0 - 3.5 m high	300.00	1.98	37.13	40.47	300.00	nr	377.59
3.5 - 4.0 m high	350.00	1.60	30.00	43.26	350.00	nr	423.26
4.0 - 4.5 m high	400.00	2.38	44.55	48.56	400.00	nr	493.11
4.5 - 5.0 m high	450.00	2.97	55.69	72.98	450.00	nr	578.67
5.0 - 6.0 m high	550.00	3.96	74.25	80.94	550.00	nr	705.19
6.0 - 7.0 m high	750.00	4.50	84.38	99.65	750.00	nr	934.02
Tree planting; "Airpot" Container grown trees; semi-mature and mature trees; Deepdale Trees Ltd; planting and back filling; planted by tele handler or by crane; delivery included; all other operations priced separately							
Semi mature trees indicative prices							
40 - 45 cm girth	550.00	4.00	75.00	48.43	550.00	nr	673.43
45 - 50 cm girth	750.00	4.00	75.00	48.43	750.00	nr	873.43
55 - 60 cm girth	1000.00	6.00	112.50	48.43	1000.00	nr	1160.93
60 - 70 cm girth	2000.00	7.00	131.25	65.58	2000.00	nr	2196.83
70 - 80 cm girth	3000.00	7.50	140.63	82.73	3000.00	nr	3223.36
80 - 90 cm girth	4000.00	8.00	150.00	96.86	4000.00	nr	4246.86

Q PAVING/PLANTING/FENCING/SITE FURNITURE

Item Excluding site overheads and profit	PC £	Labour hours	Labour £	Plant £	Material £	Unit	Total rate £
Tree planting; root balled trees; **advanced nursery stock and** **semi-mature; Lorenz von Ehren GMbH;** **Euro prices converted to £Stg at the** **rate £1 = Euro 1.47)**							
"Acer platanoides Emerald Queen"; including backfilling with excavated material (other operations not included)							
16 - 18 cm girth	57.82	1.30	24.30	3.75	57.82	nr	85.87
18 - 20 cm girth	74.83	1.60	30.00	3.75	74.83	nr	108.58
20 - 25 cm girth	95.24	4.50	84.38	18.12	95.24	nr	197.74
25 - 30 cm girth	112.24	6.00	112.50	22.50	112.24	nr	247.24
30 - 35 cm girth	183.67	11.00	206.25	30.00	183.67	nr	419.92
"Aesculus briotti"; including backfilling with excavated material (other operations not included)							
16 - 18 cm girth	68.03	1.30	24.30	3.75	68.03	nr	96.08
18 - 20 cm girth	81.63	1.60	30.00	3.75	81.63	nr	115.38
20 - 25 cm girth	102.04	4.50	84.38	18.12	102.04	nr	204.54
25 - 30 cm girth	149.66	6.00	112.50	22.50	149.66	nr	284.66
30 - 35 cm girth	204.08	11.00	206.25	30.00	204.08	nr	440.33
"Prunus avium Flora Plena"; including backfilling with excavated material (other operations not included)							
16 - 18 cm girth	51.02	1.30	24.30	3.75	51.02	nr	79.07
18 - 20 cm girth	61.22	1.60	30.00	3.75	61.22	nr	94.97
20 - 25 cm girth	74.83	4.50	84.38	18.12	74.83	nr	177.33
25 - 30 cm girth	119.05	6.00	112.50	22.50	119.05	nr	254.05
30 - 35 cm girth	204.08	11.00	206.25	26.25	204.08	nr	436.58
"Quercus palustris - Pin Oak"; including backfilling with excavated material (other operations not included)							
16 - 18 cm girth	61.22	1.30	24.30	3.75	61.22	nr	89.27
18 - 20 cm girth	68.03	1.60	30.00	3.75	68.03	nr	101.78
20 - 25 cm girth	91.84	4.50	84.38	18.12	91.84	nr	194.34
25 - 30 cm girth	121.09	6.00	112.50	22.50	121.09	nr	256.09
30 - 35 cm girth	193.88	11.00	206.25	30.00	193.88	nr	430.13
"Tilia cordata Green Spire"; including backfilling with excavated material (other operations not included)							
16 - 18 cm girth	40.82	1.30	24.30	3.75	40.82	nr	68.87
18 - 20 cm girth	51.02	1.60	30.00	3.75	51.02	nr	84.77
20 - 25 cm girth	61.22	4.50	84.38	18.12	61.22	nr	163.72
25 - 30 cm girth; 5 x transplanted 4.0 - 5.0 m tall	85.03	6.00	112.50	22.50	85.03	nr	220.03
30 - 35 cm girth; 5 x transplanted 5.0 - 7.0 m tall	170.07	11.00	206.25	30.00	170.07	nr	406.32
Tree planting; root balled trees; **semi-mature and mature trees; Lorenz** **von Ehren GMbH; Euro prices converted** **to £Stg at the rate £1 = Euro 1.47);** **planting and back filling; planted by tele** **handler or by crane; delivery included;** **all other operations priced separately**							
Semi mature trees							
40 - 45 cm girth	476.19	8.00	150.00	70.50	476.19	nr	696.69
45 - 50 cm girth	612.24	8.00	150.00	95.85	612.24	nr	858.09
55 - 60 cm girth	952.39	10.00	187.50	115.13	952.39	nr	1255.01
67 - 70 cm girth	1292.52	15.00	281.25	189.40	1292.52	nr	1763.17
75 - 80 cm girth	2380.95	18.00	337.50	187.08	2380.95	nr	2905.53
80 - 90 cm girth	3265.31	18.00	337.50	187.08	3265.31	nr	3789.89
Tree planting; tree protection - General Preamble: Care must be taken to ensure that tree grids and guards are removed when trees grow beyond the specified diameter of guard.							

Q PAVING/PLANTING/FENCING/SITE FURNITURE

Item Excluding site overheads and profit	PC £	Labour hours	Labour £	Plant £	Material £	Unit	Total rate £
Q31 PLANTING - cont'd							
Tree planting; tree protection							
Crowders Nurseries; "Crowders Tree Tube"; olive green							
1200 mm high x 80 x 80 mm	0.76	0.07	1.25	-	0.76	nr	**2.01**
stakes; 1500 mm high for "Crowders Tree Tube"; driving into ground	0.50	0.05	0.94	-	0.50	nr	**1.44**
Crowders Nurseries; galvanized welded mesh tree guards; nailing to tree stakes (tree stakes not included)							
840 mm high x 200 mm diameter	12.25	0.25	4.69	-	12.25	nr	**16.94**
1200 mm high x 250 mm diameter	10.50	0.33	6.25	-	10.50	nr	**16.75**
1200 mm high x 300 mm diameter	11.50	0.33	6.25	-	11.50	nr	**17.75**
1800 mm high x 300 mm diameter	13.00	0.33	6.25	-	13.00	nr	**19.25**
Expandable plastic tree guards; including 25 mm softwood stakes							
500 mm high	0.51	0.17	3.12	-	0.51	nr	**3.63**
1.00 m high	0.58	0.17	3.12	-	0.58	nr	**3.70**
J. Toms Ltd; "Spiral Guard" perforated PVC guards; for trees 10 to 40 mm diameter; white, black or grey							
450 mm	0.18	0.03	0.63	-	0.18	nr	**0.81**
600 mm	0.20	0.03	0.63	-	0.20	nr	**0.83**
750 mm	0.27	0.03	0.63	-	0.27	nr	**0.90**
J. Toms Ltd; Plastic Mesh Treeguards supplied in 50m rolls							
13 mm x 13 mm small mesh; roll width 60 cm; black	0.80	0.04	0.67	-	1.13	nr	**1.80**
13 mm x 13 mm small mesh; roll width 120 cm; black	1.13	0.06	1.10	-	1.57	nr	**2.67**
13 mm x 13 mm small mesh; roll width 60 cm; brown	0.70	0.06	1.10	-	1.03	nr	**2.13**
13 mm x 13 mm small mesh; roll width 120 cm; brown	1.46	0.06	1.10	-	1.94	nr	**3.04**
ready made tree guard in dark green fine plastic mesh, supplied flat packed	0.58	0.06	1.10	-	0.91	nr	**2.01**
Tree guards of 3 nr 2.40 m x 100 mm stakes; driving 600 mm into firm ground; bracing with timber braces at top and bottom; including 3 strands barbed wire	0.27	1.00	18.75	-	0.27	nr	**19.02**
English Woodlands; strimmer guard in heavy duty black plastic, 225 mm high	2.07	0.07	1.25	-	2.07	nr	**3.32**
Tubex Ltd; "Standard Treeshelter" inclusive of 25 mm stake; prices shown for quantities of 500 nr							
0.6 m high	0.43	0.05	0.94	-	0.60	each	**1.54**
0.75 m high	0.53	0.05	0.94	-	0.74	each	**1.68**
1.2 m high	1.81	0.05	0.94	-	2.10	each	**3.03**
1.5 m high	1.08	0.05	0.94	-	1.42	each	**2.36**
Tubex Ltd; "Shrubshelter" inclusive of 25 mm stake; prices shown for quantities of 500 nr							
"Ecostart" shelter for forestry transplants and seedlings,	0.50	0.07	1.25	-	0.50	nr	**1.75**
0.6 m high	0.96	0.07	1.25	-	1.13	nr	**2.38**
0.75 m high	1.37	0.07	1.25	-	1.58	nr	**2.83**
Tree guards; Netlon							
Extra over trees for spraying with antidessicant spray; "Wiltpruf"							
selected standards; standards; light standards	2.57	0.20	3.75	-	2.57	nr	**6.32**
standards; heavy standards	4.29	0.25	4.69	-	4.29	nr	**8.98**

Q PAVING/PLANTING/FENCING/SITE FURNITURE

Item Excluding site overheads and profit	PC £	Labour hours	Labour £	Plant £	Material £	Unit	Total rate £
Hedges							
Excavating trench for hedges; depositing soil							
alongside trench; by machine							
300 mm deep x 300 mm wide	-	0.03	0.56	0.27	-	m	0.83
300 mm deep x 450 mm wide	-	0.05	0.84	0.41	-	m	1.25
Excavating trench for hedges; depositing soil							
alongside trench; by hand							
300 mm deep x 300 mm wide	-	0.12	2.25	-	-	m	2.25
300 mm deep x 450 mm wide	-	0.23	4.22	-	-	m	4.22
Setting out; notching out; excavating trench;							
breaking up subsoil to minimum depth 300 mm							
minimum 400 mm deep	-	0.25	4.69	-	-	m	4.69
Hedge planting; including backfill with excavated							
topsoil; PC £0.31/nr							
single row; 200 mm centres	1.55	0.06	1.17	-	1.55	m	2.72
single row; 300 mm centres	1.03	0.06	1.04	-	1.03	m	2.08
single row; 400 mm centres	0.78	0.04	0.78	-	0.78	m	1.56
single row; 500 mm centres	0.62	0.03	0.62	-	0.62	m	1.24
double row; 200 mm centres	3.10	0.17	3.13	-	3.10	m	6.23
double row; 300 mm centres	2.07	0.13	2.50	-	2.07	m	4.57
double row; 400 mm centres	1.55	0.08	1.56	-	1.55	m	3.11
double row; 500 mm centres	1.24	0.07	1.25	-	1.24	m	2.49
Extra over hedges for incorporating manure; at 1							
m³ per 30 m	0.83	0.03	0.47	-	0.83	m	1.30
Topiary							
Clipped topiary; Lorenz von Ehren							
GmbH; German field grown clipped and							
transplanted as detailed; planted to							
plantpit; including backfilling with							
excavated material and TPMC (Euro							
prices converted to £Stg at the rate £1							
= Euro 1.47)							
Buxus sempervirens (Box); Balls							
300 diameter, 3x transplanted; container grown							
or rootballed	9.52	0.25	4.69	-	14.91	nr	19.60
500 diameter, 4x transplanted, wire rootballed	33.33	1.50	28.13	5.47	71.90	nr	105.50
900 diameter, 5x transplanted; wire rootballed	163.27	2.75	51.56	5.47	226.83	nr	283.86
1300 diameter, 6x transplanted; wire rootballed	646.26	3.10	58.13	6.56	707.54	nr	772.23
Buxus sempervirens; (Box); Pyramids							
500 high, 3x transplanted; container grown or							
rootballed	14.29	1.50	28.13	5.47	52.86	nr	86.46
900 high, 4x transplanted; wire rootballed	119.05	2.75	51.56	5.47	182.61	nr	239.64
1300 high, 5x transplanted, wire rootballed	340.14	3.10	58.13	6.56	401.42	nr	466.11
Buxus sempervirens; (Box); Truncated Pyramids							
500 high, 3x transplanted; container grown or							
rootballed	61.22	1.50	28.13	5.47	99.79	nr	133.39
900 high, 4x transplanted; wire rootballed	132.65	2.75	51.56	5.47	196.21	nr	253.24
1300 high, 5x transplanted, wire rootballed	748.30	3.10	58.13	6.56	809.58	nr	874.27
Buxus sempervirens (Box); Cubes							
500 square, 4x transplanted; rootballed	68.03	1.50	28.13	5.47	106.60	nr	140.20
900 square, 5x transplanted; wire rootballed	306.12	2.75	51.56	5.47	369.68	nr	426.71
Taxus Baccata (Yew); Balls							
500 diameter, 4x transplanted; rootballed	68.03	1.50	28.13	5.47	106.60	nr	140.20
900 diameter, 5x transplanted; wire rootballed	136.05	2.75	51.56	5.47	199.61	nr	256.64
1300 diameter, 7x transplanted, wire rootballed	459.18	3.10	58.13	6.56	520.46	nr	585.15
Taxus Baccata (Yew); Cones							
800 high, 4x transplanted; wire rootballed	42.18	1.50	28.13	5.47	80.75	nr	114.35
1500 high, 5x transplanted; wire rootballed	159.86	2.00	37.50	5.47	206.94	nr	249.91
2500 high, 7x transplanted; wire rootballed	544.22	3.10	58.13	6.56	605.50	nr	670.19
Taxus Baccata (Yew); Cubes							
500 square, 4x transplanted; rootballed	61.22	1.50	28.13	5.47	99.79	nr	133.39
900 square, 5x transplanted; wire rootballed	255.10	2.75	51.56	5.47	318.66	nr	375.69

Q PAVING/PLANTING/FENCING/SITE FURNITURE

Item Excluding site overheads and profit	PC £	Labour hours	Labour £	Plant £	Material £	Unit	Total rate £
Q31 PLANTING - cont'd							
Clipped topiary; Lorenz von Ehren GmbH - cont'd							
Taxus Baccata (Yew); Pyramids							
900 high, 4x transplanted; wire rootballed	91.84	1.50	28.13	5.47	130.41	nr	**164.01**
1500 high, 5x transplanted; wire rootballed	265.31	3.10	58.13	6.56	326.59	nr	**391.28**
2500 high, 7x transplanted; wire rootballed	544.22	4.00	75.00	10.94	624.17	nr	**710.11**
Carpinus Betulus (Common Hornbeam); Columns, round base							
800 wide, 2000 high, 4x transplanted, wire rootballed	136.05	3.10	58.13	6.56	197.33	nr	**262.02**
800 wide, 2750 high, 4x transplanted, wire rootballed	217.69	4.00	75.00	10.94	297.64	nr	**383.58**
Carpinus Betulus (Common Hornbeam); Columns, square base							
800 wide, 2000 high, 4x transplanted, wire rootballed	170.07	3.10	58.13	6.56	231.35	nr	**296.04**
800 wide, 2750 high, 4x transplanted, wire rootballed	285.71	4.00	75.00	10.94	365.66	nr	**451.60**
Carpinus Betulus (Common Hornbeam); Cones							
3m high, 5x transplanted; wire rootballed	122.45	4.00	75.00	10.94	202.40	nr	**288.34**
4m high, 6x transplanted; wire rootballed	442.18	5.00	93.75	10.94	536.33	nr	**641.02**
Carpinus Betulus 'Fastigiata'; Pyramids							
4m high, 6x transplanted; wire rootballed	442.18	-	-	-	442.18	nr	**442.18**
7m high, 7x transplanted; wire rootballed	1700.68	-	-	-	1700.68	nr	**1700.68**
Shrub planting - General							
Preamble: For preparation of planting areas see "Cultivation" at the beginning of the section on planting.							
Shrub planting							
Setting out; selecting planting from holding area; loading to wheel barrows; planting as plan or as directed; distance from holding area maximum 50 m; plants 2 - 3 litre containers							
plants in groups of 100 nr minimum	-	0.01	0.22	-	-	nr	**0.22**
plants in groups of 10 - 100 nr	-	0.02	0.31	-	-	nr	**0.31**
plants in groups of 3 - 5 nr	-	0.03	0.47	-	-	nr	**0.47**
single plants not grouped	-	0.04	0.75	-	-	nr	**0.75**
Forming planting holes; in cultivated ground (cultivating not included); by mechanical auger; trimming holes by hand; depositing excavated material alongside holes							
250 diameter	-	0.03	0.62	0.04	-	nr	**0.66**
250 x 250 mm	-	0.04	0.75	0.06	-	nr	**0.81**
300 x 300 mm	-	0.08	1.41	0.08	-	nr	**1.49**
Hand excavation; forming planting holes; in cultivated ground (cultivating not included); depositing excavated material alongside holes							
100 mm x 100 mm x 100 mm deep; with mattock or hoe	-	0.01	0.13	-	-	nr	**0.13**
250 mm x 250 mm x 300 mm deep	-	0.04	0.75	-	-	nr	**0.75**
300 mm x 300 mm x 300 mm deep	-	0.06	1.04	-	-	nr	**1.04**
400 mm x 400 mm x 400 mm deep	-	0.13	2.34	-	-	nr	**2.34**
500 mm x 500 mm x 500 mm deep	-	0.25	4.69	-	-	nr	**4.69**
600 mm x 600 mm x 600 mm deep	-	0.43	8.12	-	-	nr	**8.12**
900 mm x 900 mm x 600 mm deep	-	1.00	18.75	-	-	nr	**18.75**
1.00 m x 1.00 m x 600 mm deep	-	1.23	23.06	-	-	nr	**23.06**
1.25 m x 1.25 m x 600 mm deep	-	1.93	36.19	-	-	nr	**36.19**

Q PAVING/PLANTING/FENCING/SITE FURNITURE

Item Excluding site overheads and profit	PC £	Labour hours	Labour £	Plant £	Material £	Unit	Total rate £
Hand excavation; forming planting holes; in uncultivated ground; depositing excavated material alongside holes							
100 mm x 100 mm x 100 mm deep with mattock or hoe	-	0.03	0.47	-	-	nr	0.47
250 mm x 250 mm x 300 mm deep	-	0.06	1.04	-	-	nr	1.04
300 mm x 300 mm x 300 mm deep	-	0.06	1.17	-	-	nr	1.17
400 mm x 400 mm x 400 mm deep	-	0.25	4.69	-	-	nr	4.69
500 mm x 500 mm x 500 mm deep	-	0.33	6.10	-	-	nr	6.10
600 mm x 600 mm x 600 mm deep	-	0.55	10.31	-	-	nr	10.31
900 mm x 900 mm x 600 mm deep	-	1.25	23.44	-	-	nr	23.44
1.00 m x 1.00 m x 600 mm deep	-	1.54	28.83	-	-	nr	28.83
1.25 m x 1.25 m x 600 mm deep	-	2.41	45.23	-	-	nr	45.23
Bare foot planting; to planting holes (forming holes not included); including backfilling with excavated material (bare root plants not included)							
bare root 1+1; 30 - 90 mm high	-	0.02	0.31	-	-	nr	0.31
bare root 1+2; 90 - 120 mm high	-	0.02	0.31	-	-	nr	0.31
Containerised planting; to planting holes (forming holes not inlcuded); including backfilling with excavated material (shrub or ground cover not included)							
9 cm pot	-	0.01	0.19	-	-	nr	0.19
2 litre container	-	0.02	0.38	-	-	nr	0.38
3 litre container	-	0.02	0.42	-	-	nr	0.42
5 litre container	-	0.03	0.62	-	-	nr	0.62
10 litre container	-	0.05	0.94	-	-	nr	0.94
15 litre container	-	0.07	1.25	-	-	nr	1.25
20 litre container	-	0.08	1.56	-	-	nr	1.56
Shrub planting; 2 litre containerised plants; in cultivated ground (cultivating not included); PC £1.85/nr							
average 2 plants per m^2	-	0.06	1.05	-	3.70	m^2	4.75
average 3 plants per m^2	-	0.08	1.57	-	5.55	m^2	7.13
average 4 plants per m^2	-	0.11	2.10	-	7.40	m^2	9.50
average 6 plants per m^2	-	0.17	3.15	-	11.10	m^2	14.25
Extra over shrubs for stakes	0.50	0.02	0.31	-	0.50	nr	0.81
Composted bark soil conditioners; 20 m^3 loads; on beds by mechanical loader; spreading and rotavating into topsoil; by machine							
50 mm thick	131.00	-	-	8.09	131.00	100 m^2	139.09
100 mm thick	262.00	-	-	12.33	275.10	100 m^2	287.43
150 mm thick	393.00	-	-	16.75	412.65	100 m^2	429.40
200 mm thick	524.00	-	-	21.09	550.20	100 m^2	571.29
Mushroom compost; 25 m^3 loads; delivered not further than 25m from location; cultivating into topsoil by pedestrian operated machine							
50 mm thick	90.50	2.86	53.57	1.78	90.50	100 m^2	145.85
100 mm thick	181.00	6.05	113.42	1.78	181.00	100 m^2	296.20
150 mm thick	271.50	8.90	166.97	1.78	271.50	100 m^2	440.24
200 mm thick	362.00	12.90	241.97	1.78	362.00	100 m^2	605.74
Mushroom compost; 25 m^3 loads; on beds by mechanical loader; spreading and rotavating into topsoil by tractor drawn rotavator							
50 mm thick	90.50	-	-	8.09	90.50	100 m^2	98.59
100 mm thick	181.00	-	-	12.33	181.00	100 m^2	193.33
150 mm thick	271.50	-	-	16.75	271.50	100 m^2	288.25
200 mm thick	362.00	-	-	21.09	362.00	100 m^2	383.09
Manure; 20 m^3 loads; delivered not further than 25m from location; cultivating into topsoil by pedestrian operated machine							
50 mm thick	-	2.86	53.57	1.78	160.00	100 m^2	215.35
100 mm thick	250.00	6.05	113.42	1.78	262.50	100 m^2	377.70
150 mm thick	375.00	8.90	166.97	1.78	393.75	100 m^2	562.49
200 mm thick	500.00	12.90	241.97	1.78	525.00	100 m^2	768.74

Q PAVING/PLANTING/FENCING/SITE FURNITURE

Item Excluding site overheads and profit	PC £	Labour hours	Labour £	Plant £	Material £	Unit	Total rate £
Q31 PLANTING - cont'd							
Shrub planting - cont'd							
Fertilizers (7:7:7); PC £0.44/kg; to beds; by hand							
35 g/m^2	1.56	0.17	3.13	-	1.56	100 m^2	**4.68**
50 g/m^2	2.22	0.17	3.13	-	2.22	100 m^2	**5.35**
70 g/m^2	3.11	0.17	3.13	-	3.11	100 m^2	**6.24**
Fertilizers; "Enmag" PC £1.68/kg slow release fertilizer; to beds; by hand							
35 g/m^2	5.88	0.17	3.13	-	5.88	100 m^2	**9.01**
50 g/m^2	8.40	0.17	3.13	-	8.40	100 m^2	**11.53**
70 g/m^2	12.60	0.17	3.13	-	12.60	100 m^2	**15.73**
Note: For machine incorporation of fertilizers and soil conditioners see "Cultivation".							
Herbaceous and groundcover planting							
Herbaceous plants; PC £1.13/nr; including forming planting holes in cultivated ground (cultivating not included); backfilling with excavated material; 1 litre containers							
average 4 plants per m^2 - 500 mm centres	-	0.09	1.75	-	4.52	m^2	**6.27**
average 6 plants per m^2 - 408 mm centres	-	0.14	2.63	-	6.78	m^2	**9.41**
average 8 plants per m^2 - 354 mm centres	-	0.19	3.50	-	9.04	m^2	**12.54**
Note: For machine incorporation of fertilizers and soil conditioners see "Cultivation".							
Plant support netting; Bridport Gundry; on 50 mm diameter stakes; 750 mm long; driving into ground at 1.50 m centres							
green extruded plastic mesh; 125 mm square	0.38	0.04	0.75	-	0.38	m^2	**1.13**
Bulb planting							
Bulbs; including forming planting holes in cultivated area (cultivating not included); backfilling with excavated material							
small	13.00	0.83	15.62	-	13.00	100 nr	**28.62**
medium	22.00	0.83	15.62	-	22.00	100 nr	**37.62**
large	25.00	0.91	17.05	-	25.00	100 nr	**42.05**
Bulbs; in grassed area; using bulb planter; including backfilling with screened topsoil or peat and cut turf plug							
small	13.00	1.67	31.25	-	13.00	100 nr	**44.25**
medium	22.00	1.67	31.25	-	22.00	100 nr	**53.25**
large	25.00	2.00	37.50	-	25.00	100 nr	**62.50**
Aquatic planting							
Aquatic plants; in prepared growing medium in pool; plant size 2 - 3 litre containerised (plants not included)	-	0.04	0.75	-	-	nr	**0.75**
Operations after planting							
Initial cutting back to shrubs and hedge plants; including disposal of all cuttings	-	1.00	18.75	-	-	100 m^2	**18.75**
Mulch; Melcourt Industries Ltd; "Bark Nuggets®"; to plant beds; delivered in 25 m^3 loads; maximum distance 25 m							
50 mm thick	1.97	0.03	0.55	-	2.06	m^2	**2.61**
75 mm thick	3.32	0.07	1.25	-	3.49	m^2	**4.74**
100 mm thick	4.43	0.09	1.67	-	4.65	m^2	**6.32**
Mulch; Melcourt Industries Ltd; "Amenity Bark"; to plant beds; delivered in 80 m^3 loads; maximum distance 25 m							
50 mm thick	0.93	0.04	0.83	-	0.97	m^2	**1.81**
75 mm thick	1.39	0.07	1.25	-	1.46	m^2	**2.71**
100 mm thick	1.85	0.09	1.67	-	1.95	m^2	**3.61**

Q PAVING/PLANTING/FENCING/SITE FURNITURE

Item Excluding site overheads and profit	PC £	Labour hours	Labour £	Plant £	Material £	Unit	Total rate £
Mulch; Melcourt Industries Ltd; "Amenity Bark"; to plant beds; delivered in 25 m³ loads; maximum distance 25 m							
50 mm thick	1.58	0.04	0.83	-	1.66	m²	2.49
75 mm thick	2.37	0.07	1.25	-	2.49	m²	3.74
100 mm thick	3.15	0.09	1.67	-	3.31	m²	4.98
Mulch; Melcourt Industries Ltd; "Amenity Bark"; to plant beds; delivered in 25 m³ loads; maximum distance 25 m							
50 mm thick	1.58	0.03	0.55	-	1.66	m²	2.20
75 mm thick	2.37	0.04	0.82	-	2.49	m²	3.31
100 mm thick	3.15	0.06	1.09	-	3.31	m²	4.41
Selective herbicides; in accordance with manufacturer's instructions; PC £24.54/ litre; by machine; application rate							
30 ml/m²	-	-	-	0.29	0.74	100 m²	1.03
35 ml/m²	-	-	-	0.29	0.86	100 m²	1.15
40 ml/m²	-	-	-	0.29	0.98	100 m²	1.27
50 ml/m²	-	-	-	0.29	1.23	100 m²	1.52
3.00 l /ha	-	-	-	29.17	73.64	ha	102.81
3.50 l/ha	-	-	-	29.17	85.91	ha	115.08
4.00 l/ha	-	-	-	29.17	98.18	ha	127.35
Selective herbicides; in accordance with manufacturer's instructions; PC £24.54/ litre; by hand; application rate							
30 ml/m²	-	0.17	3.13	-	0.74	100 m²	3.86
35 ml/m²	-	0.17	3.13	-	0.86	100 m²	3.98
40 ml/m²	-	0.17	3.13	-	0.98	100 m²	4.11
50 ml/m²	-	0.17	3.13	-	1.23	100 m²	4.35
3.00 l/ha	-	16.67	312.50	-	73.64	ha	386.14
3.50 l/ha	-	16.67	312.50	-	85.91	ha	398.41
4.00 l/ha	-	16.67	312.56	-	98.18	ha	410.74
General herbicides; in accordance with manufacturer's instructions; "Knapsack" spray application							
"Casoron G (residual)"; at 1 kg/125 m²	2.82	0.05	0.94	-	3.10	100 m²	4.04
"Spasor"; at 50 ml/100 m²	0.70	0.05	0.94	-	0.70	100 m²	1.64
"Roundup Pro"; at 40 ml /100 m²	0.37	0.05	0.94	-	0.37	100 m²	1.31
"Super Verdone"; at 100 ml/100 m²	0.60	0.05	0.94	-	0.60	100 m²	1.53
Fertilizers; in top 150 mm of topsoil at 35 g/m²							
fertilizer (18:0:0+Mg+Fe)	6.88	0.12	2.30	-	6.88	100 m²	9.17
"Enmag"	5.88	0.12	2.30	-	6.17	100 m²	8.47
fertilizer (7:7:7)	1.56	0.12	2.30	-	1.56	100 m²	3.85
fertilizer (20:10:10)	2.20	0.12	2.30	-	2.20	100 m²	4.50
"Superphosphate"	1.46	0.12	2.30	-	1.53	100 m²	3.83
"Hoof and Horn"	2.01	0.12	2.30	-	2.11	100 m²	4.41
"Bone meal"	2.79	0.12	2.30	-	2.93	100 m²	5.23
Fertilizers; in top 150 mm of topsoil at 70 g/m²							
fertilizer (18:0:0+Mg+Fe)	13.76	0.12	2.30	-	13.76	100 m²	16.05
"Enmag"	11.76	0.12	2.30	-	12.35	100 m²	14.64
fertilizer (7:7:7)	3.11	0.12	2.30	-	3.11	100 m²	5.41
fertilizer (20:10:10)	4.40	0.12	2.30	-	4.40	100 m²	6.70
"Superphosphate"	2.92	0.12	2.30	-	3.07	100 m²	5.36
"Hoof and Horn"	4.02	0.12	2.30	-	4.22	100 m²	6.52
"Bone meal"	5.58	0.12	2.30	-	5.86	100 m²	8.16
Maintenance operations (Note: the following rates apply to aftercare maintenance executed as part of a landscaping contract only)							
Weeding and hand forking planted areas; including disposing weeds and debris on site; areas maintained weekly	-	-	0.08	-	-	m²	0.08
Weeding and hand forking planted areas; including disposing weeds and debris on site; areas maintained monthly	-	0.01	0.19	-	-	m²	0.19

Q PAVING/PLANTING/FENCING/SITE FURNITURE

Item Excluding site overheads and profit	PC £	Labour hours	Labour £	Plant £	Material £	Unit	Total rate £
Q31 PLANTING - cont'd							
Maintenance operations - cont'd							
Extra over weeding and hand forking planted areas for disposing excavated material off site; to tip not exceeding 13 km; mechanically loaded							
slightly contaminated	-	-	-	1.65	13.00	m³	**14.65**
rubbish	-	-	-	1.65	13.00	m³	**14.65**
inert material	-	-	-	1.10	7.50	m³	**8.60**
Mulch; Melcourt Industries Ltd; "Bark Nuggets®"; to plant beds; delivered in 80 m³ loads; maximum distance 25 m							
50 mm thick	1.56	0.03	0.50	-	1.64	m²	**2.15**
75 mm thick	2.35	0.04	0.76	-	2.47	m²	**3.22**
100 mm thick	3.13	0.05	1.01	-	3.29	m²	**4.30**
Mulch; Melcourt Industries Ltd; "Bark Nuggets®"; to plant beds; delivered in 25 m³ loads; maximum distance 25 m							
50 mm thick	2.21	0.03	0.55	-	2.33	m²	**2.87**
75 mm thick	3.32	0.04	0.82	-	3.49	m²	**4.31**
100 mm thick	4.43	0.06	1.09	-	4.65	m²	**5.74**
Mulch; Melcourt Industries Ltd; "Amenity Bark"; to plant beds; delivered in 80 m³ loads; maximum distance 25 m							
50 mm thick	0.93	0.03	0.55	-	0.97	m²	**1.52**
75 mm thick	1.39	0.04	0.82	-	1.46	m²	**2.28**
100 mm thick	1.85	0.04	0.69	-	1.95	m²	**2.64**
Mulch; Melcourt Industries Ltd; "Amenity Bark"; to plant beds; delivered in 25 m³ loads; maximum distance 25 m							
50 mm thick	1.58	0.03	0.55	-	1.66	m²	**2.20**
75 mm thick	2.37	0.04	0.82	-	2.49	m²	**3.31**
100 mm thick	3.15	0.06	1.09	-	3.31	m²	**4.41**
Selective herbicides in accordance with manufacturer's instructions; PC £24.54/ litre; by machine; application rate							
30 ml/m²	-	-	-	0.29	0.74	100 m²	**1.03**
35 ml/m²	-	-	-	0.29	0.86	100 m²	**1.15**
40 ml/m²	-	-	-	0.29	0.98	100 m²	**1.27**
50 ml/m²	-	-	-	0.29	1.23	100 m²	**1.52**
3.00 l/ha	-	-	-	29.17	73.64	ha	**102.81**
3.50 l/ha	-	-	-	29.17	85.91	ha	**115.08**
4.00 l/ha	-	-	-	29.17	98.18	ha	**127.35**
Selective herbicides; in accordance with manufacturer's instructions; PC £24.54 / litre; by hand; application rate							
30 ml/m²	-	0.17	3.13	-	0.74	100 m²	**3.86**
35 ml/m²	-	0.17	3.13	-	0.86	100 m²	**3.98**
40 ml/m²	-	0.17	3.13	-	0.98	100 m²	**4.11**
50 ml/m²	-	0.17	3.13	-	1.23	100 m²	**4.35**
3.00 l/ha	-	16.67	312.50	-	73.64	ha	**386.14**
3.50 l/ha	-	16.67	312.50	-	85.91	ha	**398.41**
4.00 l/ha	-	16.67	312.51	-	98.18	ha	**410.69**
General herbicides; in accordance with manufacturer's instructions; "Knapsack" spray application							
"Casoron G (residual)"; at 1 kg/125 m²	2.82	0.05	0.94	-	3.10	100 m²	**4.04**
"Spasor"; at 50 ml/100 m²	0.70	0.05	0.94	-	0.70	100 m²	**1.64**
"Roundup Pro"; at 40 ml/100 m²	0.37	0.05	0.94	-	0.37	100 m²	**1.31**
"Super Verdone"; at 100 ml/100 m²	0.60	0.05	0.94	-	0.60	100 m²	**1.53**
Fertilizers; at 35 g/m²							
fertilizer (18:0:0+Mg +Fe)	6.88	0.12	2.30	-	6.88	100 m²	**9.17**
"Enmag"	5.88	0.12	2.30	-	6.17	100 m²	**8.47**
fertilizer (7:7:7)	1.56	0.12	2.30	-	1.56	100 m²	**3.85**

Q PAVING/PLANTING/FENCING/SITE FURNITURE

Item Excluding site overheads and profit	PC £	Labour hours	Labour £	Plant £	Material £	Unit	Total rate £
fertilizer (20:10:10)	2.20	0.12	2.30	-	2.20	100 m²	4.50
"Superphosphate"	1.46	0.12	2.30	-	1.53	100 m²	3.83
"Hoof and Horn"	2.01	0.12	2.30	-	2.11	100 m²	4.41
"Bone meal"	2.79	0.12	2.30	-	2.93	100 m²	5.23
Fertilizers; at 70 g/m²							
fertilizer (18:0:0+Mg +Fe)	13.76	0.12	2.30	-	13.76	100 m²	16.05
"Enmag"	11.76	0.12	2.30	-	12.35	100 m²	14.64
fertilizer (7:7:7)	3.11	0.12	2.30	-	3.11	100 m²	5.41
fertilizer (20:10:10)	4.40	0.12	2.30	-	4.40	100 m²	6.70
"Superphosphate"	2.92	0.12	2.30	-	3.07	100 m²	5.36
"Hoof and Horn"	4.02	0.12	2.30	-	4.22	100 m²	6.52
"Bone meal"	5.58	0.12	2.30	-	5.86	100 m²	8.16
Treating plants with growth retardant (maleic hydrazide); at rate of 16 l/ha; in accordance with manufacturer's instructions							
"Burtolin" spray; at 1.50 litres per 5 l water	1.06	0.03	0.62	-	1.06	nr	1.68
Watering planting; evenly; at a rate of 5 litre/m²							
using hand-held watering equipment	-	0.25	4.69	-	-	100 m²	4.69
using sprinkler equipment and with sufficient water pressure to run 1 nr 15 m radius sprinkler	-	0.14	2.61	-	-	100 m²	2.61
using movable spray lines powering 3 nr sprinkler heads with a radius of 15 m and allowing for 60% overlap (irrigation machinery costs not included)	-	0.02	0.29	-	-	100 m²	0.29
Forestry planting							
Deep ploughing rough ground to form planting ridges at							
2.00 m centres	-	0.63	11.72	7.30	-	100 m²	19.02
3.00 m centres	-	0.59	11.03	6.87	-	100 m²	17.90
4.00 m centres	-	0.40	7.50	4.67	-	100 m²	12.17
Notching plant forestry seedlings; "T" or "L" notch	27.00	0.75	14.06	-	27.00	100 nr	41.06
Turf planting forestry seedlings	27.00	2.00	37.50	-	27.00	100 nr	64.50
Selective herbicides; in accordance with manufacturer's instructions; PC £24.54 / litre; by hand; application rate							
30 ml/100m²	-	0.17	3.13	-	0.74	100 m²	3.86
35 ml/100m²	-	0.17	3.13	-	0.86	100 m²	3.98
40 ml/100m²	-	0.17	3.13	-	0.98	100 m²	4.11
50 ml/100m²	-	0.17	3.13	-	1.23	100 m²	4.35
3.00 l/ha	-	16.67	312.50	-	73.64	ha	386.14
3.50 l/ha	-	16.67	312.50	-	85.91	ha	398.41
4.00 l/ha	-	16.67	312.50	-	98.18	ha	410.68
Selective herbicides; in accordance with manufacturer's instructions; PC £24.54 / litre; by machine; application rate							
30 ml/100m²	-	-	-	0.29	0.74	100 m²	1.03
35 ml/100m²	-	-	-	0.29	0.86	100 m²	1.15
40 ml/100m²	-	-	-	0.29	0.98	100 m²	1.27
50 ml/100m²	-	-	-	0.29	1.23	100 m²	1.52
3.00 l/ha	-	-	-	29.17	73.64	ha	102.81
3.50 l/ha	-	-	-	29.17	85.91	ha	115.08
4.00 l/ha	-	-	-	29.17	98.18	ha	127.35
General herbicides; in accordance with manufacturer's instructions; "Knapsack" spray application							
"Casoron G (residual)"; at 1 kg/125 m²	2.82	0.05	0.94	-	3.10	100 m²	4.04
"Spasor"; at 50 ml/100 m²	0.70	0.05	0.94	-	0.70	100 m²	1.64
"Roundup Pro"; at 40 ml/100 m²	0.37	0.05	0.94	-	0.37	100 m²	1.31
"Super Verdone"; at 100 ml/100 m²	0.60	0.05	0.94	-	0.60	100 m²	1.53
Fertilizers; at 35 g/m²							
fertilizer (18:0:0+Mg + Fe)	6.88	0.12	2.30	-	6.88	100 m²	9.17
"Enmag"	5.88	0.12	2.30	-	6.17	100 m²	8.47
fertilizer (7:7:7)	1.56	0.12	2.30	-	1.56	100 m²	3.85
fertilizer (20:10:10)	2.20	0.12	2.30	-	2.20	100 m²	4.50
"Superphosphate"	1.46	0.12	2.30	-	1.53	100 m²	3.83
"Hoof and Horn"	2.01	0.12	2.30	-	2.11	100 m²	4.41
"Bone meal"	2.79	0.12	2.30	-	2.93	100 m²	5.23

Q PAVING/PLANTING/FENCING/SITE FURNITURE

Item Excluding site overheads and profit	PC £	Labour hours	Labour £	Plant £	Material £	Unit	Total rate £
Q31 PLANTING - cont'd							
Forestry planting - cont'd							
Fertilizers; at 70 g/m^2							
fertilizer (18:0:0+Mg + Fe)	13.76	0.12	2.30	-	13.76	100 m^2	**16.05**
"Enmag"	11.76	0.12	2.30	-	12.35	100 m^2	**14.64**
fertilizer (7:7:7)	3.11	0.12	2.30	-	3.11	100 m^2	**5.41**
fertilizer (20:10:10)	4.40	0.12	2.30	-	4.40	100 m^2	**6.70**
"Superphosphate"	2.92	0.12	2.30	-	3.07	100 m^2	**5.36**
"Hoof and Horn"	4.02	0.12	2.30	-	4.22	100 m^2	**6.52**
"Bone meal"	5.58	0.12	2.30	-	5.86	100 m^2	**8.16**
Tree tubes; to young trees	126.00	0.30	5.63	-	126.00	100 nr	**131.63**
Cleaning and weeding around seedlings; once	-	0.50	9.38	-	-	100 nr	**9.38**
Treading in and firming ground around seedlings planted; at 2500 per ha after frost or other ground disturbance; once	-	0.33	6.25	-	-	100 nr	**6.25**
Beating up initial planting; once (including supply of replacement seedlings at 10% of original planting)	2.70	0.25	4.69	-	2.70	100 nr	**7.39**
Work to existing planting							
Control shoots lichen and suckers on street trees; Rhone Poulenc							
"Burtolin" spray at 1.50 litres per 5 l	1.06	0.03	0.62	-	1.06	nr	**1.68**
Cutting and trimming ornamental hedges; to specified profiles; including cleaning out hedge bottoms; hedge cut 2 occasions per annum; by hand							
up to 2.00 m high	-	0.03	0.62	-	0.19	m	**0.81**
2.00 - 4.00 m high	-	0.05	0.94	1.59	0.38	m	**2.90**
Cutting and laying field hedges; including stakes and ethering; removing or burning all debris (Note: Rate at which work executed varies greatly with width and height of hedge; a typical hedge could be cut and laid at rate of 7 m run per man day)	-	1.00	18.75	-	1.53	m	**20.28**
labour charge per man day (specialist daywork)	-	1.00	18.75	-	-	m	**18.75**
Trimming field hedges; to specified heights and shapes							
using cutting bar	-	5.00	93.75	-	-	100 m	**93.75**
using flail	-	0.20	3.75	2.55	-	100 m	**6.30**
Q35 LANDSCAPE MAINTENANCE							
Preamble; Long-term Landscape maintenance Maintenance on long-term contracts differs in cost from that of maintenance as part of a landscape contract. In this section the contract period is generally 3 - 5 years. Staff are generally allocated to a single project only and therefore productivity is higher whilst overhead costs are lower. Labour costs in this section are lower than the costs used in other parts of the book. Machinery is assumed to be leased over a 5 year period and written off over the same period. The costs of maintenance and consumables for the various machinery types have been included in the information that follows. Finance costs for the machinery have not been allowed for. The rates shown below are for machines working in unconfined contiguous areas. Users should adjust the times and rates if working in smaller spaces or spaces with obstructions.							

Q PAVING/PLANTING/FENCING/SITE FURNITURE

Item Excluding site overheads and profit	PC £	Labour hours	Labour £	Plant £	Material £	Unit	Total rate £
Grass Cutting - tractor mounted equipment; Keith Banyard Grounds Maintenance Ltd							
Using multiple-gang mower with cylindrical cutters; contiguous areas such as playing fields and the like larger than 3000 m²							
3 gang; 2.13 m cutting width	-	0.01	0.16	0.10	-	100m²	**0.26**
5 gang; 3.40 m cutting width	-	0.01	0.14	0.13	-	100 m²	**0.27**
7 gang; 4.65 m cutting width	-	0.01	0.09	0.09	-	100 m²	**0.18**
Using multiple-gang mower with cylindrical cutters; non-contiguous areas such as verges and general turf areas							
3 gang	-	0.02	0.31	0.10	-	100m²	**0.41**
5 gang	-	0.01	0.20	0.13	-	100 m²	**0.33**
Using multiple-rotary mower with vertical drive shaft and horizontally rotating bar or disc cutters; contiguous areas larger than 3000 m²							
Cutting grass, overgrowth or the like using flail mower or reaper	-	0.02	0.30	0.22	-	100 m²	**0.52**
Collection of arisings							
22 cuts per year	-	0.02	0.25	0.21	-	100 m²	**0.46**
18 cuts per year	-	0.02	0.33	0.21	-	100 m²	**0.54**
12 cuts per year	-	0.03	0.50	0.21	-	100 m²	**0.71**
4 cuts per year	-	0.07	1.00	0.21	-	100 m²	**1.21**
Disposal of arisings							
22 cuts per year	-	0.01	0.12	0.04	0.45	100 m²	**0.61**
18 cuts per year	-	0.01	0.15	0.05	0.60	100 m²	**0.80**
12 cuts per year	-	0.01	0.10	0.13	1.08	100 m²	**1.31**
4 cuts per year	-	0.02	0.30	0.76	1.80	100 m²	**2.86**
Cutting grass, overgrowth or the like; using tractor-mounted side-arm flail mower; in areas inaccessible to alternative machine; on surface							
not exceeding 30 deg from horizontal	-	0.02	0.35	0.28	-	100 m²	**0.63**
30 deg to 50 deg from horizontal	-	0.05	0.70	0.28	-	100 m²	**0.98**
Grass Cutting - 'Ride-On' self-propelled equipment; Norris and Gardener Grounds Maintenance							
Using ride-on multiple-cylinder mower							
3 gang; 2.13 m cutting width	-	0.01	0.16	0.10	-	100m²	**0.26**
5 gang; 3.40 m cutting width	-	0.01	0.10	0.13	-	100 m²	**0.23**
Using ride-on multiple-rotary mower with horizontally rotating bar, disc or chain cutters							
cutting width 1.52 m	-	0.01	0.19	0.15	-	100 m²	**0.34**
cutting width 1.82 m	-	0.01	0.16	0.11	-	100 m²	**0.27**
cutting width 2.97 m	-	0.01	0.10	0.12	-	100 m²	**0.22**
Add for using grass box/collector for							
removal and depositing of arisings	-	0.04	0.57	0.05	-	100 m²	**0.62**
Grass Cutting - pedestrian operated equipment							
Using cylinder lawn mower fitted with not less than five cutting blades, front and rear rollers; on surface not exceeding 30 deg from horizontal; arisings let fly; width of cut							
51 cm	-	0.06	0.91	0.17	-	100 m	**1.08**
61 cm	-	0.05	0.76	0.15	-	100 m	**0.91**
71 cm	-	0.04	0.66	0.16	-	100 m	**0.82**
91 cm	-	0.03	0.51	0.16	-	100 m	**0.67**
Using rotary self-propelled mower; width of cut							
45 cm	-	0.04	0.59	0.06	-	100 m²	**0.65**
81 cm	-	0.02	0.33	0.06	-	100 m²	**0.39**
91 cm	-	0.02	0.29	0.06	-	100 m²	**0.35**
120 cm	-	0.01	0.22	0.23	-	100 m²	**0.45**

Q PAVING/PLANTING/FENCING/SITE FURNITURE

Item Excluding site overheads and profit	PC £	Labour hours	Labour £	Plant £	Material £	Unit	Total rate £
Q35 LANDSCAPE MAINTENANCE - cont'd							
Grass Cutting - pedestrian operated equipment - cont'd							
Add for using grass box for collecting and depositing arisings							
removing and depositing arisings	-	0.05	0.75	-	-	100 m²	**0.75**
Add for 30 to 50 deg from horizontal	-	-	-	-	-	-	**33%**
Add for slopes exceeding 50 deg	-	-	-	-	-	-	**100%**
Cutting grass or light woody undergrowth; using trimmer with nylon cord or metal disc cutter; on surface							
not exceeding 30 deg from horizontal	-	0.20	3.00	0.27	-	100 m²	**3.27**
30 -50 deg from horizontal	-	0.40	6.00	0.54	-	100 m²	**6.54**
exceeding 50 deg from horizontal	-	0.50	7.50	0.68	-	100 m²	**8.18**
Grass Cutting - collecting arisings							
Extra over for tractor drawn and self-propelled machinery using attached grass boxes							
Depositing arisings							
22 cuts per year	-	0.05	0.75	-	-	100 m²	**0.75**
18 cuts per year	-	0.08	1.13	-	-	100 m²	**1.13**
12 cuts per year	-	0.10	1.50	-	-	100 m²	**1.50**
4 cuts per year	-	0.25	3.75	-	-	100 m²	**3.75**
Disposing arisings							
22 cuts per year	-	0.01	0.12	0.02	0.05	100 m²	**0.19**
18 cuts per year	-	0.01	0.15	0.03	0.06	100 m²	**0.23**
12 cuts per year	-	0.01	0.22	0.04	0.09	100 m²	**0.35**
4 cuts per year	-	0.04	0.65	0.12	0.28	100 m²	**1.05**
Harrowing							
Harrowing grassed area with							
drag harrow	-	0.01	0.19	0.12	-	100 m²	**0.31**
chain or light flexible spiked harrow	-	0.02	0.25	0.12	-	100 m²	**0.37**
Scarifying mechanical							
A Plant Hire Co Ltd							
Sisis ARP4, including grass collection box, towed by tractor; area scarified annually	-	0.02	0.35	0.48	-	100 m²	**0.83**
Sisis ARP4, including grass collection box, towed by tractor; area scarified two years previously	-	0.03	0.42	0.58	-	100 m²	**1.00**
Pedestrian operated self-powered equipment	-	0.07	1.05	0.65	-	100 m²	**1.70**
Add for disposal of arisings	-	0.03	0.38	0.95	9.00	100 m²	**10.33**
Scarifying by hand							
hand implement	-	0.50	7.50	-	-	100 m²	**7.50**
add for disposal of arisings	-	0.03	0.38	0.95	9.00	100 m²	**10.33**
Rolling							
Rolling grassed area; equipment towed by tractor; once over, using							
smooth roller	-	0.01	0.21	0.15	-	100 m²	**0.36**
Turf Aeration							
By machine; A Plant Hire Co Ltd							
Vertidrain turf aeration equipment towed by tractor to effect a minimum penetration of 100 to 250 mm at 100 mm centres	-	0.04	0.58	1.21	-	100 m²	**1.79**
Ryan GA 30; self-propelled turf aerating equipment; to effect a minimum penetration of 100 mm at varying centres	-	0.04	0.58	1.40	-	100 m²	**1.99**
Groundsman; pedestrian operated; self-powered solid or slitting tine turf aerating equipment to effect a minimum penetration of 100 mm	-	0.14	2.10	1.33	-	100 m²	**3.43**
Cushman core harvester; self propelled; for collection of arisings	-	0.02	0.29	0.62	-	100 m²	**0.91**

Q PAVING/PLANTING/FENCING/SITE FURNITURE

Item Excluding site overheads and profit	PC £	Labour hours	Labour £	Plant £	Material £	Unit	Total rate £
By hand							
hand fork; to effect a minimum penetration of 100 mm and spaced 150 mm apart	-	1.33	20.00	-	-	100 m²	20.00
hollow tine hand implement; to effect a minimum penetration of 100 mm and spaced 150 mm apart	-	2.00	30.00	-	-	100 m²	30.00
collection of arisings by hand	-	3.00	45.00	-	-	100 m²	45.00
Turf areas; Surface treatments and Top dressing; Boughton Loam Ltd							
Apply screened topdressing to grass surfaces; spread using Tru-Lute							
Sand soil mixes 90/10 to 50/50	0.09	-	0.03	0.03	0.09	m²	0.15
Apply screened soil 3 mm, Kettering loam to goal mouths and worn areas							
20 mm thick	0.82	0.01	0.15	-	0.99	m²	1.14
10 mm thick	0.41	0.01	0.15	-	0.49	m²	0.64
Leaf Clearance; clearing grassed area of leaves and other extraneous debris							
Using equipment towed by tractor							
Large grassed areas with perimeters of mature trees such as sports fields and amenity areas	-	0.01	0.19	0.27	-	100 m²	0.46
Large grassed areas containing ornamental trees and shrub beds	-	0.03	0.38	0.38	-	100 m²	0.76
Using pedestrian operated mechanical equipment and blowers							
Grassed areas with perimeters of mature trees such as sports fields and amenity areas	-	0.04	0.60	0.09	-	100 m²	0.69
Grassed areas containing ornamental trees and shrub beds	-	0.10	1.50	0.24	-	100 m²	1.74
Verges	-	0.07	1.00	0.16	-	100 m²	1.16
By hand							
Grassed areas with perimeters of mature trees such as sports fields and amenity areas	-	0.05	0.75	0.09	-	100 m²	0.84
grassed areas containing ornamental trees and shrub beds	-	0.08	1.25	0.15	-	100 m²	1.40
verges	-	1.00	15.00	1.86	-	100 m²	16.86
Removal of arisings							
Areas with perimeters of mature trees	-	0.01	0.10	0.08	0.72	100 m²	0.90
Areas containing ornamental trees and shrub beds	-	0.02	0.30	0.32	1.80	100 m²	2.42
Litter Clearance							
Collection and disposal of litter from grassed area	-	0.01	0.15	-	0.05	100 m²	0.20
Collection and disposal of litter from isolated grassed area not exceeding 1000 m²	-	0.04	0.60	-	0.05	100 m²	0.65
Edge Maintenance							
Maintain edges where lawn abuts pathway or hard surface using							
Strimmer	-	0.01	0.08	0.01	-	m	0.08
Shears	-	0.02	0.25	-	-	m	0.25
Maintain edges where lawn abuts plant bed using							
Mechanical edging tool	-	0.01	0.10	0.03	-	m	0.13
Shears	-	0.01	0.17	-	-	m	0.17
Half moon edging tool	-	0.02	0.30	-	-	m	0.30
Tree Guards, Stakes and Ties etc.							
Adjusting existing tree tie	-	0.03	0.50	-	-	nr	0.50
Taking up single or double tree stake and ties; removing and disposing	-	0.05	0.94	-	-	nr	0.94

Q PAVING/PLANTING/FENCING/SITE FURNITURE

Item Excluding site overheads and profit	PC £	Labour hours	Labour £	Plant £	Material £	Unit	Total rate £
Q35 LANDSCAPE MAINTENANCE - cont'd							
Pruning Shrubs							
Trimming ground cover planting							
Soft groundcover; vinca ivy and the like	-	1.00	15.00	-	-	100 m^2	**15.00**
Woody groundcover; cotoneaster and the like	-	1.50	22.50	-	-	100 m^2	**22.50**
Pruning massed shrub border (measure ground area)							
Shrub beds pruned annually	-	0.01	0.15	-	-	m^2	**0.15**
Shrub beds pruned hard every 3 years	-	0.03	0.42	-	-	m^2	**0.42**
Cutting off dead heads							
Bush or standard rose	-	0.05	0.75	-	-	nr	**0.75**
Climbing rose	-	0.08	1.25	-	-	nr	**1.25**
Pruning Roses							
Bush or standard rose	-	0.05	0.75	-	-	nr	**0.75**
Climbing rose or rambling rose; tying in as required	-	0.07	1.00	-	-	nr	**1.00**
Pruning Ornamental Shrub; height before pruning (increase these rates by 50% if pruning work has not been executed during the previous two years)							
Not exceeding 1m	-	0.04	0.60	-	-	nr	**0.60**
1 to 2 m	-	0.06	0.83	-	-	nr	**0.83**
Exceeding 2 m	-	0.13	1.88	-	-	nr	**1.88**
Removing excess growth etc from face of building etc.; height before pruning							
Not exceeding 2 m	-	0.03	0.43	-	-	nr	**0.43**
2 to 4 m	-	0.05	0.75	-	-	nr	**0.75**
4 to 6 m	-	0.08	1.25	-	-	nr	**1.25**
6 to 8 m	-	0.13	1.88	-	-	nr	**1.88**
8 to 10 m	-	0.14	2.14	-	-	nr	**2.14**
Removing epicormic growth from base of shrub or trunk and base of tree; any height; any diameter; number of growths							
Not exceeding 10	-	0.05	0.75	-	-	nr	**0.75**
10 to 20	-	0.07	1.00	-	-	nr	**1.00**
Beds Borders and Planters							
Lifting							
bulbs	-	0.50	7.50	-	-	100 nr	**7.50**
tubers or corms	-	0.40	6.00	-	-	100 nr	**6.00**
established herbaceous plants; hoeing and depositing for replanting	-	2.00	30.00	-	-	100 nr	**30.00**
Temporary staking and tying in herbaceous plant	-	0.03	0.50	-	0.10	nr	**0.59**
Cutting down spent growth of herbaceous plant; clearing arisings							
unstaked	-	0.02	0.30	-	-	nr	**0.30**
staked; not exceeding 4 stakes per plant; removing stakes and putting into store	-	0.03	0.38	-	-	nr	**0.38**
Hand weeding							
newly planted areas	-	2.00	30.00	-	-	100 m^2	**30.00**
established areas	-	0.50	7.50	-	-	100 m^2	**7.50**
Removing grasses from groundcover areas	-	3.00	45.05	-	-	100 m^2	**45.05**
Hand digging with fork; not exceeding 150 mm Deep; breaking down lumps; leaving surface with a medium tilth	-	1.33	20.00	-	-	100 m^2	**20.00**
Hand digging with fork or spade to an average depth of 230 mm; breaking down lumps; leaving surface with a medium tilth	-	2.00	30.00	-	-	100 m^2	**30.00**
Hand hoeing; not exceeding 50 mm deep; leaving surface with a medium tilth	-	0.40	6.00	-	-	100 m^2	**6.00**
Hand raking to remove stones etc.; breaking down lumps; leaving surface with a fine tilth prior to planting	-	0.67	10.00	-	-	100 m^2	**10.00**

Q PAVING/PLANTING/FENCING/SITE FURNITURE

Item Excluding site overheads and profit	PC £	Labour hours	Labour £	Plant £	Material £	Unit	Total rate £
Hand weeding; planter, window box; not exceeding 1.00 m²							
ground level box	-	0.05	0.75	-	-	nr	0.75
box accessed by stepladder	-	0.08	1.25	-	-	nr	1.25
Spreading only compost, mulch or processed bark to a depth of 75 mm							
on shrub bed with existing mature planting	-	0.09	1.36	-	-	m²	1.36
recently planted areas	-	0.07	1.00	-	-	m²	1.00
groundcover and herbaceous areas	-	0.08	1.13	-	-	m²	1.13
Clearing cultivated area of leaves, litter and other extraneous debris; using hand implement							
weekly maintenance	-	0.13	1.88	-	-	100 m²	1.88
daily maintenance	-	0.02	0.25	-	-	100 m²	0.25
Bedding							
Lifting							
bedding plants; hoeing and depositing for disposal	-	3.00	45.00	-	-	100 m²	45.00
Hand digging with fork; not exceeding 150 mm deep; breaking down lumps; leaving surface with a medium tilth	-	0.75	11.25	-	-	100 m²	11.25
Hand weeding							
newly planted areas	-	2.00	30.00	-	-	100 m²	30.00
established areas	-	0.50	7.50	-	-	100 m²	7.50
Hand digging with fork or spade to an average depth of 230 mm; breaking down lumps; leaving surface with a medium tilth	-	0.50	7.50	-	-	100 m²	7.50
Hand hoeing; not exceeding 50 mm deep; leaving surface with a medium tilth	-	0.40	6.00	-	-	100 m²	6.00
Hand raking to remove stones etc; breaking down lumps; leaving surface with a fine tilth prior to planting	-	0.67	10.00	-	-	100 m²	10.00
Hand weeding; planter, window box; not exceeding 1.00 m²							
ground level box	-	0.05	0.75	-	-	nr	0.75
box accessed by stepladder	-	0.08	1.25	-	-	nr	1.25
Spreading only; compost, mulch or processed bark to a depth of 75 mm							
on shrub bed with existing mature planting	-	0.09	1.36	-	-	m²	1.36
recently planted areas	-	0.07	1.00	-	-	m²	1.00
groundcover and herbaceous areas	-	0.08	1.13	-	-	m²	1.13
Collecting bedding from nursery	-	3.00	45.00	12.73	-	100 m²	57.73
Setting out							
mass planting single variety	-	0.13	1.88	-	-	m²	1.88
pattern	-	0.33	5.00	-	-	m²	5.00
Planting only							
massed bedding plants	-	0.20	3.00	-	-	m²	3.00
Clearing cultivated area of leaves, litter and other extraneous debris; using hand implement							
weekly maintenance	-	0.13	1.88	-	-	100 m²	1.88
daily maintenance	-	0.02	0.25	-	-	100 m²	0.25
Irrigation and watering							
Hand held hosepipe; flow rate 25 litres per minute; irrigation requirement							
10 litres/m²	-	0.74	11.05	-	-	100 m²	11.05
15 litres/m²	-	1.10	16.50	-	-	100 m²	16.50
20 litres/m²	-	1.46	21.95	-	-	100 m²	21.95
25 litres/m²	-	1.84	27.55	-	-	100 m²	27.55
Hand held hosepipe; flow rate 40 litres per minute; irrigation requirement							
10 litres/m²	-	0.46	6.93	-	-	100 m²	6.93
15 litres/m²	-	0.69	10.39	-	-	100 m²	10.39
20 litres/m²	-	0.91	13.70	-	-	100 m²	13.70
25 litres/m²	-	1.15	17.19	-	-	100 m²	17.19

Q PAVING/PLANTING/FENCING/SITE FURNITURE

Item Excluding site overheads and profit	PC £	Labour hours	Labour £	Plant £	Material £	Unit	Total rate £
Q35 LANDSCAPE MAINTENANCE - cont'd							
Hedge Cutting; field hedges cut once or twice annually							
Trimming sides and top using hand tool or hand held mechanical tools							
not exceeding 2 m high	-	0.10	1.50	0.14	-	10 m^2	**1.64**
2 to 4 m high	-	0.33	5.00	0.45	-	10 m^2	**5.45**
Hedge Cutting; ornamental							
Trimming sides and top using hand tool or hand held mechanical tools							
not exceeding 2 m high	-	0.13	1.88	0.17	-	10 m^2	**2.04**
2 to 4 m high	-	0.50	7.50	0.68	-	10 m^2	**8.18**
Hedge Cutting; reducing width; hand tool or hand held mechanical tools							
not exceeding 2 m high							
Average depth of cut not exceeding 300 mm	-	0.10	1.50	0.14	-	10 m^2	**1.64**
Average depth of cut 300 to 600 mm	-	0.83	12.50	1.13	-	10 m^2	**13.63**
Average depth of cut 600 to 900 mm	-	1.25	18.75	1.70	-	10 m^2	**20.45**
2 to 4 m high							
Average depth of cut not exceeding 300 mm	-	0.03	0.38	0.03	-	10 m^2	**0.41**
Average depth of cut 300 to 600 mm	-	0.13	1.88	0.17	-	10 m^2	**2.04**
Average depth of cut 600 to 900 mm	-	2.50	37.50	3.40	-	10 m^2	**40.90**
4 to 6 m high							
Average depth of cut not exceeding 300 mm	-	0.10	1.50	0.14	-	m^2	**1.64**
Average depth of cut 300 to 600 mm	-	0.17	2.50	0.23	-	m^2	**2.73**
Average depth of cut 600 to 900 mm	-	0.50	7.50	0.68	-	m^2	**8.18**
Hedge Cutting; reducing width; tractor mounted hedge cutting equipment							
Not exceeding 2 m high							
Average depth of cut not exceeding 300 mm	-	0.04	0.60	0.62	-	10 m^2	**1.22**
Average depth of cut 300 to 600 mm	-	0.05	0.75	0.78	-	10 m^2	**1.53**
Average depth of cut 600 to 900 mm	-	0.20	3.00	3.11	-	10 m^2	**6.11**
2 to 4 m high							
Average depth of cut not exceeding 300 mm	-	0.01	0.19	0.19	-	10 m^2	**0.38**
Average depth of cut 300 to 600 mm	-	0.03	0.38	0.39	-	10 m^2	**0.76**
Average depth of cut 600 to 900 mm	-	0.02	0.30	0.31	-	10 m^2	**0.61**
Hedge Cutting; reducing height; hand tool or hand held mechanical tools							
Not exceeding 2 m high							
Average depth of cut not exceeding 300 mm	-	0.07	1.00	0.09	-	10 m^2	**1.09**
Average depth of cut 300 to 600 mm	-	0.13	2.00	0.18	-	10 m^2	**2.18**
Average depth of cut 600 to 900 mm	-	0.40	6.00	0.54	-	10 m^2	**6.54**
2 to 4 m high							
Average depth of cut not exceeding 300 mm	-	0.03	0.50	0.05	-	m^2	**0.54**
Average depth of cut 300 to 600 mm	-	0.07	1.00	0.09	-	m^2	**1.09**
Average depth of cut 600 to 900 mm	-	0.20	3.00	0.27	-	m^2	**3.27**
4 to 6 m high							
Average depth of cut not exceeding 300 mm	-	0.07	1.00	0.09	-	m^2	**1.09**
Average depth of cut 300 to 600 mm	-	0.13	1.88	0.17	-	m^2	**2.04**
Average depth of cut 600 to 900 mm	-	0.25	3.75	0.34	-	m^2	**4.09**
Hedge Cutting; removal and disposal of arisings							
Sweeping up and depositing arisings							
300 mm cut	-	0.05	0.75	-	-	10 m^2	**0.75**
600 mm cut	-	0.20	3.00	-	-	10 m^2	**3.00**
900 mm cut	-	0.40	6.00	-	-	10 m^2	**6.00**
Chipping arisings							
300 mm cut	-	0.02	0.30	0.11	-	10 m^2	**0.41**
600 mm cut	-	0.08	1.25	0.46	-	10 m^2	**1.71**
900 mm cut	-	0.20	3.00	1.11	-	10 m^2	**4.11**

Q PAVING/PLANTING/FENCING/SITE FURNITURE

Item Excluding site overheads and profit	PC £	Labour hours	Labour £	Plant £	Material £	Unit	Total rate £
Disposal of unchipped arisings							
300 mm cut	-	0.02	0.25	0.48	1.20	10 m²	**1.93**
600 mm cut	-	0.03	0.50	0.95	1.80	10 m²	**3.25**
900 mm cut	-	0.08	1.25	2.39	4.50	10 m²	**8.14**
Disposal of chipped arisings							
300 mm cut	-	-	0.05	0.16	1.80	10 m²	**2.01**
600 mm cut	-	0.02	0.25	0.16	3.60	10 m²	**4.01**
900 mm cut	-	0.03	0.50	0.32	9.00	10 m²	**9.82**
Herbicide applications; CDA (Controlled droplet application); chemical application via low pressure specialised wands to landscape planting; application to maintain 1.00 m diameter clear circles (0.79m²) around new planting							
Rigby Taylor Ltd; Roundup Pro Green; Glyphosate; enhanced movement Glyphosate; application rate 15 litre/ha.							
plants at 1.50 m centres; 4444 nr/ha	31.56	18.52	347.25	-	34.72	ha	**381.97**
plants at 1.75 m centres; 3265 nr/ha	23.21	13.60	255.00	-	25.54	ha	**280.54**
plants at 2.00 m centres; 2500 nr/ha	17.76	10.42	195.38	-	19.54	ha	**214.91**
mass spraying	90.00	10.00	187.50	-	99.00	ha	**286.50**
Bayer Environmental Science; Vanquish Biactive; enhanced movement Glyphosate; application rate 15 litre/ ha.							
plants at 1.50 m centres; 4444 nr/ha	73.38	18.52	347.25	-	80.71	ha	**427.96**
plants at 1.75 m centres; 3265 nr/ha	53.85	13.60	255.00	-	59.23	ha	**314.23**
plants at 2.00 m centres; 2500 nr/ha	41.29	10.42	195.38	-	45.42	ha	**240.80**
mass spraying	209.25	10.00	187.50	-	230.18	ha	**417.68**
Xanadu; glyphosate and diuron for control of emerged weeds at a single application per season; application rate 20 l/ha							
plants at 1.50 m centres; 4444 nr/ha	98.28	18.52	347.25	-	108.11	ha	**455.36**
plants at 1.75 m centres; 3265 nr/ha	72.10	13.60	255.00	-	79.31	ha	**334.31**
plants at 2.00 m centres; 2500 nr/ha	55.30	10.42	195.38	-	60.83	ha	**256.20**
mass spraying	280.00	10.00	187.50	-	308.00	ha	**495.50**
Herbicide applications; standard backpack spray applicators; application to maintain 1.00 m diameter clear circles (0.79m²) around new planting							
Rigby Taylor Ltd; Roundup Pro Green; Glyphosate; enhanced movement Glyphosate; application rate 5 litre/ ha.							
plants at 1.50 m centres; 4444 nr/ha	16.10	24.69	462.94	-	17.71	ha	**480.65**
plants at 1.75 m centres; 3265 nr/ha	11.78	18.14	340.13	-	12.95	ha	**353.08**
plants at 2.00 m centres; 2500 nr/ha	9.09	13.89	260.44	-	9.99	ha	**270.43**
mass spraying	46.00	13.00	243.75	-	50.60	ha	**294.35**
Q40 FENCING AND GATES							
Protective fencing							
Cleft chestnut rolled fencing; fixing to 100 mm diameter chestnut posts; driving into firm ground at 3 m centres							
900 mm high	3.11	0.11	2.00	-	5.77	m	**7.77**
1200 mm high	3.65	0.16	3.00	-	6.34	m	**9.34**
1500 mm high; 3 strand	5.80	0.21	4.00	-	8.30	m	**12.31**
Enclosures; Earth Anchors							
"Rootfast" anchored galvanized steel enclosures post ref ADP 20-1000; 1000 mm high x 20 mm diameter with ref AA25-750 socket and							
padlocking ring	32.00	0.10	1.88	-	34.24	nr	**36.11**
steel cable, orange plastic coated	1.25	-	0.04	-	1.25	m	**1.29**

Q PAVING/PLANTING/FENCING/SITE FURNITURE

Item Excluding site overheads and profit	PC £	Labour hours	Labour £	Plant £	Material £	Unit	Total rate £
Q40 FENCING AND GATES - cont'd							
Boundary fencing; strained wire and wire mesh; AVS Fencing Supplies Ltd							
Strained wire fencing; concrete posts only at 2750 mm centres, 610 mm below ground; excavating holes; filling with concrete; replacing topsoil; disposing surplus soil off site							
900 mm high	2.69	0.56	10.42	0.50	3.97	m	**14.88**
1200 mm high	3.36	0.56	10.42	0.50	4.64	m	**15.56**
1400 mm high	3.85	0.56	10.42	0.50	5.13	m	**16.04**
1800 mm high	4.58	0.78	14.58	1.24	5.86	m	**21.69**
2400 mm high	6.33	1.67	31.25	0.50	7.61	m	**39.36**
Extra over strained wire fencing for concrete straining posts with one strut; posts and struts 610 mm below ground; struts cleats, stretchers, winders, bolts, and eye bolts; excavating holes; filling to within 150 mm of ground level with concrete (1:12) - 40 mm aggregate; replacing topsoil; disposing surplus soil off site							
900 mm high	21.63	0.67	12.56	1.12	48.10	nr	**61.78**
1200 mm high	24.04	0.67	12.50	1.12	51.05	nr	**64.68**
1400 mm high	27.46	0.67	12.50	1.12	58.05	nr	**71.68**
1800 mm high	34.23	0.67	12.50	1.12	62.85	nr	**76.48**
2400 mm high	46.96	0.83	15.62	1.33	82.50	nr	**99.45**
Extra over strained wire fencing for concrete straining posts with two struts; posts and struts 610 mm below ground; excavating holes; filling to within 150 mm of ground level with concrete (1:12) - 40 mm aggregate; replacing topsoil; disposing surplus soil off site							
900 mm high	31.20	0.91	17.06	1.44	70.49	nr	**88.99**
1200 mm high	34.75	0.85	15.94	1.12	74.94	nr	**92.00**
1400 mm high	39.84	0.91	17.02	1.12	90.54	nr	**108.69**
1800 mm high	49.69	0.85	15.94	1.12	96.45	nr	**113.51**
2400 mm high	46.96	0.67	12.50	1.33	98.90	nr	**112.73**
Strained wire fencing; galvanised steel angle posts only at 2750 mm centres; 610 mm below ground; driving in							
900 mm high; 40 mm x 40 mm x 5 mm	4.14	0.03	0.57	-	4.14	m	**4.71**
1200 mm high; 40 mm x 40 mm x 5 mm	4.86	0.03	0.62	-	4.86	m	**5.49**
1400 mm high; 40 mm x 40 mm x 5 mm	5.80	0.06	1.14	-	5.80	m	**6.93**
1800 mm high; 40 mm x 40 mm x 5 mm	7.05	0.07	1.36	-	7.05	m	**8.41**
2400 mm high; 45 mm x 45 mm x 5 mm	8.55	0.12	2.27	-	8.55	m	**10.82**
Galvanised steel angle straining posts with one strut for strained wire fencing; setting in concrete							
900 mm high; 50 mm x 50 mm x 6 mm	-	-	-	-	42.41	nr	**42.41**
1200 mm high; 50 mm x 50 mm x 6 mm	-	-	-	-	59.32	nr	**59.32**
1400 mm high; 50 mm x 50 mm x 6 mm	-	-	-	-	68.95	nr	**68.95**
1800 mm high; 50 mm x 50 mm x 6 mm	-	-	-	-	78.40	nr	**78.40**
2400 mm high; 60 mm x 60 mm x 6 mm	-	-	-	-	93.60	nr	**93.60**
Galvanised steel straining posts with two struts for strained wire fencing; setting in concrete							
900 mm high; 50 mm x 50 mm x 6 mm	-	-	-	-	63.46	nr	**63.46**
1200 mm high; 50 mm x 50 mm x 6 mm	-	-	-	-	87.32	nr	**87.32**
1400 mm high; 50 mm x 50 mm x 6 mm	-	-	-	-	103.38	nr	**103.38**
1800 mm high; 50 mm x 50 mm x 6 mm	-	-	-	-	125.69	nr	**125.69**
2400 mm high; 60 mm x 60 mm x 6 mm	-	-	-	-	156.00	nr	**156.00**
Strained wire; to posts (posts not included); 3 mm galvanized wire; fixing with galvanized stirrups							
900 mm high; 2 wire	0.22	0.03	0.62	-	0.31	m	**0.93**
1200 mm high; 3 wire	0.33	0.05	0.87	-	0.42	m	**1.29**
1400 mm high; 3 wire	0.33	0.05	0.87	-	0.42	m	**1.29**
1800 mm high; 3 wire	0.33	0.05	0.87	-	0.42	m	**1.29**

Q PAVING/PLANTING/FENCING/SITE FURNITURE

Item Excluding site overheads and profit	PC £	Labour hours	Labour £	Plant £	Material £	Unit	Total rate £
Barbed wire; to posts (posts not included); 3 mm galvanised wire; fixing with galvanised stirrups							
900 mm high; 2 wire	0.29	0.07	1.25	-	0.38	m	**1.63**
1200 mm high; 3 wire	0.43	0.09	1.75	-	0.52	m	**2.27**
1400 mm high; 3 wire	0.43	0.09	1.75	-	0.52	m	**2.27**
1800 mm high; 3 wire	0.43	0.09	1.75	-	0.52	m	**2.27**
Chain link fencing; AVS Fencing Supplies Ltd; to strained wire and posts priced separately; 3 mm galvanised wire; 51 mm mesh; galvanized steel components; fixing to line wires threaded through posts and strained with eye-bolts; posts (not included)							
900 mm high	4.58	0.07	1.25	-	4.67	m	**5.92**
1200 mm high	4.58	0.07	1.25	-	4.67	m	**5.92**
1800 mm high	7.77	0.10	1.88	-	7.86	m	**9.73**
2400 mm high	7.77	0.10	1.88	-	7.86	m	**9.73**
Chain link fencing; to strained wire and posts priced separately; 3.15 mm plastic coated galvanized wire (wire only 2.50 mm); 51 mm mesh; galvanised steel components; fencing with line wires threaded through posts and strained with eye-bolts; posts (not included) (Note: plastic coated fencing can be cheaper than galvanised finish as wire of a smaller cross-sectional area can be used)							
900 mm high	1.98	0.07	1.25	-	2.16	m	**3.41**
1200 mm high	2.63	0.07	1.25	-	2.81	m	**4.06**
1400 mm high	3.29	0.07	1.25	-	3.47	m	**4.72**
1800 mm high	3.95	0.13	2.34	-	4.67	m	**7.01**
Extra over strained wire fencing for cranked arms and galvanized barbed wire							
1 row	2.37	0.02	0.31	-	2.37	m	**2.68**
2 row	2.51	0.05	0.94	-	2.51	m	**3.45**
3 row	2.66	0.05	0.94	-	2.66	m	**3.60**
Field fencing; Jacksons Fencing; welded wire mesh; fixed to posts and straining wires measured separately							
Cattle fence; 1100m high 114 x 300 mm at bottom to 230 x 300 mm at top	0.84	0.10	1.88	-	1.04	m	**2.91**
Sheep fence; 900 mm high; 140 x 300 mm at bottom to 230 x 300 mm at top	0.76	0.10	1.88	-	0.96	m	**2.83**
Deer Fence; 1900 mm high; 89 x 150 mm at bottom to 267 x 300 mm at top	3.04	0.13	2.34	-	3.24	m	**5.58**
Extra for concreting in posts	-	-	-	-	1.70	nr	**1.70**
Extra for straining post	8.82	0.75	14.06	-	8.82	nr	**22.88**
Rabbit netting; Jacksons Fencing; timber stakes; peeled kiln dried pressure treated; pointed; 1.8 m posts driven 900 mm into ground at 3 m centres (line wires and netting priced separately)							
75 -100 mm stakes	1.62	0.25	4.69	-	1.62	m	**6.31**
Corner posts or straining posts 150 mm diameter 2.3 m high set in concrete; centres to suit local conditions or changes of direction							
1 strut	10.44	1.00	18.75	3.20	16.51	each	**38.46**
2 strut	12.38	1.00	18.75	3.20	18.45	each	**40.40**
Strained wire; to posts (posts not included); 3 mm galvanized wire; fixing with galvanized stirrups							
900 mm high; 2 wire	0.22	0.03	0.62	-	0.31	m	**0.93**
1200 mm high; 3 wire	0.33	0.05	0.87	-	0.42	m	**1.29**

Q PAVING/PLANTING/FENCING/SITE FURNITURE

Item Excluding site overheads and profit	PC £	Labour hours	Labour £	Plant £	Material £	Unit	Total rate £
Q40 FENCING AND GATES - cont'd							
Rabbit netting; 31 mm 19 gauge 1050 **high netting fixed to posts line wires and** **straining posts or corner posts all priced** **separately**							
900 high turned in	0.70	0.04	0.75	-	0.71	m	**1.46**
900 high buried 150 mm in trench	0.70	0.08	1.56	-	0.71	m	**2.27**
Boundary fencing; strained wire and wire **mesh; Jacksons Fencing** Tubular chain link fencing; galvanized; plastic-coated; 60.3 mm diameter posts at 3.0 m centres; setting 700 mm into ground; choice of ten mesh colours; including excavating holes; backfilling and removing surplus soil; with top rail only							
900 mm high	-	-	-	-	-	m	**23.10**
1200 mm high	-	-	-	-	-	m	**24.35**
1800 mm high	-	-	-	-	-	m	**27.01**
2000 mm high	-	-	-	-	-	m	**28.64**
Tubular chain link fencing; galvanised; plastic coated; 60.3 mm diameter posts at 3.0 m centres; cranked arms and 3 lines barbed wire; setting 700 mm into ground; including excavating holes; backfilling and removing surplus soil; with top rail only							
2000 mm high	-	-	-	-	-	m	**29.99**
1800 mm high	-	-	-	-	-	m	**29.41**
Boundary fencing; Steelway-Fensecure Ltd "Classic" 2 rail tubular fencing; top and bottom with strecher bars and straining wires in between; comprising 60.3 mm tubular posts at 3.00 m centres; setting in concrete; 35 mm top rail tied with aluminium and steel fittings; 50 mm x 50 mm x 355/2.5 mm PVC coated chain link; all components galvanized and coated in green nylon							
964 mm high	-	-	-	-	-	m	**28.29**
1269 mm high	-	-	-	-	-	m	**29.99**
1574 mm high	-	-	-	-	-	m	**34.81**
1878 mm high	-	-	-	-	-	m	**35.49**
2188 mm high	-	-	-	-	-	m	**37.37**
2458 mm high	-	-	-	-	-	m	**41.87**
2948 mm high	-	-	-	-	-	m	**53.78**
3562 mm high	-	-	-	-	-	m	**55.88**
End Posts; Classic range 60.3 mm diameter; setting in concrete							
964 mm high	-	-	-	-	-	nr	**40.56**
1269 mm high	-	-	-	-	-	nr	**43.48**
1574 mm high	-	-	-	-	-	nr	**47.47**
1878 mm high	-	-	-	-	-	nr	**52.56**
2188 mm high	-	-	-	-	-	nr	**57.02**
2458 mm high	-	-	-	-	-	nr	**66.03**
2948 mm high	-	-	-	-	-	nr	**74.91**
3562 mm high	-	-	-	-	-	nr	**88.08**
Corner Posts; 60.3 mm diameter; setting in concrete							
964 mm high	-	-	-	-	-	nr	**32.29**
1269 mm high	-	-	-	-	-	nr	**15.99**
1574 mm high	-	-	-	-	-	nr	**36.81**
1878 mm high	-	-	-	-	-	nr	**37.49**
2188 mm high	-	-	-	-	-	nr	**43.87**
2458 mm high	-	-	-	-	-	nr	**46.37**
2948 mm high	-	-	-	-	-	nr	**55.78**
3562 mm high	-	-	-	-	-	nr	**62.38**

Q PAVING/PLANTING/FENCING/SITE FURNITURE

Item Excluding site overheads and profit	PC £	Labour hours	Labour £	Plant £	Material £	Unit	Total rate £
Boundary fencing; McArthur Group							
"Paladin" Welded mesh colour coated green							
fencing; fixing to metal posts at 2.975 m centres							
with manufacturer's fixings; setting 600 mm deep							
in firm ground; including excavating holes;							
backfilling and removing surplus excavated							
material; includes 10 year product guarantee							
1800 mm high	57.20	0.66	12.38	-	57.20	m	69.58
2000 mm high	65.98	0.66	12.38	-	65.98	m	78.35
2400 mm high	6.47	0.66	12.38	-	75.07	m	87.44
Extra over welded galvanised plastic coated							
mesh fencing for concreting in posts	7.95	0.11	2.08	-	7.95	m	10.03
Boundary fencing; H. Langdon & Son							
"Orsogril" rectangular steel bar mesh fence							
panels; pleione pattern; bolting to 60 mm x 8 mm							
uprights at 2 m centres; mesh 62 mm x 66 mm;							
setting in concrete							
930 mm high panels	-	-	-	-	-	m	105.86
1326 mm high panels	-	-	-	-	-	m	146.53
1722 mm high panels	-	-	-	-	-	m	176.22
Langdons; "Orsogril" rectangular steel bar mesh							
fence panels; pleione pattern; bolting to 8 mm x							
80 mm uprights at 2 m centres; mesh 62 mm x 66							
mm; setting in concrete							
930 mm high panel	-	-	-	-	-	m	108.45
1326 mm high panel	-	-	-	-	-	m	142.98
1722 mm high panel	-	-	-	-	-	m	181.73
"Orsogril" rectangular steel bar mesh fence							
panels; sterope pattern; bolting to 60 mm x 80							
mm uprights at 2.00 m centres; mesh 62 mm x							
132 mm; setting in concrete paving (paving not							
included)							
930 mm high panels	-	-	-	-	-	m	113.57
1326 mm high panels	-	-	-	-	-	m	148.86
1722 mm high panels	-	-	-	-	-	m	187.22
Security fencing; Jacksons Fencing							
"Barbican" galvanized steel paling fencing; on							
60 mm x 60 mm posts at 3 m centres; setting in							
concrete							
1250 mm high	-	-	-	-	-	m	59.21
1500 mm high	-	-	-	-	-	m	63.38
2000 mm high	-	-	-	-	-	m	73.07
2500 mm high	-	-	-	-	-	m	85.87
Gates; to match "Barbican" galvanised steel							
paling fencing							
width 1m	-	-	-	-	-	nr	778.03
width 2 m	-	-	-	-	-	nr	787.55
width 3 m	-	-	-	-	-	nr	808.96
width 4 m	-	-	-	-	-	nr	836.32
width 8 m	-	-	-	-	-	pair	1704.77
width 9m	-	-	-	-	-	pair	2055.71
width 10 m	-	-	-	-	-	pair	2336.48
Security fencing; Jacksons Fencing;							
intruder guards							
"Viper Spike Intruder Guards"; to existing							
structures, including fixing bolts							
ref Viper 1; 40 mm x 5 mm x 1.1 m long with base							
plate	24.70	1.00	18.75	-	24.70	nr	43.45
ref Viper 3; 160 mm x 190 mm wide; U shape; to							
prevent intruders climbing pipes	31.05	0.50	9.38	-	31.05	nr	40.42

Q PAVING/PLANTING/FENCING/SITE FURNITURE

Item Excluding site overheads and profit	PC £	Labour hours	Labour £	Plant £	Material £	Unit	Total rate £
Q40 FENCING AND GATES - cont'd							
Security fencing; Jacksons Fencing "Razor Barb Concertina"; spiral wire security barriers; fixing to 600 mm steel ground stakes							
ref 3275; 450 mm diameter roll - medium barb; galvanized	1.95	0.02	0.38	-	2.07	m	**2.45**
ref 3276; 730 mm diameter roll - medium barb; galvanized	2.49	0.02	0.38	-	2.61	m	**2.99**
ref 3277; 950 mm diameter roll - medium barb; galvanized	2.50	0.02	0.38	-	2.63	m	**3.00**
3 lines "Barbed Tape" Medium Barb barbed tape; on 50 mm x 50 mm mild steel angle posts; setting in concrete							
ref 3283; Barbed Tape medium barb; galvanized	0.72	0.21	3.95	-	12.15	m	**16.10**
5 lines "Barbed Tape" Medium Barb barbed tape; on 50 mm x 50 mm mild steel angle posts; setting in concrete							
ref 3283; Barbed Tape medium barb; galvanized	1.20	0.22	4.17	-	12.63	m	**16.79**
Trip rails; metal; Broxap Streetscene double row 48.3 mm internal diameter mild steel tubular rails with sleeved joints; to cast iron posts 525 mm above ground mm long; setting in concrete at 1.20 m centres; all standard painted							
BX 201; Cast iron with single rail 42 mm diameter rail	54.83	1.00	18.75	-	60.46	m	**79.21**
22 mm diameter mild steel rails with ferrule joints; fixing through holes in 44 mm x 13 mm mild steel standards 700 mm long; setting in concrete at 1.20 m centres; priming	22.66	1.33	25.00	-	22.93	m	**47.93**
Heavy duty; Anti Ram 540 high with 100 x 100 mm rail; Birdsmouth style	40.65	1.00	18.75	-	40.92	m	**59.67**
Trip rails; Birdsmouth; Jacksons Fencing Diamond rail fencing (Birdsmouth); posts 100 x 100 mm softwood planed at 1.35 m centres set in 1:3:6 concrete; rail 75 x 75 nominal secured with galvanized straps nailed to posts							
Posts 900 mm (600 above ground) at 1.35 m centres	6.32	0.50	9.38	0.21	7.97	m	**17.56**
Posts 1.20 mm (900 above ground) at 1.35 m centres	7.22	0.50	9.38	0.21	8.87	m	**18.46**
Posts 900 mm (600 above ground) at 1.80 m centres	5.30	0.38	7.05	0.16	6.56	m	**13.77**
Posts 1.20 mm (900 above ground) at 1.80m centres	5.98	0.38	7.05	0.16	7.22	m	**14.43**
Trip rails; Townscape Products Ltd Hollow steel section knee rails; galvanised; 500 mm high; setting in concrete							
1000 mm bays	80.22	2.00	37.50	-	86.71	m	**124.21**
1200 mm bays	16.90	2.00	37.50	-	75.28	m	**112.77**
Timber fencing; AVS Fencing Supplies Ltd Palisade fencing; 22 mm x 75 mm softwood vertical palings with flat tops; nailing to 3 nr 50 mm x 100 mm horizontal softwood rails; housing into 100 mm x 100 mm softwood posts with weathered tops at 3.00 m centres; setting in concrete; all treated timber							
1800 mm high	17.38	0.47	8.76	-	21.27	m	**30.03**

Q PAVING/PLANTING/FENCING/SITE FURNITURE

Item Excluding site overheads and profit	PC £	Labour hours	Labour £	Plant £	Material £	Unit	Total rate £
Palisade fencing; 22 mm x 75 mm oak vertical palings with flat tops; nailing to 3 nr 50 mm x 125 mm horizontal oak rails; housing into 100 mm x 100 mm oak posts with weathered tops at 3.00 m centres; setting in concrete							
1800 mm high	33.83	0.47	8.75	-	37.72	m	**46.47**
Post-and-rail fencing; 3 nr 90 mm x 38 mm softwood horizontal rails; fixing with galvanized nails to 150 mm x 75 mm softwood posts; including excavating and backfilling into firm ground at 1.80 m centres; all treated timber							
1200 mm high	9.86	0.35	6.57	-	9.91	m	**16.47**
Post-and-rail fencing; 3 nr 90 mm x 38 mm oak horizontal rails; fixing with galvanised nails to 150 mm x 75 mm oak posts; including excavating and backfilling into firm ground at 1.80 m centres							
1200 mm high	24.72	0.35	6.57	-	24.77	m	**31.34**
Morticed post-and-rail fencing; 3 nr horizontal 90 mm x 38 mm softwood rails; fixing with galvanized nails; 90 mm x 38 mm softwood centre prick posts; to 150 mm x 75 mm softwood posts; including excavating and backfilling into firm ground at 2.85 m centres; all treated timber							
1200 mm high	8.39	0.35	6.57	-	8.44	m	**15.01**
1350 mm high five rails	10.06	0.40	7.50	-	10.11	m	**17.61**
Cleft rail fencing; oak or chestnut adze tapered rails 2.80 m long; morticed into joints; to 125 mm x 100 mm softwood posts 1.95 m long; including excavating and backfilling into firm ground at 2.80 m centres							
two rails	12.91	-	-	-	12.91	m	**12.91**
three rails	15.56	0.28	5.25	-	15.56	m	**20.81**
four rails	18.21	0.35	6.57	-	18.21	m	**24.78**
Close boarded fencing; 2 nr softwood rails; 150 x 25 mm gravel boards; 2 ex 100 x 22 softwood pales lapped 13 mm; to concrete posts; including excavating and backfilling into firm ground at 3.00 m centres							
900 mm high	10.70	0.47	8.76	-	11.98	m	**20.74**
1050 mm high	7.43	0.47	8.76	-	13.71	m	**22.47**
1200 mm high	8.03	0.47	8.76	-	14.31	m	**23.07**
Close boarded fencing; 3 nr softwood rails; 150 x 25 mm gravel boards; 89 mm x 19 mm softwood pales lapped 13 mm; to concrete posts; including excavating and backfilling into firm ground at 3.00 m centres							
1350 mm high	15.17	0.56	10.50	-	16.45	m	**26.96**
1650 mm high	16.43	0.56	10.50	-	17.71	m	**28.21**
1800 mm high	16.67	0.56	10.50	-	17.95	m	**28.45**
Close boarded fencing; 3 nr softwood rails; 150 x 25 mm gravel boards; 89 mm x 19 mm softwood pales lapped 13 mm; to softwood posts 100 x 100 mm; including excavating and backfilling into firm ground at 3.00 m centres							
1350 mm high	11.54	0.56	10.50	-	12.82	m	**23.32**
1650 mm high	12.53	0.56	10.50	-	13.81	m	**24.31**
1800 mm high	12.96	0.56	10.50	-	14.24	m	**24.75**
Close boarded fencing; 3 nr softwood rails; 150 x 25 mm gravel boards; 89 mm x 19 mm softwood pales lapped 13 mm; to oak posts; including excavating and backfilling into firm ground at 3.00 m centres							
1350 mm high	9.79	0.56	10.50	-	15.66	m	**26.17**
1650 mm high	16.00	0.56	10.50	-	17.28	m	**27.79**
1800 mm high	16.75	0.56	10.50	-	18.03	m	**28.54**

Q PAVING/PLANTING/FENCING/SITE FURNITURE

Item Excluding site overheads and profit	PC £	Labour hours	Labour £	Plant £	Material £	Unit	Total rate £
Q40 FENCING AND GATES - cont'd							
Timber fencing; AVS Fencing Supplies Ltd - cont'd							
Close boarded fencing; 3 nr softwood rails; 150 x 25 mm gravel boards; 89 mm x 19 mm softwood pales lapped 13 mm; to oak posts; including excavating and backfilling into firm ground at 2.47 m centres							
1800 mm high	18.07	0.68	12.71	-	19.35	m	32.05
Close boarded fencing - all oak; 2 nr 76 mm x 38 mm rectangular rails; 152 mm x 25 mm gravel boards; 89 mm x 19 mm oak featheredge boards lapped 13 mm; to oak posts; including excavating and backfilling into firm ground at 2.47 m centres							
1070 mm high	7.85	0.56	10.50	-	32.77	m	43.27
1800 mm high	38.51	0.56	10.50	-	39.79	m	50.29
Extra over close boarded fencing for 65 mm x 38 mm oak cappings	4.99	0.05	0.94	-	4.99	nr	5.93
"Hit and miss" horizontal rail fencing; 87 mm x 38 mm top and bottom rails; 100 mm x 22 mm vertical boards arranged alternately on opposite side of rails; to 100 mm x 100 mm posts; including excavating and backfilling into firm ground; setting in concrete at 1.8 m centres							
treated softwood; 1800 mm high	20.69	1.33	25.00	-	23.25	m	48.25
treated softwood; 2000 mm high	21.53	1.33	25.00	-	26.89	m	51.89
primed softwood; 1800 mm high	20.69	1.67	31.25	-	23.63	m	54.88
primed softwood; 2000 mm high	21.53	1.67	31.25	-	27.26	m	58.52
Screen fencing in horizontal wavy edge overlap panels; 1.83 m wide; to 100 mm x 100 mm preservative treated softwood posts; including excavating and backfilling into firm ground							
900 mm high	6.67	0.47	8.76	-	11.90	m	20.66
1200 mm high	6.74	0.47	8.76	-	12.49	m	21.25
1500 mm high	6.78	0.56	10.53	-	13.94	m	24.48
1800 mm high	6.80	0.56	10.53	-	14.44	m	24.98
Screen fencing for oak in lieu of softwood posts							
900 mm high	5.23	-	-	-	5.23	m	5.23
1200 mm high	6.54	-	-	-	6.54	m	6.54
1500 mm high	7.85	-	-	-	7.85	m	7.85
1800 mm high	9.11	-	-	-	9.11	m	9.11
Extra over screen fencing for 300 mm high trellis tops; slats at 100 mm centres; including additional length of posts	3.08	0.10	1.88	-	3.08	m	4.96
Timber trellis; Landscapes by Design; **pressure treated softwood panels** Timber section 38 x 19 mm at 120 mm centres at 45, 55 or 90 deg angle; framed and weathered; panel size 1.8 x 1.8 m							
fixed to timber freestanding posts 100 x 100 mm	-	-	-	-	-	m	63.00
fixed to wall on 38 mm batten	-	-	-	-	-	m	58.80
Extra for post finials	-	-	-	-	-	nr	10.08
extra for angled top to each trellis panel at 55 deg projecting 450 mm above the top frame of each panel	-	-	-	-	-	nr	12.60
extra for scalloped radius top depression 300 mm into each panel	-	-	-	-	-	nr	15.75
extra for painting with Sadolin	-	-	-	-	-	m	12.60

Q PAVING/PLANTING/FENCING/SITE FURNITURE

Item Excluding site overheads and profit	PC £	Labour hours	Labour £	Plant £	Material £	Unit	Total rate £
Concrete fencing							
Panel fencing; to precast concrete posts; in 2 m bays; setting posts 600 mm into ground; sandfaced finish							
900 mm high	12.93	0.25	4.69	-	14.50	m	19.19
1200 mm high	17.24	0.25	4.69	-	18.81	m	23.50
Panel fencing; to precast concrete posts; in 2 m bays; setting posts 750 mm into ground; sandfaced finish							
1500 mm high	21.55	0.33	6.25	-	23.12	m	29.37
1800 mm high	25.87	0.36	6.82	-	27.43	m	34.25
2100 mm high	30.18	0.40	7.50	-	31.74	m	39.24
2400 mm high	35.92	0.40	7.50	-	37.49	m	44.99
Extra over concrete panel fencing for aggregate faced one side	8.20	-	-	-	8.20	m²	8.20
Windbreak fencing							
Fencing; English Woodlands; "Tensar Shade and Shelter Netting" windbreak fencing; green; to 100 mm diameter treated softwood posts; setting 450 mm into ground; fixing with 50 mm x 25 mm treated softwood battens nailed to posts, including excavating and backfilling into firm ground; setting in concrete at 3 m centres							
1200 mm high	1.24	0.16	2.92	-	4.48	m	7.39
1800 mm high	1.76	0.16	2.92	-	5.00	m	7.91
Ball stop fencing; Steelway-Fensecure Ltd							
Ball Stop Net; 30 x 30 mm netting fixed to 60.3 diameter 12 mm solid bar lattice galvanised dual posts, top, middle and bottom rails							
4.5 m high	-	-	-	-	-	m	74.19
5 .0 m high	-	-	-	-	-	m	84.99
6.0 m high	-	-	-	-	-	m	93.55
7.0 m high	-	-	-	-	-	m	101.11
8.0 m high	-	-	-	-	-	m	109.05
9.0 m high	-	-	-	-	-	m	117.31
10.0 m high	-	-	-	-	-	m	125.95
Ball Stop Net; Corner posts							
4.5 m high	-	-	-	-	-	m	51.41
5.0 m high	-	-	-	-	-	m	56.27
6.0 m high	-	-	-	-	-	m	87.19
7.0 m high	-	-	-	-	-	m	77.87
8.0 m high	-	-	-	-	-	m	88.67
9.0 m high	-	-	-	-	-	m	99.47
10 m high	-	-	-	-	-	m	110.27
Ball Stop Net; End posts							
4.5 m high	-	-	-	-	-	m	42.77
5.0 m high	-	-	-	-	-	m	47.63
6.0 m high	-	-	-	-	-	m	58.43
7.0 m high	-	-	-	-	-	m	69.23
8.0 m high	-	-	-	-	-	m	80.03
9.0 m high	-	-	-	-	-	m	90.83
10 m high	-	-	-	-	-	m	102.71
Railings; H. Langdon & Son							
Mild steel bar railings of balusters at 115 mm centres welded to flat rail top and bottom; bays 2.00 m long; bolting to 51 mm x 51 mm hollow square section posts; setting in concrete							
galvanized; 900 mm high	-	-	-	-	-	m	57.00
galvanized; 1200 mm high	-	-	-	-	-	m	69.05
galvanized; 1500 mm high	-	-	-	-	-	m	76.73
galvanized; 1800 mm high	-	-	-	-	-	m	92.06

Q PAVING/PLANTING/FENCING/SITE FURNITURE

Item Excluding site overheads and profit	PC £	Labour hours	Labour £	Plant £	Material £	Unit	Total rate £
Q40 FENCING AND GATES - cont'd							
Railings; H. Langdon & Son - cont'd							
primed; 900 mm high	-	-	-	-	-	m	53.72
primed; 1200 mm high	-	-	-	-	-	m	65.23
primed; 1500 mm high	-	-	-	-	-	m	72.12
primed; 1800 mm high	-	-	-	-	-	m	85.94
Mild steel blunt top railings of 19 mm balusters at 130 mm centres welded to bottom rail; passing through holes in top rail and welded; top and bottom rails 40 mm x 10 mm; bolting to 51 mm x 51 mm hollow square section posts; setting in concrete							
galvanized; 800 mm high	-	-	-	-	-	m	61.38
galvanized; 1000 mm high	-	-	-	-	-	m	64.45
galvanized; 1300 mm high	-	-	-	-	-	m	76.73
galvanized; 1500 mm high	-	-	-	-	-	m	82.87
primed; 800 mm high	-	-	-	-	-	m	59.08
primed; 1000 mm high	-	-	-	-	-	m	61.38
primed; 1300 mm high	-	-	-	-	-	m	72.90
primed; 1500 mm high	-	-	-	-	-	m	78.28
Railings; traditional pattern; 16 mm diameter verticals at 127 mm intervals with horizontal bars near top and bottom; balusters with spiked tops; 51 x 20 mm standards; including setting 520 mm into concrete at 2.75 m centres							
primed; 1200 mm high	-	-	-	-	-	m	69.05
primed; 1500 mm high	-	-	-	-	-	m	78.28
primed; 1800 mm high	-	-	-	-	-	m	84.40
Interlaced bow-top mild steel railings; traditional park type; 16 mm diameter verticals at 80 mm intervals, welded at bottom to 50 x 10 mm flat and slotted through 38 mm x 8 mm top rail to form hooped top profile; 50 mm x 10 mm standards; setting 560 mm into concrete at 2.75 m centres							
galvanized; 900 mm high	-	-	-	-	-	m	104.36
galvanized; 1200 mm high	-	-	-	-	-	m	127.38
galvanized; 1500 mm high	-	-	-	-	-	m	151.93
galvanized; 1800 mm high	-	-	-	-	-	m	176.48
primed; 900 mm high	-	-	-	-	-	m	101.28
primed; 1200 mm high	-	-	-	-	-	m	122.77
primed; 1500 mm high	-	-	-	-	-	m	145.80
primed; 1800 mm high	-	-	-	-	-	m	168.80
Metal Estate Fencing; Jacksons Fencing; mild steel flat bar angular fencing; galvanized; main posts at 5.00 m centres; fixed with 1:3:6 concrete 430 deep; intermediate posts fixed with fixing claw at 1.00 m centres							
5 nr plain flat or round bar							
1200 high; under 500 m	44.10	0.28	5.21	-	46.34	m	51.54
1200 high; over 500 m	37.80	0.28	5.21	-	40.04	m	45.24
Five bar horizontal steel fencing; 1.20 m high; 38 mm x 10 mm joiner standards at 4.50 m centres; 38 mm x 8 mm intermediate standards at 900 mm centres; including excavating and backfilling into firm ground; setting in concrete							
galvanized	-	-	-	-	-	m	81.50
primed	-	-	-	-	-	m	67.00
Extra over five bar horizontal steel fencing for 76 mm diameter end and corner posts							
galvanized	-	-	-	-	-	each	89.10
primed	-	-	-	-	-	each	52.14

Q PAVING/PLANTING/FENCING/SITE FURNITURE

Item Excluding site overheads and profit	PC £	Labour hours	Labour £	Plant £	Material £	Unit	Total rate £
H. Langdon & Son; Five bar horizontal steel fencing with mild steel copings; 38 mm square hollow section mild steel standards at 1.80 m centres; including excavating and backfilling into firm ground; setting in concrete							
galvanized; 900 mm high	-	-	-	-	-	m	106.66
galvanized; 1200 mm high	-	-	-	-	-	m	112.20
galvanized; 1500 mm high	-	-	-	-	-	m	120.12
galvanized; 1800 mm high	-	-	-	-	-	m	128.04
primed; 900 mm high	-	-	-	-	-	m	94.64
primed; 1200 mm high	-	-	-	-	-	m	97.15
primed; 1500 mm high	-	-	-	-	-	m	102.43
primed; 1800 mm high	-	-	-	-	-	m	110.22
Gates - General							
Preamble: Gates in fences; see specification for fencing, as gates in traditional or proprietary fencing systems are usually constructed of the same materials and finished as the fencing itself.							
Gates, hardwood; Longlyf Timber Products Ltd							
Hardwood entrance gate, five bar diamond braced, curved hanging stile, planed iroko; fixed to 150 mm x 150 mm softwood posts; inclusive of hinges and furniture							
ref 1100 040; 0.9 m wide	165.17	5.00	93.75	-	219.05	nr	312.80
ref 1100 041; 1.2 m wide	176.73	5.00	93.75	-	230.61	nr	324.36
ref 1100 042; 1.5 m wide	229.56	5.00	93.75	-	283.44	nr	377.19
ref 1100 043; 1.8 m wide	242.95	5.00	93.75	-	296.83	nr	390.58
ref 1100 044; 2.1 m wide	266.14	5.00	93.75	-	320.02	nr	413.77
ref 1100 045; 2.4 m wide	281.39	5.00	93.75	-	335.27	nr	429.02
ref 1100 047; 3.0 m wide	305.92	5.00	93.75	-	359.80	nr	453.55
ref 1100 048; 3.3 m wide	319.52	5.00	93.75	-	373.40	nr	467.15
ref 1100 049; 3.6 m wide	330.44	5.00	93.75	-	384.32	nr	478.07
Hardwood field gate, five bar diamond braced, planed iroko; fixed to 150 mm x 150 mm softwood posts; inclusive of hinges and furniture							
ref 1100 100; 0.9 m wide	119.11	4.00	75.00	-	172.99	nr	247.99
ref 1100 101; 1.2 m wide	124.26	4.00	75.00	-	178.14	nr	253.14
ref 1100 102; 1.5 m wide	129.42	4.00	75.00	-	183.30	nr	258.30
ref 1100 103; 1.8 m wide	165.84	4.00	75.00	-	219.72	nr	294.72
ref 1100 104; 2.1 m wide	204.24	4.00	75.00	-	258.12	nr	333.12
ref 1100 105; 2.4 m wide	216.36	4.00	75.00	-	270.24	nr	345.24
ref 1100 106; 2.7 m wide	228.53	4.00	75.00	-	282.41	nr	357.41
ref 1100 108; 3.3 m wide	252.87	4.00	75.00	-	306.75	nr	381.75
ref 1100 109; 3.6 m wide	265.65	4.00	75.00	-	319.53	nr	394.53
Gates, softwood; Jacksons Fencing							
Timber field gates, including wrought iron ironmongery; five bar type; diamond braced; 1.80 m high; to 200 mm x 200 mm posts; setting 750 mm into firm ground							
treated softwood; width 2400 mm	70.06	10.00	187.50	-	152.28	nr	339.78
treated softwood; width 2700 mm	76.81	10.00	187.50	-	159.03	nr	346.53
treated softwood; width 3000 mm	80.01	10.00	187.50	-	162.22	nr	349.72
treated softwood; width 3300 mm	85.86	10.00	187.50	-	168.07	nr	355.57
Featherboard garden gates, including ironmongery; to 100 mm x 120 mm posts; 1 nr diagonal brace							
treated softwood; 1.0m x 1.2m high	14.31	3.00	56.25	-	80.90	nr	137.15
treated softwood; 1.0m x 1.5m high	18.90	3.00	56.25	-	87.11	nr	143.36
treated softwood; 1.0m x 1.8m high	64.13	3.00	56.25	-	92.96	nr	149.21

Q PAVING/PLANTING/FENCING/SITE FURNITURE

Item Excluding site overheads and profit	PC £	Labour hours	Labour £	Plant £	Material £	Unit	Total rate £
Q40 FENCING AND GATES - cont'd							
Gates, softwood; Jacksons Fencing - cont'd							
Picket garden gates, including ironmongery; to							
match picket fence; width 1000 mm; to 100 mm x							
120 mm posts; 1 nr diagonal brace							
treated softwood; 950 mm high	57.83	3.00	56.25	-	79.82	nr	136.07
treated softwood; 1200 mm high	60.48	3.00	56.25	-	82.47	nr	138.72
treated softwood; 1800 mm high	69.84	3.00	56.25	-	98.67	nr	154.92
Gates, tubular steel; Jacksons Fencing							
Tubular mild steel field gates, including							
ironmongery; diamond braced; 1.80 m high; to							
tubular steel posts; setting in concrete							
galvanized; width 3000 mm	87.58	5.00	93.75	-	161.66	nr	255.41
galvanized; width 3300 mm	92.67	5.00	93.75	-	166.76	nr	260.51
galvanized; width 3600 mm	97.97	5.00	93.75	-	172.05	nr	265.80
galvanized; width 4200 mm	108.52	5.00	93.75	-	182.61	nr	276.35
Gates, sliding; Jacksons Fencing							
"Sliding Gate"; including all galvanized rails and							
vertical rail infill panels; special guide and							
shutting frame posts (Note: foundations installed							
by suppliers)							
access width 4.00 m; 1.5m high gates	-	-	-	-	-	nr	3729.55
access width 4.00 m; 2.0m high gates	-	-	-	-	-	nr	3781.90
access width 4.00 m; 2.5m high gates	-	-	-	-	-	nr	3833.05
access width 6.00 m; 1.5m high gates	-	-	-	-	-	nr	4261.32
access width 6.00 m; 2.0m high gates	-	-	-	-	-	nr	4324.38
access width 6.00 m; 2.5m high gates	-	-	-	-	-	nr	4385.05
access width 8.00 m; 1.5m high gates	-	-	-	-	-	nr	4856.15
access width 8.00 m; 2.0m high gates	-	-	-	-	-	nr	4928.71
access width 8.00 m; 2.5m high gates	-	-	-	-	-	nr	4997.72
access width 10.00 m; 2.0m high gates	-	-	-	-	-	nr	5988.70
access width 10.00 m; 2.5m high gates	-	-	-	-	-	nr	6077.92
Stiles and kissing gates							
Stiles; Jacksons Fencing; "Jacksons							
Nr 2007" stiles; 2 nr posts; setting into firm							
ground; 3 nr rails; 2 nr treads	65.16	3.00	56.25	-	75.76	nr	132.01
Kissing gates; Jacksons Fencing; in							
galvanised metal bar; fixing to fencing posts							
(posts not included); 1.65 m x 1.30 m x 1.00 m							
high	168.75	5.00	93.75	-	168.75	nr	262.50
Pedestrian guard rails and barriers							
Mild steel pedestrian guard rails; Broxap							
Streetscene; to BS 3049:1976; 1.00 m high with							
150 mm toe space; to posts at 2.00 m centres,							
galvanized finish							
vertical bar infill panel	26.00	2.00	37.50	-	32.07	m	69.57
vertical bar infill with 200 mm visibility gap at top	29.00	2.00	37.50	-	35.07	m	72.57

Q PAVING/PLANTING/FENCING/SITE FURNITURE

Item Excluding site overheads and profit	PC £	Labour hours	Labour £	Plant £	Material £	Unit	Total rate £
Q50 SITE/STREET FURNITURE/EQUIPMENT							
Standing Stones; CED Ltd; erect standing stones; vertical height above ground; in concrete base; including excavation setting in concrete to 1/3 depth and crane offload into position							
Purple schist							
1.00 m high	47.50	1.50	28.13	14.13	62.41	nr	**104.65**
1.25 m high	95.00	1.50	28.13	14.13	104.29	nr	**146.54**
1.50 m high	152.00	2.00	37.50	16.95	163.85	nr	**218.30**
2.00 m high	247.00	3.00	56.25	28.25	281.24	nr	**365.74**
2.50 m high	380.00	1.50	28.13	56.50	446.37	nr	**531.00**
Cattle Grids - General							
Preamble: Cattle grids are not usually prefabricated, owing to their size and weight. Specification must take into account the maximum size and weight of vehicles likely to cross the grid. Drainage and regular clearance of the grid is essential. Warning signs should be erected on both approaches to the grid. For the cost of a complete grid and pit installation see "Approximate Estimates" section (book only).							
Cattle grids; Jacksons Fencing							
Cattle grids; supply only							
to carry 12 tonnes evenly distributed; 2.90 m x 2.36 m grid; galvanized	604.93	-	-	-	604.93	nr	**604.93**
to carry 12 tonnes evenly distributed; 3.66 m x 2.50 m grid; galvanized	1054.63	-	-	-	1054.63	nr	**1054.63**
Barriers - General							
Preamble: The provision of car and lorry control barriers may form part of the landscape contract. Barriers range from simple manual counterweighted poles to fully automated remote-control security gates, and the exact degree of control required must be specified. Complex barriers may need special maintenance and repair.							
Barriers: Autopa Ltd							
Manually operated pole barriers; counterbalance; to tubular steel supports; bolting to concrete foundation (foundation not included); aluminium boom; various finishes							
clear opening up to 3.00 m	670.00	6.00	112.50	-	690.49	nr	**802.99**
clear opening up to 4.00 m	740.00	6.00	112.50	-	760.49	nr	**872.99**
clear opening 5.00	805.00	6.00	112.50	-	825.49	nr	**937.99**
clear opening 6.00	885.00	6.00	112.50	-	905.49	nr	**1017.99**
clear opening 7.00	975.00	6.00	112.50	-	995.49	nr	**1107.99**
catch pole; arm rest for all manual barriers	110.00	-	-	-	112.56	nr	**112.56**
Electrically operated pole barriers; enclosed fan-cooled motor; double worm reduction gear; overload clutch coupling with remote controls; aluminium boom; various finishes (exclusive of electrical connections by electrician)							
Autopa AU 3.0, 3m boom	1800.00	6.00	112.50	-	1820.49	nr	**1932.99**
Autopa AU 4.5, 4.5m boom with catchpost	1920.00	6.00	112.50	-	1940.49	nr	**2052.99**
Autopa AU 6.0, 6.0m boom with catchpost	1995.00	6.00	112.50	-	2015.49	nr	**2127.99**
Vehicle crash barriers - General							
Preamble: See Department of Environment Technical Memorandum BE5.							

Prices for Measured Works

Q PAVING/PLANTING/FENCING/SITE FURNITURE

Item Excluding site overheads and profit	PC £	Labour hours	Labour £	Plant £	Material £	Unit	Total rate £
Q50 SITE/STREET FURNITURE/EQUIPMENT - cont'd							
Vehicle crash barriers Steel corrugated beams; untensioned; Broxap Streetscene; effective length 3.20 m 310 mm deep x 85 mm corrugations							
steel posts; Z-section; roadside posts	66.56	0.33	6.25	0.70	70.47	m	**77.42**
steel posts; Z-section; off highway	60.94	0.33	6.25	0.70	64.84	m	**71.80**
steel posts RSJ 760 high; for anchor fixing	75.31	0.33	6.25	-	79.81	m	**86.06**
steel posts RSJ 560 high; for anchor fixing to car parks	-	0.33	6.25	-	77.63	m	**83.87**
extra over for curved rail 6.00 m radius	25.50	-	-	-	25.50	m	**25.50**
Bases for street furniture Excavating; filling with no-fines cement:aggregate (1:12); bases for street furniture							
300 mm x 450 mm x 500 mm deep	-	0.75	14.06	-	6.66	nr	**20.72**
300 mm x 600 mm x 500 mm deep	-	1.11	20.81	-	8.88	nr	**29.69**
300 mm x 900 mm x 500 mm deep	-	1.67	31.31	-	13.32	nr	**44.63**
1750 mm x 900 mm x 300 mm deep	-	2.33	43.69	-	46.60	nr	**90.29**
2000 mm x 900 mm x 300 mm deep	-	2.67	50.06	-	53.25	nr	**103.32**
2400 mm x 900 mm x 300 mm deep	-	3.20	60.00	-	63.90	nr	**123.91**
2400 mm x 1000 mm x 300 mm deep	-	3.56	66.75	-	71.01	nr	**137.76**
Precast concrete flags; to BS 7263; to concrete bases (not included); bedding and jointing in cement:mortar (1:4)							
450 mm x 600 mm x 50 mm	7.81	1.17	21.87	-	13.00	m²	**34.88**
Precast concrete paving blocks; to concrete bases (not included); bedding in sharp sand; butt joints							
200 mm x 100 mm x 65 mm	7.83	0.50	9.38	-	9.73	m²	**19.11**
200 mm x 100 mm x 80 mm	8.74	0.50	9.38	-	10.55	m²	**19.92**
Engineering paving bricks; to concrete bases (not included); bedding and jointing in sulphate-resisting cement:lime:sand mortar (1:1:6)							
over 300 mm wide	10.50	0.56	10.53	-	19.85	m²	**30.38**
Edge restraints to pavings; haunching in concrete (1:3:6)							
200 mm x 300 mm	-	0.10	1.88	-	4.97	m	**6.84**
Bases; Earth Anchors Ltd "Rootfast" ancillary anchors; ref A1; 500 mm long 25 mm diameter and top strap ref F2; including bolting to site furniture (site furniture not included)	8.50	0.50	9.38	-	8.50	set	**17.88**
installation tool for above	21.00	-	-	-	21.00	nr	**21.00**
"Rootfast" ancillary anchors; ref A4; heavy duty 40 mm square fixed head anchors; including bolting to site furniture (site furniture not included)	-	0.33	6.25	-	-	set	**6.25**
"Rootfast" ancillary anchors; F4; vertical socket; including bolting to site furniture (site furniture not included)	11.75	0.05	0.94	-	11.75	set	**12.69**
"Rootfast" ancillary anchors; ref F3; horizontal socket; including bolting to site furniture (site furniture not included)	12.90	0.05	0.94	-	12.90	set	**13.84**
installation tools for the above	59.00	-	-	-	59.00	nr	**59.00**
"Rootfast" ancillary anchors; ref A3 Anchored bases; including bolting to site furniture (site furniture not included)	48.00	0.33	6.25	-	48.00	set	**54.25**
installation tools for the above	59.00	-	-	-	59.00	nr	**59.00**
Furniture/equipment - General Preamble: The following items include fixing to manufacturer's instructions; holding down bolts or other fittings and making good (excavating, backfilling and tarmac, concrete or paving bases not included).							

Q PAVING/PLANTING/FENCING/SITE FURNITURE

Item Excluding site overheads and profit	PC £	Labour hours	Labour £	Plant £	Material £	Unit	Total rate £
Dog waste bins; all-steel							
Earth Anchors Ltd							
HG45A; 45l; earth anchored; post mounted	159.00	0.33	6.25	-	159.00	nr	**165.25**
HG45A; 45l; as above with pedal operation	187.00	0.33	6.25	-	187.00	nr	**193.25**
Dog waste bins; cast iron							
Bins; Furnitubes International Ltd							
ref PED 701; "Pedigree"; post mounted cast iron							
dog waste bins; 1250 mm total height above							
ground; 400 mm square bin	603.00	0.75	14.06	-	609.07	nr	**623.13**
Litter bins; precast concrete in textured							
white or exposed aggregate finish; with							
wire baskets and drainage holes							
Bins; Marshalls Plc							
"Boulevard 700" concrete circular litter bin	205.00	0.50	9.38		205.00	nr	**214.38**
Bins; Neptune Outdoor Furniture Ltd							
ref SF 16 - 42l	174.00	0.50	9.38	-	174.00	nr	**183.38**
ref SF 14 - 100l	242.00	0.50	9.38	-	242.00	nr	**251.38**
Bins; Townscape Products Ltd							
"Sutton"; 750 mm high x 500 mm diameter, 70							
litre capacity including GRP canopy	154.22	0.33	6.25	-	264.42	nr	**270.67**
"Braunton"; 750 mm high x 500 mm diameter, 70							
litre capacity including GRP canopy	163.49	1.50	28.13	-	273.69	nr	**301.81**
Litter bins; metal; stove-enamelled							
perforated metal for holder and container							
Bins; Townscape Products Ltd							
"Metro"; 440 x 420 x 800 high; 62 litre capacity	530.56	0.33	6.25	-	530.56	nr	**536.81**
"Voltan" Large Round; 460 diameter x 780 high;							
56 litre capacity	456.56	0.33	6.25	-	456.56	nr	**462.81**
"Voltan" Small Round with pedestal; 410							
diameter x 760 high; 31 litre capacity	442.06	0.33	6.25	-	442.06	nr	**448.31**
Litter bins; all-steel							
Bins; Marshall Plc							
"MSF Central"; Steel litter bin	250.00	1.00	18.75	-	250.00	nr	**268.75**
"MSF Central"; Stainless steel litter bin	485.00	0.67	12.50	-	485.00	nr	**497.50**
Bins; Earth Anchors Ltd							
"Ranger", 100l Litter bin, 107l, pedestal mounted	697.00	0.50	9.38	-	697.00	nr	**706.38**
"Big Ben", 82l, steel frame and liner, Vyflex							
coated finish, earth anchored	254.00	1.00	18.75	-	254.00	nr	**272.75**
"Beau", 42l, steel frame and liner, Vyflex coated							
finish, earth anchored	204.00	1.00	18.75	-	204.00	nr	**222.75**
Bins; Townscape Products Ltd							
"Baltimore Major" with GRP canopy; 560							
diameter x 960 high, 140 litre capacity	766.62	1.00	18.75	-	766.62	nr	**785.37**
Litter bins; cast Iron							
Bins; Marshalls Plc							
"MSF5501" Heritage, cast iron litter bin	465.00	1.00	18.75	4.69	465.00	nr	**488.44**
Bins; Furnitubes International Ltd							
"Wave Bin"; ref WVB 440; free standing 55 litre							
liners; 440 mm diameter x 850 mm high	410.00	0.50	9.38	-	413.78	nr	**423.15**
"Wave Bin" ref WVB 520; free standing 85 litre							
liners; 520 mm diameter x 850 mm high; cast iron							
plinth	428.00	0.50	9.38	-	431.78	nr	**441.15**
"Covent Garden"; ref COV 702; side opening,							
500 mm square x 1050 mm high; 105 litre							
capacity	443.50	0.50	9.38	-	449.84	nr	**459.22**
"Covent Garden"; ref COV 803; side opening,							
500 mm diameter x 1025 mm high; 85 litre							
capacity	393.00	0.50	9.38	-	399.34	nr	**408.72**
"Covent Garden"; ref COV 912; open top; 500							
mm A/F octagonal x 820 mm high; 85 litre							
capacity	235.00	0.50	9.38	-	241.34	nr	**250.72**

Q PAVING/PLANTING/FENCING/SITE FURNITURE

Item Excluding site overheads and profit	PC £	Labour hours	Labour £	Plant £	Material £	Unit	Total rate £
Q50 SITE/STREET FURNITURE/EQUIPMENT - cont'd							
Litter bins; cast Iron - cont'd							
"Albert"; ref ALB 800; open top; 400 mm diameter x 845 mm high; 55 litre capacity	372.00	0.50	9.38	-	378.34	nr	387.72
Bins; Broxap Streetscene							
"Chester"; pedestal mounted	297.00	1.00	18.75	-	303.34	nr	322.09
Bins; Townscape Products Ltd							
"York Major"; 650 diameter x 1060 high, 140 litre capacity	771.04	1.00	18.75	-	772.82	nr	791.57
Litter bins; timber faced; hardwood slatted casings with removable metal litter containers; ground or wall fixing							
Bins; SMP Playgrounds Ltd							
ref LBC22; 560 mm high	463.00	1.00	18.75	-	471.23	nr	489.98
Bins; Lister Lutyens Co Ltd							
"Monmouth"; 675 mm high x 450 mm wide; freestanding	104.00	0.33	6.25	-	104.00	nr	110.25
"Monmouth"; 675 mm high x 450 mm wide; bolting to ground (without legs)	96.00	1.00	18.75	-	104.23	nr	122.98
Bins; Woodscape Ltd							
Square, 580 mm x 580 mm x 950 mm high, with lockable lid	575.00	0.50	9.38	-	575.00	nr	584.38
Round, 580 mm diameter by 950 mm high, with lockable lid	570.00	0.50	9.38	-	570.00	nr	579.38
Plastic litter and grit bins; glassfibre reinforced polyester grit bins; yellow body; hinged lids							
Bins; Wybone Ltd; Victoriana glass fibre, cast iron effect litter bins, including lockable liner							
ref LBV/2; 521 mm x 521 mm x 673 mm high; open top; square shape; 0.078 m^3 capacity	190.26	1.00	18.75	-	196.60	nr	215.35
ref LVC/3; 457 mm diameter x 648 mm high; open top; drum shape; 0.084 m^3 capacity; with lockable liner	207.56	1.00	18.75	-	213.90	nr	232.65
Bins; Amberol Ltd; floor standing double walled to accept stabilizing ballast							
"Enviro Bin" ; 150 litre	168.00	0.33	6.25	-	168.00	nr	174.25
"Enviro Bin" ; 90 litre	-	0.33	6.25	-	126.00	nr	132.25
"Westminster " hooded; 90 litre	110.00	0.33	6.25	-	110.00	nr	116.25
"Westminster" 90 litre	85.00	0.33	6.25	-	85.00	nr	91.25
Bins; Amberol Ltd; pole mounted							
Envirobin 50 litre top emptying with liner	75.00	0.50	9.38	-	75.00	nr	84.38
Grit Bins; Furnitubes International Ltd							
Q11 Grit and Salt Bin; Yellow Glass Fibre; Hinged Lid; 310 litre	273.00	-	-	-	273.00	ea	273.00
Q6 Grit and Salt Bin; Yellow Glass Fibre; Hinged Lid; 170 litre	243.00	-	-	-	243.00	ea	243.00
Lifebuoy stations							
Lifebuoy stations; Earth Anchors Ltd							
"Rootfast" lifebuoy station complete with post ref AP44; SOLAS approved lifebouys 590 mm and lifeline	153.00	2.00	37.50	-	162.49	nr	199.99
installation tool for above	-	-	-	-	75.00	nr	75.00

Q PAVING/PLANTING/FENCING/SITE FURNITURE

Item Excluding site overheads and profit	PC £	Labour hours	Labour £	Plant £	Material £	Unit	Total rate £
Outdoor seats **CED Ltd; Stone bench; "Sinuous bench";** **Stone type bench to organic "S"** **pattern; laid to concrete base; 500 high** **x 500 wide (not included)**							
2.00 m long	1000.00	4.00	75.00	-	1000.00	nr	1075.00
5.00 m long	2500.00	8.00	150.00	-	2500.00	nr	2650.00
10.00 m long	5000.00	11.00	206.25	-	5000.00	nr	5206.25
Outdoor seats; concrete framed - **General** Preamble: Prices for the following concrete framed seats and benches with hardwood slats include for fixing (where necessary) by bolting into existing paving or concrete bases (bases not included) or building into walls or concrete foundations (walls and foundations not included).							
Outdoor seats; concrete framed Outdoor seats; Townscape Products Ltd							
"Oxford" benches; 1800 mm x 430 mm x 440 mm	224.95	2.00	37.50	-	237.09	nr	274.59
"Maidstone" seats; 1800 mm x 610 mm x 785 mm	336.40	2.00	37.50	-	348.54	nr	386.04
Outdoor seats; concrete Outdoor seats; Marshalls Plc							
"Boulevard 2000" concrete seat	590.00	2.00	37.50	-	609.70	nr	647.20
Outdoor seats; metal framed - General Preamble: Metal framed seats with hardwood backs to various designs can be bolted to ground anchors.							
Outdoor seats; metal framed Outdoor seats; Furnitubes International Ltd							
ref NS 6; "Newstead"; steel standards with iroko slats; 1.80 m long	238.00	2.00	37.50	-	257.70	nr	295.20
ref NEB 6; "New Forest Single Bench"; cast iron standards with iroko slats; 1.83 m long	242.00	2.00	37.50	-	261.70	nr	299.20
ref EA 6; "Eastgate"; cast iron standards with iroko slats; 1.86 m long	240.00	2.00	37.50	-	259.70	nr	297.20
ref NE 6; "New Forest Seat"; cast iron standards with iroko slats; 1.83 m long	384.00	2.00	37.50	-	403.70	nr	441.20
Outdoor seats; Columbia Cascade Ltd; perforated metal bench seats, range of 8 nr colours							
ref 2604-6; back-to-back metal benches; 1800 mm x 1500 mm x 860 mm high	1295.00	3.00	56.25	-	1307.68	nr	1363.93
Outdoor seats; Orchard Street Furniture Ltd; "Bramley"; broad iroko slats to steel frame							
1.20 m long	161.44	2.00	37.50	-	174.12	nr	211.62
1.80 m long	198.94	2.00	37.50	-	211.62	nr	249.12
2.40 m long	229.79	2.00	37.50	-	242.47	nr	279.97
Outdoor seats; Orchard Street Furniture Ltd; "Laxton"; narrow iroko slats to steel frame							
1.20 m long	182.45	2.00	37.50	-	195.13	nr	232.63
1.80 m long	224.20	2.00	37.50	-	236.88	nr	274.38
2.40 m long	254.79	2.00	37.50	-	267.47	nr	304.97
Outdoor seats; Orchard Street Furniture Ltd; "Lambourne"; iroko slats to cast iron frame							
1.80 m long	473.94	2.00	37.50	-	486.62	nr	524.12
2.40 m long	557.71	2.00	37.50	-	570.39	nr	607.89
Outdoor seats; SMP Playgrounds Ltd; "Datchet", steel and timber							
backless benches; ref OFDB6; 1.80 m long	310.00	2.00	37.50	-	322.68	nr	360.18
bench seats with backrests; ref OFDS6; 1.80 m long	418.00	2.00	37.50	-	430.68	nr	468.18

Q PAVING/PLANTING/FENCING/SITE FURNITURE

Item Excluding site overheads and profit	PC £	Labour hours	Labour £	Plant £	Material £	Unit	Total rate £
Q50 SITE/STREET FURNITURE/EQUIPMENT - cont'd							
Outdoor seats; metal framed - cont'd							
Outdoor seats; Broxap Streetscene; metal and timber							
"Eastgate"	378.00	2.00	37.50	-	390.68	nr	428.18
"Serpent"	398.00	2.00	37.50	-	410.68	nr	448.18
"Rotherham"	566.00	2.00	37.50	-	578.68	nr	616.18
Outdoor seats; Earth Anchors Ltd; "Forest-Saver", steel frame and recycled slats							
bench, 1.8m	182.00	1.00	18.75	-	182.00	nr	200.75
seat, 1.8m	283.00	1.00	18.75	-	283.00	nr	301.75
Outdoor seats; Earth Anchors Ltd; "Evergreen", ci frame and recycled slats							
bench, 1.8m	385.00	1.00	18.75	-	385.00	nr	403.75
seat, 1.8m	467.00	1.00	18.75	-	467.00	nr	485.75
Outdoor seats; all-steel							
Outdoor seats; Marshalls Plc							
"MSF Central" Stainless Steel seat	1000.00	0.50	9.38	-	1000.00	nr	1009.38
"MSF Central" Steel seat	500.00	2.00	37.50	-	500.00	nr	537.50
"MSF 502" Cast Iron Heritage seat	468.00	2.00	37.50	-	468.00	nr	505.50
Outdoor seats; Earth Anchors Ltd, "Ranger"							
bench, 1.8m	259.00	1.00	18.75	-	259.00	nr	277.75
seat, 1.8m	419.00	1.00	18.75	-	419.00	nr	437.75
Outdoor seats; all timber							
Outdoor seats; Geometric Furniture Ltd; Teak bench with back and armrests							
ref 8204; two seater; 1.38 m	328.90	2.00	37.50	-	348.60	nr	386.10
ref 8207; two seater; 1.50 m	339.37	2.00	37.50	-	359.07	nr	396.57
ref 8206; three seater; 1.85 m	614.45	2.00	37.50	-	634.15	nr	671.65
ref 8202; three seater; 1.95 m	460.46	2.00	37.50	-	480.16	nr	517.66
ref 8205; four seater; 2.40 m	738.53	2.00	37.50	-	758.23	nr	795.73
Outdoor seats; Lister Lutyens Co Ltd; "Mendip"; teak							
1.524 m long	416.00	1.00	18.75	-	439.49	nr	458.24
1.829 m long	431.00	1.00	18.75	-	454.49	nr	473.24
2.438 m long (inc. centre leg)	577.00	1.00	18.75	-	600.49	nr	619.24
Outdoor seats; Lister Lutyens Co Ltd; "Sussex"; hardwood							
1.5 m	173.00	2.00	37.50	-	196.49	nr	233.99
Outdoor seats; Woodscape Ltd; solid hardwood							
seat type "3" with back; 2.00 m long; freestanding	640.00	2.00	37.50	-	663.49	nr	700.99
seat type "3" with back; 2.00 m long; building in	680.00	4.00	75.00	-	703.49	nr	778.49
seat type "4" with back; 2.00 m long; fixing to wall	370.00	3.00	56.25	-	381.34	nr	437.59
seat type "5" with back; 2.50 m long; freestanding	750.00	2.00	37.50	-	773.49	nr	810.99
seat type "4"; 2.00 m long; fixing to wall	295.00	2.00	37.50	-	318.49	nr	355.99
seat type "5" with back; 2.00 m long; freestanding	655.00	2.00	37.50	-	678.49	nr	715.99
seat type "4" with back; 2.50 m long; fixing to wall	435.00	3.00	56.25	-	446.34	nr	502.59
seat type "5" with back; building in	695.00	2.00	37.50	-	718.49	nr	755.99
seat type "5" with back; 2.50 m long; building in	790.00	3.00	56.25	-	813.49	nr	869.74
bench type "1"; 2.00 m long; freestanding	505.00	2.00	37.50	-	528.49	nr	565.99
bench type "1"; 2.00 m long; building in	545.00	4.00	75.00	-	568.49	nr	643.49
bench type "2"; 2.00 m long; freestanding	480.00	2.00	37.50	-	503.49	nr	540.99
bench type "2"; 2.00 m long; building in	520.00	4.00	75.00	-	543.49	nr	618.49
bench type "2"; 2.50 m long; freestanding	545.00	2.00	37.50	-	568.49	nr	605.99
bench type "2"; 2.50 m long; building in	585.00	4.00	75.00	-	608.49	nr	683.49
bench type "2"; 2.00 m long overall; curved to 5 m radius, building in	680.00	4.00	75.00	-	703.49	nr	778.49

Q PAVING/PLANTING/FENCING/SITE FURNITURE

Item Excluding site overheads and profit	PC £	Labour hours	Labour £	Plant £	Material £	Unit	Total rate £
Outdoor seats; Orchard Street Furniture Ltd; "Allington"; all iroko							
1.20 m long	254.26	2.00	37.50	-	277.75	nr	315.25
1.80 m long	292.29	2.00	37.50	-	315.78	nr	353.28
2.40 m long	355.59	2.00	37.50	-	379.08	nr	416.58
Outdoor seats; SMP Playgrounds Ltd; "Lowland range"; all timber							
bench seats with backrest; ref OFDS6; 1.50 m long	418.00	2.00	37.50	-	430.68	nr	468.18
Outdoor seats; tree benches/seats							
Tree bench; Neptune Outdoor Furniture Ltd, "Beaufort" Hexagonal, Timber							
SF34-15A, 1500 mm diameter	800.00	0.50	9.38	-	800.00	nr	809.38
Tree seat; Neptune Outdoor Furniture Ltd, "Beaufort" Hexagonal, Timber, with back							
SF32-10A, 720 mm diameter	960.00	0.50	9.38	-	960.00	nr	969.38
SF32-20A, 1720 mm diameter	1122.00	0.50	9.38	-	1122.00	nr	1131.38
Street furniture ranges							
Townscape Products Ltd; "Belgrave"; natural grey concrete							
bollards; 250 mm diameter x 500 mm high	71.20	1.00	18.75	-	74.39	nr	93.14
seats; 1800 mm x 600 mm x 736 mm high	262.00	2.00	37.50	-	262.00	nr	299.50
Picnic benches - General							
Preamble: The following items include for fixing to ground in to manufacturer's instructions or concreting in.							
Picnic tables and benches							
Picnic tables and benches; Broxap Streetscene "Eastgate" picnic unit	548.00	2.00	37.50	-	560.68	nr	598.18
Picnic tables and benches; Woodscape Ltd							
Table and benches built in, 2 m long	1635.00	2.00	37.50	-	1647.68	nr	1685.18
Market Prices of Containers							
Plant containers; terracotta							
Capital Garden Products Ltd							
Large Pot LP63; weathered terracotta - 1170 x 1600 dia.	-	-	-	-	412.00	nr	412.00
Large Pot LP38; weathered terracotta - 610 x 970 dia.	-	-	-	-	255.00	nr	255.00
Large Pot LP23; weathered terracotta - 480 x 580 dia.	-	-	-	-	136.00	nr	136.00
Manhole Cover Planter 3022; 760 x 560 x 210 high	-	-	-	-	94.00	nr	94.00
Apple Basket 2215; 380 x 560 dia.	-	-	-	-	97.00	nr	97.00
Indian style Shimmer Pot 2322; 585 x 560 dia.	-	-	-	-	135.00	nr	135.00
Indian style Shimmer Pot 1717; 430 x 430 dia.	-	-	-	-	93.00	nr	93.00
Indian style Shimmer Pot 1314; 330 x 355 dia.	-	-	-	-	71.00	nr	71.00
Plant containers; faux lead							
Capital Garden Products Ltd							
Trough 2508 'Tudor Rose'; 620 x 220 x 230 high	-	-	-	-	53.00	nr	53.00
Tub 2004 'Elizabethan'; 510 mm square	-	-	-	-	84.00	nr	84.00
Tub 1513 'Elizabethan'; 380 mm square	-	-	-	-	48.00	nr	48.00
Tub 1601 'Tudor Rose'; 420 x 400 dia.	-	-	-	-	61.00	nr	61.00
Plant containers; window boxes							
Capital Garden Products Ltd							
Adam 5401; 1370 x 270 x 210 h	-	-	-	-	82.00	nr	82.00
Wheatsheaf WH54; 1370 x 270 x 210 h	-	-	-	-	89.00	nr	89.00
Oakleaf OAK24; 610 x 230 x 240 h	-	-	-	-	51.00	nr	51.00
Swag 2402; 610 x 200 x 210 h	-	-	-	-	53.00	nr	53.00

Q PAVING/PLANTING/FENCING/SITE FURNITURE

Item Excluding site overheads and profit	PC £	Labour hours	Labour £	Plant £	Material £	Unit	Total rate £
Q50 SITE/STREET FURNITURE/EQUIPMENT - cont'd							
Plant containers; timber							
Plant containers; Hardwood; Neptune Outdoor Furniture Ltd							
'Beaufort' T38-4D, 1500 x 1500 x 900 mm high	-	-	-	-	896.00	nr	**896.00**
'Beaufort' T38-3C, 1000 x 1500 x 700 mm high	-	-	-	-	667.00	nr	**667.00**
'Beaufort' T38-2A, 1000 x 500 x 500 mm high	-	-	-	-	348.00	nr	**348.00**
'Kara' T42-4D, 1500 x 1500 x 900 mm high	-	-	-	-	967.00	nr	**967.00**
'Kara' T42-3C, 1000 x 1500 x 700 mm high	-	-	-	-	720.00	nr	**720.00**
'Kara' T42-2A, 1000 x 500 x 500 mm high	-	-	-	-	376.00	nr	**376.00**
Plant containers; Hardwood; Woodscape Ltd							
Square, 900 x 900 x 420 mm high	-	-	-	-	270.00	nr	**270.00**
Measured Works							
Plant containers; precast concrete							
Plant containers; Marshalls Plc							
"Boulevard 700" circular base and ring	440.00	2.00	37.50	16.50	440.00	nr	**494.00**
"Boulevard 1200" circular base and ring	531.00	1.00	18.75	16.50	531.00	nr	**566.25**
Cycle holders							
Cycle stands; Marshalls Plc							
"Sheffield" steel cycle stand; RCS1	44.00	0.50	9.38	-	44.00	nr	**53.38**
"Sheffield" stainless steel cycle stand; RSCS1	121.00	0.50	9.38	-	121.00	nr	**130.38**
Cycle holders; George Fischer Sales; "Velop A" galvanized steel							
ref R; fixing to wall or post; making good	29.00	1.00	18.75	-	41.68	nr	**60.43**
ref SR(V); fixing in ground; making good	31.50	1.00	18.75	-	44.18	nr	**62.93**
Cycle holders; Townscape Products Ltd							
"Guardian" cycle holders; tubular steel frame; setting in concrete; 1250 mm x 550 mm x 775 mm high; making good	251.70	1.00	18.75	-	264.38	nr	**283.13**
"Penny" cycle stands; 600 mm diameter; exposed aggregate bollards with 8 nr cycle holders; in galvanized steel; setting in concrete; making good	501.79	1.00	18.75	-	514.47	nr	**533.22**
Cycle holders; Broxap Streetscene							
"Neath" cycle rack; ref BX/MW/AG; for 6 nr cycles; semi-vertical; galvanised and polyester powder coated; 1.320 m wide x 2.542 m long x 1.80 m high	345.00	10.00	187.50	-	370.36	nr	**557.86**
"Premier Senior" combined shelter and rack; ref BX/MW/AW; for 10 nr cycles; horizontal; galvanised only; 2.13 m wide x 3.05 m long x 2.15 m high	689.00	10.00	187.50	-	714.36	nr	**901.86**
"Toast Rack" double sided free standing cycle rack; ref BX/MW/GH; for 10 nr cycles; galvanised and polyester coated; 3.250 m long	424.00	4.00	75.00	-	434.24	nr	**509.24**
Directional signage; cast aluminium							
Signage; Furnitubes International Ltd							
ref FFL 1; "Lancer"; cast aluminium finials	61.00	0.07	1.25	-	61.00	nr	**62.25**
ref FAA IS; arrow end type cast aluminium directional arms, single line; 90 mm wide	160.00	2.00	37.50	-	160.00	nr	**197.50**
ref FAA ID; arrow end type cast aluminium directional arms; double line; 145 mm wide	195.00	0.13	2.50	-	195.00	nr	**197.50**
ref FAA IS; arrow end type cast aluminium directional arms; treble line; 200 mm wide	233.00	0.20	3.75	-	233.00	nr	**236.75**
ref FCK1 211 G; "Kingston"; composite standard root columns	345.00	2.00	37.50	-	360.53	nr	**398.03**
Excavating; for bollards and barriers; by hand							
Holes for bollards							
400 mm x 400 mm x 400 mm; disposing off site	-	0.42	7.81	-	0.58	nr	**8.39**
600 mm x 600 mm x 600 mm; disposing off site	-	0.97	18.19	-	1.62	nr	**19.81**

Q PAVING/PLANTING/FENCING/SITE FURNITURE

Item Excluding site overheads and profit	PC £	Labour hours	Labour £	Plant £	Material £	Unit	Total rate £
Concrete bollards - General Preamble: Precast concrete bollards are available in a very wide range of shapes and sizes. The bollards listed here are the most commonly used sizes and shapes; manufacturer's catalogues should be consulted for the full range. Most manufacturers produce bollards to match their suites of street furniture, which may include planters, benches, litter bins and cycle stands. Most parallel sided bollards can be supplied in removable form, with a reduced shank, precast concrete socket and lifting hole to permit removal with a bar.							
Concrete bollards Marshalls Plc; cylinder; straight or tapered; 200 mm - 400 mm diameter; plain grey concrete; setting into firm ground (excavating and backfilling not included)							
"Bridgford" 915 mm high above ground	59.51	2.00	37.50	-	69.86	nr	**107.36**
Marshalls Plc; cylinder; straight or tapered; 200 mm - 400 mm diameter; "Beadalite" reflective finish; setting into firm ground (excavating and backfilling not included)							
"Wexham" concrete bollard; exposed silver grey	123.50	2.00	37.50	-	129.07	nr	**166.57**
"Wexham Major" concrete bollard; exposed silver grey	199.00	2.00	37.50	-	209.35	nr	**246.85**
Precast concrete verge markers; various shapes; 450 mm high							
plain grey concrete	31.48	1.00	18.75	-	35.46	nr	**54.21**
white concrete	33.01	1.00	18.75	-	36.99	nr	**55.74**
exposed aggregate	34.95	1.00	18.75	-	38.93	nr	**57.68**
Other bollards Removable parking posts; Dee-Organ Ltd "Spacekeeper"; ref 3014203; folding plastic coated galvanized steel parking posts with key; reflective bands; including 300 mm x 300 mm x 300 mm concrete foundations and fixing bolts;							
850 mm high	114.22	2.00	37.50	-	125.85	nr	**163.35**
"Spacesaver"; ref 1508101; hardwood removable bollards including key and base-plate; 150 mm x 150 mm; 600 mm high	116.86	2.00	37.50	-	120.93	nr	**158.43**
"Spacesaver" ref 1506101; hardwood removable bollards including key and base-plate; 150 mm x 150 mm; 800 mm high	130.04	2.00	37.50	-	134.11	nr	**171.61**
Removable parking posts; Marshalls Plc							
"RT\RD4" Domestic telescopic bollard	150.00	2.00	37.50	-	155.57	nr	**193.07**
"RT\R8" Heavy Duty Telescopic bollard	223.00	2.00	37.50	-	228.57	nr	**266.07**
Plastic Bollards; Marshalls Plc							
"Lismore" 3 ring recycled plastic bollard	78.00	2.00	37.50	-	83.57	nr	**121.07**
Cast iron bollards - General Preamble: The following bollards are particularly suitable for conservation areas. Logos for civic crests can be incorporated to order.							

Q PAVING/PLANTING/FENCING/SITE FURNITURE

Item Excluding site overheads and profit	PC £	Labour hours	Labour £	Plant £	Material £	Unit	Total rate £
Q50 SITE/STREET FURNITURE/EQUIPMENT - cont'd							
Cast iron bollards							
Bollards; Furnitubes International Ltd (excavating and backfilling not included)							
"Doric Round"; 920 mm high x 170 mm diameter	81.00	2.00	37.50	-	86.30	nr	123.80
"Gunner round"; 750 mm high x 165 mm diameter	52.00	2.00	37.50	-	57.30	nr	94.80
"Manchester Round"; 975 mm high; 225 mm square base	87.00	2.00	37.50	-	92.30	nr	129.80
"Cannon"; 1140 mm x 210 mm diameter	111.00	2.00	37.50	-	116.30	nr	153.80
"Kenton"; heavy duty galvanized steel; 900 mm high; 350 mm diameter	199.00	2.00	37.50	-	204.30	nr	241.80
Bollards; Marshalls Plc (excavating and backfilling not included)							
"MSF103" Cast Iron bollard - "Small Manchester"	125.00	2.00	37.50	-	130.57	nr	168.07
"MSF102" Cast Iron bollard - "Manchester"	135.00	2.00	37.50	-	140.57	nr	178.07
Cast iron bollards with rails - General							
Preamble: The following cast iron bollards are suitable for conservation areas.							
Cast iron bollards with rails							
Cast iron posts with steel tubular rails; Broxap Streetscene; setting into firm ground (excavating not included)							
"Sheffield"; 450 mm high; one rail, type A	62.00	1.50	28.13	-	64.56	nr	92.69
"Morecambe"; 1149 mm high above ground; four rails, type C	122.00	1.00	18.75	-	124.56	nr	143.31
"Mersey"; 1085 mm high above ground; two rails, type D	121.00	1.50	28.13	-	123.56	nr	151.69
"Promenade"; 1150 mm high above ground; square; three rails, type C	122.00	1.50	28.13	-	124.56	nr	152.69
"Type A" mild steel tubular rail including connector	4.67	0.03	0.62	-	6.06	m	6.68
"Type C" mild steel tubular rail including connector	5.83	0.03	0.62	-	7.84	m	8.47
Steel Bollards							
Steel Bollards; Marshalls Plc; (excavating and backfilling not included)							
"SSB0" Stainless Steel bollard; 101 x 1250	95.00	2.00	37.50	-	100.57	nr	138.07
"RS00" Stainless steel bollard; 114 x 1500	130.00	2.00	37.50	-	135.57	nr	173.07
"RB119" - Brunel" Steel Bollard; 168 x 1500	119.00	2.00	37.50	-	129.35	nr	166.85
Timber bollards							
Woodscape Ltd; Durable Hardwood							
RP 250/1500; 250 mm diameter x 1500 mm long	190.60	1.00	18.75	-	190.83	nr	209.58
SP 250/1500; 250 mm square x 1500 mm long	190.60	1.00	18.75	-	190.83	nr	209.58
SP 150/1200; 150 mm square x 1200 mm long	68.70	1.00	18.75	-	68.93	nr	87.68
SP 125/750; 125 mm square x 750 mm long	37.90	1.00	18.75	-	38.13	nr	56.88
RP 125/750; 125 mm diameter x 750 mm long	37.90	1.00	18.75	-	38.13	nr	56.88
Deterrent bollards							
Semi-mountable vehicle deterrent and kerb protection bollards; Furnitubes International Ltd (excavating and backfilling not included)							
"Bell decorative"	261.00	2.00	37.50	-	268.54	nr	306.04
"Half bell"	302.00	2.00	37.50	-	309.54	nr	347.04
"Full bell"	409.00	2.00	37.50	-	416.54	nr	454.04
"Three quarter bell"	397.00	2.00	37.50	-	404.54	nr	442.04

Q PAVING/PLANTING/FENCING/SITE FURNITURE

Item Excluding site overheads and profit	PC £	Labour hours	Labour £	Plant £	Material £	Unit	Total rate £
Security Bollards							
Security bollards; Furnitubes International Ltd							
(excavating and backfilling not included)							
"Gunner"; reinforced with steel insert and tie							
bars; 750 mm high above ground; 600 mm below							
ground	77.00	2.00	37.50	-	84.95	nr	122.45
"Burr Bloc Type 6" removable steel security							
bollard; 750 mm high above ground; 410 mm x							
285 mm	425.00	2.00	37.50	-	442.89	nr	480.39
Tree grilles; Cast Iron							
Cast iron tree grilles; Furnitubes International Ltd							
ref GS 1070 Greenwich; two part; 1000 mm							
square; 700 mm diameter tree hole	76.00	2.00	37.50	-	76.00	nr	113.50
ref GC 1270 Greenwich; two part, 1200 mm							
diameter; 700 mm diameter tree hole	97.00	2.00	37.50	-	97.00	nr	134.50
ref GCF 1026 Greenwich; steel tree grille frame							
for GC 1045, one part	171.50	2.00	37.50	-	171.50	nr	209.00
ref GSF 1226 Greenwich; steel tree grille frame							
for GC 1270 and GC 1245, one part	178.50	2.00	37.50	-	178.50	nr	216.00
Cast iron tree grilles; Marshalls Plc							
"Heritage" cast iron grille plus frame; 1m x 1m	336.96	3.00	56.25	-	347.98	nr	404.23
Cast iron tree grilles; Townscape Products Ltd							
"Baltimore" 1200 mm square x 460 mm diameter							
tree hole	362.46	2.00	37.50	-	362.46	nr	399.96
"Baltimore" Hexagonal, maximum width 1440							
mm nominal, 600 mm diameter tree hole	583.21	2.00	37.50	-	583.21	nr	620.71
Note: Care must be taken to ensure that tree							
grids and guards are removed when trees grow							
beyond the specified diameter of guard.							
Playground equipment - General							
Preamble: The range of equipment							
manufactured or available in the UK is so great							
that comprehensive coverage would be							
impossible, especially as designs, specifications							
and prices change fairly frequently. The following							
information should be sufficient to give guidance							
to anyone designing or equipping a playground.							
In comparing prices note that only outline							
specification details are given here and that other							
refinements which are not mentioned may be the							
reason for some difference in price between two							
apparently identical elements. The fact that a							
particular manufacturer does not appear under							
one item heading does not necessarily imply that							
he does not make it. Landscape designers are							
advised to check that equipment complies with							
ever more stringent safety standards before							
specifying.							
Playground equipment - Installation							
The rates below include for installation of the							
specified equipment by the manufacturers. Most							
manufactures will offer an option to install the							
equipment supplied by them.							
Play systems; Kompan Ltd "Galaxy";							
multiple play activity systems for							
non-prescribed play; for children 6-14							
years; galvanized steel and high density							
polyethylene							
Propys GXY 910; 21 different play activities	-	-	-	-	-	nr	15500.00
Adara GXY 906; 14 different play activities	-	-	-	-	-	nr	12600.00
Sirius GXY 8011; 10 different play activities	-	-	-	-	-	nr	7800.00

Q PAVING/PLANTING/FENCING/SITE FURNITURE

Item Excluding site overheads and profit	PC £	Labour hours	Labour £	Plant £	Material £	Unit	Total rate £
Q50 SITE/STREET FURNITURE/EQUIPMENT - cont'd							
Sports and social areas; multi-use games **areas (MUGA); Kompan Ltd; Enclosed** **sports areas; complete with surfacing** **boundary and goals and targets;** **galvanized steel framework with high** **density polyethylene panels, galvanized** **steel goals and equipment; surfacing** **priced separately**							
Pitch Complete; suitable for multiple ball sports; suitable for use with natural artificial or hard landscape surfaces; fully enclosed including 2 nr end sports walls							
FRE 2202 End sports wall multigoal only	-	-	-	-	-	nr	8800.00
FRE 2000 12.0 x 20.0 m	-	-	-	-	-	nr	21760.00
FRE 2116 19.00 x 36.00	-	-	-	-	-	nr	32320.00
FRE 3000 Meeting point shelter or social area inclusive of 2 nr modular S benches, 3.920 m x 1.40 m	-	-	-	-	-	nr	3575.00
Swings - General							
Preamble: Prices for the following vary considerably. Those given represent the middle of the range and include multiple swings with tubular steel frames and timber or tyre seats; ground fixing and priming only.							
Swings							
Swings; Wicksteed Leisure Ltd							
traditional swings; 1850 mm high; 1 bay; 2 seat	-	-	-	-	-	nr	1748.00
traditional swings; 1850 mm high; 2 bay; 4 seat	-	-	-	-	-	nr	2811.00
traditional swings; 2450 mm high; 1 bay; 2 seat	-	-	-	-	-	nr	1690.00
traditional swings; 2450 mm high; 2 bay; 4 seat	-	-	-	-	-	nr	2665.00
traditional swings; 3050 mm high; 1 bay; 2 seat	-	-	-	-	-	nr	1793.00
traditional swings; 3050 mm high; 2 bay; 4 seat	-	-	-	-	-	nr	2795.00
single arch swing; cradle safety seat; 1850 mm high	-	-	-	-	-	nr	1376.00
double arch swing; cradle safety seats; 1850 mm high	-	-	-	-	-	nr	1772.00
single arch swing; flat rubber safety seat; 2450 mm high	-	-	-	-	-	nr	1365.00
double arch swing; flat rubber safety seats; 2450 mm high	-	-	-	-	-	nr	1674.00
Swings; Lappset UK Ltd							
ref 020414; swing frame with two flat seats	-	-	-	-	-	nr	1215.00
Swings; Kompan Ltd							
ref M947-52; double swings	-	-	-	-	-	nr	1600.00
ref M948-52; double swings	-	-	-	-	-	nr	1650.00
ref M951; "Sunflower" swings	-	-	-	-	-	nr	1220.00
Slides							
Slides; Wicksteed Leisure Ltd							
"Pedestal" slides; 3.40 m	-	-	-	-	-	nr	2425.00
"Pedestal" slides; 4.40 m	-	-	-	-	-	nr	3157.00
"Pedestal" slides; 5.80 m	-	-	-	-	-	nr	3675.00
"Embankment" slides; 3.40 m	-	-	-	-	-	nr	1859.00
"Embankment" slides; 4.40 m	-	-	-	-	-	nr	2416.00
"Embankment" slides; 5.80 m	-	-	-	-	-	nr	3075.00
"Embankment" slides; 7.30 m	-	-	-	-	-	nr	3907.00
"Embankment" slides; 9.10 m	-	-	-	-	-	nr	4894.00
"Embankment" slides; 11.00 m	-	-	-	-	-	nr	5853.00
Slides; Lappset UK Ltd							
ref 142015; slide	-	-	-	-	-	nr	2350.00
ref 141115; "Jumbo" slide	-	-	-	-	-	nr	3150.00

Q PAVING/PLANTING/FENCING/SITE FURNITURE

Item Excluding site overheads and profit	PC £	Labour hours	Labour £	Plant £	Material £	Unit	Total rate £
Slides; Kompan Ltd							
ref M351; slides	-	-	-	-	-	nr	2080.00
ref M322; slide and cave	-	-	-	-	-	nr	2440.00
Moving equipment - General							
Preamble: The following standard items of playground equipment vary considerably in quality and price; the following prices are middle of the range.							
Moving equipment							
Roundabouts; Wicksteed Leisure Ltd							
"Turnstile"	-	-	-	-	-	nr	677.00
"Speedway" (without restrictor)	-	-	-	-	-	nr	2665.00
"Spiro Whirl" (without restrictor)	-	-	-	-	-	nr	3035.00
Roundabouts; Kompan Ltd							
Supernova GXY916 multifunctional spinning and balancing disc; capacity approximately 15 children	-	-	-	-	-	nr	3135.00
Seesaws							
Seesaws; Lappset UK Ltd							
ref 010300; seesaws	-	-	-	-	-	nr	810.00
ref 010237; seesaws	-	-	-	-	-	nr	1665.00
Seesaws; Wicksteed Leisure Ltd							
"Seesaw" (non-bump)	-	-	-	-	-	nr	1966.00
"Jolly Gerald" (non-bump)	-	-	-	-	-	nr	2121.00
"Rocking Rockette" (with motion restrictor)	-	-	-	-	-	nr	2846.00
"Rocking Horse" (with motion restrictor)	-	-	-	-	-	nr	3248.00
Play sculptures - General							
Preamble: Many variants on the shapes of playground equipment are available, simulating spacecraft, trains, cars, houses etc., and these designs are frequently changed. The basic principles remain constant but manufacturer's catalogues should be checked for the latest styles.							
Play sculptures							
Wooden animals; SMP Playgrounds Ltd							
pig	-	-	-	-	-	nr	451.00
donkey	-	-	-	-	-	nr	471.50
rhino	-	-	-	-	-	nr	635.50
giraffe	-	-	-	-	-	nr	645.75
camel	-	-	-	-	-	nr	666.25
hippo	-	-	-	-	-	nr	973.75
elephant	-	-	-	-	-	nr	1271.00
Climbing equipment and play structures; General							
Preamble: Climbing equipment generally consists of individually designed modules. Play structures generally consist of interlinked and modular pieces of equipment and sculptures. These may consist of climbing, play and skill based modules, nets and various other activities. Both are set into either safety surfacing or defined sand pit areas.The equipment below outlines a range from various manufacturers. Individual catalogues should be consulted in each instance. Safety areas should be allowed round all equipment.							

Q PAVING/PLANTING/FENCING/SITE FURNITURE

Item Excluding site overheads and profit	PC £	Labour hours	Labour £	Plant £	Material £	Unit	Total rate £
Q50 SITE/STREET FURNITURE/EQUIPMENT - cont'd							
Climbing equipment and play structures							
Climbing equipment and play structures; Kompan Ltd							
ref MQ1002; "Mosaic" combination system	-	-	-	-	-	nr	7650.00
ref OK1103; "Oasis" combination system	-	-	-	-	-	nr	6030.00
ref MQ1001; "Mosaiq"	-	-	-	-	-	nr	7200.00
ref MQ2004; "Mosaic" combination system	-	-	-	-	-	nr	9530.00
ref OK3100; "Oasis" combination system	-	-	-	-	-	nr	8520.00
ref M480 "Castle"	-	-	-	-	-	nr	24420.00
Climbing equipment; Wicksteed Leisure Ltd							
"Junior Commando Bridge"	-	-	-	-	-	nr	1027.00
"Fantasy Funrun - Under Starters Orders" set of 12 units	-	-	-	-	-	nr	13025.00
Climbing equipment; Lappset UK Ltd							
ref 120254; "Climbing Frame"	-	-	-	-	-	nr	2520.00
ref 122457; "Playhouse"	-	-	-	-	-	nr	3000.00
ref 120100; "Activity Tower"	-	-	-	-	-	nr	10850.00
ref 122124; "Tower and Climbing Frame"	-	-	-	-	-	nr	10300.00
SMP Playgrounds Ltd							
Action Pack - Cape Horn; 8 module	-	-	-	-	-	nr	7544.00
Spring equipment							
Spring based equipment for 1 - 8 year olds; Kompan Ltd							
ref M101; "Crazy Hen"	-	-	-	-	-	nr	530.00
ref M128; "Crazy Daisy"	-	-	-	-	-	nr	730.00
ref M141; "Crazy Seesaw"	-	-	-	-	-	nr	1360.00
ref M155; "Quartet Seesaw"	-	-	-	-	-	nr	1030.00
ref M199; "Crazy Springboard"	-	-	-	-	-	nr	1820.00
Spring based equipment for under 12's; Lappset UK Ltd							
ref 010443; "Rocking Horse"	-	-	-	-	-	nr	1070.00
Sand pits							
Market Prices							
Play Pit sand; Boughton Loam Ltd	-	-	-	-	46.04	m³	46.04
Kompan Ltd							
" Basic 500" 2900 mm x 1570 mm x 310 mm deep	-	-	-	-	-	nr	810.00
Flagpoles							
Flagpoles; Harrison External Display Systems; in glass fibre; smooth white finish; terylene halyards with mounting accessories; setting in concrete; to manufacturers recommendations (excavating not included)							
6.00 m high, plain	159.00	3.00	56.25	-	179.49	nr	235.74
6.00 m high, hinged baseplate, external halyard	198.00	4.00	75.00	-	218.49	nr	293.49
6.00 m high, hinged baseplate, internal halyard	303.00	4.00	75.00	-	323.49	nr	398.49
10.00 m high, plain	315.00	5.00	93.75	-	335.49	nr	429.24
10.00 m high, hinged baseplate, external halyard	771.00	5.00	93.75	-	791.49	nr	885.24
10.00 m high, hinged baseplate, internal halyard	448.00	5.00	93.75	-	468.49	nr	562.24
15.00 m high, plain	721.00	6.00	112.50	-	741.49	nr	853.99
15.00 m high, hinged baseplate, external halyard	771.00	6.00	112.50	-	791.49	nr	903.99
15.00 m high, hinged baseplate, internal halyard	970.00	6.00	112.50	-	990.49	nr	1102.99
Flagpoles; Harrison External Display Systems; tapered hollow steel section; base plate flange; galvanized; white painted finish; lowering gear; including bolting to concrete (excavating not included)							
7.00 m high	131.00	3.00	56.25	-	150.42	nr	206.67
10.00 m high	315.00	3.50	65.63	-	337.29	nr	402.92
13.00 m high	342.00	6.00	112.50	-	368.53	nr	481.03

Q PAVING/PLANTING/FENCING/SITE FURNITURE

Item Excluding site overheads and profit	PC £	Labour hours	Labour £	Plant £	Material £	Unit	Total rate £
Flagpoles; Harrison External Display Systems; in glass fibre; smooth white finish; terylene halyards for wall mounting							
3.0 m pole	180.00	2.00	37.50	-	180.00	nr	217.50
Flagpoles; Harrison External Display Systems; parallel galvanised steel banner flagpole - guide price							
6 m high	1000.00	4.00	75.00	-	1020.49	nr	1095.49
Sports equipment; Edwards Sports Products Ltd							
Tennis courts; steel posts and fixings for hardcourt							
Round	198.00	1.00	18.75	-	198.00	set	216.75
Square	209.00	1.00	18.75	-	209.00	set	227.75
Tennis nets; not including posts or fixings							
ref 5021; "Matchplay"	65.00	3.00	56.25	-	71.83	set	128.08
ref 5075; "Club"	49.50	3.00	56.25	-	56.33	set	112.58
ref 5001; "Championship"	77.00	3.00	56.25	-	83.83	set	140.08
Aluminum football goal posts with sockets; senior; including nets							
ref 2026; "Club" nets	638.00	4.00	75.00	-	644.83	set	719.83
ref 2037; "Medium Club" nets	625.00	0.14	2.70	-	814.70	set	817.40
Steel football goal posts with sockets; ref 2760; including nets							
ref 2026; "Club" net	463.00	4.00	75.00	-	469.83	set	544.83
ref 2037; "Medium Club" net	455.00	4.00	75.00	-	461.83	set	536.83
Junior aluminium football goal posts with sockets; including 2 mm nets	555.00	4.00	75.00	-	561.83	set	636.83
Upright one-piece tubular steel rugby goal post with sockets							
ref 2800; 7.30 m high	480.00	6.00	112.50	-	498.21	set	610.71
Telescopic tubular steel rugby goal post with sockets							
ref 2802; 10.70 m high	690.00	6.00	112.50	-	708.21	set	820.71
Hockey goal posts with sockets; including "Club" nets							
ref 2851; steel posts	230.00	0.14	2.70	-	372.27	set	374.98
backboards 450 mm	296.00	-	-	-	296.00	set	296.00
backboards 150 mm practice	110.00	-	-	-	110.00	set	110.00
Football mini goal complete with posts, net and fittings							
12' x 6'; aluminium and steel	397.00	4.00	75.00	-	403.83	set	478.83
12' x 6'; all steel	352.00	3.00	56.25	-	358.83	set	415.08
Clothes line fittings							
Rotary outdoor clothes line; Hills Industries Ltd; mild steel tube; to concrete base (base not included)							
"Airdry"; ref 4/40	16.10	2.00	37.50	-	19.51	nr	57.01
"Airdry"; ref 3/30	12.69	2.00	37.50	-	16.10	nr	53.60
"Portadry"; ref 3/35	19.13	2.00	37.50	-	22.54	nr	60.04
"Supadry"; ref 4/50	43.56	2.00	37.50	-	46.97	nr	84.47
"Supadry"; ref 3/50	39.05	2.00	37.50	-	42.46	nr	79.96
"Builders Special"	70.99	2.00	37.50	-	74.40	nr	111.90
"Handiline"; ref 4/45	39.29	2.00	37.50	-	42.70	nr	80.20
"Handiline"; ref 3/35	25.42	2.00	37.50	-	28.83	nr	66.33
Pair of precast concrete clothes posts 2.65 m long; 125 mm x 125 mm at base tapering to 75 mm; including setting in concrete (1:2:4); base size 300 mm x 450 mm x 600 mm deep;							
including excavating and disposing off site	60.12	3.00	56.25	-	72.26	nr	128.51

R DISPOSAL SYSTEMS

Item Excluding site overheads and profit	PC £	Labour hours	Labour £	Plant £	Material £	Unit	Total rate £
R12 DRAINAGE BELOW GROUND							
Silt pits and inspection chambers							
Excavating pits; starting from ground level; by machine							
maximum depth not exceeding 1.00 m	-	0.50	9.38	16.50	-	m³	**25.88**
maximum depth not exceeding 2.00 m	-	0.50	9.38	24.75	-	m³	**34.13**
maximum depth not exceeding 4.00 m	-	0.50	9.38	49.50	-	m³	**58.88**
Disposal of excavated material; depositing on site in permanent spoil heaps; average 50 m	-	0.04	0.78	1.61	-	m³	**2.39**
Filling to excavations; obtained from on site spoil heaps; average thickness not exceeding 0.25 m	-	0.13	2.50	4.40	-	m³	**6.90**
Surface treatments; compacting; bottoms of excavations	-	0.05	0.94	-	-	m²	**0.94**
Earthwork support; distance between opposing faces not exceeding 2.00 m							
maximum depth not exceeding 1.00 m	-	0.20	3.75	-	4.51	m³	**8.26**
maximum depth not exceeding 2.00 m	-	0.30	5.63	-	4.51	m³	**10.13**
maximum depth not exceeding 4.00 m	-	0.67	12.50	-	1.78	m³	**14.27**
Silt pits and inspection chambers; in situ concrete							
Beds; plain in situ concrete; 11.50 N/mm² - 40 mm aggregate							
thickness not exceeding 150 mm	-	1.00	18.75	-	94.85	m³	**113.60**
thickness 150 mm - 450 mm	-	0.67	12.50	-	94.85	m³	**107.35**
Benchings in bottoms; plain in situ concrete; 25.50 N/mm² - 20 mm aggregate							
thickness 150 mm - 450 mm	-	2.00	37.50	-	94.85	m³	**132.35**
Isolated cover slabs; reinforced in situ concrete; 21.00 N/mm² - 20 mm aggregate							
thickness not exceeding 150 mm	94.85	4.00	75.00	-	94.85	m³	**169.85**
Fabric reinforcement; BS 4483; A193 (3.02 kg/m²) in cover slabs	2.05	0.06	1.17	-	2.25	m²	**3.43**
Formwork to reinforced in situ concrete; isolated cover slabs							
soffits; horizontal	-	3.28	61.50	-	4.63	m²	**66.13**
height not exceeding 250 mm	-	0.97	18.19	-	1.97	m	**20.16**
Silt pits and inspection chambers; precast concrete units							
Precast concrete inspection chamber units; BS 5911; bedding, jointing and pointing in cement mortar (1:3); 600 mm x 450 mm internally							
600 mm deep	30.00	6.00	112.50	-	34.64	nr	**147.14**
900 mm deep	37.00	7.00	131.25	-	41.81	nr	**173.06**
Drainage chambers; 1200 mm x 750 mm reducing to 600 mm x 600 mm; no base unit; depth of invert							
1050 mm deep	198.00	9.00	168.75	-	209.43	nr	**378.18**
1650 mm deep	286.00	11.00	206.25	-	303.52	nr	**509.77**
2250 mm deep	374.00	12.50	234.38	-	396.31	nr	**630.69**
Cover slabs for chambers or shaft sections; heavy duty							
900 mm diameter internally	34.00	0.67	12.49	-	34.00	nr	**46.49**
1050 mm diameter internally	41.00	2.00	37.50	11.00	41.00	nr	**89.50**
1200 mm diameter internally	46.60	1.00	18.75	11.00	46.60	nr	**76.35**
1500 mm diameter internally	90.00	1.00	18.75	11.00	90.00	nr	**119.75**
1800 mm diameter internally	108.00	2.00	37.50	24.75	108.00	nr	**170.25**
Brickwork							
Walls to manholes; common bricks; PC £200.00 /1000; in cement mortar (1:3)							
one brick thick	24.00	3.00	56.25	-	32.98	m²	**89.23**
one and a half brick thick	36.00	4.00	75.00	-	49.47	m²	**124.47**
two brick thick projection of footing or the like	48.00	4.80	90.00	-	65.96	m²	**155.96**

R DISPOSAL SYSTEMS

Item Excluding site overheads and profit	PC £	Labour hours	Labour £	Plant £	Material £	Unit	Total rate £
Walls to manholes; engineering bricks; PC £260.00 /1000; in cement mortar (1:3)							
one brick thick	31.20	3.00	56.25	-	40.54	m²	96.79
one and a half brick thick	46.80	4.00	75.00	-	60.81	m²	135.81
two brick thick projection of footing or the like	62.40	4.80	90.00	-	81.08	m²	171.08
Extra over common or engineering bricks in any mortar for fair face; flush pointing as work proceeds; English bond walls or the like	-	0.13	2.50	-	-	m²	2.50
In situ finishings; cement:sand mortar (1:3); steel trowelled; 13 mm one coat work to manhole walls; to brickwork or blockwork base; over 300 mm wide	-	0.80	15.00	-	2.59	m²	17.59
Building into brickwork; ends of pipes; making good facings or renderings							
small	-	0.20	3.75	-	-	nr	3.75
large	-	0.30	5.63	-	-	nr	5.63
extra large	-	0.40	7.50	-	-	nr	7.50
extra large; including forming ring arch cover	-	0.50	9.38	-	-	nr	9.38
Inspection Chambers; Hepworth Plc							
Polypropylene up to 600 mm deep							
300 mm diameter; 960 mm deep; heavy duty round covers and frames; double seal recessed with 6 nr 110 mm outlets/inlets	141.49	3.00	56.25	-	143.98	nr	200.23
Polypropylene up to 1200 mm deep							
475 mm diameter; 1030 mm deep; heavy duty round covers and frames; double seal recessed with 5 nr 110 mm outlets/inlets	38.47	4.00	75.00	-	193.28	nr	268.28
Cast iron inspection chambers; to BS 437; St Gobain Pipelines Plc, drainage systems bolted flat covers; bedding in cement mortar (1:3); mechanical coupling joints							
100 mm x 100 mm							
one branch each side	147.80	2.20	41.25	-	150.49	nr	191.74
two branches each side	269.23	2.40	45.00	-	270.52	nr	315.52
100 mm x 150 mm							
one branch each side	187.52	1.80	33.75	-	187.78	nr	221.53
150 mm x 150 mm							
one branch	193.30	1.50	28.13	-	193.56	nr	221.68
one branch each side	222.87	2.75	51.56	-	223.13	nr	274.69
Step irons; to BS 1247; Ashworth Ltd drainage systems; malleable cast iron; galvanized; building into joints							
General purpose pattern; for one brick walls	3.62	0.17	3.19	-	3.62	nr	6.80
Best quality vitrified clay half section channels; Hepworth Plc; bedding and jointing in cement:mortar (1:2)							
Channels; straight							
100 mm	4.16	0.80	15.00	-	5.97	m	20.97
150 mm	6.94	1.00	18.75	-	8.74	m	27.49
225 mm	15.59	1.35	25.31	-	17.39	m	42.70
300 mm	31.99	1.80	33.75	-	33.79	m	67.54
Bends; 15, 30, 45 or 90 degrees							
100 mm bends	3.58	0.75	14.06	-	4.49	nr	18.55
150 mm bends	6.20	0.90	16.88	-	7.55	nr	24.43
225 mm bends	24.05	1.20	22.50	-	25.85	nr	48.35
300 mm bends	73.53	1.10	20.63	-	75.79	nr	96.42

R DISPOSAL SYSTEMS

Item Excluding site overheads and profit	PC £	Labour hours	Labour £	Plant £	Material £	Unit	Total rate £
R12 DRAINAGE BELOW GROUND - cont'd							
Best quality vitrified clay three quarter section channels; Hepworth Plc; bedding and jointing in cement:mortar (1:2)							
Branch bends; 115, 140 or 165 degrees; left or right hand							
100 mm	3.58	0.75	14.06	-	4.49	nr	**18.55**
150 mm	6.20	0.90	16.88	-	7.55	nr	**24.43**
Intercepting traps							
Vitrified clay; inspection arms; brass stoppers; iron levers; chains and staples; galvanized; staples cut and pinned to brickwork; cement:mortar (1:2) joints to vitrified clay pipes and channels; bedding and surrounding in concrete; 11.50 N/mm^2 - 40 mm aggregate; cutting and fitting brickwork; making good facings							
100 mm inlet; 100 mm outlet	53.22	3.00	56.25	-	78.08	nr	**134.33**
150 mm inlet; 150 mm outlet	76.73	2.00	37.50	-	106.47	nr	**143.97**
Excavating trenches; using 3 tonne tracked excavator; to receive pipes; grading bottoms; earthwork support; filling with excavated material to within 150 mm of finished surfaces and compacting; completing fill with topsoil; disposal of surplus soil							
Services not exceeding 200 mm nominal size							
average depth of run not exceeding 0.50 m	0.78	0.12	2.25	0.99	0.94	m	**4.17**
average depth of run not exceeding 0.75 m	0.78	0.16	3.05	1.37	0.94	m	**5.36**
average depth of run not exceeding 1.00 m	0.78	0.28	5.31	2.40	0.94	m	**8.64**
average depth of run not exceeding 1.25 m	0.78	0.38	7.19	3.22	0.78	m	**11.19**
Granular beds to trenches; lay granular material , to trenches excavated separately, to receive pipes (not included)							
300 wide x 100 thick							
reject sand	-	0.05	0.94	0.23	1.65	m	**2.81**
reject gravel	-	0.05	0.94	0.23	1.31	m	**2.48**
shingle 40 mm aggregate	-	0.05	0.94	0.23	2.94	m	**4.10**
sharp sand	2.92	0.05	0.94	0.23	3.21	m	**4.38**
300 wide x 150 thick							
reject sand	-	0.08	1.41	0.34	2.48	m	**4.22**
reject gravel	-	0.08	1.41	0.34	1.97	m	**3.71**
shingle 40 mm aggregate	-	0.08	1.41	0.34	4.41	m	**6.15**
sharp sand	4.38	0.08	1.41	0.34	4.82	m	**6.56**
Excavating trenches; using 3 tonne tracked excavator; to receive pipes; grading bottoms; earthwork support; filling with imported granular material type 2 and compacting; disposal of surplus soil							
Services not exceeding 200 mm nominal size							
average depth of run not exceeding 0.50 m	4.63	0.09	1.63	0.69	5.98	m	**8.30**
average depth of run not exceeding 0.75 m	6.95	0.11	2.03	0.86	8.97	m	**11.87**
average depth of run not exceeding 1.00 m	9.27	0.14	2.62	1.13	11.97	m	**15.71**
average depth of run not exceeding 1.25 m	11.58	0.23	4.29	1.93	14.96	m	**21.17**

R DISPOSAL SYSTEMS

Item Excluding site overheads and profit	PC £	Labour hours	Labour £	Plant £	Material £	Unit	Total rate £
Excavating trenches, using 3 tonne **tracked excavator, to receive pipes;** **grading bottoms; earthwork support;** **filling with concrete, ready mixed ST2;** **disposal of surplus soil**							
Services not exceeding 200 mm nominal size							
average depth of run not exceeding 0.50 m	9.71	0.11	2.00	0.36	11.30	m	**13.66**
average depth of run not exceeding 0.75 m	14.56	0.13	2.44	0.45	16.94	m	**19.83**
average depth of run not exceeding 1.00 m	19.41	0.17	3.13	0.60	22.60	m	**26.32**
average depth of run not exceeding 1.25 m	24.27	0.23	4.22	0.90	28.25	m	**33.37**
Earthwork Support; providing support to **opposing faces of excavation; moving** **along as work proceeds; A Plant Acrow**							
Maximum depth not exceeding 2.00 m							
distance between opposing faces not exceeding 2.00 m	-	0.80	15.00	16.95	-	m	**31.95**
Clay pipes and fittings: to BS **EN295:1:1991; Hepworth Plc;** **Supersleeve**							
100 mm clay pipes; polypropylene slip coupling; in trenches (trenches not included)							
laid straight	2.20	0.25	4.69	-	3.81	m	**8.50**
short runs under 3.00 m	2.20	0.31	5.86	-	3.81	m	**9.67**
Extra over 100 mm clay pipes for							
bends; 15-90 degree; single socket	6.99	0.25	4.69	-	6.99	nr	**11.67**
junction; 45 or 90 degree; double socket	15.26	0.25	4.69	-	15.26	nr	**19.95**
slip couplings polypropylene	2.59	0.08	1.56	-	2.59	nr	**4.15**
gully with "P" trap; 100 mm; 154 mm x 154 mm plastic grating	64.44	1.00	18.75	-	85.44	nr	**104.19**
150 mm vitrified clay pipes; polypropylene slip coupling; in trenches (trenches not included)							
laid straight	8.77	0.30	5.63	-	8.77	m	**14.40**
short runs under 3.00 m	6.75	0.33	6.25	-	6.75	m	**13.00**
Extra over 150 mm vitrified clay pipes for							
bends; 15 - 90 degree	9.77	0.28	5.25	-	14.48	nr	**19.73**
junction; 45 or 90 degree; 100 x 150	-	0.40	7.50	-	28.09	nr	**35.59**
junction; 45 or 90 degree; 150 x 150	20.49	0.40	7.50	-	29.91	nr	**37.41**
slip couplings polypropylene	4.71	0.05	0.94	-	4.71	nr	**5.65**
taper pipe 100 -150 mm	14.69	0.50	9.38	-	14.69	nr	**24.07**
taper pipe 150 -225 mm	36.06	0.50	9.38	-	36.06	nr	**45.44**
socket adaptor; connection to traditional pipes and fittings	9.47	0.33	6.19	-	14.18	nr	**20.37**
Accessories in clay							
150 mm access pipe	-	-	-	-	118.57	nr	**118.57**
150 mm rodding eye	47.97	0.50	9.38	-	52.11	nr	**61.49**
gully with "P" traps; 150 mm; 154 mm x 154 mm plastic grating	64.44	1.00	18.75	-	70.81	nr	**89.56**
PVC-u pipes and fittings; to BS EN1401; **Wavin Plastics Ltd; OsmaDrain**							
110 mm PVC-u pipes; in trenches (trenches not included)							
laid straight	11.12	0.08	1.50	-	11.12	m	**12.62**
short runs under 3.00 m	11.12	0.12	2.25	-	12.51	m	**14.76**
Extra over 110 mm PVC-u pipes for							
bends; short radius	21.68	0.25	4.69	-	21.68	nr	**26.37**
bends; long radius	37.59	0.25	4.69	-	37.59	nr	**42.28**
junctions; equal; double socket	25.87	0.25	4.69	-	25.87	nr	**30.56**
slip couplings	12.54	0.25	4.69	-	12.54	nr	**17.23**
adaptors to clay	21.51	0.50	9.38	-	21.60	nr	**30.98**
160 mm PVC-u pipes; in trenches (trenches not included)							
laid straight	25.37	0.08	1.50	-	25.37	m	**26.87**
short runs under 3.00 m	44.16	0.12	2.25	-	47.36	m	**49.61**

R DISPOSAL SYSTEMS

Item Excluding site overheads and profit	PC £	Labour hours	Labour £	Plant £	Material £	Unit	Total rate £
R12 DRAINAGE BELOW GROUND - cont'd							
PVC-u pipes and fittings; to BS EN1401; **Wavin Plastics Ltd; OsmaDrain** - cont'd							
Extra over 160 mm PVC-u pipes for							
socket bend double 90 or 45 degrees	71.42	0.20	3.75	-	71.42	nr	**75.17**
socket bend double 15 or 30 degrees	66.50	0.20	3.75	-	66.50	nr	**70.25**
socket bend single 87.5 or 45 degrees	47.69	0.20	3.75	-	47.69	nr	**51.44**
socket bend single 15 or 30 degrees	42.76	0.20	3.75	-	42.76	nr	**46.51**
bends; short radius	56.36	0.25	4.69	-	78.04	nr	**82.73**
bends; long radius	126.41	0.25	4.69	-	126.41	nr	**131.10**
junctions; single	138.90	0.33	6.25	-	138.90	nr	**145.15**
pipe coupler	28.18	0.05	0.94	-	28.18	nr	**29.12**
slip couplings PVC-u	15.24	0.05	0.94	-	15.24	nr	**16.18**
adaptors to clay	58.55	0.50	9.38	-	58.64	nr	**68.02**
level invert reducer	25.00	0.50	9.38	-	25.00	nr	**34.38**
spiggot	53.18	0.20	3.75	-	53.18	nr	**56.93**
Accessories in PVC-u							
110 mm screwed access cover	-	-	-	-	18.61	nr	**18.61**
110 mm rodding eye	44.33	0.50	9.38	-	48.47	nr	**57.85**
gully with "P" traps; 110 mm; 154 mm x 154 mm grating	43.58	1.00	18.75	-	45.23	nr	**63.98**
Kerbs; to gullies; in one course Class B engineering bricks; to 4 nr sides; rendering in cement:mortar (1:3); dished to gully gratings	2.08	1.00	18.75	-	2.96	nr	**21.71**
Gullies Concrete; Hepworth Plc Concrete road gullies; to BS 5911; trapped with rodding eye and stoppers; 450 mm diameter x 1.07 m deep	22.00	6.00	112.50	-	44.86	nr	**157.36**
Gullies Vitrified Clay; Hepworth Plc; **bedding in concrete; 11.50 N/mm² - 40** **mm aggregate** Vitrified clay yard gullies (mud); trapped; domestic duty (up to 1 tonne)							
RGP5; 100 mm outlet; 100 mm dia; 225 mm internal width 585 mm internal depth	70.78	3.50	65.63	-	71.31	nr	**136.94**
RGP7; 150 mm outlet; 100 mm dia; 225 mm internal width 585 mm internal depth	70.78	3.50	65.63	-	71.31	nr	**136.94**
Vitrified clay yard gullies (mud); trapped; medium duty (up to 5 tonnes)							
100 mm outlet; 100 mm dia; 225 mm internal width 585 mm internal depth	100.05	3.50	65.63	-	100.58	nr	**166.21**
150 mm outlet; 100 mm dia; 225 mm internal width 585 mm internal depth	101.56	3.50	65.63	-	102.10	nr	**167.72**
Combined filter and silt bucket for Yard gullies 225 mm wide	24.88	-	-	-	24.88	nr	**24.88**
Vitrified clay road gullies; trapped with rodding eye							
100 mm outlet; 300 mm internal dia, 600 mm internal depth	62.13	3.50	65.63	-	62.66	nr	**128.28**
150 mm outlet; 300 mm internal dia, 600 mm internal depth	63.62	3.50	65.63	-	64.15	nr	**129.77**
150 mm outlet; 400 mm internal dia, 750 mm internal depth	73.78	3.50	65.63	-	74.31	nr	**139.94**
150 mm outlet; 450 mm internal dia, 900 mm internal depth	-	3.50	65.63	-	100.36	nr	**165.99**
Hinged Gratings and frames for gullies; alloy							
135 mm for 100 mm dia gully	-	-	-	-	9.12	nr	**9.12**
193 mm for 150 mm dia gully	-	-	-	-	15.76	nr	**15.76**
120 mm x 120 mm	-	-	-	-	5.60	nr	**5.60**
150 mm x 150 mm	-	-	-	-	10.14	nr	**10.14**
230 mm x 230 mm	-	-	-	-	18.54	nr	**18.54**
316 mm x 316 mm	-	-	-	-	49.18	nr	**49.18**

R DISPOSAL SYSTEMS

Item Excluding site overheads and profit	PC £	Labour hours	Labour £	Plant £	Material £	Unit	Total rate £
Hinged Gratings and frames for gullies; Cast Iron;							
265 mm for 225 mm dia gully	-	-	-	-	31.51	nr	**31.51**
150 mm x 150 mm	-	-	-	-	10.14	nr	**10.14**
230 mm x 230 mm	-	-	-	-	18.54	nr	**18.54**
316 mm x 316 mm	-	-	-	-	49.18	nr	**49.18**
Universal Gully Trap PVC-u; Wavin							
Plastics Ltd; OsmaDrain system; bedding							
in concrete; 11.50 N/mm^2 - 40 mm							
aggregate							
Universal gully fitting; comprising gully trap only							
110 mm outlet; 110 mm dia; 205 mm internal							
depth	18.39	3.50	65.63	-	18.92	nr	**84.55**
Vertical inlet hopper c\w plastic grate							
272 x 183 mm	24.71	0.25	4.69	-	24.71	nr	**29.40**
Sealed access hopper							
110 x 110 mm	47.24	0.25	4.69	-	47.24	nr	**51.93**
Universal Gully PVC-u; Wavin Plastics							
Ltd; OsmaDrain system; accessories to							
Universal gully trap							
Hoppers; backfilling with clean granular material;							
tamping; surrounding in lean mix concrete							
plain hopper with 110 spigot 150 mm long	20.20	0.40	7.50	-	20.55	nr	**28.05**
vertical inlet hopper with 110 spigot 150 mm long	24.71	0.40	7.50	-	24.71	nr	**32.21**
sealed access hopper with 110 spigot 150 mm							
long	47.24	0.40	7.50	-	47.24	nr	**54.74**
plain hopper; solvent weld to trap	13.80	0.40	7.50	-	13.80	nr	**21.30**
vertical inlet hopper; solvent wed to trap	20.87	0.40	7.50	-	20.87	nr	**28.37**
sealed access cover; PVC-u	24.06	0.10	1.88	-	24.06	nr	**25.93**
Gullies PVC-u; Wavin Plastics Ltd;							
OsmaDrain system; bedding in concrete;							
11.50 N/mm^2 - 40 mm aggregate							
Bottle Gully; providing access to the drainage							
system for cleaning							
bottle gully; 228 x 228 x 317 mm deep	41.89	0.50	9.38	-	42.39	nr	**51.76**
sealed access cover; PVC-u 217 x 217 mm	31.01	0.10	1.88	-	31.01	nr	**32.89**
grating; ductile iron 215 x 215 mm	23.90	0.10	1.88	-	23.90	nr	**25.77**
bottle gully riser; 325 mm	5.56	0.50	9.38	-	6.64	nr	**16.01**
Yard Gully; trapped 300 mm diameter 600 mm							
deep; including catchment bucket and ductile							
iron cover and frame, medium duty loading							
305 mm diameter 600 mm deep	270.49	2.50	46.88	-	273.69	nr	**320.56**
Kerbs to Gullies							
One course Class B engineering bricks to 4 nr							
sides; rendering in cement:mortar (1:3); dished to							
gully gratings							
150 mm x 150 mm	1.04	0.33	6.25	-	1.62	nr	**7.87**
Access Covers and Frames							
Loading Information; Note - Groups 5 - 6 are not							
covered within the scope of this book. Readers							
should refer to the Spons Civil Engineering Price							
Book							
Group 1 - Class A15; Pedestrian access only - 1.5							
tonne maximum weight							
Group 2 - Class B125; Car parks and pedestrian							
areas with occasional vehicle use - 12.5 tonne							
maximum weight limit							
Group 3 - Class B250; Gully tops in areas							
extending more than 500 mm from the kerb into							
the carriageway							
Group 4 - Class D400; Carriageways of roads							

R DISPOSAL SYSTEMS

Item Excluding site overheads and profit	PC £	Labour hours	Labour £	Plant £	Material £	Unit	Total rate £
R12 DRAINAGE BELOW GROUND - cont'd							
Access covers and frames; BSEN124;							
St Gobain Pipelines Plc; bedding frame							
in cement mortar (1:3); cover in grease							
and sand; light duty; clear opening							
sizes; base size shown in brackets							
Group 1; solid top single seal; Ductile Iron; plastic frame							
Pedestrian; 450 mm diameter (525) x 35 deep	12.76	3.00	56.25	-	15.22	nr	71.47
Driveway; 450 mm diameter (550) x 35 deep	15.64	3.00	56.25	-	18.10	nr	74.35
Group 1; solid top single seal; Ductile Iron; Steel frame							
450 mm x 450 mm (565 x 565) x 35 deep	33.75	1.50	28.13	-	35.64	nr	63.77
450 mm x 600 mm (710 x 510) x 35 deep	42.14	1.80	33.75	-	44.78	nr	78.53
600 mm x 600 mm (710 x 710) x 35 deep	31.50	2.00	37.50	-	34.88	nr	72.38
Group 1; coated; double seal solid top; Fabricated Steel							
450 mm x 450 mm	24.00	1.50	28.13	-	25.89	nr	54.02
600 mm x 450 mm	25.50	1.80	33.75	-	28.14	nr	61.89
600 mm x 600 mm	31.50	2.00	37.50	-	34.88	nr	72.38
Group 2; Single seal solid top							
450 diameter x 41 deep	30.04	1.50	28.13	-	31.93	nr	60.05
600 diameter x 75 deep	57.52	1.50	28.13	-	59.41	nr	87.53
600 x 450 (710x 560) x 41 deep	42.14	1.50	28.13	-	44.03	nr	72.16
600 x 600 (711 x 711) x 44 deep	45.15	1.50	28.13	-	47.04	nr	75.17
600 x 600 (755 x 755) x 75 deep	55.90	1.50	28.13	-	57.79	nr	85.91
Recessed Covers and Frames; Bripave; Galvanized; Single seal; clear opening sizes shown in brackets							
450 x 450 x 93 (596 x 596) 1.5 tonne load	62.73	1.50	28.13	-	64.62	nr	92.75
450 x 450 x 93 (596 x 596) 5 tonne load	70.11	1.50	28.13	-	72.00	nr	100.13
450 x 450 x 93 (596 x 596) 11 tonne load	87.75	1.50	28.13	-	89.64	nr	117.77
600 x 450 x 93 (746 x 596) 1.5 tonne load	176.41	2.00	37.50	-	178.30	nr	215.80
600 x 450 x 93 (746 x 596) 5 tonne load	189.62	2.00	37.50	-	191.51	nr	229.01
600 x 450 x 93 (746 x 596) 11 tonne load	195.14	2.00	37.50	-	197.03	nr	234.53
600 x 600 x 93 (746 x 596) 1.5 tonne load	184.24	2.33	43.69	-	187.62	nr	231.31
600 x 600 x 93 (746 x 596) 5 tonne load	193.19	2.33	43.69	-	196.58	nr	240.26
600 x 600 x 93 (746 x 596) 11 tonne load	202.87	2.33	43.69	-	206.25	nr	249.94
750 x 600 x 93 (896 x 746) 1.5 tonne load	209.96	3.00	56.25	-	211.85	nr	268.10
750 x 600 x 93 (896 x 746) 5 tonne load	233.47	3.00	56.25	-	235.37	nr	291.62
750 x 600 x 93 (896 x 746) 11 tonne load	255.08	3.00	56.25	-	256.97	nr	313.22
Extra over to the above for double seal to covers							
450 x 450 x 93 (596 x 596)	-	-	-	-	22.50	nr	22.50
600 x 450 x 93 (746 x 596)	-	-	-	-	26.25	nr	26.25
600 x 600 x 93 (746 x 596)	-	-	-	-	30.00	nr	30.00
750 x 600 x 93 (896 x 746)	-	-	-	-	33.75	nr	33.75
Access Covers and Frames; Jones of							
Oswestry; bedding frame in cement							
mortar (1:3); cover in grease and sand;							
clear opening sizes							
Access Covers and frames; "Suprabloc"; to paved areas; filling with blocks cut and fitted to match surrounding paving; BS standard size manholes available							
pedestrian weight; 300 mm x 300 mm	81.05	2.50	46.88	-	87.27	nr	134.15
pedestrian weight; 450 mm x 450 mm	100.06	2.50	46.88	-	100.06	nr	146.94
pedestrian weight; 450 mm x 600 mm	110.79	2.50	46.88	-	117.27	nr	164.15
light vehicular weight; 300 mm x 300 mm	83.19	2.40	45.00	-	89.41	nr	134.41
light vehicular weight; 450 mm x 450 mm	102.70	3.00	56.25	-	109.83	nr	166.08
light vehicular weight; 450 mm x 600 mm	113.71	4.00	75.00	-	119.93	nr	194.93
heavy vehicular weight; 300 mm x 300 mm	85.32	3.00	56.25	-	91.80	nr	148.05
heavy vehicular weight; 450 mm x 600 mm	105.33	3.00	56.25	-	112.46	nr	168.71
heavy vehicular weight; 600 mm x 600 mm	116.63	4.00	75.00	-	127.00	nr	202.00

R DISPOSAL SYSTEMS

Item Excluding site overheads and profit	PC £	Labour hours	Labour £	Plant £	Material £	Unit	Total rate £
Extra over "Suprabloc" manhole frames and covers for							
filling recessed manhole covers with brick paviors; PC £305.00 /1000	12.20	1.00	18.75	-	13.92	m^2	**32.67**
filling recessed manhole covers with vehicular paving blocks; PC £7.83 /m^2	7.83	0.75	14.06	-	8.33	m^2	**22.39**
filling recessed manhole covers with concrete paving flags; PC £9.71/m^2	9.71	0.35	6.56	-	10.01	m^2	**16.57**
R13 LAND DRAINAGE							
Ditching; clear silt and bottom ditch not exceeding 1.50 m deep; strim back vegetation; disposing to spoil heaps; by machine							
Up to 1.50 m wide at top	-	6.00	112.50	27.00	-	100 m	**139.50**
1.50 - 2.50 m wide at top	-	8.00	150.00	36.00	-	100 m	**186.00**
2.50 - 4.00 m wide at top	-	9.00	168.75	40.50	-	100 m	**209.25**
Ditching; clear only vegetation from ditch not exceeding 1.50 m deep; disposing to spoil heaps; by strimmer							
Up to 1.50 m wide at top	-	5.00	93.75	6.80	-	100 m	**100.55**
1.50 - 2.50 m wide at top	-	6.00	112.50	12.24	-	100 m	**124.74**
2.50 - 4.00 m wide at top	-	7.00	131.25	19.04	-	100 m	**150.29**
Ditching; clear silt from ditch not exceeding 1.50 m deep; trimming back vegetation; disposing to spoil heaps; by hand							
Up to 1.50 m wide at top	-	15.00	281.25	6.80	-	100 m	**288.05**
1.50 - 2.50 m wide at top	-	27.00	506.25	12.24	-	100 m	**518.49**
2.50 - 4.00 m wide at top	-	42.00	787.50	19.04	-	100 m	**806.54**
Ditching; excavating and forming ditch and bank to given profile (normally 45 degrees); in loam or sandy loam; by machine							
Width 300 mm							
depth 600 mm	-	3.70	69.38	33.30	-	100 m	**102.67**
depth 900 mm	-	5.20	97.50	46.80	-	100 m	**144.30**
depth 1200 mm	-	7.20	135.00	64.80	-	100 m	**199.80**
depth 1500 mm	-	9.40	176.25	74.03	-	100 m	**250.28**
Width 600 mm							
depth 600 mm	-	-	-	89.51	-	100 m	**89.51**
depth 900 mm	-	-	-	132.82	-	100 m	**132.82**
depth 1200 mm	-	-	-	180.47	-	100 m	**180.47**
depth 1500 mm	-	-	-	259.88	-	100 m	**259.88**
Width 900 mm							
depth 600 mm	-	-	-	315.00	-	100 m	**315.00**
depth 900 mm	-	-	-	555.00	-	100 m	**555.00**
depth 1200 mm	-	-	-	745.80	-	100 m	**745.80**
depth 1500 mm	-	-	-	924.00	-	100 m	**924.00**
Width 1200 mm							
depth 600 mm	-	-	-	495.00	-	100 m	**495.00**
depth 900 mm	-	-	-	735.00	-	100 m	**735.00**
depth 1200 mm	-	-	-	973.50	-	100 m	**973.50**
depth 1500 mm	-	-	-	1247.40	-	100 m	**1247.40**
Width 1500 mm							
depth 600 mm	-	-	-	610.50	-	100 m	**610.50**
depth 900 mm	-	-	-	930.60	-	100 m	**930.60**
depth 1200 mm	-	-	-	1237.50	-	100 m	**1237.50**
depth 1500 mm	-	-	-	1551.00	-	100 m	**1551.00**
Extra for ditching in clay	-	-	-	-	-	-	**20%**

R DISPOSAL SYSTEMS

Item Excluding site overheads and profit	PC £	Labour hours	Labour £	Plant £	Material £	Unit	Total rate £
R13 LAND DRAINAGE - cont'd							
Ditching; excavating and forming ditch							
and bank to given profile (normal 45							
degrees); in loam or sandy loam; by hand							
Width 300 mm							
depth 600 mm	-	36.00	675.00	-	-	100 m	**675.00**
depth 900 mm	-	42.00	787.50	-	-	100 m	**787.50**
depth 1200 mm	-	56.00	1050.00	-	-	100 m	**1050.00**
depth 1500 mm	-	70.00	1312.50	-	-	100 m	**1312.50**
Width 600 mm							
depth 600 mm	-	56.00	1050.00	-	-	100 m	**1050.00**
depth 900 mm	-	84.00	1575.00	-	-	100 m	**1575.00**
depth 1200 mm	-	112.00	2100.00	-	-	100 m	**2100.00**
depth 1500 mm	-	140.00	2625.00	-	-	100 m	**2625.00**
Width 900 mm							
depth 600 mm	-	84.00	1575.00	-	-	100 m	**1575.00**
depth 900 mm	-	126.00	2362.50	-	-	100 m	**2362.50**
depth 1200 mm	-	168.00	3150.00	-	-	100 m	**3150.00**
depth 1500 mm	-	210.00	3937.50	-	-	100 m	**3937.50**
Width 1200 mm							
depth 600 mm	-	112.00	2100.00	-	-	100 m	**2100.00**
depth 900 mm	-	168.00	3150.00	-	-	100 m	**3150.00**
depth 1200 mm	-	224.00	4200.00	-	-	100 m	**4200.00**
depth 1500 mm	-	280.00	5250.00	-	-	100 m	**5250.00**
Extra for ditching in clay	-	-	-	-	-	-	**50%**
Extra for ditching in compacted soil	-	-	-	-	-	-	**90%**
Piped ditching							
Jointed concrete pipes; to BS 5911 pt.3;							
including bedding, haunching and topping with							
150 mm concrete; 11.50 N/mm^2 - 40 mm							
aggregate; to existing ditch							
300 mm diameter	11.44	0.67	12.50	6.60	28.79	m	**47.89**
450 mm diameter	17.93	0.67	12.50	6.60	43.96	m	**63.06**
600 mm diameter	26.87	0.67	12.50	6.60	63.82	m	**82.92**
900 mm diameter	69.68	1.00	18.75	9.90	87.93	m	**116.58**
Jointed concrete pipes; to BS 5911 pt.1 class S;							
including bedding, haunching and topping with							
150 mm concrete; 11.50 N/mm^2 - 40 mm							
aggregate; to existing ditch							
1200 mm diameter	110.89	1.00	18.75	9.90	134.77	m	**163.42**
Extra over jointed concrete pipes for bends to 45							
deg	91.52	0.67	12.50	8.25	98.80	nr	**119.55**
Extra over jointed concrete pipes for single							
junctions 300 mm dia	78.00	0.67	12.50	8.25	85.28	nr	**106.03**
Extra over jointed concrete pipes for single							
junctions 450 mm dia	97.00	0.67	12.50	7.50	104.28	nr	**124.28**
Extra over jointed concrete pipes for single							
junctions 600 mm dia	105.92	0.67	12.50	7.50	113.20	nr	**133.20**
Extra over jointed concrete pipes for single							
junctions 900 mm dia	256.06	0.67	12.50	7.50	263.34	nr	**283.34**
Extra over jointed concrete pipes for single							
junctions 1200 mm dia	379.71	0.67	12.50	7.50	386.99	nr	**406.99**
Concrete road gullies; to BS 5911; trapped;							
cement:mortar (1:2) joints to concrete pipes;							
bedding and surrounding in concrete; 11.50							
N/mm^2 - 40 mm aggregate; 450 mm diameter x							
1.07 m deep; rodding eye; stoppers	22.00	6.00	112.50	-	38.26	nr	**150.76**

R DISPOSAL SYSTEMS

Item Excluding site overheads and profit	PC £	Labour hours	Labour £	Plant £	Material £	Unit	Total rate £
Mole drainage; White Horse Contractors Ltd							
Drain by mole plough; 50 mm diameter mole set							
at depth of 450 mm in parallel runs							
1.20 m centres	-	-	-	-	-	100 m²	11.58
1.20 m centres	-	-	-	-	-	ha	576.87
1.50 m centres	-	-	-	-	-	100 m²	10.64
1.50 m centres	-	-	-	-	-	ha	521.64
2.00 m centres	-	-	-	-	-	100 m²	9.68
2.00 m centres	-	-	-	-	-	ha	479.13
2.50 m centres	-	-	-	-	-	100 m²	8.60
2.50 m centres	-	-	-	-	-	ha	426.69
3.00 m centres	-	-	-	-	-	100 m²	7.52
3.00 m centres	-	-	-	-	-	ha	383.79
Drain by mole plough; 75 mm diameter mole set							
at depth of 450 mm in parallel runs							
1.20 m centres	-	-	-	-	-	100 m²	13.83
1.20 m centres	-	-	-	-	-	ha	691.30
1.50 m centres	-	-	-	-	-	100 m²	12.60
1.50 m centres	-	-	-	-	-	ha	622.11
2.00 m centres	-	-	-	-	-	100 m²	12.17
2.00 m centres	-	-	-	-	-	ha	589.02
2.50 m centres	-	-	-	-	-	100 m²	10.17
2.50 m centres	-	-	-	-	-	ha	510.12
3.00 m centres	-	-	-	-	-	100 m²	10.64
3.00 m centres	-	-	-	-	-	ha	443.38
Sand slitting; White Horse Contractors Ltd							
Drainage slits; at 2.00 m centres; using slitting							
machine; backfilling to 75 mm of surface with pea							
gravel; blind with sharp sand							
250 mm depth	-	-	-	-	-	100 m²	26.18
300 mm depth	-	-	-	-	-	100 m²	29.41
400 mm depth	-	-	-	-	-	100 m²	35.36
450 mm depth	-	-	-	-	-	100 m²	37.98
250 mm depth	-	-	-	-	-	ha	6727.50
300 mm depth	-	-	-	-	-	ha	6986.25
400 mm depth	-	-	-	-	-	ha	7245.00
Trenchless drainage system; White							
Horse Contractors Ltd; insert perforated							
pipes by means of laser graded deep							
plough machine; backfill with gravel (not							
included)							
laterals 60 mm							
depth 700 mm	-	-	-	-	-	m	1.78
main 100 mm							
depth 900 mm	-	-	-	-	-	m	2.91
main 160 mm							
depth 1000 mm	-	-	-	-	-	m	3.73
Gravel backfill							
laterals 60 mm	-	-	-	-	-	m	3.95
main 110 mm	-	-	-	-	-	m	5.90
main 160 mm	-	-	-	-	-	m	7.29

Prices for Measured Works

R DISPOSAL SYSTEMS

Item Excluding site overheads and profit	PC £	Labour hours	Labour £	Plant £	Material £	Unit	Total rate £
R13 LAND DRAINAGE - cont'd							
Agricultural drainage; calculation table							
For calculation of drainage per hectare, the							
following table can be used; rates show the							
lengths of drains per unit and not the value							
Lateral drains							
4.00 m centres	-	-	-	-	-	m/ha	2500.00
6.00 m centres	-	-	-	-	-	m/ha	1670.00
8.00 m centres	-	-	-	-	-	m/ha	1250.00
10.00 m centres	-	-	-	-	-	m/ha	1000.00
15.00 m centres	-	-	-	-	-	m/ha	650.00
25.00 m centres	-	-	-	-	-	m/ha	400.00
30.00 m centres	-	-	-	-	-	m/ha	330.00
Main drains (per hectare)							
1 nr (at 100 m centres)	-	-	-	-	-	m/ha	100.00
2 nr (at 50 m centres)	-	-	-	-	-	m/ha	200.00
3 nr (at 33.333 m centres)	-	-	-	-	-	m/ha	300.00
4 nr (at 25 m centres)	-	-	-	-	-	m/ha	400.00
Agricultural drainage; excavating							
Removing 150 mm depth of topsoil; 300 mm							
wide; depositing beside trench; by machine	-	1.50	28.13	13.50	-	100 m	41.63
Removing 150 mm depth of topsoil; 300 mm							
wide; depositing beside trench; by hand	-	8.00	150.00	-	-	100 m	150.00
Disposing on site; to spoil heaps; by machine							
not exceeding 100 m distance	-	0.07	1.24	2.34	-	m³	3.58
average 100 - 150 m distance	-	0.08	1.49	2.81	-	m³	4.29
average 150 - 200 m distance	-	0.09	1.74	3.28	-	m³	5.02
Removing excavated material from site to tip not							
exceeding 13 km; mechanically loaded							
excavated material and clean hardcore rubble	-	-	-	1.10	7.50	m³	8.60
Agricultural drainage; White Horse							
Contractors Ltd; excavating trenches							
(with minimum run of 500 m) by trenching							
machine; including disposing subsoil to							
spoil heaps not exceeding 100 m							
Width 150 mm							
depth 450 mm	-	5.04	94.50	84.30	-	100 m	178.80
depth 600 mm	-	5.79	108.50	98.90	-	100 m	207.40
depth 750 mm	-	6.53	122.43	113.37	-	100 m	235.80
Width 225 mm							
depth 450 mm	-	4.67	87.66	79.87	-	100 m	167.53
depth 600 mm	-	6.60	123.75	118.55	-	100 m	242.30
depth 750 mm	-	7.42	139.22	136.26	-	100 m	275.48
Width 300 mm							
depth 600 mm	-	7.33	137.44	138.66	-	100 m	276.09
depth 750 mm	-	8.17	153.19	159.15	-	100 m	312.33
depth 900 mm	-	9.00	168.75	179.25	-	100 m	348.00
depth 1000 mm	-	9.72	182.25	195.06	-	100 m	377.31
Width 375 mm							
depth 600 mm	-	8.67	162.56	166.38	-	100 m	328.94
depth 750 mm	-	9.58	179.63	189.60	-	100 m	369.23
depth 900 mm	-	10.50	196.88	213.21	-	100 m	410.09
depth 1000 mm	-	11.78	220.88	238.60	-	100 m	459.47

R DISPOSAL SYSTEMS

Item Excluding site overheads and profit	PC £	Labour hours	Labour £	Plant £	Material £	Unit	Total rate £
Agricultural drainage; excavating for drains; by backacter excavator JCB C3X Sitemaster; including disposing spoil to spoil heaps not exceeding 100 m							
Width 150 mm							
depth 450 mm	-	0.50	9.38	85.50	-	100 m	**94.88**
depth 600 mm	-	0.67	12.50	108.50	-	100 m	**121.00**
depth 700 mm	-	0.78	14.63	129.42	-	100 m	**144.04**
depth 900 mm	-	1.00	18.75	138.00	-	100 m	**156.75**
Width 225 mm							
depth 450 mm	-	0.75	14.06	95.25	-	100 m	**109.31**
depth 600 mm	-	1.00	18.75	121.50	-	100 m	**140.25**
depth 700 mm	-	1.17	21.94	144.63	-	100 m	**166.57**
depth 900 mm	-	1.50	28.13	174.00	-	100 m	**202.13**
depth 1000 mm	-	1.67	31.31	230.13	-	100 m	**261.44**
Width 300 mm							
depth 450 mm	-	1.00	18.75	105.99	-	100 m	**124.74**
depth 700 mm	-	1.56	29.25	164.79	-	100 m	**194.04**
depth 900 mm	-	2.00	37.50	211.65	-	100 m	**249.15**
depth 1000 mm	-	2.22	41.63	235.08	-	100 m	**276.70**
depth 1200 mm	-	2.67	50.06	282.33	-	100 m	**332.39**
depth 1500 mm	-	3.33	62.44	352.62	-	100 m	**415.06**
Width 375 mm							
depth 450 mm	-	1.25	23.44	132.24	-	100 m	**155.68**
depth 700 mm	-	1.94	36.38	205.68	-	100 m	**242.06**
depth 900 mm	-	2.50	46.88	264.48	-	100 m	**311.36**
depth 1000 mm	-	2.78	52.13	293.88	-	100 m	**346.00**
depth 1200 mm	-	3.33	62.44	352.62	-	100 m	**415.06**
depth 1500 mm	-	4.17	78.19	440.82	-	100 m	**519.01**
Width 450 mm							
depth 450 mm	-	1.50	28.13	173.01	-	100 m	**201.13**
depth 700 mm	-	2.33	43.69	269.07	-	100 m	**312.76**
depth 900 mm	-	3.00	56.25	346.02	-	100 m	**402.27**
depth 1000 mm	-	3.33	62.44	384.63	-	100 m	**447.07**
depth 1200 mm	-	4.00	75.00	461.58	-	100 m	**536.58**
depth 1500 mm	-	5.00	93.75	576.81	-	100 m	**670.56**
depth 2000 mm	-	6.67	125.06	769.32	-	100 m	**894.38**
depth 2500 mm	-	8.33	156.19	961.44	-	100 m	**1117.63**
Width 600 mm							
depth 450 mm	-	6.63	124.31	258.57	-	100 m	**382.88**
depth 600 mm	-	8.84	165.75	344.76	-	100 m	**510.51**
depth 700 mm	-	10.31	193.31	402.09	-	100 m	**595.40**
depth 900 mm	-	11.05	207.19	430.95	-	100 m	**638.14**
depth 1000 mm	-	14.73	276.19	574.47	-	100 m	**850.66**
depth 1200 mm	-	17.68	331.50	689.52	-	100 m	**1021.02**
depth 1500 mm	-	22.10	414.38	861.90	-	100 m	**1276.28**
depth 2000 mm	-	29.46	552.38	1148.94	-	100 m	**1701.32**
depth 2500 mm	-	36.83	690.56	1436.37	-	100 m	**2126.93**
depth 3000 mm	-	36.83	690.56	1436.37	-	100 m	**2126.93**
Width 900 mm							
depth 450 mm	-	6.63	124.31	258.57	-	100 m	**382.88**
depth 600 mm	-	8.84	165.75	344.76	-	100 m	**510.51**
depth 700 mm	-	10.31	193.31	402.09	-	100 m	**595.40**
depth 900 mm	-	11.05	207.19	430.95	-	100 m	**638.14**
depth 1000 mm	-	14.73	276.19	574.47	-	100 m	**850.66**
depth 1200 mm	-	17.68	331.50	689.52	-	100 m	**1021.02**
depth 1500 mm	-	22.10	414.38	861.90	-	100 m	**1276.28**
depth 2000 mm	-	29.46	552.38	1148.94	-	100 m	**1701.32**
depth 2500 mm	-	36.83	690.56	1436.37	-	100 m	**2126.93**
depth 3000 mm	-	36.83	690.56	1436.37	-	100 m	**2126.93**

Prices for Measured Works

R DISPOSAL SYSTEMS

Item Excluding site overheads and profit	PC £	Labour hours	Labour £	Plant £	Material £	Unit	Total rate £
R13 LAND DRAINAGE - cont'd							
Agricultural drainage; excavating for							
drains; by 7 tonne tracked excavator;							
including disposing spoil to spoil heaps							
not exceeding 100 m							
Width 600 mm							
depth 450 mm	-	4.49	84.19	91.76	-	100 m	**175.95**
depth 600 mm	-	5.99	112.31	122.42	-	100 m	**234.73**
depth 700 mm	-	6.99	131.06	142.86	-	100 m	**273.92**
depth 900 mm	-	8.98	168.38	183.53	-	100 m	**351.90**
depth 1000 mm	-	9.98	187.13	203.97	-	100 m	**391.09**
depth 1200 mm	-	11.98	224.63	244.84	-	100 m	**469.47**
depth 1500 mm	-	14.97	280.69	305.95	-	100 m	**586.64**
depth 2000 mm	-	19.97	374.44	408.14	-	100 m	**782.57**
depth 2500 mm	-	24.96	468.00	973.44	-	100 m	**1441.44**
depth 3000 mm	-	29.95	561.56	612.10	-	100 m	**1173.67**
Width 700 mm							
depth 450 mm	-	5.24	98.25	107.09	-	100 m	**205.34**
depth 600 mm	-	6.99	131.06	142.86	-	100 m	**273.92**
depth 700 mm	-	8.15	152.81	166.57	-	100 m	**319.38**
depth 900 mm	-	10.48	196.50	214.19	-	100 m	**410.69**
depth 1000 mm	-	11.65	218.44	238.10	-	100 m	**456.53**
depth 1200 mm	-	13.98	262.13	285.72	-	100 m	**547.84**
depth 1500 mm	-	17.47	327.56	357.04	-	100 m	**684.61**
depth 2000 mm	-	23.29	436.69	475.99	-	100 m	**912.68**
depth 2500 mm	-	29.11	545.85	1135.68	-	100 m	**1681.53**
depth 3000 mm	-	34.94	655.13	714.09	-	100 m	**1369.21**
Width 900 mm							
depth 600 mm	-	9.89	185.44	202.13	-	100 m	**387.56**
depth 700 mm	-	11.54	216.38	235.85	-	100 m	**452.22**
depth 900 mm	-	14.84	278.25	303.44	-	100 m	**581.69**
depth 1000 mm	-	16.48	309.00	336.81	-	100 m	**645.81**
depth 1200 mm	-	19.78	370.88	404.25	-	100 m	**775.13**
depth 1500 mm	-	24.73	463.69	505.42	-	100 m	**969.11**
depth 2000 mm	-	32.97	618.19	673.82	-	100 m	**1292.01**
depth 2500 mm	-	41.21	772.69	1607.19	-	100 m	**2379.88**
depth 3000 mm	-	49.45	927.19	1010.63	-	100 m	**1937.82**
Width 1000 mm							
depth 600 mm	-	11.64	218.25	237.89	-	100 m	**456.14**
depth 700 mm	-	13.59	254.81	277.75	-	100 m	**532.56**
depth 900 mm	-	17.47	327.56	357.04	-	100 m	**684.61**
depth 1000 mm	-	19.41	363.94	396.69	-	100 m	**760.63**
depth 1200 mm	-	23.29	436.69	475.99	-	100 m	**912.68**
depth 1500 mm	-	29.11	545.81	594.94	-	100 m	**1140.75**
depth 2000 mm	-	38.81	727.69	793.18	-	100 m	**1520.87**
depth 2500 mm	-	48.52	909.75	1892.28	-	100 m	**2802.03**
depth 3000 mm	-	58.22	1091.63	1189.87	-	100 m	**2281.50**
Width 1500 mm							
depth 600 mm	-	17.47	327.56	357.04	-	100 m	**684.61**
depth 700 mm	-	20.38	382.13	416.52	-	100 m	**798.64**
depth 900 mm	-	26.20	491.25	535.46	-	100 m	**1026.71**
depth 1000 mm	-	29.11	545.81	594.94	-	100 m	**1140.75**
depth 1200 mm	-	34.93	654.94	713.88	-	100 m	**1368.82**
depth 1500 mm	-	43.67	818.81	892.51	-	100 m	**1711.32**
depth 2000 mm	-	58.22	1091.63	1189.87	-	100 m	**2281.50**
depth 2500 mm	-	72.78	1364.63	2838.42	-	100 m	**4203.05**
depth 3000 mm	-	87.33	1637.44	1784.81	-	100 m	**3422.24**

R DISPOSAL SYSTEMS

Item Excluding site overheads and profit	PC £	Labour hours	Labour £	Plant £	Material £	Unit	Total rate £
Agricultural drainage; excavating for **drains; by hand, including disposing** **spoil to spoil heaps not exceeding 100 m**							
Width 150 mm							
depth 450 mm	-	22.03	413.06	-	-	100 m	**413.06**
depth 600 mm	-	29.38	550.88	-	-	100 m	**550.88**
depth 700 mm	-	34.27	642.56	-	-	100 m	**642.56**
depth 900 mm	-	44.06	826.13	-	-	100 m	**826.13**
Width 225 mm							
depth 450 mm	-	33.05	619.69	-	-	100 m	**619.69**
depth 600 mm	-	44.06	826.13	-	-	100 m	**826.13**
depth 700 mm	-	51.41	963.94	-	-	100 m	**963.94**
depth 900 mm	-	66.10	1239.38	-	-	100 m	**1239.38**
depth 1000 mm	-	73.44	1377.00	-	-	100 m	**1377.00**
Width 300 mm							
depth 450 mm	-	44.06	826.13	-	-	100 m	**826.13**
depth 600 mm	-	58.75	1101.56	-	-	100 m	**1101.56**
depth 700 mm	-	68.54	1285.13	-	-	100 m	**1285.13**
depth 900 mm	-	88.13	1652.44	-	-	100 m	**1652.44**
depth 1000 mm	-	97.92	1836.00	-	-	100 m	**1836.00**
Width 375 mm							
depth 450 mm	-	55.08	1032.75	-	-	100 m	**1032.75**
depth 600 mm	-	73.44	1377.00	-	-	100 m	**1377.00**
depth 700 mm	-	85.68	1606.50	-	-	100 m	**1606.50**
depth 900 mm	-	110.16	2065.50	-	-	100 m	**2065.50**
depth 1000 mm	-	122.40	2295.00	-	-	100 m	**2295.00**
Width 450 mm							
depth 450 mm	-	66.10	1239.38	-	-	100 m	**1239.38**
depth 600 mm	-	88.13	1652.44	-	-	100 m	**1652.44**
depth 700 mm	-	102.82	1927.88	-	-	100 m	**1927.88**
depth 900 mm	-	132.19	2478.56	-	-	100 m	**2478.56**
depth 1000 mm	-	146.88	2754.00	-	-	100 m	**2754.00**
Width 600 mm							
depth 450 mm	-	88.13	1652.44	-	-	100 m	**1652.44**
depth 600 mm	-	117.50	2203.13	-	-	100 m	**2203.13**
depth 700 mm	-	137.09	2570.44	-	-	100 m	**2570.44**
depth 900 mm	-	176.26	3304.88	-	-	100 m	**3304.88**
depth 1000 mm	-	195.84	3672.00	-	-	100 m	**3672.00**
Width 900 mm							
depth 450 mm	-	132.19	2478.56	-	-	100 m	**2478.56**
depth 600 mm	-	176.26	3304.88	-	-	100 m	**3304.88**
depth 700 mm	-	205.63	3855.56	-	-	100 m	**3855.56**
depth 900 mm	-	264.38	4957.13	-	-	100 m	**4957.13**
depth 1000 mm	-	293.76	5508.00	-	-	100 m	**5508.00**
Earthwork Support; moving along as **work proceeds**							
Maximum depth not exceeding 2.00 m							
distance between opposing faces not exceeding							
2.00 m	-	0.80	15.00	16.95	-	m	**31.95**
Agricultural drainage; pipe laying							
Hepworth; Agricultural clay drain pipes; to BS							
1196; 300 mm length; butt joints; in straight runs							
75 mm diameter	285.05	8.00	150.00	-	292.17	100 m	**442.17**
100 mm diameter	490.18	9.00	168.75	-	502.43	100 m	**671.18**
150 mm diameter	1004.33	10.00	187.50	-	1029.44	100 m	**1216.94**
Extra over clay drain pipes for filter-wrapping							
pipes with "Terram" or similar filter fabric							
"Terram 700"	0.20	0.04	0.75	-	0.20	m^2	**0.95**
"Terram 1000"	0.20	0.04	0.75	-	0.20	m^2	**0.95**
Junctions between drains in clay pipes							
75 mm x 75 mm	10.36	0.25	4.69	-	10.58	nr	**15.27**
100 mm x 100 mm	12.94	0.25	4.69	-	13.31	nr	**17.99**
100 mm x 150 mm	15.94	0.25	4.69	-	16.48	nr	**21.17**

R DISPOSAL SYSTEMS

Item Excluding site overheads and profit	PC £	Labour hours	Labour £	Plant £	Material £	Unit	Total rate £
R13 LAND DRAINAGE - cont'd							
Wavin Plastics Ltd; flexible plastic							
perforated pipes in trenches (not							
included); to a minimum depth of 450 mm							
(couplings not included)							
"OsmaDrain"; flexible plastic perforated pipes in							
trenches (not included); to a minimum depth of							
450 mm (couplings not included)							
60 mm diameter; available in 150 m coil	54.66	2.00	37.50	-	56.03	100 m	93.53
80 mm diameter; available in 100 m coil	82.57	2.00	37.50	-	84.63	100 m	122.13
100 mm diameter; available in 100 m coil	145.38	2.00	37.50	-	149.01	100 m	186.51
160 mm diameter; available in 35 m coil	351.24	2.00	37.50	-	360.02	100 m	397.52
"WavinCoil"; plastic pipe junctions							
60 mm x 60 mm	3.14	0.05	0.94	-	3.14	nr	4.08
80 mm x 80 mm	3.44	0.05	0.94	-	3.44	nr	4.38
100 mm x 100 mm	3.85	0.05	0.94	-	3.85	nr	4.79
80 mm x 60 mm	3.39	0.05	0.94	-	3.39	nr	4.33
100 mm x 60 mm	3.62	0.05	0.94	-	3.62	nr	4.56
100 mm x 80 mm	3.67	0.05	0.94	-	3.67	nr	4.61
125 mm x 60 mm	3.99	0.05	0.94	-	3.99	nr	4.93
125 mm x 100 mm	4.14	0.05	0.94	-	4.14	nr	5.08
160 mm x 160 mm	9.94	0.05	0.94	-	9.94	nr	10.88
"WavinCoil"; couplings for flexible pipes							
60 mm diameter	1.23	0.03	0.62	-	1.23	nr	1.85
80 mm diameter	1.42	0.03	0.62	-	1.42	nr	2.04
100 mm diameter	1.57	0.03	0.62	-	1.57	nr	2.19
160 mm diameter	2.12	0.03	0.62	-	2.12	nr	2.74
Market prices of backfilling materials							
Sand	15.00	-	-	-	18.00	m³	18.00
Gravel rejects	12.50	-	-	-	13.75	m³	13.75
Topsoil; allowing for 20% settlement	12.48	-	-	-	14.98	m³	14.98
Agricultural drainage; backfilling trench							
after laying pipes with gravel rejects or							
similar; blind filling with ash or sand;							
topping with 150 mm topsoil from dumps							
not exceeding 100 m; by machine							
Width 150 mm							
depth 450 mm	-	3.30	61.88	53.40	88.37	100 m	203.65
depth 600 mm	-	4.30	80.63	69.90	119.20	100 m	269.72
depth 750 mm	-	4.96	93.00	80.79	135.25	100 m	309.04
depth 900 mm	-	6.30	118.13	102.90	172.75	100 m	393.77
Width 225 mm							
depth 450 mm	-	4.95	92.81	80.10	136.74	100 m	309.66
depth 600 mm	-	6.45	120.94	104.85	178.87	100 m	404.66
depth 750 mm	-	7.95	149.06	129.60	221.12	100 m	499.78
depth 900 mm	-	9.45	177.19	154.35	263.24	100 m	594.78
Width 375 mm							
depth 450 mm	-	8.25	154.69	133.50	227.81	100 m	516.00
depth 600 mm	-	10.75	201.56	174.75	298.06	100 m	674.38
depth 750 mm	-	13.25	248.44	216.00	368.44	100 m	832.88
depth 900 mm	-	15.75	295.31	257.25	438.69	100 m	991.25
Agricultural drainage; backfilling trench							
after laying pipes with gravel rejects or							
similar, blind filling with ash or sand,							
topping with 150 mm topsoil from dumps							
not exceeding 100 m; by hand							
Width 150 mm							
depth 450 mm	-	18.63	349.31	-	82.75	100 m	432.07
depth 600 mm	-	24.84	465.75	-	119.20	100 m	584.95
depth 750 mm	-	31.05	582.19	-	107.17	100 m	689.35
depth 900 mm	-	37.26	698.63	-	172.75	100 m	871.37

R DISPOSAL SYSTEMS

Item Excluding site overheads and profit	PC £	Labour hours	Labour £	Plant £	Material £	Unit	Total rate £
Width 225 mm							
depth 450 mm	-	27.94	523.88	-	136.74	100 m	**660.62**
depth 600 mm	-	37.26	698.63	-	178.87	100 m	**877.49**
depth 750 mm	-	46.57	873.19	-	221.12	100 m	**1094.31**
depth 900 mm	-	55.89	1047.94	-	263.24	100 m	**1311.18**
Width 375 mm							
depth 450 mm	-	46.57	873.19	-	227.81	100 m	**1101.00**
depth 600 mm	-	61.10	1145.63	-	298.06	100 m	**1443.69**
depth 750 mm	-	77.63	1455.56	-	368.44	100 m	**1824.00**
depth 900 mm	-	93.15	1746.56	-	438.69	100 m	**2185.25**
Catchwater or french drains; 100 mm **diameter non-coilable perforated plastic** **pipes; to BS4962; including straight** **jointing; pipes laid with perforations** **uppermost; lining trench; wrapping pipes** **with filter fabric**							
Width 300 mm							
depth 450 mm	995.35	9.10	170.63	34.65	1372.36	100 m	**1577.64**
depth 600 mm	998.81	9.84	184.50	46.86	1502.34	100 m	**1733.70**
depth 750 mm	1002.43	10.60	198.75	59.40	1632.51	100 m	**1890.66**
depth 900 mm	1005.97	11.34	212.63	71.61	1762.58	100 m	**2046.81**
depth 1000 mm	1008.33	11.84	222.00	79.86	1849.29	100 m	**2151.15**
depth 1200 mm	1013.05	12.84	240.75	96.36	2022.72	100 m	**2359.83**
Width 450 mm							
depth 450 mm	1000.66	10.44	195.75	86.46	1567.47	100 m	**1849.68**
depth 600 mm	1005.97	11.34	212.63	71.61	1762.58	100 m	**2046.81**
depth 750 mm	1011.28	12.46	233.63	90.09	1957.68	100 m	**2281.40**
depth 900 mm	1016.59	13.60	255.00	108.90	2152.79	100 m	**2516.69**
depth 1000 mm	1020.13	14.34	268.88	121.11	2282.86	100 m	**2672.84**
depth 1200 mm	1027.21	15.84	297.00	145.86	2543.00	100 m	**2985.86**
depth 1500 mm	1037.83	18.10	339.38	183.15	2933.21	100 m	**3455.74**
depth 2000 mm	1055.53	21.84	409.50	244.86	3583.57	100 m	**4237.93**
Width 600 mm							
depth 450 mm	1005.97	11.24	210.75	71.61	1762.58	100 m	**2044.94**
depth 600 mm	1013.05	12.84	240.75	96.36	2022.72	100 m	**2359.83**
depth 750 mm	1020.13	14.34	268.88	121.11	2282.86	100 m	**2672.84**
depth 900 mm	1027.21	15.84	297.00	145.86	2543.00	100 m	**2985.86**
depth 1000 mm	1033.95	16.84	315.75	162.36	2708.31	100 m	**3186.42**
depth 1200 mm	1041.37	18.84	353.25	195.36	3063.28	100 m	**3611.89**
depth 1500 mm	1055.53	21.84	409.50	469.86	3583.57	100 m	**4462.93**
depth 2000 mm	1079.13	19.84	372.00	442.86	4450.70	100 m	**5265.56**
Width 900 mm							
depth 450 mm	1016.59	13.60	255.00	108.90	2152.79	100 m	**2516.69**
depth 600 mm	1027.21	7.92	148.50	145.86	2543.00	100 m	**2837.36**
depth 750 mm	1037.83	9.05	169.69	183.15	2933.21	100 m	**3286.05**
depth 900 mm	1048.45	10.17	190.69	220.11	3323.42	100 m	**3734.22**
depth 1000 mm	1055.53	10.92	204.75	244.86	3583.57	100 m	**4033.18**
depth 1200 mm	1069.69	12.42	232.88	294.36	4103.85	100 m	**4631.08**
depth 1500 mm	1090.93	14.67	275.06	368.61	4884.27	100 m	**5527.94**
depth 2000 mm	1126.33	18.92	354.75	567.36	6184.98	100 m	**7107.09**
depth 2500 mm	1161.73	22.17	415.69	691.11	7485.69	100 m	**8592.48**
depth 3000 mm	1197.13	25.92	486.00	739.86	8786.39	100 m	**10012.25**

R DISPOSAL SYSTEMS

Item Excluding site overheads and profit	PC £	Labour hours	Labour £	Plant £	Material £	Unit	Total rate £
R13 LAND DRAINAGE - cont'd							
Catchwater or french drains; 160 mm **diameter non-coilable perforated plastic** **pipes; to BS4962; including straight** **jointing; pipes laid with perforations** **uppermost; lining trench; wrapping pipes** **with filter fabric**							
Width 600 mm							
depth 450 mm	1709.39	11.20	210.00	69.30	2463.56	100 m	**2742.86**
depth 600 mm	1716.47	12.70	238.13	94.05	2723.70	100 m	**3055.88**
depth 750 mm	1723.55	14.20	266.25	118.80	2983.84	100 m	**3368.89**
depth 900 mm	1730.63	15.70	294.38	143.55	3243.98	100 m	**3681.91**
depth 1000 mm	1735.35	16.70	313.13	160.05	3417.41	100 m	**3890.59**
depth 1200 mm	1744.79	18.70	350.63	193.15	3764.27	100 m	**4308.04**
depth 1500 mm	1758.95	21.70	406.88	242.55	4284.55	100 m	**4933.97**
depth 2000 mm	1782.55	26.70	500.63	325.05	5151.69	100 m	**5977.36**
depth 2500 mm	1806.15	31.70	594.38	407.55	6018.82	100 m	**7020.75**
depth 3000 mm	1829.75	36.70	688.13	490.05	6885.96	100 m	**8064.14**
Width 900 mm							
depth 450 mm	1016.59	13.60	255.00	108.90	2152.79	100 m	**2516.69**
depth 600 mm	1027.21	15.84	297.00	145.86	2543.00	100 m	**2985.86**
depth 750 mm	1037.83	18.10	339.38	183.15	2933.21	100 m	**3455.74**
depth 900 mm	1048.45	20.34	381.38	220.11	3323.42	100 m	**3924.91**
depth 1000 mm	1055.53	21.84	409.50	244.86	3583.57	100 m	**4237.93**
depth 1200 mm	1069.69	24.84	465.75	294.36	4103.85	100 m	**4863.96**
depth 1500 mm	1090.93	29.34	550.13	368.61	4884.27	100 m	**5803.01**
depth 2000 mm	1126.33	36.84	690.75	492.36	6184.98	100 m	**7368.09**
depth 2500 mm	1161.73	44.34	831.38	616.11	7485.69	100 m	**8933.17**
depth 3000 mm	1197.13	51.84	972.00	739.86	8786.39	100 m	**10498.25**
Catchwater or french drains; Exxon **Chemical Geopolymers; "Filtram" filter** **drain; in trenches (trenches not** **included); comprising filter fabric, liquid** **conducting core and 110 mm PVC-u** **slitpipes; all in accordance with** **manufacturer's instructions; backfilling** **with clean broken stone 40 mm minimum;** **top 300 mm of fill to be gravel rejects or** **quarry waste**							
Width 600 mm							
depth 1000 mm	1031.93	16.84	315.75	162.36	2716.43	100 m	**3194.54**
depth 1200 mm	1041.37	18.84	353.25	195.36	3063.28	100 m	**3611.89**
depth 1500 mm	1055.53	21.84	409.50	469.86	3583.57	100 m	**4462.93**
depth 2000 mm	1079.13	19.84	372.00	442.86	4450.70	100 m	**5265.56**
Width 900 mm							
depth 450 mm	1016.59	13.60	255.00	108.90	2152.79	100 m	**2516.69**
depth 600 mm	1027.21	15.84	297.00	145.86	2543.00	100 m	**2985.86**
depth 750 mm	1037.83	18.10	339.38	183.15	2933.21	100 m	**3455.74**
depth 900 mm	1048.45	20.34	381.38	220.11	3323.42	100 m	**3924.91**
depth 1000 mm	1055.53	21.84	409.50	244.86	3583.57	100 m	**4237.93**
depth 1200 mm	1069.69	24.84	465.75	294.36	4103.85	100 m	**4863.96**
depth 1500 mm	1090.93	29.34	550.13	368.61	4884.27	100 m	**5803.01**
depth 2000 mm	1126.33	37.84	709.50	567.36	6184.98	100 m	**7461.84**
depth 2500 mm	1161.73	44.34	831.38	691.11	7485.69	100 m	**9008.17**
depth 3000 mm	1197.13	51.84	972.00	814.86	8786.39	100 m	**10573.25**
Outfalls							
Reinforced concrete outfalls to water course; flank walls; for 150 mm drain outlets; overall dimensions							
900 mm x 1050 mm x 900 mm high	-	-	-	-	-	m	**424.00**

R DISPOSAL SYSTEMS

Item Excluding site overheads and profit	PC £	Labour hours	Labour £	Plant £	Material £	Unit	Total rate £
Soakaway design based on BS EN 752-4. Flat rate hourly rainfall = 50 mm /hr and assumes 100 impermeability of the run-off area. A storage capacity of the soakaway should be 1/3 of the hourly rainfall; Formulae for calculating soakaway depths are provided in the publications mentioned below and in the memoranda section of this publication. The design of soakaways is dependent on amongst other factors, soil conditions, permeability, groundwater level and runoff. The definitive documents for design of soakaways are CIRIA 156 and BRE Digest 365 dated September 1991; The suppliers of the systems below will assist through their technical divisions. Excavation earthwork support of pits and disposal not included.							
Excavating; mechanical							
To reduce levels							
maximum depth not exceeding 1.00 m; JCB sitemaster	-	0.05	0.94	1.65	-	m^3	**2.59**
maximum depth not exceeding 1.00 m; 360 Tracked excavator	-	0.04	0.75	1.65	-	m^3	**2.40**
maximum depth not exceeding 2.00 m; 360 Tracked excavator	-	0.06	1.13	2.48	-	m^3	**3.60**
Disposal; mechanical							
Excavated material; off site; to tip; mechanically loaded (JCB)							
inert	-	0.04	0.78	1.25	14.85	m^3	**16.88**
Insitu concrete ring beam foundations to base of soakaway; 300 mm wide x 250 deep; poured on or against earth or unblinded hardcore							
internal diameters of rings							
900 mm	-	4.00	75.00	-	19.32	nr	**94.32**
1200 mm	23.41	4.50	84.38	-	25.76	nr	**110.13**
1500 mm	29.27	5.00	93.75	-	32.19	nr	**125.94**
2400 mm	46.83	5.50	103.13	-	51.51	nr	**154.64**
Concrete soakaway rings; Milton Pipes Ltd; perforations and step irons to concrete rings at manufacturers recommended centres; placing of concrete ring soakaways to insitu concrete ring beams (1:3:6) (not included); filling and surrounding base with gravel 225 deep (not included)							
Ring diameter 900 mm							
1.00 deep; volume 636 litres	36.00	1.25	23.44	18.05	114.05	nr	**155.54**
1.50 deep; volume 954 litres	54.00	1.65	30.94	47.64	168.85	nr	**247.44**
2.00 deep; volume 1272 litres	72.00	1.65	30.94	47.64	223.65	nr	**302.24**
Ring diameter 1200 mm							
1.00 deep; volume 1131 litres	50.00	3.00	56.25	43.31	140.11	nr	**239.67**
1.50 deep; volume 1696 litres	75.00	4.50	84.38	64.97	206.21	nr	**355.55**
2.00 deep; volume 2261 litres	100.00	4.50	84.38	64.97	272.31	nr	**421.65**
2.50 deep; volume 2827 litres	125.00	6.00	112.50	86.63	308.31	nr	**507.44**
Ring diameter 1500 mm							
1.00 deep; volume 1767 litres	86.00	4.50	84.38	43.31	197.77	nr	**325.46**
1.50 deep; volume 2651 litres	129.00	6.00	112.50	60.64	290.47	nr	**463.61**
2.00 deep; volume 3534 litres	172.00	6.00	112.50	60.64	383.17	nr	**556.31**
2.50 deep; volume 4418 litres	125.00	7.50	140.63	108.28	385.87	nr	**634.78**

Prices for Measured Works

R DISPOSAL SYSTEMS

Item Excluding site overheads and profit	PC £	Labour hours	Labour £	Plant £	Material £	Unit	Total rate £
R13 LAND DRAINAGE - cont'd							
Concrete soakaway rings - cont'd							
Ring diameter 2400 mm							
1.00 deep; volume 4524 litres	284.00	6.00	112.50	43.31	434.47	nr	590.28
1.50 deep; volume 6786 litres	426.00	7.50	140.63	60.64	643.37	nr	844.63
2.00 deep; volume 9048 litres	568.00	7.50	140.63	60.64	856.57	nr	1057.83
2.50 deep; volume 11310 litres	710.00	9.00	168.75	108.28	1065.47	nr	1342.50
Extra over for							
250 mm depth chamber ring	-	-	-	-	-	-	100%
500 mm depth chamber ring	-	-	-	-	-	-	50%
Cover slabs to soakaways							
heavy duty precast concrete							
900 diameter	51.50	1.00	18.75	21.66	51.50	nr	91.91
1200 diameter	69.50	1.00	18.75	21.66	69.50	nr	109.91
1500 diameter	109.50	1.00	18.75	21.66	109.50	nr	149.91
2400 diameter	442.50	1.00	18.75	21.66	442.50	nr	482.91
Step irons to concrete chamber rings	22.00	-	-	-	22.00	m	22.00
Extra over soakaways for filter wrapping with a proprietary filter membrane							
900 mm diameter x 1.00 m deep	1.11	1.00	18.75	-	1.39	nr	20.14
900 mm diameter x 2.00 m deep	2.23	1.50	28.13	-	2.78	nr	30.91
1050 mm diameter x 1.00 m deep	1.30	1.50	28.13	-	1.62	nr	29.75
1050 mm diameter x 2.00 m deep	2.60	2.00	37.50	-	3.24	nr	40.74
1200 mm diameter x 1.00 m deep	1.48	2.00	37.50	-	1.85	nr	39.35
1200 mm diameter x 2.00 m deep	2.97	2.50	46.88	-	3.71	nr	50.58
1500 mm diameter x 1.00 m deep	1.85	2.50	46.88	-	2.32	nr	49.19
1500 mm diameter x 2.00 m deep	3.70	2.50	46.88	-	4.62	nr	51.50
1800 mm diameter x 1.00 m deep	2.22	3.00	56.25	-	2.78	nr	59.03
1800 mm diameter x 2.00 m deep	4.44	3.25	60.94	-	5.56	nr	66.49
Gravel surrounding to concrete ring soakaway							
40 mm aggregate backfilled to vertical face of soakaway wrapped with geofabric (not included)							
250 thick	28.00	0.20	3.75	8.66	28.00	m³	40.41
Backfilling to face of soakaway; carefully compacting as work proceeds							
arising from the excavations							
average thickness exceeding 0.25 m; depositing in layers 150 mm maximum thickness	-	0.03	0.62	4.33	-	m³	4.96
"Aquacell" soakaway; Wavin Plastics Ltd; preformed polypropylene soakaway infiltration crate units; to trenches; surrounded by geotextile and 40 mm aggregate laid 100 thick. in trenches (excavation, disposal and backfilling not included)							
1.00 x 500 x 400; internal volume 190 litres							
4 crates; 2.00 x 1.00 x 400; 760 litres	263.20	1.60	30.00	13.86	305.87	nr	349.73
8 crates; 2.00 x 1.00 x 800; 1520 litres	526.40	3.20	60.00	17.82	578.27	nr	656.09
12 crates; 6.00 x 500 x 800; 2280 litres	789.60	4.80	90.00	36.96	885.92	nr	1012.88
16 crates; 4.00 x 1.00 x 800; 3040 litres	1052.80	6.40	120.00	33.00	1139.92	nr	1292.92
20 crates; 5.00 x 1.00 x 800; 3800 litres	1316.00	8.00	150.00	32.67	1412.83	nr	1595.50
30 crates; 15.00 x 1.00 x 400; 5700 litres	1974.00	12.00	225.00	89.10	2201.89	nr	2515.99
60 crates; 15.00 x 1.00 x 800; 11400 litres	3948.00	20.00	375.00	116.49	4260.46	nr	4751.95
Geofabric surround to Aquacell units; Terram Ltd							
"Terram" synthetic fibre filter fabric; to face of concrete rings (not included); anchoring whilst backfilling (not included)							
"Terram 1000", 0.70 mm thick; mean water flow 50 litre/m²/s	0.39	0.05	0.94	-	0.47	m²	1.41

S PIPED SUPPLY SYSTEMS

Item Excluding site overheads and profit	PC £	Labour hours	Labour £	Plant £	Material £	Unit	Total rate £
S10 COLD WATER							
Blue MDPE polythene pipes; type 50; for **cold water services; with compression** **fittings; bedding on 100 mm DOT type 1** **granular fill material**							
Pipes							
20 mm diameter	0.29	0.08	1.50	-	3.96	m	5.46
25 mm diameter	0.36	0.08	1.50	-	4.03	m	5.53
32 mm diameter	0.89	0.08	1.50	-	4.57	m	6.07
50 mm diameter	1.54	0.10	1.88	-	5.24	m	7.12
60 mm diameter	3.33	0.10	1.88	-	7.08	m	8.95
Hose union bib taps; to BS 5412; including fixing to wall; making good surfaces							
15 mm	10.56	0.75	14.06	-	12.34	nr	26.40
22 mm	14.90	0.75	14.06	-	17.57	nr	31.63
Stopcocks; to BS 1010; including fixing to wall; making good surfaces							
15 mm	6.26	0.75	14.06	-	8.04	nr	22.10
22 mm	9.76	0.75	14.06	-	11.54	nr	25.60
Standpipes; to existing 25 mm water mains							
1.00 m high	-	-	-	-	-	nr	225.00
Hose junction bib taps; to standpipes							
19 mm	-	-	-	-	-	nr	80.00
S14 IRRIGATION							
Quality Irrigation Ltd; Main or ring main **supply**							
Excavate and lay mains supply pipe 32 mm to supply irrigated area							
32 mm MDPE	-	-	-	-	-	m	4.00
25 mm MDPE	-	-	-	-	-	m	3.00
20 mm LDPE	-	-	-	-	-	m	1.50
Install low voltage electrical cable for station or solenoid control with main supply							
12 core cable	-	-	-	-	-	m	1.50
16 core cable	-	-	-	-	-	m	2.50
20 core cable	-	-	-	-	-	m	3.10
Quality Irrigation Ltd; Supply irrigation **infrastructure for irrigation system**							
Header tank and submersible pump and pressure stat							
2270 litre (500 gallon 25 mm / 10 days to 1500 m^2)	-	-	-	-	-	nr	2250.00
4540 litre (1000 gallon 25 mm / 10 days to 3500 m^2)	-	-	-	-	-	nr	4050.00
9080 litre (2000 gallon 25 mm / 10 days to 7000 m^2)	-	-	-	-	-	nr	5250.00
Electric multistation controllers with rain cut out							
6 station controller	-	-	-	-	-	nr	560.00
12 station controller	-	-	-	-	-	nr	760.00
24 station controller	-	-	-	-	-	nr	1200.00
Solenoid valve; 25 mm with chamber; extra over for each active station	-	-	-	-	-	nr	80.00

S PIPED SUPPLY SYSTEMS

Item Excluding site overheads and profit	PC £	Labour hours	Labour £	Plant £	Material £	Unit	Total rate £
S14 IRRIGATION - cont'd							
Quality Irrigation Ltd; Station consisting of multiple sprinklers; inclusive of all trenching wiring and connections to ring main or main supply							
Gear drive sprinklers placed to provide head to head (100%) overlap; average 4 sprinklers per station; inclusive of trenching and supply pipework from ring main; price per sprinkler							
300 mm pop-up; max 10.5 m centres; 350 m^2 average cover	-	-	-	-	-	nr	155.00
100 mm pop-up; max 15 m centres; 350 m^2 maximum cover	-	-	-	-	-	nr	125.00
Sprinkler on fixed riser; max 15 m centres; 350 m^2 maximum cover	-	-	-	-	-	nr	115.00
Pop-up Sprays; installed to 80% spray overlap; (average 6 sprays per station); price per spray							
300 mm; 4.6m spray radius; 3.7 m centres; 66 m^2 maximum cover	-	-	-	-	-	nr	91.00
100 mm; 4.6m spray radius; 3.7 m centres; 66 m^2 maximum cover	-	-	-	-	-	nr	71.00
Extra over for supply to each area treated by mini sprays	-	-	-	-	-	nr	55.00
Mini Sprayers: installed to 60% spray overlap; (average 30 sprays per station) price per spray							
1.50 m radius; 7.0 m^2 maximum cover	-	-	-	-	-	nr	8.00
3.00 m radius; 28 m^2 maximum cover	-	-	-	-	-	nr	9.00
Extra over for supply to each area treated by mini sprays	-	-	-	-	-	nr	55.00
Commissioning and testing of Irrigation system							
per station	-	-	-	-	-	nr	30.00
Annual maintenance costs of Irrigation system							
Call out charge per visit	-	-	-	-	-	nr	250.00
extra over per station	-	-	-	-	-	nr	10.00
Leaky Pipe Systems Ltd; "Leaky Pipe"; moisture leaking pipe irrigation system.							
Main supply pipe inclusive of machine excavation; exclusive of connectors							
20 mm LDPE Polytubing	0.84	0.05	0.94	0.57	0.86	m	2.37
16 mm LDPE Polytubing	0.56	0.05	0.94	0.57	0.57	m	2.08
Water filters and cartridges							
No 10; 20 mm	-	-	-	-	41.90	nr	41.90
Big Blue and RR30 cartridge; 25 mm	-	-	-	-	107.20	nr	107.20
Water filters and pressure regulator sets							
Complete assemblies							
No 10; Flow rate 3.1 - 82 litres per minute	-	-	-	-	71.50	nr	71.50
Leaky pipe hose; placed 150 mm sub surface for turf irrigation							
Distance between laterals 350 mm; excavation and backfilling priced separately							
LP12L low leak	3.65	0.04	0.75	-	3.65	m^2	4.40
LP12H high leak	3.16	0.04	0.75	-	3.16	m^2	3.91
LP12UH ultra high leak	3.85	0.04	0.75	-	3.85	m^2	4.60
Leaky pipe hose; laid to surface for landscape Irrigation; distance between laterals 600 mm							
LP12L low leak	2.12	0.03	0.47	-	2.18	m^2	2.65
LP12H high leak	1.84	0.03	0.47	-	1.89	m^2	2.36
LP12UH ultra high leak	2.24	0.03	0.47	-	2.30	m^2	2.77

S PIPED SUPPLY SYSTEMS

Item Excluding site overheads and profit	PC £	Labour hours	Labour £	Plant £	Material £	Unit	Total rate £
Leaky pipe hose; laid to surface for landscape Irrigation; distance between laterals 900 mm							
LP12L low leak	1.42	0.02	0.31	-	1.46	m²	1.77
LP12H high leak	1.23	0.02	0.31	-	1.26	m²	1.58
LP12UH ultra high leak	1.50	0.02	0.31	-	1.54	m²	1.85
Leaky pipe hose; laid to surface for tree irrigation laid around circumference of tree pit							
LP12L low leak	1.96	0.13	2.34	-	2.15	nr	4.50
LP12H high leak	1.70	0.13	2.34	-	1.87	nr	4.21
LP12UH ultra high leak	2.07	0.13	2.34	-	2.27	nr	4.62
Accessories							
Automatic multi-station controller stations inclusive of connections	289.00	2.00	80.00	-	289.00	nr	369.00
Solenoid valves inclusive of wiring and connections to a multi-station controller; nominal distance from controller 25 m	48.00	0.50	9.38	-	398.00	nr	407.38

S15 FOUNTAINS/WATER FEATURES

Lakes and ponds - General
Preamble: The pressure of water against a
retaining wall or dam is considerable, and where
water retaining structures form part of the design
of water features, the landscape architect is
advised to consult a civil engineer. Artificially
contained areas of water in raised reservoirs over
25,000 m³ have to be registered with the local
authority, and their dams will have to be covered
by a civil engineer's certificate of safety.

Typical linings - General
Preamble: In addition to the traditional methods
of forming the linings of lakes and ponds in
puddled clay or concrete, there are a number of
lining materials available. They are mainly used
for reservoirs but can also help to form
comparatively economic water features
especially in soil which is not naturally water
retentive. Information on the construction of
traditional clay puddle ponds can be obtained
from the British Trust for Conservation Volunteers,
36 St. Mary's Street, Wallingford, Oxfordshire.
OX10 0EU. Tel: (01491) 39766. The cost of
puddled clay ponds depends on the availability of
suitable clay, the type of hand or machine labour
that can be used, and the use to which the pond
is to be put.

Item	PC	Labour	Labour	Plant	Material	Unit	Total
Lake liners; Fairwater Ltd; to evenly graded surface of excavations (excavating not included); all stones over 75 mm; removing debris; including all welding and jointing of liner sheets							
Geotextile underlay; inclusive of spot welding to prevent dragging							
to water features	-	-	-	-	-	m²	1.90
to lakes or large features	-	-	-	-	-	1000m²	1900.00
Butyl rubber liners; "Varnamo" inclusive of site vulcanising							
0.75 mm thick	-	-	-	-	-	m²	6.66
0.75 mm thick	-	-	-	-	-	1000m²	6600.00
1.00 mm thick	-	-	-	-	-	m²	7.59
1.00 mm thick	-	-	-	-	-	1000m²	7590.00

S PIPED SUPPLY SYSTEMS

Item Excluding site overheads and profit	PC £	Labour hours	Labour £	Plant £	Material £	Unit	Total rate £
S15 FOUNTAINS/WATER FEATURES - cont'd							
Lake liners; Landline Ltd; "Landflex" or "Alkorplan" geomembranes; to prepared surfaces (surfaces not included); all joints fully welded; installation by Landline employees							
"Landflex HC" polyethylene geomembranes							
0.50 mm thick	-	-	-	-	-	1000m^2	**2940.00**
0.75 mm thick	-	-	-	-	-	1000m^2	**3360.00**
1.00 mm thick	-	-	-	-	-	1000m^2	**3780.00**
1.50 mm thick	-	-	-	-	-	1000m^2	**4300.00**
2.00 mm thick	-	-	-	-	-	1000m^2	**4780.00**
"Alkorplan PVC" geomembranes							
0.80 mm thick	-	-	-	-	-	1000 m	**4720.00**
1.20 mm thick	-	-	-	-	-	1000 m	**7090.00**
Lake liners; Rawell Water Control Systems Ltd; to prepared graded surfaces (surfaces not included); maximum slope 2 horizontal:1 vertical; in accordance with manufacturer's instructions							
"Rawmat - type P" bentonite sheets; 200 mm lap joints; spreading 150 mm thick screened approved topsoil; "Rawmat" to be bedded in anchor trenches round perimeter (excavation and material to trenches not included)	914.80	4.00	75.00	14.44	1097.76	100 m^2	**1187.20**
Lake liners; Monarflex							
Polyethylene lake and reservoir lining system; welding on site by Monarflex technicians (surface preparation and backfilling not included)							
"Blackline"; 500 micron	-	-	-	-	-	100 m^2	**331.00**
"Blackline"; 750 micron	-	-	-	-	-	100 m^2	**385.00**
"Blackline"; 1000 micron	-	-	-	-	-	100 m^2	**445.00**
Operations over surfaces of lake liners							
Dug ballast; evenly spread over excavation already brought to grade							
150 mm thick	499.50	2.00	37.50	28.88	549.45	100 m^2	**615.83**
200 mm thick	666.00	3.00	56.25	43.31	732.60	100 m^2	**832.16**
300 mm thick	999.00	3.50	65.63	50.53	1098.90	100 m^2	**1215.06**
Imported topsoil; evenly spread over excavation							
100 mm thick	173.20	1.50	28.13	21.66	207.84	100 m^2	**257.62**
150 mm thick	259.80	2.00	37.50	28.88	311.76	100 m^2	**378.13**
200 mm thick	346.40	3.00	56.25	43.31	415.68	100 m^2	**515.24**
Blinding existing subsoil with 50 mm sand	146.05	1.00	18.75	28.88	160.66	100 m^2	**208.28**
Topsoil from excavation; evenly spread over excavation							
100 mm thick	-	-	-	21.66	-	100 m^2	**21.66**
200 mm thick	-	-	-	28.88	-	100 m^2	**28.88**
300 mm thick	-	-	-	43.31	-	100 m^2	**43.31**
Extra over for screening topsoil using a "Powergrid screener; removing debris	-	-	-	4.33	0.38	m^3	**4.71**
Waterfall construction; Fairwater Ltd							
Stone placed on top of butyl liner; securing with concrete and dressing to form natural rock pools and edgings							
Portland stone m^3 rate	-	-	-	-	-	m^3	**583.30**
Portland stone tonne rate	-	-	-	-	-	tonne	**323.00**

S PIPED SUPPLY SYSTEMS

Item Excluding site overheads and profit	PC £	Labour hours	Labour £	Plant £	Material £	Unit	Total rate £
Balancing tank; blockwork construction ; inclusive of recirculation pump and pond level control; 110 mm balancing pipe to pond; waterproofed with butyl rubber membrane; pipework mains water top-up and overflow							
450 x 600 x 1000 mm	-	-	-	-	-	nr	1213.15
extra for pump							
2000 gallons per hour; submersible		-	-	-	-	nr	185.25
Ornamental pools - General							
Preamble: Small pools may be lined with one of the materials mentioned under lakes and ponds, or may be in rendered brickwork, puddled clay or, for the smaller sizes, fibreglass Most of these tend to be cheaper than waterproof concrete. Basic prices for various sizes of concrete pools are given in the approximate estimates section (book only). Prices for excavation, grading, mass concrete, and precast concrete retaining walls are given in the relevant sections. The manufacturers should be consulted before specifying the type and thickness of pool liner, as this depends on the size, shape and proposed use of the pool. The manufacturer's recommendation on foundations and construction should be followed.							
Ornamental pools							
Pool liners; to 50 mm sand blinding to excavation (excavating not included); all stones over 50 mm; removing debris from surfaces of excavation; including all welding and jointing of liner sheets							
black polythene; 1000 gauge	-	-	-	-	-	m²	2.26
blue polythene; 1000 gauge	-	-	-	-	-	m²	2.52
coloured PVC; 1500 gauge	-	-	-	-	-	m²	3.06
black PVC; 1500 gauge	-	-	-	-	-	m²	2.61
black butyl; 0.75 mm thick	-	-	-	-	-	m²	6.33
black butyl; 1.00 mm thick	-	-	-	-	-	m²	7.21
black butyl; 1.50 mm thick	-	-	-	-	-	m²	24.69
Fine gravel; 100 mm; evenly spread over area of pool; by hand	2.81	0.13	2.34	-	2.81	m²	5.15
Selected topsoil from excavation; 100 mm; evenly spread over area of pool; by hand	-	0.13	2.34	-	-	m²	2.34
Extra over selected topsoil for spreading imported topsoil over area of pool; by hand	17.32	-	-	-	20.78	m³	20.78
Pool surrounds and ornament; **Haddonstone Ltd; Portland Bath or** **Terracotta cast stone**							
Pool surrounds; installed to pools or water feature construction priced separately; surrounds and copings to 112.5 mm internal brickwork							
C4HSKVP half small pool surround; internal diameter 1780 mm; kerb features continuous moulding enriched with ovolvo and palmette designs; inclusive of plinth and integral conch shell vases flanked by dolphins	782.98	16.00	300.00	-	795.32	nr	1095.32
C4SKVP small pool surround as above but with full circular construction; internal diameter 1780 mm	1565.96	48.00	900.00	-	1580.92	nr	2480.92
C4MKVP medium pool surround; internal diameter 2705 mm; inclusive of plinth and integral vases	2348.94	48.00	900.00	-	2380.25	nr	3280.25
C4XLKVP extra large pool surround; internal diameter 5450 mm	4697.87	140.00	2625.00	-	4734.98	nr	7359.98

S PIPED SUPPLY SYSTEMS

Item Excluding site overheads and profit	PC £	Labour hours	Labour £	Plant £	Material £	Unit	Total rate £
S15 FOUNTAINS/WATER FEATURES - cont'd							
Pool surrounds and ornament; **Haddonstone Ltd** - cont'd							
Pool centre pieces and fountains; inclusive of plumbing and pumps							
HC350 Lotus bowl; 1830 wide with C1700 triple dolphin fountain; HD2900 Doric pedestal	2421.28	8.00	150.00	-	2423.87	nr	2573.87
C251 Gothic Fountain and Gothic Upper Base A350; freestanding fountain	655.32	4.00	75.00	-	657.91	nr	732.91
HC521 Romanesque Fountain; freestanding bowl with self-circulating fountain; filled with cobbles; 815 mm diameter; 348 mm high	302.13	2.00	37.50	-	352.72	nr	390.22
C300 Lion fountain 610 mm high on fountain base C305; 280 mm high	178.72	2.00	37.50	-	181.31	nr	218.81
Wall Fountains, Watertanks and **Fountains; inclusive of installation** **drainage, automatic top up, pump, and** **balancing tank** Capital Garden Products Ltd							
Lion Wall Fountain F010; 970 x 940 x 520	-	4.00	75.00	-	1765.00	nr	1840.00
Dolphin Wall Fountain F001; 740 x 510	-	4.00	75.00	-	1626.00	nr	1701.00
Dutch Master Wall Fountain F012; 740 x 410	-	4.00	75.00	-	1625.00	nr	1700.00
James II Watertank 2801; 730 x 730 x 760 h - 405 litres	-	4.00	75.00	-	1752.00	nr	1827.00
James II Fountain 4901 bp; 710 x 1300 x 1440 h - 655 litres	-	4.00	75.00	-	1627.00	nr	1702.00
Marble Wall Fountains; Architectural **Heritage Ltd, inclusive of installation** **drainage, automatic top up, pump, and** **balancing tank** Breccia Pernice Marble Wall Fountain supported by two Dolphins, 1770 high, 1000 wide, 640 deep	16000.00	4.00	75.00	-	17545.00	nr	17620.00
'The River God', Verona Marble wall mask fountain, 890 high, 810 wide, 230 deep	8000.00	4.00	75.00	-	9545.00	nr	9620.00
Fountain kits; typical prices of **submersible units comprising fountain** **pumps, fountain nozzles, underwater** **spotlights, nozzle extension armatures,** **underwater terminal boxes and electrical** **control panels** Single aerated white foamy water columns; ascending jet 70 mm diameter; descending water up to four times larger; jet height adjustable between 1.00 m and 1.70 m	3600.00	-	-	-	4170.00	nr	4170.00
Single aerated white foamy water columns, ascending jet 110 mm diameter; descending water up to four times larger; jet height adjustable between 1.50 m and 3.00 m	6300.00	-	-	-	7180.02	nr	7180.02

V ELECRICAL SUPPLY/POWER/LIGHTING SYSTEMS

Item Excluding site overheads and profit	PC £	Labour hours	Labour £	Plant £	Material £	Unit	Total rate £
V41 STREET AREA FLOODLIGHTING							
Street area floodlighting - Generally Preamble: There are an enormous number of luminaires available which are designed for small scale urban and garden projects. The designs are continually changing and the landscape designer is advised to consult the manufacturer's latest catalogue. Most manufacturers supply light fittings suitable for column, bracket, bulkhead, wall or soffit mounting. Highway lamps and columns for trafficked roads are not included in this section as the design of highway lighting is a very specialised subject outside the scope of most landscape contracts. The IP reference number refers to the waterproof properties of the fitting; the higher the number the more waterproof the fitting. Most items can be fitted with time clocks or PIR controls.							
Market Prices of Lamps Lamps							
70 w HQIT-S	-	-	-	-	9.22	nr	**9.22**
70 w HQIT	-	-	-	-	18.87	nr	**18.87**
70 w SON	-	-	-	-	4.22	nr	**4.22**
70 w SONT	-	-	-	-	4.22	nr	**4.22**
100 w SONT	-	-	-	-	6.34	nr	**6.34**
150 w SONT	-	-	-	-	7.71	nr	**7.71**
28 w 2D	-	-	-	-	22.00	nr	**22.00**
100 w GLS\E27	-	-	-	-	2.07	nr	**2.07**
Bulkhead and canopy fittings; including **fixing to wall and light fitting (lamp, final** **painting, electric wiring, connections or** **related fixtures such as switch gear and** **time clock mechanisms not included** **unless otherwise indicated)** Bulkhead and canopy fittings; Louis Poulsen "Nyhavn Wall" small domed top conical shade with rings; copper wall lantern; finished untreated copper to achieve verdigris finish; also available in white aluminium; 310 mm diameter shade; with wall mounting arm; to IP 44	377.44	0.50	9.38	-	380.11	nr	**389.49**
Bulkhead and canopy fittings; Sugg Lighting "Princess" IP54 backlamp; 375 mm x 229 mm; copper frame; stove painted in black; with chimney and lampholder	205.05	0.50	9.38	-	207.72	nr	**217.09**
"Victoria" IP54 backlamp; 502 mm x 323 mm; copper frame; polished copper finish; with chimney, door and lampholder	274.33	0.50	9.38	-	277.00	nr	**286.38**
"Palace" IP54 backlamp; 457 mm x 321 mm; copper frame; stove painted black finish; with chimney, door and lampholder	273.74	0.50	9.38	-	276.41	nr	**285.79**
"Windsor" IP54 backlamp; 650 mm x 306 mm; copper frame; polished and lacquered finish; with door and lampholder	267.96	0.50	9.38	-	270.63	nr	**280.00**
"Windsor" IP54 gas backlamp, 650 mm x 306 mm, copper frame, polished and lacquered finish, hinged door, double inverted cluster mantle with permanent pilot and mains solenoid	480.00	0.50	9.38	-	482.67	nr	**492.05**

V ELECTRICAL SUPPLY/POWER/LIGHTING SYSTEMS

Item Excluding site overheads and profit	PC £	Labour hours	Labour £	Plant £	Material £	Unit	Total rate £
V41 STREET AREA FLOODLIGHTING - cont'd							
Floodlighting; ground, wall or pole **mounted; including fixing, (lamp, final** **painting, electric wiring, connections or** **related fixtures such as switch gear and** **time clock mechanisms not included** **unless otherwise indicated)** Floodlighting; Louis Poulsen - Landscape Division							
"Compact Flood"; ref FL-01; multi purpose ground/spike mounted wide angle floodlight; 70 w HIT or SON; to IP65	238.70	1.00	18.75	-	244.19	nr	262.94
"Morph 38"; ref SP-07; mains voltage spotlight; 60/80/120 w PAR 38; to IP54	44.81	1.00	18.75	-	50.30	nr	69.05
Floodlight Accessories; Louis Poulsen - Landscape Division							
"ES" earth spike for Compact Flood	39.25	-	-	-	39.25	nr	39.25
"L" Louvre for Compact Flood	46.68	-	-	-	46.68	nr	46.68
"L38" Louvre for Morph 38	20.21	-	-	-	20.21	nr	20.21
"C " Cowl for Compact Flood	29.71	-	-	-	29.71	nr	29.71
"BD" Barn doors for Compact Flood	64.43	-	-	-	64.43	nr	64.43
Large area/pitch floodlighting; CU Phosco ref FL444 1000 w SON-T; floodlights with lamp and loose gear; narrow asymmetric beam	350.00	1.00	40.00	-	350.00	nr	390.00
ref FL444 2.0 kw MBIOS; floodlight with lamp and loose gear; projector beam	519.70	1.00	40.00	-	519.70	nr	559.70
Large area floodlighting; CU Phosco ref FL345/G/250S; floodlight with lamp and integral gear	288.20	1.00	40.00	-	288.20	nr	328.20
ref FL345/G/400 MBI; floodlight with lamp and integral gear	308.00	1.00	40.00	-	308.00	nr	348.00
Small area floodlighting; Sugg Lighting; floodlight for feature lighting; clear or toughened glass							
"Scenario"; Lamp 150 w HPS-T	451.80	1.00	40.00	-	451.80	nr	491.80
"Scenario"; Lamp 150 w HQI-T	451.80	1.00	40.00	-	451.80	nr	491.80
Spotlights for uplighting and for **illuminating signs and notice boards,** **statuary and other features); ground,** **wall or pole mounted; including fixing,** **light fitting and priming (lamp, final** **painting, electric wiring, connections or** **related fixtures such as switch gear and** **time clock mechanisms not included); all** **mains voltage (240 v) unless otherwise** **stated** Spotlights; Louis Poulsen - Landscape Division							
"Maxispotter" miniature solid copper spotlight designed to patternate and age naturally; with mounting bracket; and internal anti glare louvre; low voltage halogen reflector; 20/35/50 w; to IP56	74.03	1.00	18.75	-	79.52	nr	98.27
"WeeBee Spot"; ref SP-05 for Low voltage halogen reflector; 20/35/50 w; c/w integral transformer and wall mounting box; to IP65	99.35	1.00	18.75	-	104.84	nr	123.59
Spotlight Accessories; Louis Poulsen - Landscape Division							
"ES"; earth spike for WeeBee Spot	15.59	-	-	-	15.59	nr	15.59
Spotlighters, uplighters and cowl lighting; Marlin Lighting; TECHNO - SHORT ARM; head adjustable 130 deg rotation 350 deg; projection 215 mm on 210 mm base plate; 355 high; integral gear; PG16 cable gland; black							
ref SIM 3518-09; 150 w CDMT/HQIT	363.96	1.00	40.00	-	363.96	nr	403.96
ref SIM 3517-09; 70 w CDMT/HQIT	327.62	1.00	40.00	-	327.62	nr	367.62

V ELECTRICAL SUPPLY/POWER/LIGHTING SYSTEMS

Item Excluding site overheads and profit	PC £	Labour hours	Labour £	Plant £	Material £	Unit	Total rate £
Recessed uplighting; including walk/drive over fully recessed uplighting; excavating, ground fixing, concreting in and making good surfaces (electric wiring, connections or related fixtures such as switch gear and time-clock mechanisms not included unless otherwise stated) (Note: transformers will power multiple lights dependent on the distance between the light units); all mains voltage (240 v) unless otherwise stated							
Recessed uplighting; Louis Poulsen - Landscape Division; "WeeBee Up"; ultra-compact recessed halogen uplight in diecast aluminium and S/Steel top plate and toughened safety glass; 20/35 w; complete with installation sleeve; low voltage requires transformer; to IP67							
ref BU-01	97.25	2.00	37.50	-	102.74	nr	**140.24**
Recessed uplighting; Louis Poulsen - Landscape Division; Pharo uplighters; 266 mm diameter; diecast aluminium with stainless steel top plate 10 mm toughened safety glass; 2000 kg drive over; integral control gear; to IP67							
BU-Pharo; HIT Metal Halide; white light; spot or flood or wall wash distribution	322.36	1.50	28.13	-	327.85	nr	**355.97**
HSE-E High pressure sodium; golden light; spot or flood or wall wash distribution	339.42	1.50	28.13	-	344.91	nr	**373.03**
HME-Mercury Vapour; cool white; flood distribution	295.79	1.50	28.13	-	301.28	nr	**329.40**
Tungsten halogen 150 w; flood distribution	284.10	1.50	28.13	-	289.59	nr	**317.71**
TC-T Compact fluorescent; white low power consumption; flood distribution	311.43	1.50	28.13	-	316.92	nr	**345.04**
Accessories for Pharo uplighters; Louis Poulsen - Landscape Division							
"RG" Rockguard	39.10	-	-	-	39.10	nr	**39.10**
"IS" Stainless steel installation sleeve	59.01	-	-	-	59.01	nr	**59.01**
"L" Anti glare louvre (internal tilt)	24.04	-	-	-	24.04	nr	**24.04**
Recessed uplighting; Louis Poulsen - Landscape Division; Nimbus 125-150 mm diameter uplighters; manufactured from diecast aluminium with stainless steel top plate 8 mm toughened safety glass; 2000 kg drive over; integral control gear; to IP67							
BU-02; 95 mm deep; 20/35/50 w low voltage halogen; requires remote transformer	103.81	2.00	37.50	-	109.30	nr	**146.80**
BU-03; 172 mm deep; 20/35/50 w low voltage halogen; with integral transformer	140.97	2.00	37.50	-	146.46	nr	**183.96**
Accessories for Nimbus uplighters; Louis Poulsen - Landscape Division							
"IS" installation sleeve	34.41	-	-	-	34.41	nr	**34.41**
Wall recessed; Marlin Lighting; "EOS" range; integral gear; asymmetric reflector; toughened reeded glass; to IP55							
ref SIM 4629-09; 10 w TCD "MINI EOS" 145 x 90 mm	88.63	1.00	40.00	-	88.63	nr	**128.63**
ref SIM 4639-09; "MEGA EOS" Wall recessing box	14.97	0.25	10.00	-	14.97	nr	**24.97**
ref SIM 4619-09; 26 w TCD "RECTANGULAR EOS" 270 x 145 mm	145.15	1.00	40.00	-	145.15	nr	**185.15**
ref SIM 4532-12; "RECTANGULAR EOS" Wall recessing box	14.97	0.25	10.00	-	14.97	nr	**24.97**
ref SIM 4628-09; 12 v 20 w QT9 "MINI EOS" 145 x 90 mm	132.85	1.00	40.00	-	132.85	nr	**172.85**
ref SIM 4623-12; "MINI EOS" Wall recessing box	9.02	0.25	10.00	-	9.02	nr	**19.02**

V ELECTRICAL SUPPLY/POWER/LIGHTING SYSTEMS

Item Excluding site overheads and profit	PC £	Labour hours	Labour £	Plant £	Material £	Unit	Total rate £
V41 STREET AREA FLOODLIGHTING - cont'd							
Underwater Lighting Underwater Lighting; Louis Poulsen - Landscape Division							
UW-05 "Minipower" cast bronze underwater floodlight c\w mounting bracket; 50 w low voltage reflector lamp; requires remote transformer	185.40	1.00	40.00	-	185.40	nr	**225.40**
Low-level lighting; positioned to provide glare free light wash to pathways steps and terraces; including forming post holes, concreting in, making good to surfaces (final painting, electric wiring, connections or related fixtures such as switch gear and time clock mechanisms not included) Low-level lighting; Louis Poulsen - Landscape Division							
"Sentry"; ref AM-08; single-sided bollard type sculptural pathway light; aluminium; 670 mm high; for 35 w HIT; to IP65	344.21	2.00	37.50	-	349.70	nr	**387.20**
"Footliter"; ref GL-05; low voltage pathlighter; 311 mm high; 6.00 m light distribution; in aluminium or solid copper; c\w earth spike; requires remote transformer; 20 w halogen; to IP44; copper finish	73.21	2.00	37.50	-	78.70	nr	**116.20**
"Bricklight"; ref ST-03; for 9 w PL lamp; diecast aluminium for recessing to walls; conforms to standard brick size; to IP65	54.64	2.00	37.50	-	60.13	nr	**97.63**
Lighted bollards; including excavating, ground fixing, concreting in and making good surfaces (lamp, final painting, electric wiring, connections or related fixtures such as switch gear and time-clock mechanisms not included) (Note: all illuminated bollards must be earthed); heights given are from ground level to top of bollards Lighted bollards; Louis Poulsen							
"Orbiter" Vandal resistant; head of cast aluminium; domed top; anti-glare rings; pole extruded aluminium; diffuser clear UV stabilised polycarbonate; powder coated; 1040 mm high; 255 mm diameter; with root or base plate; IP44	452.54	2.50	46.88	-	458.61	nr	**505.49**
"Waterfront" solidly proportioned; head of cast silumin; domed top; symmetrical distribution; pole extruded aluminium sandblasted or painted white; internal diffuser clear UV stabilised polycarbonate; 865 mm high; diameter 260 mm; IP55	500.85	2.50	46.88	-	506.92	nr	**553.80**
"Bysted" concentric louvred bollard; head cast iron; post COR-TEN steel; externally untreated to provide natural aging effect of uniform oxidised red surface finish; internal painted white; lamp diffuser rings of clear polycarbonate; 1130 mm high 280 mm diameter; to IP44	711.89	2.50	46.88	-	717.96	nr	**764.84**
"Centurion" heavy duty bollard; 2 sided light distribution from stepped reflector; cast aluminium; 900 mm high; width 240 x 160 mm; to IP44	570.76	2.50	46.88	-	576.83	nr	**623.71**

V ELECTRICAL SUPPLY/POWER/LIGHTING SYSTEMS

Item Excluding site overheads and profit	PC £	Labour hours	Labour £	Plant £	Material £	Unit	Total rate £
"Nyhavn" split level overhanging canopy type luminaire head of cast silumin; diameter 565 mm; sandblasted or powder coated; diffuser clear UV stabilized polycarbonate; post of extruded aluminium; diameter 220 mm; 1235 mm high; to IP44	593.25	2.50	46.88	-	599.32	nr	646.20
Lighted bollards; Woodscape Ltd							
Illuminated bollard; in "Very durable hardwood"; integral die cast aluminium lighting unit; 165 x 165 x 1.00 high	209.40	2.50	46.88	-	221.26	nr	268.13
Reproduction street lanterns and columns; including excavating, concreting in, backfilling and disposing of spoil, or fixing to ground or wall, making good surfaces, light fitting and priming (final painting, electric wiring, connections or related fixtures such as switch gear and time-clock mechanisms not included) (Note: lanterns up to 14 in and columns up to 7 ft are suitable for residential lighting)							
Reproduction street lanterns; Sugg Lighting; reproduction lanterns hand made to original designs; all with ES lamp holder							
"Westminster" IP54 hexagonal hinged top lantern with door, two piece folded polycarbonate glazing; Lamp 100 w HQI T; integral photo electric cell							
"Small" 1016 mm high x 356 mm wide	364.98	0.50	9.38	-	364.98	nr	374.36
"Large" 1124 mm high x 760 mm wide	669.17	0.50	9.38	-	669.17	nr	678.54
"Guildhall" IP54; handcrafted copper frame; clear polycarbonate glazing circular tapered lantern with hemispherical top; integral photo electric cell; with door							
"small" 711 mm high x 330 mm wide	358.73	0.50	9.38	-	361.40	nr	370.78
"medium" 1150 mm high x 432 mm wide	492.86	0.50	9.38	-	495.53	nr	504.91
"large" 1370 mm high x 550 mm wide	660.33	0.50	9.38	-	663.00	nr	672.38
"Grosvenor" circular lantern with door; copper frame; polished copper finish; polycarbonate glazing							
"small" 790 mm x 330 mm; IP54	445.50	0.50	9.38	-	448.17	nr	457.55
"medium" 1080 mm x 435 mm; IP65	405.08	0.50	9.38	-	407.75	nr	417.13
"Classic Globe" IP54 lantern with hinged outer frame; cast aluminium frame; black polyester powder coating							
"medium" 965 mm x 483 mm	404.25	0.50	9.38	-	406.92	nr	416.30
"Windsor" lantern; copper frame; polished copper finish							
"small" 905 x 356 mm; IP54; with door	281.82	0.50	9.38	-	284.49	nr	293.87
"small" 905 x 356 mm; IP65; without door	280.67	0.50	9.38	-	283.34	nr	292.72
"medium" 1124 mm x 420 mm; IP54; with door	306.08	0.50	9.38	-	308.75	nr	318.13
"large" 1124 mm x 470 mm; IP54; with door	347.66	0.50	9.38	-	350.33	nr	359.71
"medium" gas lantern, 1124 mm x 420 mm; IP54; with door; integral solenoid and pilot	589.00	0.50	9.38	-	591.67	nr	601.04
Reproduction brackets and suspensions; Sugg lighting							
Iron Brackets							
"Bow" bracket - 6'0"	525.73	2.00	37.50	-	528.40	nr	565.90
"Ornate" iron bracket - large	235.88	2.00	37.50	-	238.55	nr	276.05
"Ornate" iron bracket - medium	226.79	2.00	37.50	-	229.46	nr	266.96
"Swan neck" iron bracket - large	204.36	2.00	37.50	-	207.03	nr	244.53
"Swan neck" iron bracket - medium	115.82	2.00	37.50	-	118.49	nr	155.99

V ELECTRICAL SUPPLY/POWER/LIGHTING SYSTEMS

Item Excluding site overheads and profit	PC £	Labour hours	Labour £	Plant £	Material £	Unit	Total rate £
V41 STREET AREA FLOODLIGHTING - cont'd							
Reproduction brackets and suspensions; **Sugg lighting** - cont'd							
Cast Brackets							
"Universal" cast bracket	157.06	2.00	37.50	-	159.73	nr	**197.23**
"Abbey" bracket	87.32	2.00	37.50	-	89.99	nr	**127.49**
"Short Abbey" bracket	73.38	2.00	37.50	-	76.05	nr	**113.55**
"Plaza" cast bracket	77.62	2.00	37.50	-	80.29	nr	**117.79**
Base mountings							
"Universal" pedestal	143.80	2.00	37.50	-	146.47	nr	**183.97**
"Universal" plinth	78.23	2.00	37.50	-	80.90	nr	**118.40**
Reproduction lighting columns							
Reproduction lighting columns; Sugg Lighting							
"Harborne" C11; fabricated iron/steel heavy duty post with integral cast root; 3 - 5 m	637.91	8.00	150.00	-	645.50	nr	**795.50**
"Aylesbury" C12; base fabricated heavy gauge 89 mm aluminium post; 3 - 5 m	1029.03	8.00	150.00	-	1029.03	nr	**1179.03**
"Cannonbury" C13; fabricated heavy gauge 89 mm aluminium post; 3 - 5 m	591.82	8.00	150.00	-	591.82	nr	**741.82**
Standard column C14; rooted British Steel 168/89 mm embellished post; 5 - 8 m	318.36	8.00	150.00	-	318.36	nr	**468.36**
"Large Constitution Hill" C22X; cast aluminium, steel cored post with extended spigot; 4.3 m	2595.90	6.00	112.50	-	2595.90	nr	**2708.40**
"Cardiff" C29; cast aluminium post; 3.5 m	1334.64	6.00	112.50	-	1334.64	nr	**1447.14**
"Seven Dials" C36; cast aluminium rooted post; 3.8 m	898.64	8.00	150.00	-	898.64	nr	**1048.64**
"Royal Exchange" C42; traditional cast aluminium post; welded multi arm construction; 2.1 m	1579.01	6.00	112.50	-	1579.01	nr	**1691.51**
Precinct lighting lanterns; ground, wall **or pole mounted; including fixing and** **light fitting (lamps, poles, brackets, final** **painting, electric wiring, connections or** **related fixtures such as switch gear and** **time clock mechanisms not included)**							
Precinct lighting lanterns; Louis Poulsen							
"Nyhavn Park" side entry 90 degree curved arm mounted; steel rings over domed top; housing cast silumin sandblasted with integral gear; protected by UV stabilized clear polycarbonate diffuser; IP55	519.84	1.00	18.75	-	519.84	nr	**538.59**
"Kipp"; bottom entry pole-top; shot blasted or lacquered Hanover design award; hinged diffuser for simple maintenance; indirect lighting technique ensures low glare; IP55	281.14	1.00	18.75	-	281.14	nr	**299.89**
"Bjarne Bech"; luminaire head of galvanized sheet steel; domed head over cylindrical light shield; internal reflector surfaces painted white; IP23	287.50	1.00	18.75	-	287.50	nr	**306.25**
Precinct lighting lanterns; Marlin Lighting; "Sphereline"; clear or opal; vandal resistant; integral HPF gear; black finish							
ref 8031 XX; 300 mm diameter opal polycarbonate globe; 10/13w TCD; to IP44	62.49	1.00	18.75	-	62.49	nr	**81.24**
Precinct lighting lanterns; Marlin Lighting; "Sphereline"; accessories and fittings							
ref 8030 PB26 "Boulevard" pole top 2 globe bracket	340.13	0.50	20.00	-	340.13	nr	**360.13**

V ELECTRICAL SUPPLY/POWER/LIGHTING SYSTEMS

Item Excluding site overheads and profit	PC £	Labour hours	Labour £	Plant £	Material £	Unit	Total rate £
Precinct lighting lanterns; Sugg Lighting							
Juno Dome; IP66; molded GRP body; graphite finish	305.00	1.00	18.75	-	305.00	nr	323.75
Juno Cone; IP66; molded GRP body; graphite finish	305.00	1.00	18.75	-	305.00	nr	323.75
Sharkon; IP65; cast aluminium body; silver and blue finish	282.59	1.00	18.75	-	282.59	nr	301.34
Precinct lighting columns; including excavating, concreting, backfilling and disposing of spoil (final painting, electric wiring, connections or related fixtures such as switch gear and time-clock mechanisms not included)							
Precinct lighting columns; Marlin Lighting; "Skyline" tubular steel poles							
ref SPR 750 A6; waisted; 5.0 m high	209.61	6.00	112.50	-	239.11	nr	351.61
ref SPR 730 A6; waisted; 3.0 m high	175.62	6.00	112.50	-	205.12	nr	317.62
Transformers for low voltage lighting; Louis Poulsen - Landscape Division; distance from electrical supply 25 m; trenching and backfilling measured separately							
Boxed Transformer							
50 Va for maximum 50 w of lamp	52.32	2.00	80.00	-	73.43	nr	153.43
100 Va for maximum 100 w of lamp	60.77	2.00	80.00	-	81.89	nr	161.89
150 Va for maximum 150 w of lamp	64.30	2.00	80.00	-	85.41	nr	165.41
200 Va for maximum 200 w of lamp	75.78	2.00	80.00	-	96.90	nr	176.90
250 Va for maximum 50 w of lamp	80.14	2.00	80.00	-	101.26	nr	181.26
Buried transformers							
50 Va for maximum 50 w of lamp	60.72	2.00	80.00	-	81.84	nr	161.84
Installation; electric cable in trench 500 mm deep (trench not included); all in accordance with IEE regulations							
Light duty 600 volt grade armoured							
2 core 2.5 mm	1.20	0.01	0.09	-	1.20	m	1.29
3 core 2.5 mm	1.44	0.01	0.09	-	1.44	m	1.53
4 core 2.5 mm	1.67	0.01	0.09	-	1.67	m	1.76
Twin and earth PVC cable in plastic conduit (conduit not included)							
2.50 mm2	0.14	0.01	0.19	-	0.14	m	0.33
4.00 mm2	0.43	0.01	0.19	-	0.43	m	0.62
16 mm heavy gauge high impact PVC conduit	3.08	0.03	0.62	-	3.08	m	3.70
Main switch and fuse unit; 30 A	-	-	-	-	37.92	nr	37.92
Weatherproof switched socket; 13 A	-	2.00	80.00	-	15.38	nr	95.38

DAVIS LANGDON

EUROPE & MIDDLE EAST
office locations

ENGLAND

DAVIS LANGDON

LONDON
Mid City Place
71 High Holborn
London WC1V 6QS
Tel: (020) 7061 7000
Fax: (020) 7061 7061
Email: neill.morrison@davislangdon.com

BIRMINGHAM
75-77 Colmore Row
Birmingham
B3 2HD
Tel: (0121) 710 1100
Fax: (0121) 710 1399
Email: david.daly@davislangdon.com

BRISTOL
St Lawrence House
29/31 Broad Street
Bristol BS1 2HF
Tel: (0117) 927 7832
Fax: (0117) 925 1350
Email: alan.francis@davislangdon.com

CAMBRIDGE
36 Storey's Way
Cambridge
CB3 ODT
Tel: (01223) 351 258
Fax: (01223) 321 002
Email: laurence.brett@davislangdon.com

LEEDS
No 4 The Embankment
Victoria Wharf
Sovereign Street
Leeds LS1 4BA
Tel: (0113) 243 2481
Fax: (0113) 242 4601
Email: duncan.sissons@davislangdon.com

LIVERPOOL
Cunard Building
Water Street
Liverpool L3 1JR
Tel: (0151) 236 1992
Fax: (0151) 227 5401
Email: andrew.stevenson@davislangdon.com

MAIDSTONE
11 Tower View
Kings Hill
West Malling
Kent ME19 4UY
Tel: (01732) 840 429
Fax: (01732) 842 305
Email: nick.leggett@davislangdon.com

MANCHESTER
Cloister House
Riverside
New Bailey Street
Manchester M3 5AG
Tel: (0161) 819 7600
Fax: (0161) 819 1818
Email: paul.stanion@davislangdon.com

MILTON KEYNES
Everest House
Rockingham Drive
Linford Wood
Milton Keynes
MK14 6LY
Tel: (01908) 304 700
Fax: (01908) 660 059
Email: kevin.sims@davislangdon.com

NORWICH
63 Thorpe Road
Norwich NR1 1UD
Tel: (01603) 628 194
Fax: (01603) 615 928
Email: michael.ladbrook@davislangdon.com

OXFORD
Avalon House
Marcham Road
Abingdon
Oxford OX14 1TZ
Tel: (01235) 555 025
Fax: (01235) 554 909
Email: paul.coomber@davislangdon.com

PETERBOROUGH
Clarence House
Minerva Business Park
Lynchwood
Peterborough PE2 6FT
Tel: (01733) 362 000
Fax: (01733) 230 875
Email: stuart.bremner@davislangdon.com

PLYMOUTH
1 Ensign House
Parkway Court
Longbridge Road
Plymouth PL6 8LR
Tel: (01752) 827 444
Fax: (01752) 221 219
Email: gareth.steventon@davislangdon.com

SOUTHAMPTON
Brunswick House
Brunswick Place
Southampton SO15 2AP
Tel: (023) 8033 3438
Fax: (023) 8022 6099
Email: chris.tremellen@davislangdon.com/
peter.boote@davislangdon.com

**DAVIS LANGDON
LEGAL SUPPORT**
Mid City Place
71 High Holborn
London WC1V 6QS
Tel: (020) 7061 7000
Fax: (020) 7061 7061
Email: mark.hackett@davislangdon.com

**DAVIS LANGDON
CONSULTANCY**
Mid City Place
71 High Holborn
London WC1V 6QS
Tel: (020) 7061 7007
Fax: (020) 7061 7005
Email: john.connaughton@davislangdon.com

**DAVIS LANGDON
SCHUMANN SMITH**
Southgate House
St Georges Way
Stevenage
Hertfordshire SG1 1HG
Tel: (01438) 742 642
Fax: (01438) 742 632
Email: nick.schumann@schumannsmith.com

**DAVIS LANGDON
MOTT GREEN & WALL**
Mid City Place
71 High Holborn
London WC1V 6QS
Tel: (020) 7061 7777
Fax: (020) 7061 7009
Email: general@mottgreenwall.co.uk

**DAVIS LANGDON
CROSHER & JAMES**
Mid City Place
71 High Holborn
London WC1V 6QS
Tel: (020) 7061 7077
Fax: (020) 7061 7078
Email: tony.llewellyn@crosherjames.com

BIRMINGHAM
102 New Street
Birmingham B2 4HQ
Tel: (0121) 632 3600
Fax: (0121) 632 3601
Email: clive.searle@crosherjames.com

CARDIFF
4 Piershead Street
Capital Waterside
Cardiff
CF10 4QP
Tel: (029) 2049 7497
Fax: (029) 2049 7111
Email: michael.murraym@crosherjames.com

EDINBURGH
39 Melville Street
Edinburgh
EH3 7JF
Tel: (0131) 220 4225
Fax: (0131) 220 4226
Email: ian.mcfarlane@crosherjames.com

GLASGOW
Monteith House
11 George Square
Glasgow
G2 1DY
Tel: (0141) 248 0333
Fax: (0141) 248 0313
Email: fraserk@nbwcrosherjames.com

MANCHESTER
Cloister House
Riverside
New Bailey Street
Manchester M3 5AG
Tel: (0161) 819 7600
Fax: (0161) 819 1818
Email: sharmas@nbwcrosherjames.com

SOUTHAMPTON
Brunswick House
Brunswick Place
Southampton SO15 2AP
Tel: (023) 8068 2800
Fax: (023) 8033 6360
Email: reesd@nbwcrosherjames.com

SCOTLAND

DAVIS LANGDON

GLASGOW
Monteith House
11 George Square
Glasgow G2 1DY
Tel: (0141) 248 0300
Fax: (0141) 248 0303
Email:
sam.mackenzie@davislangdon.com

EDINBURGH
39 Melville Street
Edinburgh
EH3 7JF
Tel: (0131) 240 1350
Fax: (0131) 240 1399
Email: erland.rendall@davislangdon.com

WALES

CARDIFF
4 Pierhead Street
Capital Waterside
Cardiff CF10 4QP
Tel: (029) 2049 7497
Fax: (029) 2049 7111
Email: paul.edwards@davislangdon.com

IRELAND

DAVIS LANGDON PKS

DUBLIN
24 Lower Hatch Street
Dublin 2
Ireland
Tel: (00 353 1) 676 3671
Fax: (00 353 1) 676 3672
Email: mwebb@dlpks.ie

GALWAY
Heritage Hall
Kirwan's Lane
Galway, Ireland
Tel: (00 353 91) 530 199
Fax: (00 353 91) 530 198
Email: joregan@dlpks.ie

LIMERICK
8 The Crescent
Limerick
Ireland
Tel: (00 353 61) 318 870
Fax: (00 353 61) 318 871
Email: cbarry@dlpks.ie

SPAIN

DAVIS LANGDON EDETCO

BARCELONA
C/Muntaner, 479, 12"
Barcelona 08021
Spain
Tel: (00 34 93) 418 6899
Fax: (00 34 93) 211 0003
Email: fmonells@barcelona.edetco.com

GIRONA
C/Salt 10
Girona 17005
Spain
Tel: (00 34 97) 223 8000
Fax: (00 34 97) 224 2661
Email: girona@girona.edetco.com

FRANCE

DAVIS LANGDON
5 Rue St Germain l'Auxerrois
75001 Paris
France
Tel: (00 33 1) 5340 9480
Fax: (00 33 1) 5340 9481
Email: andrew.richardson@dleparis.com

POLAND

DAVIS LANGDON
Warsaw Trade Tower
ul. Chlodna 51, 26th Floor
00-867 Warsaw, Poland
Tel: (00 48 22) 455 39 00
Fax: (00 48 22) 455 39 01
Email: warsaw@davislangdon-polska.pl

RUSSIA

DAVIS LANGDON
Office 5
Myasnitskaya
Moscow, 101000
Russia
Tel: (00 7 095) 933 7810
Fax: (00 7 095) 933 7811
Email: stephen.thomas@davislangdon.com

MIDDLE EAST

DAVIS LANGDON
PO Box 13-5422-Shouran
Beirut
Lebanon
Tel: (00 9611) 780 1t1
Fax: (00 9611) 809 045
Email: DLL.MI@cyberia.net.lb

ARABIAN GULF

DAVIS LANGDON

BAHRAIN
3rd Floor Building 256
Road No 3605
Area No 336
PO Box 640, Manama
State of Bahrain
Arabian Gulf
Tel: (00 973) 1782 7567
Fax: (00 973) 1772 8257
Email: david.galbraith@davislangdon-
bahrain.com

UNITED ARAB EMIRATES
PO Box 7856
Office 410
Oud Metha Office Building
Dubai, UAE
Tel: (00 9714) 32 42 919
Fax: (00 9714) 32 42 838
Email: neil.taylor@davislangdon-dubai.com

QATAR
PO Box 3206, Doha
State of Qatar
Tel: (00 974) 4580 150
Fax: (00 974) 4697 905
Email: david.craig@davislangdon-qatar.com

EGYPT
35 Misr Helwan Road
Maadi 11431
Cairo
Egypt
Tel: (00 20 2) 526 2319
Fax: (00 20 2) 527 1338
Email: dlegypt@link.net

Specialist Service Lines
Project Management | Cost Management | Management Consulting | Legal Support | Specification Consulting | Engineering Services | Property Tax & Finance

Specialist Sectors
Arts | Commercial Offices | Distribution | Education | Food Processing | Health | Heritage | Hotels & Leisure | Industrial | Infrastructure | Public Buildings | Regeneration | Residential | Retail | Sports | Transportation

Davis Langdon LLP is a member firm of Davis Langdon & Seah International, with offices throughout Europe and the Middle East, Asia, Australasia, Africa and the USA

Spon's Estimating Costs Guide to Minor Landscaping, Gardening and External Works

Bryan Spain

SPON'S
ESTIMATING COST GUIDE TO
MINOR LANDSCAPING,
GARDENING AND
EXTERNAL WORKS

BRYAN SPAIN

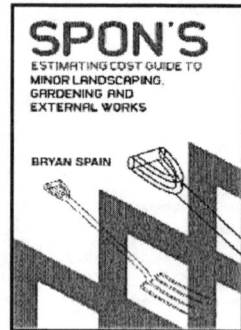

Especially written for contractors and small businesses carrying out small works, Spon's Estimating Costs Guide to Small Landscaping Work, Gardening and External Works contains accurate information on thousands of rates, each broken down to labour, material overheads and profit.

This is the first book to focus on garden maintenance work on blocks of flats and individual houses, schools and sports fields, garden makeovers, laying patios and paths, small land drainage schemes on farmland, and small-scale local authority maintenance work.

More than just a price book, it gives easy-to-read, professional advice on setting up and running a business including help on producing estimates faster, and keeping estimates accurate and competitive.

Suitable for any size firm from one-man-band to established business, this book contains valuable commercial and cost information that contractors can't afford to be without.

March 2005: 216x138 mm: 304 pages
PB: 0-415-34410-7: £24.99

To Order: Tel: +44 (0) 1264 343071 Fax: +44 (0) 1264 343005, or
Post: Taylor and Francis Customer Services, Thomson Publishing Services, Cheriton House, Andover, Hants, SP10 5BE, UK Email: book.orders@tandf.co.uk

For a complete listing of all our titles visit:
www.sponpress.com

Taylor & Francis
Taylor & Francis Group plc

Part 6

Approximate Estimates

Prices in this section are based upon the Prices for Measured Work, but allow for incidentals which would normally be measured separately in a Bill of Quantities. They do not include for Preliminaries which are priced elsewhere in this book.

Items shown as sub-contract or specialist rates would normally include the specialists overhead and profit. All other items which could fall within the scope of works of general landscape and external works contractors would not include profit

Based on current commercial rates, profits of 15% to 35% may be added to these rates to indicate the likely "with profit" values of the tasks below. The variation quoted above is dependent on the sector in which the works are taking place – domestic, public or commercial.

ESSENTIAL READING FROM TAYLOR AND FRANCIS

The Dynamic Landscape
Naturalistic Planting in an Urban Context

Edited by
Nigel Dunnett and James Hitchmough

The last quarter of the twentieth-century witnessed a burgeoning of interest in ecological or naturally inspired use of vegetation in the designed landscape. More recently a strong aesthetic element has been added to what was formerly a movement aimed at creating nature-like landscapes.

This book advances a fusion of scientific and ecological planting design philosophy that can address the need for more sustainable designed landscapes. It is a major statement on the design, implementation and management of ecologically-inspired landscape vegetation. With contributions from people at the forefront of developments in this field, in both Europe and North America, it provides a valuable synthesis of current thinking.

Contents: 1. Introduction 2. The Historical Development of Naturalistic Planting 3. A Contemporary Overview of Naturalistic Planting 4. The Dynamic Nature of Plant Communities 5. A Naturalistic Design Process 6. Herbaceous Plantings 7. Exploring Woodland Design 8. Wetlands and Water Bodies 9. Communicating Naturalistic Plantings: Plans and Specifications 10. The Creative Management of Naturalistic Plantings 11. The Social and Cultural Context of Naturalistic Plantings. Index.

June 2004: 238x225mm: 336 pages
20 tables, 40 line drawings, 67 b+w photos and 85 colour photos
HB: 0-415-25620-8: £60.00

To Order: Tel: +44 (0) 1264 343071 Fax: +44 (0) 1264 343005, or
Post: Taylor and Francis Customer Services, Thomson Publishing Services, Cheriton House, Andover, Hants, SP10 5BE, UK Email: book.orders@tandf.co.uk

For a complete listing of all our titles visit:
www.sponpress.com

Taylor & Francis
Taylor & Francis Group plc

PRELIMINARIES

Item Excluding site overheads and profit	Unit	Total rate £
CONTRACT ADMINISTRATION AND MANAGEMENT		
Prepare and maintain Health and Safety file, prepare risk and Coshh assessments, method statements, works programmes before the start of works on site; Project value:		
£35000.00	nr	212.00
£75000.00	nr	318.00
£100000.00	nr	424.00
£200000.00 - £500000.00	nr	1060.00
Contract management		
Site management by site surveyor carrying out supervision and administration duties only; inclusive of on costs		
full time management	week	1200.00
managing one other contract	week	600.00
Parking		
Parking expenses where vehicles do not park on the site area; per vehicle		
Metropolitan area city centre	week	200.00
Metropolitan area; outer areas	week	160.00
Suburban restricted parking areas	week	40.00
Congestion charging		
London only	week	40.00
SITE SETUP AND SITE ESTABLISHMENT		
Site Compounds		
Establish site administration area on reduced compacted Type 1 base; 150 mm thick; allow for initial setup of site office and welfare facilities for site manager and maximum 8 site operatives		
Temporary base or hard standing; 20 m x 10 m	200 m²	2006.80
To hardstanding area above; erect office and welfare accommodation; allow for furniture, telephone connection, and power; restrict access using temporary safety fencing to the perimeter of the site administration area		
Establishment; delivery and collection costs only		
Site office and chemical toilet; 20 m of site fencing	nr	761.60
Site office and toilet; 40 m of site fencing	nr	827.20
Site office and toilet; 60 m of site fencing	nr	892.80
Weekly hire rates of site compound equipment		
Site office 2.4 x 3.6 m; Chemical toilet; 20 m of site fencing	week	82.76
Site office 4.8 x 2.4 m; Chemical toilet; 20 m of site fencing; Armoured store 2.4 x 3.6 m	week	95.80
Site office 2.4 x 3.6 m; Chemical toilet; 40 m of site fencing	week	110.76
Site office 2.4 x 3.6 m; Chemical toilet; 60 m of site fencing	week	138.76
Removal of site compound		
Remove base; fill with topsoil and reinstate to turf on completion of site works	200 m²	1205.62

PRELIMINARIES

Item Excluding site overheads and profit	Unit	Total rate £
SITE SETUP AND SITE ESTABLISHMENT - cont'd		
Surveying and Setting out		
Note in all instances a quotation should be obtained from a surveyor. The following cost projections are based on the costs of a surveyor using EDM (Electronic distance measuring equipment where an average of 400 different points on site are recorded in a day. The surveyed points are drawn in a CAD application and the assumption used that there is one day of drawing up for each day of on-site surveying.		
Survey existing site using laser survey equipment and produce plans of existing site conditions		
Site up to 1000 to 10000 m² with sparse detail and features; boundary and level information only; average 5 surveying stations	nr	750.00
Site up to 1000 m² with dense detail such as buildings paths walls and existing vegetation; average 5 survey stations	nr	1500.00
Site up to 4000 m² with dense detail such as buildings paths walls and existing vegetation; average 10 survey stations	nr	2250.00

DEMOLITION AND SITE CLEARANCE

Item Excluding site overheads and profit	Unit	Total rate £
DEMOLITION AND SITE CLEARANCE		
Site fencing		
Supply and erect temporary protective fencing		
and remove after planting		
Cleft chestnut paling 1.20 m high 75 larch posts		
at 3 m centres	100 m	960.00
Cleft chestnut paling 1.50 m high 75 larch posts		
at 3 m centres	100 m	1174.00
"Heras" fencing		
Erection of fencing		
Collection and delivery charges		
Clear existing scrub vegetation,		
including shrubs and hedges and burn		
on site; spread ash on specified area		
by machine		
light scrub and grasses	100 m²	17.55
undergrowth brambles and heavy weed growth	100 m²	39.00
small shrubs	100 m²	39.00
As above but remove vegetation to licensed tip	100 m²	33.13
by hand	100 m²	82.50
As above but remove vegetation to licensed tip	100 m²	54.38
Fell trees on site, grub up roots, all by		
machine and remove debris to licensed		
tip		
trees 600 girth	each	215.72
trees 1.5 - 3.00 m girth	each	624.96
trees over 3.00 m girth	each	1426.39
Break up plain concrete slab and		
remove to licensed tip		
150 thick	m²	7.45
200 thick	m²	9.94
300 thick	m²	14.90
Clear away light fencing and gates (chain link,		
chestnut paling, light boarded fence or similar)		
and remove to licensed tip	100 m	111.79
Break up plain concrete slab and		
preserve arisings for hardcore; break		
down to maximum size of 200 mm x		
200 mm and transport to location		
maximum distance 50 m		
150 thick	m²	5.11
200 thick	m²	6.82
300 thick	m²	10.22
Break up reinforced concrete slab and		
remove to licensed tip		
150 thick	m²	11.81
200 thick	m²	15.76
300 thick	m²	23.62
Site Clearance - generally		
Clear away light fencing and gates (chain link,		
chestnut paling, light boarded fence or similar)		
and remove to licensed tip	100 m	111.79

GROUNDWORK

Item Excluding site overheads and profit	Unit	Total rate £
GROUNDWORK		
Cut and strip by machine turves 50 thick		
load to dumper or palette by hand, stack on site		
not exceeding 100 m travel to stack	100 m²	41.21
As above but all by hand, including barrowing	100 m²	437.49
Excavate trenches for foundations;		
trenches 225 mm deeper than specified		
foundation thickness; pour plain		
concrete foundations GEN 1 10 N/mm		
and to thickness as described; disposal		
of excavated material off site;		
dimensions of concrete foundations		
250 mm wide x 250 mm deep	m³	142.10
300 mm wide x 300 mm deep	m³	122.73
400 mm wide x 600 mm deep	m³	122.73
Prices for excavating and reducing to levels are		
for work in light or medium soils; multiplying		
factors for other soils are as follows		
clay	1.5	
compact Gravel	1.2	
soft chalk	2.0	
hard rock	3.0	
Excavating topsoil for preservation		
Remove topsoil average depth 300 mm, deposit		
in spoil heaps not exceeding 100 m travel to		
heap, treat once with paraquat-diquat weedkiller,		
turn heap once during storage		
by machine	m³	10.39
by hand	m³	110.24
Excavate to reduce levels and remove		
spoil to dump not exceeding 100 m		
travel, all by machine		
0.25 m deep average	m²	1.81
0.30 m deep	m²	2.20
1.0 m deep using 21 tonne 360 tracked		
excavator	m²	4.24
1.0 m deep using JCB	m²	6.66
As above but all by hand		
0.10 m deep average	m²	11.88
0.20 m deep average	m²	25.00
0.30 m deep average	m²	141.87
1.0 m deep average	m²	181.24
Extra for carting spoil to licensed tip offsite (20		
tonnes)	m³	16.88
Extra for carting spoil to licensed tip offsite (by		
skip - machine loaded)	m³	27.63
Extra for carting spoil to licensed tip offsite (by		
skip - hand loaded)	m³	92.50
Spread excavated material to levels in		
layers not exceeding 150 mm, using		
scraper blade or bucket		
average thickness 100 mm	m²	0.21
As above but with imported topsoil	m²	1.64
average thickness 200 mm	m²	0.41
As above but with imported topsoil	m²	3.28
average thickness 250 mm	m²	0.52
As above but with imported topsoil	m²	4.09
Extra for work to banks exceeding 30 slope	-	30%

GROUNDWORK

Item Excluding site overheads and profit	Unit	Total rate £
Rip subsoil using approved sub-soiling **machine to a depth of 600 mm below** **topsoil at 600 mm centres, all tree roots** **and debris over 150 x 150 mm to be** **removed; cultivate ground where shown** **on drawings to a depths as shown**		
100 mm	100 m²	3.49
200 mm	100 m²	3.63
300 mm	100 m²	3.87
400 mm	100 m²	5.78
TRENCHES		
Excavate trenches for foundations; **trenches 225 mm deeper than specified** **foundation thickness; pour plain** **concrete foundations GEN 1 10 N/mm** **and to thickness as described; disposal** **of excavated material off site;** **dimensions of concrete foundations**		
250 mm wide x 250 mm deep	m	10.89
300 mm wide x 300 mm deep	m	13.04
400 mm wide x 600 mm deep	m	32.02
600 mm wide x 400 mm deep	m	33.44
Ha-Ha Excavate ditch 1200 mm deep x 900 mm wide at bottom battered one side 45° slope, excavate for foundation to wall and place GEN 1 concrete foundation 150 x 500 mm, construct one and a half brick wall (Brick PC £300.00/1000) battered 10° from vertical, laid in cement:lime:sand (1:1:6) mortar in English garden wall bond, precast concrete coping weathered and throated set 150 mm above ground on high side, rake bottom and sides of ditch and seed with low maintenance grass at 35 g/m²:		
wall 600 high	m	225.70
wall 900 high	m	279.97
wall 1200 high	m	347.23
Excavate ditch 1200 mm deep x 900 mm wide at bottom battered one side 45° slope; place deer or cattle fence 1.20 m high at centre of excavated trench; rake bottom and sides of ditch and seed with low maintenance grass at 35 g/m²		
fence 1200 high	m	20.64

Approximate Estimates

GROUND STABILIZATION

Item Excluding site overheads and profit	Unit	Total rate £
GROUND STABILIZATION		
Excavate and grade banks to grade to receive stabilization treatments below; removal of excavated material not included		
By 21 tonne excavator	m³	0.28
By 7 tonne excavator	m³	9.24
By 5 tonne excavator	m³	1.71
By 3 tonne excavator	m³	2.83
Excavate sloped bank to vertical to receive retaining or stabilization treatment priced below; allow 1.00 m working space at top of bank; backfill working space and remove balance of arisings off site		
Bank height 1.00; Excavation by 5 tonne excavator		
10 degree	m	75.42
30 degree	m	25.87
45 degree	m	15.91
60 degree	m	9.71
As above but arisings moved to stockpile on site		
10 degree	m	15.83
30 degree	m	7.81
45 degree	m	5.28
60 degree	m	4.81
Bank height 2.00; Excavation by 5 tonne excavator		
10 degree	m	294.51
30 degree	m	95.53
45 degree	m	59.01
60 degree	m	36.92
As above but arisings moved to stockpile on site		
10 degree	m	55.32
30 degree	m	22.78
45 degree	m	16.81
60 degree	m	13.29
Bank height 3.00; Excavation by 21 tonne excavator		
20 degree	m	300.98
30 degree	m	191.43
45 degree	m	107.48
60 degree	m	63.04
As above but arisings moved to stockpile on site		
20 degree	m	40.18
30 degree	m	27.19
45 degree	m	12.61
60 degree	m	8.18

GROUND STABILIZATION

Item Excluding site overheads and profit	Unit	Total rate £
Bank stabilization; Erosion control mats Excavate fixing trench for erosion control mat 300 x 300 mm; backfill after placing mat selected below	m	5.30
To anchor trench above, place mat to **slope to be retained; allow extra over to** **the slope for the area of matting required** **to the anchor trench** Rake only bank to final grade; lay erosion control solution; sow with low maintenance grass at 35 g/m², spread imported topsoil to BS3882 25 mm thick incorporating medium grade sedge peat at 3 kg/m² and fertilizer at 30 g/m², water lightly		
unseeded "Eromat Light"	m²	2.76
seeded "Covamat Standard"	m²	3.36

Lay Greenfix biodegradable pre-seeded erosion
  control mat, fixed with 6 x 300 mm steel pegs at
  1.0 m centres
	m²	3.59

"Tensar" open textured erosion mat	m²	4.01
Excavate trench and lay foundation **concrete 1:3:6 600 x 300 mm deep**		
By machine	m	28.13
By hand	m	38.76
Concrete Block retaining walls; Stepoc Excavate trench 750 mm deep and lay concrete foundation 600 wide x 600 deep; construct Forticrete precast hollow concrete block wall with 450 mm below ground laid all in accordance with manufacturer's instructions; fix reinforcing bar 12 mm as work proceeds; fill blocks with concrete 1:3:6 as work proceeds		
Walls 1.00 m high		
Type 256; 400 x 225 x 256 mm	m	196.42
Type 190; 400 x 225 x 190 mm	m	186.58
As above but 1.50 m high		
Type 256; 400 x 225 x 256 mm	m	258.29
Type 190; 400 x 225 x 190 mm	m	245.05
Walls 1.50 m high as above but with foundation 1.80 m wide x 600 mm deep		
Type 256; 400 x 225 x 256 mm	m	304.49
Type 190; 400 x 225 x 190 mm	m	289.01
Walls 1.80 m high		
Type 256; 400 x 225 x 256 mm	m	329.14
Type 190; 400 x 225 x 190 mm	m	313.86
On foundation measured above, supply **and RCC precast concrete "L" shaped** **units, constructed all in accordance with** **manufacturer's instructions; backfill with** **approved excavated material compacted** **as the work proceeds**		
1250 mm high x 1000 mm wide	m	129.72
2400 mm high x 1000 mm wide	m	222.85
3750 mm high x 1000 mm wide	m	410.45

GROUND STABILIZATION

Item Excluding site overheads and profit	Unit	Total rate £
GROUND STABILIZATION - cont'd		
Timber Crib Wall		
Excavate trench to receive foundation 300 mm Deep; place plain concrete foundation 150 thick in 11.50 N/mm² concrete (sulphate-resisting cement); construct timber crib retaining wall and backfill with excavated spoil behind units		
Keller-Comtec timber crib wall system, average 4.2 m high	m	194.80
Keller-Comtec timber crib wall system, average 1.5 m high	m	140.50
Grass Concrete; bank revetments; excluding bulk earthworks		
Bring bank to final grade by machine; lay regulating layer of Type 1; lay filter membrane; lay 100 mm drainage layer of broken stone or approved hardcore 28 - 10 mm size; blind with 25 mm sharp sand; lay grass concrete surface (price does not include edgings or toe beams); fill with approved topsoil and fertilizer at 35 g/m²; seed with dwarf rye based grass seed mix		
Grasscrete in situ reinforced concrete surfacing GC 1, 100 thick	m²	68.48
Grasscrete in situ reinforced concrete surfacing GC 2, 150 thick	m²	73.22
Grassblock 103 open matrix blocks 406 x 406 x 103	m²	70.16
Gabion Walls		
Construct revetment of Maccaferri Ltd Reno mattress gabions laid on firm level ground, tightly packed with broken stone or concrete and securely wired, all in accordance with manufacturers' instructions		
gabions 6 m x 2 m x 0.17 m, one course	m²	67.30
gabions 6 m x 2 m x 0.17 m, two courses	m²	93.04
Timber Log Retaining Walls		
Excavate trench 300 wide to one third of the finished height of the retaining walls below; lay 100 mm hardcore; fix machine rounded logs set in concrete 1:3:6; remove excavated material from site; fix geofabric to rear of timber logs; backfill with previously excavated material set aside in position; all works by machine		
100 mm diameter logs		
500 mm high (constructed from 1.80 m lengths)	m	90.59
1.20 mm high (constructed from 1.80 m lengths)	m	149.48
1.60 mm high (constructed from 2.40 m lengths)	m	158.94
2.00 mm high (constructed from 3.00 m lengths)	m	209.45
As above but 150 mm diameter logs		
1.20 mm high (constructed from 1.80 m lengths)	m	172.43
1.60 mm high (constructed from 2.40 m lengths)	m	210.52
2.00 mm high (constructed from 3.00 m lengths)	m	276.12
2.40 mm high (constructed from 3.60 m lengths)	m	356.74

GROUND STABILIZATION

Item Excluding site overheads and profit	Unit	Total rate £
Geogrid soil reinforcement (Note:		
measured per metre run at the top of the		
bank)		
Backfill excavated bank; lay Tensar		
geogrid; cover with excavated material		
as work proceeds		
2.00 m high bank; geogrid at 1.00 m vertical lifts		
10 degree slope	m	**155.56**
20 degree slope	m	**84.43**
30 degree slope	m	**60.35**
2.00 m high bank; geogrid at 0.50 m vertical lifts		
45 degree slope	m	**58.50**
60 degree slope	m	**46.44**
3.00 m high bank		
10 degree slope	m	**155.56**
20 degree slope	m	**84.43**
30 degree slope	m	**60.35**
45 degree slope	m	**140.28**

IN SITU CONCRETE

Item Excluding site overheads and profit	Unit	Total rate £
IN SITU CONCRETE		
Mix concrete on site; aggregates **delivered in 20 tonne loads; deliver** **mixed concrete to location by barrow** **distance 25 m**		
1:3:6	m³	126.12
1:2:4	m³	138.15
As above but aggregates delivered in 1 tonne bags		
1:3:6	m³	160.82
1:2:4	m³	172.85
As above but ready mixed concrete		
10 N/mm²	m³	139.22
15 N/mm²	m³	144.73
As above but concrete discharged directly from ready mix lorry to required location		
10 N/mm²	m³	94.22
15 N/mm²	m³	99.73
Excavate foundation trench **mechanically, remove spoil offsite, lay** **1:3:6 site mixed concrete foundations;** **distance from mixer 25 m; depth of** **trench to be 225 mm deeper than** **foundation to allow for 3 underground** **brick courses priced separately** Foundation size		
200 mm deep x 400 mm wide	m	20.04
300 mm deep x 500 mm wide	m	30.76
400 mm deep x 400 mm wide	m	30.60
400 mm deep x 600 mm wide	m	46.04
600 mm deep x 600 mm wide	m	66.82
As above but hand excavation and disposal to spoil heap 25m by barrow and off site by grab lorry		
200 mm deep x 400 mm wide	m	37.87
300 mm deep x 500 mm wide	m	66.75
400 mm deep x 400 mm wide	m	67.10
400 mm deep x 600 mm wide	m	77.35
600 mm deep x 600 mm wide	m	141.30
Excavate foundation trench **mechanically, remove spoil offsite, lay** **ready mixed concrete GEN1 discharged** **directly from delivery lorry to location;** **depth of trench to be 225 mm deeper** **than foundation to allow for 3** **underground brick courses priced** **separately; foundation size**		
200 mm deep x 400 mm wide	m	16.35
300 mm deep x 500 mm wide	m	23.84
400 mm deep x 400 mm wide	m	23.22
400 mm deep x 600 mm wide	m	34.96
600 mm deep x 600 mm wide	m	50.21
Reinforced concrete wall to foundations **above (site mixed concrete)**		
up to 1.00 m high x 200 thick	m²	127.31
up to1.00 m high x 300 thick	m²	144.54
Reinforced concrete wall to foundations **above (ready mix concrete RC35)**		
1.00 m high x 200 thick	m²	122.61
1.00 m high x 300 thick	m²	134.39

BRICK/BLOCK WALLING

Item Excluding site overheads and profit	Unit	Total rate £
BRICK/BLOCK WALLING		
Excavate foundation trench 500 deep, remove spoil to dump off site, lay site mixed concrete foundations 1:3:6 350 x 150 thick, construct half brick wall with one brick piers at 2.0 m centres, laid in cement:lime:sand (1:1:6) mortar with flush joints, fair face one side, DPC two courses underground, engineering brick in cement:sand (1:3) mortar, coping of headers on end:		
Wall 900 high above DPC		
in engineering brick (class B) - £260.00/1000	m	157.95
in sandfaced facings - £300.00/1000	m	178.78
in rough stocks - £450.00/1000	m	187.24
Excavate foundation trench 400 deep, remove spoil to dump off site, lay GEN 1 concrete foundations 450 wide x 250 thick, construct one brick wall with one and a half brick piers at 3.0 m centres, all in English Garden Wall bond, laid in cement:lime:sand (1:1:6) mortar with flush joints, fair face one side, DPC two courses engineering brick in cement:sand (1:3) mortar, precast concrete coping 152 x 75 mm:		
Wall 900 high above DPC		
in engineering brick (class B) - £260.00/1000	m	269.39
in sandfaced facings - £300.00/1000	m	277.11
in rough stocks - £450.00/1000	m	294.09
Wall 1200 high above DPC		
in engineering brick (class B) - £260.00/1000	m	300.64
in sandfaced facings - £300.00/1000	m	309.66
in rough stocks - £450.00/1000	m	382.35
Wall 1800 high above DPC		
in engineering brick (class B) - £260.00/1000	m	373.68
in sandfaced facings - £300.00/1000	m	449.76
in rough stocks - £450.00/1000	m	478.54
Excavate foundation trench 450 deep, remove spoil to dump off site, lay GEN 1 concrete foundations 600 x 300 thick, construct one and a half brick wall with two thick brick piers at 3.0 m centres, all in English Garden Wall bond, laid in cement:lime:sand (1:1:6) mortar with flush joints, fair face one side, DPC two courses engineering brick in cement:sand (1:3) mortar, coping of headers on edge:		
Wall 900 high above DPC		
in engineering brick (class B) - £260.00/1000	m	539.70
in sandfaced facings - £300.00/1000	m	277.11
in rough stocks - £450.00/1000	m	294.09
Wall 1200 high above DPC		
in engineering brick (class B) - £260.00/1000	m	300.64
in sandfaced facings - £300.00/1000	m	309.66
in rough stocks - £450.00/1000	m	382.35
Wall 1800 high above DPC		
in engineering brick (class B) - £260.00/1000	m	373.68
in sandfaced facings - £300.00/1000	m	449.76
in rough stocks - £450.00/1000	m	478.54

BRICK/BLOCK WALLING

Item Excluding site overheads and profit	Unit	Total rate £
BRICK/BLOCK WALLING - cont'd		
Excavate foundation trench 450 deep, **remove spoil to dump off site, lay GEN 1** **concrete foundations 600 x 300 thick,** **construct wall of concrete block** Solid blocks 7 N/mm²		
100 mm thick	m²	56.23
140 mm thick	m²	62.28
100 mm blocks laid "on flat"		
215 mm thick	m²	104.02
Hollow Blocks filled with concrete		
215 mm thick	m²	85.72
Hollow blocks but with steel bar cast into the foundation		
215 mm thick	m²	99.10
Excavate foundation trench 450 deep, **remove spoil to dump off site, lay GEN 1** **concrete foundations 600 x 300 thick,** **construct wall of concrete block with** **brick face 112.5 thick to stretcher bond;** **Place stainless steel ties at 4 nr / m² of** **wall face** Solid blocks 7 N/mm² with Strecher bond brick face; Bricks PC £500.00 /1000		
100 mm thick	m²	133.48
140 mm thick	m²	139.53

ROADS AND PAVINGS

Item Excluding site overheads and profit	Unit	Total rate £
BASES FOR PAVING		
Excavate ground and reduce levels, to receive 38 mm thick slab and 25 mm mortar bed; dispose of excavated material off site; treat substrate with total herbicide		
Lay granular fill Type 1 150 thick laid to falls and compacted		
All by machine	m²	11.27
All by hand except disposal by grab	m²	37.55
Lay 1:2:4 concrete base 150 thick laid to falls		
All by machine	m²	25.29
All by hand except disposal by grab	m²	58.61
As above but with concrete base 150 deep but inclusive of 150 mm compacted hardcore		
By machine	m²	33.10
As above but inclusive of reinforcement fabric A142		
350 mm deep	m²	38.88
KERBS AND EDGINGS		
Note: Excavation is by machine unless otherwise mentioned.		
Excavate trench and construct concrete foundation 300 mm wide x 150 mm deep; lay precast concrete kerb units bedded in semi-dry concrete, slump 35 mm maximum; haunching one side; disposal of arisings off site		
Kerbs laid straight		
125 mm high x 125 mm thick bullnosed type BN	m	17.42
125 x 255 mm; ref HB2; SP	m	19.37
150 x 305 mm; ref HB1	m	23.52
Excavate and construct concrete foundation 450 mm wide x 150 mm deep; lay precast concrete kerb units bedded in semi-dry concrete, slump 35 mm maximum; haunching one side; lay channel units bedded in 1:3 lime:sand mortar, jointed in cement:sand mortar 1:3		
Kerbs; 125 mm high x 125 mm thick bullnosed type BN		
Dished channel 125 x 225 Ref CS	m	38.01
Square channel 125 x 150 Ref CS2	m	35.44
Bullnosed channel 305 x 150 Ref CBN	m	43.26
Bullnosed channel 125 x 225 Ref CS	m	114.01
Dished channel 125 x 225 Ref CS	m	114.01
On main paving base previously laid, supply and lay precast concrete channel units 305 mm wide x 150 mm deep dished type CD, bedded in 1:3 lime:sand mortar, jointed in cement:sand mortar 1:3	m	24.71
Excavate and construct concrete foundation 600 mm wide x 200 mm deep; lay channel of five courses of Class B engineering bricks to depths and falls bricks to be laid as headers along the channel, bedded in 1:3 lime:sand mortar bricks close jointed in cement:sand mortar 1:3	m	61.47

ROADS AND PAVINGS

Item Excluding site overheads and profit	Unit	Total rate £
KERBS AND EDGINGS - cont'd		
Excavate and lay precast concrete		
edging on concrete foundation 100 x		
150 deep and haunching one side 1:2:4		
including all necessary formwork; to one		
side of straight path		
Rectangular chamfered or bullnosed		
50 x 150	m	20.64
50 x 200	m	24.97
50 x 250	m	26.59
Rectangular chamfered or bullnosed as above		
but to both sides of straight paths		
50 x 150	m	43.81
50 x 200	m	50.10
50 x 250	m	53.35
Timber edgings; Softwood		
Straight		
150 x 38	m	6.64
150 x 50	m	6.84
Curved; over 5 m radius		
150 x 38	m	8.28
150 x 50	m	9.14
Curved; 4 - 5 m radius		
150 x 38	m	8.94
150 x 50	m	9.79
Curved; 3 - 4 m radius		
150 x 38	m	9.92
150 x 50	m	11.10
Curved; 1.00 - 3.00 m radius		
150 x 38	m	11.56
150 x 50	m	12.74
Timber edgings hardwood		
150 x 38	m	11.53
150 x 50	m	13.38
Brick or concrete block edge restraint;		
excavate for groundbeam, lay concrete		
1:2:4 200 wide x 150mm deep; on 50 mm		
thick sharp sand bed; lay blocks or		
bricks inclusive of haunching one side		
Blocks 200 x 100 x 60; PC £7.83 / m²; butt		
jointed		
header course	m	14.16
stretcher course	m	12.58
Bricks 215 x 112.5 x 50; PC £300.00/1000; with		
mortar joints		
header course	m	16.49
stretcher course	m	15.63
Sawn Yorkstone edgings; excavate for		
groundbeam, lay concrete 1:2:4 150 mm		
deep x 33.3% wider than the edging; on		
35 mm thick mortar bed; inclusive of		
haunching one side		
Yorkstone 50 mm thick		
100 mm wide x random lengths	m	22.62
100 mm x 100 mm	m	25.81
100 mm wide x 200 long	m	33.07
250 mm wide x random lengths	m	33.68
500 mm wide x random lengths	m	54.65

ROADS AND PAVINGS

Item Excluding site overheads and profit	Unit	Total rate £
ROADS		
Excavate weak points of excavation by hand and fill with well rammed Type 1 granular material		
100 thick	m²	9.64
200 thick	m²	12.49
Excavate road bed and dispose excavated material off site, bring to grade; lay reinforcing mesh A142 lapped and joined, lay 150 mm thick in situ reinforced concrete roadbed 21 N/mm², with 15 impregnated fibreboard expansion joints and polysulphide based sealant at 50 m centres, on 150 hardcore blinded with ash or sand, kerbs 155 x 255 to both sides, including foundations haunched one side in 11.5 N/mm² concrete with all necessary formwork		
Falls, crossfalls and cambers not exceeding 15 degrees		
4.90 m wide	m	218.89
6.10 m wide	m	296.18
7.32 m wide	m	312.78
Macadam Roadway over 1000 m²		
Excavate 350 mm for pathways or roadbed, bring to grade, lay 100 well-rolled hardcore; lay 150 mm Type 1 granular material; lay precast edgings kerbs 50 x 150 on both sides, including foundations haunched one side in 11.5 N/mm² concrete with all necessary formwork; machine lay surface of 90 mm macadam 60 mm base course and 30 mm wearing course; all disposal off site		
Macadam Roadway over 1000 m²		
1.50 m wide	m	136.40
2.00 m wide	m	165.06
3.00 m wide	m	225.92
4.00 m wide	m	286.80
5.00 m wide	m	334.53
6.00 m wide	m	408.56
7.00 m wide	m	469.44
Macadam Roadway between 400 and 1000 m²		
1.50 m wide	m	141.53
2.00 m wide	m	171.90
3.00 m wide	m	236.18
4.00 m wide	m	300.48
5.00 m wide	m	364.78
6.00 m wide	m	429.08
7.00 m wide	m	493.38

ROADS AND PAVINGS

Item Excluding site overheads and profit	Unit	Total rate £
CAR PARKS		
Excavate 350 for pathways or roadbed to receive surface 100 mm thick priced separately, bring to grade, lay 100 well-rolled hardcore; lay 150 mm Type 1 granular material; lay kerbs BS 7263; 125 x 255 on both sides, including foundations haunched one side in 11.5 N/mm² concrete with all necessary formwork		
Work to falls, crossfalls and cambers not exceeding 15 degrees		
1.50 m wide	m	120.41
2.00 m wide	m	145.54
3.00 m wide	m	193.25
4.00 m wide	m	240.98
5.00 m wide	m	288.71
6.00 m wide	m	336.44
7.00 m wide	m	384.17
As above but excavation 450 mm deep and base of Type 1 at 250 mm thick		
1.50 m wide	m	188.18
2.00 m wide	m	234.08
3.00 m wide	m	326.09
4.00 m wide	m	418.08
5.00 m wide	m	510.09
6.00 m wide	m	602.10
7.00 m wide	m	694.09
To excavated and prepared base above, lay roadbase of 40 size dense bitumen macadam to BS 4987, 70 thick, lay wearing course of 10 size dense bitumen macadam 30 thick, mark out car parking bays 5.0 m x 2.4 m with thermoplastic road paint: surfaces all mechanically laid		
per bay 5.0 m x 2.4 m	each	176.52
gangway	m²	12.43
As above but stainless steel road studs 100 x 100 to BS 873:pt.4, two per bay in lieu of thermoplastic paint		
per bay 5.0 m x 2.4 m	each	187.52
Car park as above but with interlocking concrete blocks 200 x 100 x 80 mm; grey		
per bay 5.0 m x 2.4 m	each	470.04
gangway	m²	39.17
Car park as above but with interlocking concrete blocks 200 x 100 x 80 mm; colours		
per bay 5.0 m x 2.4 m	each	486.96
gangway	m²	40.58
Car park as above but with interlocking concrete blocks 200 x 100 x 60 mm; grey		
per bay 5.0 m x 2.4 m	each	458.64
gangway	m²	38.22
Car park as above but with interlocking concrete blocks 200 x 100 x 60 mm; colours		
per bay 5.0 m x 2.4 m	each	469.80
gangway	m²	39.15

ROADS AND PAVINGS

Item Excluding site overheads and profit	Unit	Total rate £
Car park as above but with lay Grass Concrete Grasscrete in situ continuously reinforced cellular surfacing, including expansion joints at 10 m centres, fill with topsoil and peat (5:1) and fertilizer at 35 g/m², seed with dwarf rye grass at 35 g/m²		
GC2, 150 thick for HGV traffic including dust carts		
per bay 5.0 m x 2.4 m	each	345.72
gangway	m²	28.81
GC1, 100 thick for cars and light traffic		
per bay 5.0 m x 2.4 m	each	288.84
gangway	m²	24.07
CAR PARKING FOR DISABLED PEOPLE		
To excavated and prepared base above, lay roadbase of 20 size dense bitumen macadam to BS 4987, 80 thick, lay wearing course of 10 size dense bitumen macadam 30 thick, mark out car parking bays with thermoplastic road paint		
per bay 5.80 m x 3.25 m (ambulant)	each	234.31
per bay 6.55 m x 3.80 m (wheelchair)	each	309.38
gangway	m²	12.43
Car park as above but with interlocking concrete blocks 200 x 100 x 60 mm; grey, but mark out bays		
per bay 5.80 m x 3.25 m (ambulant)	each	720.45
per bay 6.55 m x 3.80 m (wheelchair)	each	951.30
gangway	m²	38.22
PLAY AREA FOR BALL GAMES		
Excavate for playground, bring to grade, lay 225 consolidated hardcore, blind with 100 type 1, lay macadam base of 40 size dense bitumen macadam to 75 thick, lay wearing course 30 thick		
Over 1000 m²	100 m²	3381.48
400 m² -1000 m²	100 m²	3554.78
Excavate for playground, bring to grade, lay 100 consolidated hardcore, blind with 100 type 1, lay macadam base of dense bitumen macadam to 50 thick, lay wearing course 20 thick		
Over 1000 m²	100 m²	2541.92
400 m² -1000 m²	100 m²	3209.78
Excavate for playground, bring to grade, lay 100 consolidated hardcore, blind with 100 type 1, lay macadam base of dense bitumen macadam to 50 thick, lay wearing course of Addagrip resin coated aggregate 3mm diameter, 6mm thick		
Over 1000 m²	100 m²	4501.92
400 m² -1000 m²	100 m²	5013.78

ROADS AND PAVINGS

Item Excluding site overheads and profit	Unit	Total rate £
INTERLOCKING BLOCK PAVING		
Edge restraint to block paving; excavate		
for groundbeam, lay concrete 1:2:4 200		
wide x 150mm deep; on 50 mm thick		
sharp sand bed; inclusive of haunching		
one side		
Blocks 200 x 100 x 60; PC £7.83 /m²; butt		
jointed		
header course	m	14.16
stretcher course	m	12.58
Bricks 215 x 112.5 x 50; PC £300.00/1000; with		
mortar joints		
header course	m	16.49
stretcher course	m	15.63
Excavate ground, treat substrate with		
total herbicide, supply and lay granular		
fill Type 1 150 thick laid to falls and		
compacted, supply and lay block pavers,		
laid on 50 compacted sharp sand,		
vibrated, joints filled with loose sand		
excluding edgings or kerbs measured		
separately		
Blocks 200 x 100 x 60		
Blocks 200 x 100 x 60	m²	51.48
200 x 100 x 80	m²	51.84
Reduce levels, lay 150 granular material		
Type 1, lay 200 x 100 x 60 vehicular		
block paving to 90 degree herringbone		
pattern, on 50 compacted sand bed,		
vibrated, jointed in sand and vibrated,		
excavate and lay precast concrete		
edging 50 x 150 to BS 7263, on		
concrete foundation 1:2:4		
1.0 m wide clear width between edgings	m	91.77
1.5 m wide clear width between edgings	m	116.99
2.0 m wide clear width between edgings	m	142.26
As above but blocks laid 45 degree		
herringbone pattern including cutting		
edging blocks		
1.0 m wide clear width between edgings	m	97.96
1.5 m wide clear width between edgings	m	123.18
2.0 m wide clear width between edgings	m	148.45
As above but all excavation by hand		
disposal off site by grab		
1.0 m wide clear width between edgings	m	182.06
1.5 m wide clear width between edgings	m	221.74
2.0 m wide clear width between edgings	m	262.01

ROADS AND PAVINGS

Item Excluding site overheads and profit	Unit	Total rate £
BRICK PAVING		
WORKS BY MACHINE		
Excavate and lay base Type 1 150 thick		
remove arisings; all by machine; lay clay		
brick paving		
200 x 100 x 50 thick; butt jointed on 50 mm		
sharp sand bed		
PC £300.00 / 1000	m²	61.10
PC £600.00 / 1000	m²	76.47
200 x 100 x 50 thick; 10 mm mortar joints on		
35mm mortar bed		
PC £300.00 / 1000	m²	75.13
PC £600.00 / 1000	m²	88.44
Excavate and lay base Type 1 250 thick		
all by machine; remove arisings; lay clay		
brick paving		
200 x 100 x 50 thick; 10 mm mortar joints on		
35mm mortar bed		
PC £300.00 / 1000	m²	81.42
Excavate and lay 150 mm readymix		
concrete base reinforced with A393		
mesh; all by machine; remove arisings;		
lay clay brick paving		
200 x 100 x 50 thick; 10 mm mortar joints on		
35mm mortar bed; running or stretcher bond		
PC £300.00 / 1000	m²	92.17
PC £600.00 / 1000	m²	105.48
200 x 100 x 50 thick; 10 mm mortar joints on		
35mm mortar bed; butt jointed; herringbone		
bond		
PC £300.00 / 1000	m²	78.14
PC £600.00 / 1000	m²	93.51
Excavate and lay base readymix		
concrete base 150 mm thick reinforced		
with A393 mesh; all by machine; remove		
arisings; lay clay brick paving		
215 x 102.5 x 50 thick; 10 mm mortar joints on		
35 mm mortar bed		
PC £300.00 / 1000; herringbone	m²	94.28
PC £600.00 / 1000; herringbone	m²	107.13
WORKS BY HAND		
Excavate and lay base Type 1 150 thick		
by hand; arisings barrowed to spoil heap		
maximum distance 25 m and removal off		
site by grab; lay clay brick paving		
200 x 100 x 50 thick; butt jointed on 50 mm		
sharp sand bed		
PC £300.00 / 1000	m²	105.76
PC £600.00 / 1000	m²	133.10
Excavate and lay 150 mm concrete		
base; 1:3:6: site mixed concrete		
reinforced with A393 mesh; remove		
arisings to stockpile and then off site by		
grab; lay clay brick paving		
215 x 102.5 x 50 thick; 10 mm mortar joints on		
35mm mortar bed		
PC £300.00 / 1000	m²	103.16
PC £600.00 / 1000	m²	130.50

ROADS AND PAVINGS

Item Excluding site overheads and profit	Unit	Total rate £
FLAG PAVING TO PEDESTRIAN AREAS		
Prices are inclusive of all mechanical excavation and disposal		
Supply and lay precast concrete flags; **excavate ground and reduce levels,** **treat substrate with total herbicide,** **supply and lay granular fill Type 1 150** **thick laid to falls and compacted** Standard precast concrete flags to BS 7263 bedded and jointed in lime:sand mortar (1:3)		
450 x 450 x 70 chamfered	m²	34.92
450 x 450 x 50 chamfered	m²	33.39
600 x 300 x 50	m²	32.43
600 x 450 x 50	m²	31.47
600 x 600 x 50	m²	29.10
750 x 600 x 50	m²	28.76
900 x 600 x 50	m²	28.25
Coloured flags bedded and jointed in lime:sand mortar (1:3)		
600 x 600 x 50	m²	32.66
450 x 450 x 70 chamfered	m²	36.84
600 x 600 x 50	m²	33.06
400 x 400 x 65	m²	41.55
750 x 600 x 50	m²	32.29
900 x 600 x 50	m²	31.27
Marshalls Saxon; textured concrete flags; reconstituted Yorkstone in colours, butt jointed bedded in lime:sand mortar (1:3)		
300 x 300 x 35	m²	59.59
450 x 450 x 50	m²	52.28
600 x 300 x 35	m²	47.44
600 x 600 x 50	m²	45.35
Tactile flags; Marshalls Plc, Blister Tactile pavings; red or buff; for blind pedestrian guidance laid to designed pattern		
450 x 450	m²	42.10
400 x 400	m²	44.79
PEDESTRIAN DETERRENT PAVING		
Excavate ground and bring to levels, **treat substrate with total herbicide,** **supply and lay granular fill Type 1 150** **mm thick laid to falls and compacted,** **supply and lay precast deterrent paving** **units bedded in lime:sand mortar (1:3)** **and jointed in lime:sand mortar (1:3)** Marshalls Plc		
Lambeth pyramidal paving 600 x 600 x 75	m²	58.04
Townscape Abbey square cobble pattern pavings; reinforced		
600 x 600 x 60	m²	55.19
Geoset chamfered studs 600 x 600 x 60	m²	50.99

Item Excluding site overheads and profit	Unit	Total rate £
IMITATION YORKSTONE PAVINGS		
Excavate ground and bring to levels,		
treat substrate with total herbicide,		
supply and lay granular fill Type 1 150		
mm thick laid to falls and compacted,		
supply and lay imitation Yorkstone		
paving laid to coursed patterns bedded		
in lime:sand mortar (1:3) and jointed in		
lime:sand mortar (1:3)		
Marshalls Heritage square or rectangular		
300 x 300 x 38	m²	62.61
600 x 300 x 30	m²	49.03
600 x 450 x 38	m²	48.33
450 x 450 x 38	m²	46.41
600 x 600 x 38	m²	43.10
As above but laid to Random rectangular		
patterns		
Various sizes selected from the above	m²	54.58
Imitation Yorkstone laid random rectangular as		
above but on concrete base 150 thick		
By machine	m²	68.60
By hand	m²	94.88
NATURAL STONE SLAB PAVING		
WORKS BY MACHINE		
Excavate ground by machine and reduce		
levels, to receive 65 mm thick slab and		
35 mm mortar bed; dispose of excavated		
material off site; treat substrate with		
total herbicide, lay granular fill Type 1		
150 thick laid to falls and compacted;		
lay to random rectangular pattern on 35		
mm mortar bed		
new riven slabs		
laid random rectangular	m²	97.12
new riven slabs; but to 150 mm plain concrete		
base		
laid random rectangular	m²	111.14
new riven slabs; disposal by grab		
laid random rectangular	m²	101.97
reclaimed Cathedral grade riven slabs		
laid random rectangular	m²	133.98
reclaimed Cathedral grade riven slabs; but to 150		
mm plain concrete base		
laid random rectangular	m²	147.20
reclaimed Cathedral grade riven slabs; disposal		
by grab		
laid random rectangular	m²	138.83
new slabs sawn 6 sides		
laid random rectangular	m²	100.98
3 sizes, laid to coursed pattern	m²	105.14
new slabs sawn 6 sides; but to 150 mm plain		
concrete base		
laid random rectangular	m²	115.00
3 sizes, laid to coursed pattern	m²	119.16
new slabs sawn 6 sides; disposal by grab		
laid random rectangular	m²	105.83
3 sizes, laid to coursed pattern	m²	109.99

ROADS AND PAVINGS

Item Excluding site overheads and profit	Unit	Total rate £
NATURAL STONE SLAB PAVING - cont'd		
WORKS BY HAND		
Excavate ground by hand and reduce		
levels, to receive 65 mm thick slab and		
35 mm mortar bed; barrow all materials		
and arisings 25 m; dispose of excavated		
material off site by grab; treat substrate		
with total herbicide, lay granular fill Type		
1 150 thick laid to falls and compacted;		
lay to random rectangular pattern on 35		
mm mortar bed		
new riven slabs		
laid random rectangular	m²	127.08
new riven slabs laid random rectangular; but to		
150 mm plain concrete base		
laid random rectangular	m²	143.66
new riven slabs; but disposal to skip		
laid random rectangular	m²	140.20
reclaimed Cathedral grade riven slabs		
laid random rectangular	m²	163.94
reclaimed Cathedral grade riven slabs; but to 150		
mm plain concrete base		
laid random rectangular	m²	180.52
reclaimed Cathedral grade riven slabs; but		
disposal to skip		
laid random rectangular	m²	177.06
new slabs sawn 6 sides		
laid random rectangular	m²	130.94
3 sizes, sawn 6 sides laid to coursed pattern	m²	134.72
new slabs sawn 6 sides; but to 150 mm plain		
concrete base		
laid random rectangular	m²	147.52
3 sizes, sawn 6 sides laid to coursed pattern	m²	151.68
GRANITE SETT PAVING - PEDESTRIAN		
Excavate ground and bring to levels, lay		
100 hardcore to falls, compacted with 5		
tonne roller, blind with compacted Type		
1 50 mm thick, lay 100 concrete 1:2:4 ,		
supply and lay granite setts 100 x 100 x		
100 mm bedded in cement:sand mortar		
(1:3) 25 mm thick minimum, close butted		
and jointed in fine sand, all excluding		
edgings or kerbs measured separately		
Setts laid to bonded pattern and jointed		
new setts 100 x 100 x 100	m²	134.13
second-hand cleaned 100 x 100 x 100	m²	150.48
Setts laid in curved pattern		
new setts 100 x 100 x 100	m²	140.32
second-hand cleaned 100 x 100 x 100	m²	156.67

ROADS AND PAVINGS

Item Excluding site overheads and profit	Unit	Total rate £
GRANITE SETT PAVING TRAFFICKED AREAS		
Excavate ground and bring to levels, lay 100 hardcore to falls, compacted with 5 tonne roller, blind with compacted Type 1 50 mm thick, lay 150 site mixed concrete 1:2:4 reinforced with steel fabric to BS 4483 ref: A 142, supply and lay granite setts 100 x 100 x 100 mm bedded in cement:sand mortar (1:3) 25 mm thick minimum, close butted and jointed in fine sand, all excluding edgings or kerbs measured separately		
Site mixed concrete		
new setts 100 x 100 x 100	m²	142.36
second-hand cleaned 100 x 100 x 100	m²	158.71
Ready mixed concrete		
new setts 100 x 100 x 100	m²	138.43
second-hand cleaned 100 x 100 x 100	m²	154.78
CONCRETE PAVING		
Pedestrian Areas		
Excavate to reduce levels, lay 100 mm Type 1, lay PAV 1 air entrained concrete, joints at max width of 6.0m cut out and sealed with sealant to BS 5212. Inclusive of all formwork and stripping		
100 mm thick	m²	66.25
Trafficked Areas		
Excavate to reduce levels, lay 150 mm Type 1 lay PAV 2 40 N/mm² air entrained concrete, reinforced with steel mesh to BS 4483 200 x 200 square at 2.22kg/m², joints at max width of 6.0m cut out and sealed with sealant to BS 5212		
150 mm thick	m²	85.49
BEACH COBBLE PAVING		
Excavate ground and bring to levels and fill with compacted Type 1 fill 100 thick, lay GEN 1 concrete base 100 thick, supply and lay cobbles individually laid by hand bedded in cement:sand mortar (1:3) 25 thick minimum, dry grout with 1:3 cement:sand grout, brush off surplus grout and water in, sponge of cobbles as work proceeds all excluding formwork edgings or kerbs measured separately		
Scottish beach cobbles 200 - 100 mm	m²	87.14
Kidney flint cobbles 100 - 75 mm	m²	95.26
Scottish beach cobbles 75 -50 mm	m²	107.13

ROADS AND PAVINGS

Item Excluding site overheads and profit	Unit	Total rate £
CONCRETE SETT PAVING		
Excavate ground and bring to levels,		
supply and lay 150 granular fill Type 1		
laid to falls and compacted, supply and		
lay setts bedded in 50 sand and		
vibrated, joints filled with dry sand and		
vibrated, all excluding edgings or kerbs		
measured separately		
Marshalls Plc; Tegula precast concrete setts		
random sizes 60 thick	m²	44.43
single size 60 thick	m²	42.24
random size 80 thick	m²	47.57
single size 80 thick	m²	53.77
Concrete setts Blanc de Bierges		
140 x 140 x 80	m²	46.67
210 x 140 x 80	m²	45.65
70 x 70 x 70	m²	47.83
GRASS CONCRETE PAVING		
Reduce levels, lay 150 mm Type 1		
granular fill compacted, supply and lay		
precast grass concrete blocks on 20		
sand and level by hand, fill blocks with		
3mm sifted topsoil and pre-seeding		
fertilizer at 50g/m², sow with perennial		
ryegrass/chewings fescue see seed at		
35g/m²		
Grass Concrete		
GB103 406 x 406 x 103	m²	34.32
GB83 406 x 406 x 83	m²	32.35
Extra for geotextile fabric underlayer	m²	0.70
Firepaths		
Excavate to reduce levels, lay 300 mm well		
rammed hardcore, blinded with 100 mm type 1,		
supply and lay Marshalls Plc "Grassguard		
180" precast grass concrete blocks on 50 sand		
and level by hand, fill blocks with sifted topsoil		
and pre-seeding fertilizer at 50 g/m²		
firepath 3.8 m wide	m	212.25
firepath 4.4 m wide	m	245.74
firepath 5.0 m wide	m	279.44
turning areas	m²	55.89
Charcon Hard Landscaping Grassgrid: 366 x 274		
x 100	m²	31.21
SLAB/BRICK PATHS		
Stepping stone path inclusive of hand		
excavation and mechanical disposal,		
100 mm Type 1 and sand blinding; slabs		
to comply with BS 7263 laid 100 mm		
apart to existing turf		
600 wide; 600 x 600 x 50 slabs		
natural finish	m	41.25
coloured	m	43.17
exposed aggregate	m	51.02
900 wide; 600 x 900 x 50 slabs		
natural finish	m	46.06
coloured	m	48.95
exposed aggregate	m	60.72

ROADS AND PAVINGS

Item Excluding site overheads and profit	Unit	Total rate £
Pathway inclusive of mechanical **excavation and disposal, 100 mm Type** **1 and sand blinding; slabs to comply** **with BS 7263 close butted**		
Straight butted path 900 wide; 600 x 900 x 50 slabs		
natural finish	m	17.96
coloured	m	19.88
exposed aggregate	m	27.73
Straight butted path 1200 wide; double row; 600 x 900 x 50 slabs, laid stretcher bond		
natural finish	m	31.55
coloured	m	35.82
exposed aggregate	m	53.25
Straight butted path 1200 wide, one row of 600 x 600 x 50, one row 600 x 900 x 50 slabs		
natural finish	m	31.04
coloured	m	34.74
exposed aggregate	m	59.44
Straight butted path 1500 wide, slabs of 600 x 900 x 50, and 600 x 600 x 50 slabs, laid to bond		
natural finish	m	38.51
coloured	m	42.50
exposed aggregate	m	59.44
Straight butted path 1800 wide, two rows of 600 x 900 x 50 slabs, laid bonded		
natural finish	m	41.82
coloured	m	53.58
exposed aggregate	m	79.95
Straight butted path 1800 wide, three rows of 600 x 900 x 50 slabs, laid stretcher bond		
natural finish	m	46.51
coloured	m	58.27
exposed aggregate	m	84.64
Brick paved paths: Bricks 215 x 112.5 x **65 with mortar joints; prices are** **inclusive of excavation, disposal off** **site, 100 hardcore, 100 1:2:4 concrete** **bed, edgings of brick 215 wide,** **haunched; all jointed in** **cement:lime:sand mortar (1:1:6).**		
Path 1015 wide laid stretcher bond; edging course of headers		
rough stocks - £400.00/1000	m	127.17
engineering brick - £280.00/1000	m	122.30
Path 1015 wide laid stack bond		
rough stocks - £400.00/1000	m	131.86
engineering brick - £280.00/1000	m	126.99
Path 1115 wide laid header bond		
rough stocks - £400.00/1000	m	138.47
engineering brick - £280.00/1000	m	133.60
Path 1330 wide laid basketweave bond		
rough stocks - £400.00/1000	m	159.70
engineering brick - £280.00/1000	m	153.31
Path 1790 wide laid basketweave bond		
rough stocks - £400.00/1000	m	207.85
engineering brick - £280.00/1000	m	153.31

ROADS AND PAVINGS

Item Excluding site overheads and profit	Unit	Total rate £
SLAB/BRICK PATHS – cont'd		
Brick paved paths; brick paviors 200 x 100 x 50 chamfered edge with butt joints; prices are inclusive of excavation, 100 mm Type 1, and 50 mm sharp sand; jointing in kiln dried sand brushed in; exclusive of edge restraints		
Path 1000 wide laid stretcher bond; edging course of headers		
rough stocks - £400.00/1000	m	118.35
engineering brick - £280.00/1000	m	122.30
Path 1330 wide laid basketweave bond		
rough stocks - £400.00/1000	m	159.70
engineering brick - £280.00/1000	m	153.31
Path 1790 wide laid basketweave bond		
rough stocks - £400.00/1000	m	207.85
engineering brick - £280.00/1000	m	153.31
GRAVEL PATHS		
Reduce levels and remove spoil to dump on site, lay 150 hardcore well rolled, lay 25 mm sand blinding and geofabric, lay 50 gravel watered and rolled, excavate and timber edge 150 x 38 to both sides of straight paths		
1.0 m wide	m	21.49
1.5 m wide	m	29.46
2.0 m wide	m	33.40
As above but Breedon Gravel		
1.0 m wide	m	27.68
1.5 m wide	m	37.36
2.0 m wide	m	47.04
BARK PAVING		
Excavate to reduce levels, remove all topsoil to dump on site, treat area with herbicide; lay 150 mm clean hardcore, blind with sand to BS 882 Grade C, lay 0.7 mm geotextile filter fabric water flow 50 l/m²/sec; supply and fix treated softwood edging boards 50 x 150 mm to bark area on hardcore base extended 150 mm beyond the bark area, boards fixed with galvanized nails to treated softwood posts 750 x 50 x 75 mm driven into firm ground at 1.0 m centres; edging boards to finish 25 mm above finished bark surface; tops of posts to be flush with edging boards and once weathered		
Supply and lay 100 mm Melcourt conifer walk chips 10-40 mm size		
1.00 m wide	m	62.94
2.00 m wide	m	105.58
3.00 m wide	m	149.46
4.00 m wide	m	190.87
Extra over for wood fibre	m²	-1.53
Extra over for hardwood chips	m²	-1.37

ROADS AND PAVINGS

Item Excluding site overheads and profit	Unit	Total rate £
FOOTPATHS		
Excavate footpath to reduce level,		
remove arisings to tip on site maximum		
distance 25 m; lay Type 1 granular fill		
100 thick, lay base course of 28 size		
dense bitumen macadam 50 thick,		
wearing course of 10 size dense bitumen		
macadam 20 thick; timber edge 150 x		
38 mm		
Areas over 1000 m²		
1.0 m wide	m	29.06
1.5 m wide	m	40.37
2.0 m wide	m	49.80
Areas 400 m² - 1000 m²		
1.0 m wide	m	32.37
1.5 m wide	m	45.34
2.0 m wide	m	67.98

SPECIAL SURFACES FOR SPORT/PLAYGROUNDS

Item Excluding site overheads and profit	Unit	Total rate £
BOWLING GREEN CONSTRUCTION; **Baylis Landscape Contractors**		
Bowling Green; Complete Excavate 300 mm deep and grade to level; excavate and install 100 mm land drain to perimeter and backfill with shingle; install 60 mm land drain to surface at 4.5 m centres backfilled with shingle; install 50 m non perforated pipe 50 m long; install 100 mm compacted and levelled drainage stone, blind with grit and sand; spread 150 mm imported 70:30 sand:soil accurately compacted and levelled; lay bowling green turf and top dress twice luted into surface; exclusive of perimeter ditches and bowls protection		
6 rink green 38.4 x 38.4 m	each	52522.50

Install "Toro" automatic bowling green irrigation
  system with pump, tank, controller, electrics,
  pipework and 8 nr "Toro 780" sprinklers		

Excluding pumphouse (optional)	each	8155.67
Supply and install to perimeter of green, "Sportsmark" preformed bowling green ditch channels		
"Ultimate Design 99" steel reinforced concrete channel, 600 mm long section	each	7739.27
"Ultimate GRC " glass reinforced concrete channel, 1.2 m long section	each	11775.99

"Ultimate Design 2001" medium density rotational
  moulded channel. 1 m long section with integral
  bowls protection
	each	11775.99

Supply and fit Bowls Protection material to rear hitting face of "Ultimate" channels 1 and 2 above		
"Curl Grass" artificial grass, 0.45 m wide	each	2174.84
"Astroturf" artificial grass, 0.45 m wide	each	1973.27
50 mm rubber bowls bumper (2 rows)	each	4482.30
Bowls protection ditch liner laid loose, 300 mm wide	each	1326.13
JOGGING TRACK		
Excavate track 250 deep; treat substrate with Casoron G4; lay filter membrane; lay 100 depth gravel waste or similar; lay 100 mm compacted gravel	100 m²	5839.04
Extra for treated softwood edging 50 x 150 on 50 x 50 posts		
both sides	m	8.32
PLAYGROUNDS		
Excavate playground area, and dispose **of arisings to tip; lay Type 1 granular fill,** **lay macadam surface two coat work 80** **thick; base course of 28 size dense** **bitumen macadam 50 thick, wearing** **course of 10 size dense bitumen** **macadam 30 thick**		
Areas over 1000 m²		
Excavation 180 mm base 100 mm thick	m²	21.59
Excavation 225 mm base 150 mm thick	m²	24.74
Areas 400 m² -1000 m²		
Excavation 180 mm base 100 mm thick	m²	24.96
Excavation 225 mm base 150 mm thick	m²	28.11

SPECIAL SURFACES FOR SPORT/PLAYGROUNDS

Item Excluding site overheads and profit	Unit	Total rate £
Excavate playground area to given **levels and falls, remove soil off site and** **backfill with compacted Type 1 granular** **fill; lay ready mixed concrete to fall 2%** **in all direction**		
base 150 mm thick; surface 100 mm thick	m²	22.50
base 150 mm thick; surface 150 mm thick	m²	28.42
SAFETY SURFACING; **Baylis Landscape Contractors**		
Excavate ground and reduce levels, to **receive 38 mm thick slab and 25 mm** **mortar bed; dispose of excavated** **material off site; treat substrate with** **total herbicide; lay granular fill Type 1** **150 thick laid to falls and compacted;** **lay macadam base 40 mm thick; supply** **and lay "Ruberflex" wet pour safety** **system to thicknesses as specified**		
All by machine except macadam by hand		
Black		
15 mm thick	100 m²	4730.82
35 mm thick	100 m²	6143.82
60 mm thick	100 m²	7449.82
Coloured		
15 mm thick	100 m²	8318.82
35 mm thick	100 m²	8862.82
60 mm thick	100 m²	9732.82
All by hand except disposal by grab		
Black		
15 mm thick	100 m²	7620.55
35 mm thick	100 m²	9033.55
60 mm thick	100 m²	10339.55
Coloured		
15 mm thick	100 m²	11208.55
35 mm thick	100 m²	11752.55
60 mm thick	100 m²	12622.55
SPECIAL SURFACES FOR SPORT/PLAYGROUNDS		
Excavate playground area to 450 depth, **lay 150 broken stone or clean hardcore,** **lay filter membrane, lay bark surface**		
Melcourt Industries		
Playbark 10/50 - 300 thick	100 m²	3629.12
Playbark 8/25 - 300 thick	100 m²	3444.12

SPECIAL SURFACES FOR SPORT/PLAYGROUNDS

Item Excluding site overheads and profit	Unit	Total rate £
SPORTSGROUND CONSTRUCTION; **Baylis Landscape Contractors**		
Plain sports pitches; site clearance, grading and drainage not included.		
Cultivate ground and grade to levels, **apply pre-seeding fertilizer at 900 kg/ha,** **apply pre-seeding selective weedkiller,** **seed in two operations with sports pitch** **type grass seed at 350 kg/ha, harrow** **and roll lightly, including initial cut; size**		
association football, senior 114 m x 72 m	each	3455.29
association football, junior 106 m x 58 m	each	2786.52
rugby union pitch 156 m x 81 m	each	5071.46
rugby league pitch 134 m x 60 m	each	3343.83
hockey pitch 95 m x 60 m	each	2563.60
shinty pitch 186 m x 96 m	each	7133.49
men's lacrosse pitch 100 m x 55 m	each	2452.14
women's lacrosse pitch 110 m x 73 m	each	3566.75
target archery ground 150 m x 50 m	each	2897.99
cricket outfield 160 m x 142 m	each	9028.32
cycle track outfield 160 m x 80 m	each	5573.04
polo ground 330 m x 220 m	each	26639.14
Cricket square; excavate to depth of **150 mm, pass topsoil through 6 mm** **screen, return and mix evenly with** **imported marl or clay loam, bring to** **accurate levels, apply pre-seeding** **fertilizer at 50 g/m , apply selective** **weedkiller, seed with cricket square type** **g grass seed at 50 g/m², rake in and roll** **lightly, erect and remove temporary** **protective chestnut fencing, allow for** **initial cut and watering three times** 22.8 m x 22.8 m	each	14489.91

PREPARATION FOR PLANTING/TURFING

Item Excluding site overheads and profit	Unit	Total rate £
CULTIVATION BY HAND		
Spread only and lightly consolidate topsoil brought from spoil heap in layers not exceeding 150, grade to specified levels, remove stones over 25mm, treat with paraquat-diquat weedkiller, all by hand		
100 mm thick	100 m²	375.07
150 mm thick	100 m²	562.61
300 mm thick	100 m²	1125.22
450 mm thick	100 m²	1687.84
As above but inclusive of loading to location by barrow maximum distance 25 m; finished topsoil depth		
100 mm thick	100 m²	825.07
150 mm thick	100 m²	1237.61
300 mm thick	100 m²	2475.22
450 mm thick	100 m²	3712.84
500 mm thick	100 m²	4123.50
600 mm thick	100 m²	4944.83
As above but loading to location by barrow maximum distance 100 m; finished topsoil depth		
100 mm thick	100 m²	843.87
150 mm thick	100 m²	1265.81
300 mm thick	100 m²	2531.62
450 mm thick	100 m²	3797.44
500 mm thick	100 m²	4217.50
600 mm thick	100 m²	5057.63
Extra to the above for incorporating mushroom compost at 50 mm/m² into the top 150 mm of topsoil; by hand	100 m²	174.00
Extra to above for imported topsoil PC 12.48 m³ allowing for 20% settlement		
100 mm thick	100 m²	149.80
150 mm thick	100 m²	224.70
300 mm thick	100 m²	449.40
450 mm thick	100 m²	674.10
500 mm thick	100 m²	749.00
600 mm thick	100 m²	898.80
750 mm thick	100 m²	1123.50
1.00 m thick	100 m²	1498.00
CULTIVATION BY MACHINE		
Treat area with systemic non selective herbicide 1 month before starting cultivation operations; rip up subsoil using subsoiling machine to a depth of 250 below topsoil at 1.20 m centres in light to medium soils, rotavate to 200 deep in two passes, cultivate with chain harrow, roll lightly, clear stones over 50mm		
by tractor	100 m²	9.09
As above but carrying out operations in clay or compacted gravel	100 m²	9.67
As above but ripping by tractor rotavation by pedestrian operated rotavator, clearance and raking by hand, herbicide application by knapsack sprayer	100 m²	38.33
As above but carrying out operations in clay or compacted gravel	100 m²	51.95

PREPARATION FOR PLANTING/TURFING

Item Excluding site overheads and profit	Unit	Total rate £
CULTIVATION BY MACHINE - cont'd		
Spread and lightly consolidate topsoil **brought from spoil heap not exceeding** **100 m, in layers not exceeding 150,** **grade to specified levels, remove stones** **over 25, treat with paraquat-diquat** **weedkiller, all by machine**		
100 mm thick	100 m²	66.26
150 mm thick	100 m²	99.73
300 mm thick	100 m²	199.46
450 mm thick	100 m²	298.82
Extra to above for imported topsoil PC **£12.48 m³ allowing for 20% settlement**		
100 mm thick	100 m²	149.80
150 mm thick	100 m²	224.70
300 mm thick	100 m²	449.40
450 mm thick	100 m²	674.10
500 mm thick	100 m²	749.00
600 mm thick	100 m²	898.80
750 mm thick	100 m²	1123.50
1.00 m thick	100 m²	1498.00
Extra for incorporating mushroom compost at 50 mm/m² into the top 150 mm of topsoil (compost delivered in 20 m³) loads		
manually spread, mechanically rotavated	100 m²	145.85
mechanically spread and rotavated	100 m²	98.59
Extra for incorporating manure at 50 mm/m² into the top 150 mm of topsoil loads		
manually spread, mechanically rotavated 20m³ loads	100 m²	215.35
mechanically spread and rotavated 60 m³ loads	100 m²	180.35

SEEDING AND TURFING

Item Excluding site overheads and profit	Unit	Total rate £
SEEDING AND TURFING		
Bring top 200 of topsoil to a fine tilth **using tractor drawn implements, remove** **stones over 25 by mechanical stone** **rake and bring to final tilth by harrow,** **apply pre-seeding fertilizer at 50 g/m²** **and work into top 50 during final** **cultivation, seed with certified grass** **seed in two operations, roll seedbed** **lightly after sowing**		
general amenity grass at 35 g/m²; BSH A3	100 m²	37.63
general amenity grass at 35 g/m²; Johnsons Taskmaster	100 m²	31.28
shaded areas; BSH A6 at 50 g/m²	100 m²	42.93
Motorway and road verges; Perryfields Pro 120 25-35 g/m²	100 m²	33.46
Bring top 200 of topsoil to a fine tilth **using pedestrian operated rotavator,** **remove stones over 25 mm, apply** **pre-seeding fertilizer at 50 g/m² and** **work into top 50 during final hand** **cultivation, seed with certified grass** **seed in two operations, rake and roll** **seedbed lightly after sowing**		
general amenity grass at 35 g/m²; BSH A3	100 m²	102.49
general amenity grass at 35 g/m²; Johnsons Taskmaster	100 m²	96.14
shaded areas; BSH A6 at 50 g/m²	100 m²	107.79
Motorway and road verges; Perryfields Pro 120 25-35 g/m²	100 m²	98.32
Extra to above for using imported topsoil spread by machine		
100 minimum depth	100 m²	237.00
150 minimum depth	100 m²	355.00
Extra for slopes over 30	-	50%
Extra to above for using imported topsoil spread by hand; maximum distance for transporting soil 100 m		
100 minimum depth	m²	583.07
150 minimum depth	m²	874.61
Extra for slopes over 30	-	50%
Extra for mechanically screening top 25 of topsoil through 6 mm screen and spreading on seedbed, debris carted to dump on site not exceeding 100 m	m³	5.65

Approximate Estimates

SEEDING AND TURFING

Item Excluding site overheads and profit	Unit	Total rate £
SEEDING AND TURFING - cont'd		
Bring top 200 of topsoil to a fine tilth, remove stones over 50, apply pre-emergent weedkiller in accordance with manufacturer's instructions, apply pre-seeding fertilizer at 50 g/m² and work into top 50 of topsoil during final cultivation, seed with certified grass seed in two operations, harrow and roll seedbed lightly after sowing		
Apply pesticide at 100 g/m², level with mechanical lute		
outfield grass at 350 kg/ha - Perryfields Pro 40	ha	8796.00
outfield grass at 350 kg/ha - Perryfields Pro 70	ha	8741.00
sportsfield grass at 300 kg/ha - Johnsons Sportsmaster	ha	8633.00
Areas not requiring pesticide or mechanical operations		
low maintenance grass at 350 kg/ha	ha	4526.00
verge mixture grass at 150 kg/ha	ha	3853.00
Extra for wild flora mixture at 30 kg/ha		
BSH WSF 75kg /ha	ha	3225.00
Extra for slopes over 30	-	50%
Cut existing turf to 1.0 x 1.0 x 0.5 m turves, roll up and move to stack not exceeding 100 m		
by pedestrian operated machine, roll up and stack by hand	100 m²	70.19
all works by hand	100 m²	203.13
Extra for boxing and cutting turves	100 m²	3.63
Bring top 200 of topsoil to a fine tilth in 2 passes, remove stones over 25 by and bring to final tilth, apply pre-seeding fertilizer at 50 g/m² and work into top 50 during final cultivation, roll turf bed lightly		
using tractor drawn implements and mechanical stone rake	m²	0.25
cultivation by pedestrian rotavator, all other operations by hand	m²	0.61
As above but bring turf from stack not exceeding 100 m, lay turves to stretcher bond using plank barrow runs, firm turves using wooden turf beater		
using tractor drawn implements and mechanical stone rake	m²	1.85
cultivation by pedestrian rotavator, all other operations by hand	m²	2.06
As above but including imported turf; Rolawn Medallion		
using tractor drawn implements and mechanical stone rake	m²	3.75
cultivation by pedestrian rotavator, all other operations by hand	m²	3.96
Extra over to all of the above for watering on two occasions and carrying out initial cut		
by ride on triple mower	100 m²	1.12
by pedestrian mower	100 m²	4.72
by pedestrian mower; box cutting	100 m²	5.28
Extra for using imported topsoil spread and graded by machine		
25 minimum depth	m²	0.67
75 minimum depth	m²	1.78
100 minimum depth	m²	2.37
150 minimum depth	m²	3.55

SEEDING AND TURFING

Item Excluding site overheads and profit	Unit	Total rate £
Extra for using imported topsoil spread and graded by hand; distance of barrow run 25 m		
25 minimum depth	m²	1.55
75 minimum depth	m²	3.86
100 minimum depth	m²	4.90
150 minimum depth	m²	7.34
Extra for work on slopes over 30 including pegging with 200 galvanized wire pins	m²	1.64
Inturf Big Roll **Supply, deliver in one consignment, fully** **prepare the area and install in Big Roll** **format Inturf 553; a turfgrass comprising** **dwarf perennial ryegrass, smooth stalked** **rneadowgrass and fescues; installation** **by tracked machine**		
Preparation by tractor drawn rotavator	m²	3.02
cultivation by pedestrian rotavator, all other operations by hand	m²	3.42
Erosion control On ground previously cultivated, bring area to level, treat with herbicide; lay 20 mm thick open texture erosion control mat with 100 mm laps, fixed with 8 x 400 mm steel pegs at 1.0 m centres; sow with low maintenance grass suitable for erosion control on slopes at 35 g/m²; spread imported topsoil 25 mm thick and fertilizer at 35 g/m², water lightly using sprinklers on two occasions	m²	5.69
As above but hand-watering by hose pipe maximum distance from mains supply 50 m	m²	5.77

AFTERCARE

Item Excluding site overheads and profit	Unit	Total rate £
AFTERCARE MAINTENANCE OF GRASSED AREAS		
Maintenance executed as part of a **landscape construction contract** **Grass cutting** Grass cutting; fine turf, using pedestrian guided machinery; arisings boxed and disposed of off site		
per occasion	100 m²	5.15
per annum 26 cuts	m²	1.34
per annum 18 cuts	m²	0.96
Grass cutting; standard turf, using self propelled 3 gang machinery		
per occasion	100 m²	0.38
per annum 26 cuts	m²	0.10
per annum 18 cuts	m²	0.07
Maintenance for one year; recreation areas, parks, amenity grass areas, using tractor drawn machinery:		
per occasion	ha	26.92
per annum 26 cuts	ha	699.92
per annum 18 cuts	ha	484.56
Aeration of turfed areas Aerate ground with spiked aerator, apply weedkiller once, apply spring/summer fertilizer once, apply autumn/winter fertilizer once, cut grass, 16 cuts, sweep up leaves twice		
as part of a landscape contract, defects liability	ha	1996.72
as part of a long term maintenance contract	ha	1564.00
LANDSCAPE MAINTENANCE		
Vegetation Control; Native Planting - **Roadside Railway or Forestry planted** **areas** **Post planting maintenance; control of** **weeds and grass; herbicide spray** **applications; maintain weed free circles** **1.00 m diameter to planting less than 5** **years old in roadside, rail or forestry** **planting environments and the like; strim** **grass to 50-75 mm; prices per occasion** **(3 applications of each operation** **normally required)** Knapsack spray application; Glyphosate; planting at		
1.50 mm centres	ha	1002.37
1.75 mm centres	ha	877.11
2.00 mm centres	ha	860.12
Maintain planted areas; control of weeds **and grass; maintain weed free circles** **1.00 m diameter to planting less than 5** **years old in roadside, rail or forestry** **planting environments and the like; strim** **surrounding grass to 50-75 mm; prices** **per occasion (3 applications of each** **operation normally required)** Herbicide spray applications; CDA (Controlled droplet application) Glyphosate and strimming; plants planted at the following centres		
1.50 mm centres	ha	903.69
1.75 mm centres	ha	877.11
2.00 mm centres	ha	860.12

AFTERCARE

Item Excluding site overheads and profit	Unit	Total rate £
Post planting maintenance; control of **weeds and grass; herbicide spray** **applications; CDA (Controlled droplet** **application); "Xanadu"Glyphosate/** **diuron; maintain weed free circles 1.00** **m diameter to planting less than 5 years** **old in roadside, rail or forestry planting** **environments and the like; strim grass to** **50-75 mm; prices per occasion (1.5** **applications of herbicide and 3 strim** **operations normally required)**		
Plants planted at the following centres		
1.50 mm centres	ha	977.08
1.75 mm centres	ha	930.88
2.00 mm centres	ha	901.41
Ornamental Shrub beds **Hand weed ornamental shrub bed during** **the growing season; planting less than** **2 years old**		
Mulched beds; weekly visits; planting centres		
600 ccs	100 m²	9.38
400 ccs	100 m²	11.25
300 ccs	100 m²	15.00
ground covers	100 m²	18.75
Mulched beds; monthly visits; planting centres		
600 ccs	100 m²	14.06
400 ccs	100 m²	18.75
300 ccs	100 m²	22.50
ground covers	100 m²	28.13
Non - mulched beds; weekly visits; planting centres		
600 ccs	100 m²	14.06
400 ccs	100 m²	15.00
300 ccs	100 m²	28.13
ground covers	100 m²	37.50
Non - mulched beds; monthly visits; planting centres		
600 ccs	100 m²	18.75
400 ccs	100 m²	22.50
300 ccs	100 m²	28.13
ground covers	100 m²	46.88
Re-mulch planting bed at the start of the **planting season; top up mulch 25 mm** **thick; Melcourt Ltd**		
Larger areas maximum distance 25 m; 80 m³ loads		
Ornamental bark mulch	100 m²	126.00
Melcourt Bark nuggets	100 m²	121.75
Amenity Bark	100 m²	89.88
Forest biomulch	100 m²	82.25
Smaller areas; maximum distance 25 m; 25 m³ loads		
Ornamental bark mulch	100 m²	158.50
Melcourt Bark nuggets	100 m²	154.25
Amenity Bark	100 m²	122.38
Forest biomulch	100 m²	114.75

PLANTING

Item Excluding site overheads and profit	Unit	Total rate £
HEDGE PLANTING		
Excavate trench for hedge 300 wide x 450 deep by machine, deposit spoil alongside, and plant hedging plants in single row at 200 centres, and backfill with excavated material incorporating organic manure at 1 m³ per 5 m³, and carry out initial cut, including delivery of plants from nursery		
bare root hedging plants		
PC - £0.30	100 m	617.06
PC - £0.60	100 m	767.06
PC - £1.00	100 m	967.06
PC - £1.20	100 m	1067.06
PC - £1.50	100 m	1217.06
As above but two rows of hedging plants at 300 centres staggered rows		
PC - £0.30	100 m	802.06
PC - £0.60	100 m	1002.06
PC - £1.00	100 m	1268.72
PC - £1.50	100 m	1602.06
As above but all by hand		
two rows of hedging plants at 300 centres staggered rows		
PC - £0.30	100 m	1009.00
PC - £0.60	100 m	1209.00
PC - £1.00	100 m	1475.66
PC - £1.50	100 m	1809.00
TREE PLANTING		
Excavate tree pit by hand, fork over bottom of pit, plant tree with roots well spread out, backfill with excavated material, incorporating organic manure at 1 m³ per 3 m³ of soil, one tree stake and two ties; tree pits square in sizes shown		
Light standard bare root tree in pit; PC £10.00		
600 x 600 deep	each	40.56
900 x 900 deep	each	54.18
Standard tree bare root tree in pit; PC £13.95		
600 x 600 deep	each	46.37
900 x 600 deep	each	55.39
900 x 900 deep	each	66.25
Standard root balled tree in pit; PC £16.00		
600 x 600 deep	each	57.87
900 x 600 deep	each	54.27
900 x 900 deep	each	77.75
Selected standard bare root tree in pit; PC £18.80		
900 x 900 deep	each	76.42
1.00 m x 1.00 m deep	each	86.92
Selected standard root ball tree in pit; PC £30.00		
900 x 900 deep	each	96.05
1.00 m x 1.00 m deep	each	138.00
Heavy standard bare root tree in pit; PC £39.50		
900 x 900 deep	each	99.10
1.00 m x 1.00 m deep	each	114.23
1.20 m x 1.00 m deep	each	135.25
Heavy standard root ball tree in pit; PC £50.00		
900 x 900 deep	each	122.87
1.00 m x 1.00 m deep	each	138.00
1.20 m x 1.00 m deep	each	159.02

PLANTING

Item Excluding site overheads and profit	Unit	Total rate £
Extra heavy standard bare root tree in pit; PC £49.00		
1.00 m x 1.00 deep	each	127.42
1.20 m x 1.00 m deep	each	180.07
1.50 m x 1.00 m deep	each	187.97
Extra heavy standard root ball tree in pit; PC £67.00		
1.00 m x 1.00 deep	each	161.17
1.20 m x 1.00 m deep	each	213.82
1.50 m x 1.00 m deep	each	221.72
Excavate tree pit by machine, fork over **bottom of pit, plant tree with roots well** **spread out, backfill with excavated** **material, incorporating organic manure at** **1 m³ per 3 m³ of soil, one tree stake and** **two ties; tree pits square in sizes shown**		
Light standard bare root tree in pit; PC £10.00		
600 x 600 deep	each	34.02
900 x 900 deep	each	45.22
Standard tree bare root tree in pit; PC £13.95		
600 x 600 deep	each	39.83
900 x 600 deep	each	42.60
900 x 900 deep	each	47.91
Standard root balled tree in pit; PC £16.00		
600 x 600 deep	each	51.33
900 x 600 deep	each	54.10
900 x 900 deep	each	59.41
Selected standard bare root tree, in pit; PC £18.80		
900 x 900 deep	each	59.79
1.00 m x 1.00 m deep	each	70.53
Selected standard root ball tree in pit; PC £30.00		
900 x 900 deep	each	96.05
1.00 m x 1.00 m deep	each	121.61
Heavy standard bare root tree, in pit; PC £39.50		
900 x 900 deep	each	99.10
1.00 m x 1.00 m deep	each	86.93
1.20 m x 1.00 m deep	each	111.65
Heavy standard root ball tree, in pit; PC £50.00		
900 x 900 deep	each	122.87
1.00 m x 1.00 m deep	each	121.61
1.20 m x 1.00 m deep	each	135.42
Extra heavy standard bare root tree in pit; PC £49.00		
1.00 m x 1.00 deep	each	111.03
1.20 m x 1.00 m deep	each	124.84
1.50 m x 1.00 m deep	each	151.36
Extra heavy standard root ball tree in pit; PC £67.00		
1.00 m x 1.00 deep	each	144.78
1.20 m x 1.00 m deep	each	158.59
1.50 m x 1.00 m deep	each	185.11

PLANTING

Item Excluding site overheads and profit	Unit	Total rate £
SEMI MATURE TREE PLANTING		
Excavate tree pit deep by machine, fork over bottom of pit, plant rootballed tree Acer platanoides "Emerald Queen" using telehandler where necessary, backfill with excavated material, incorporating Melcourt Topgrow bark/manure mixture at 1 m³ per 3 m³ of soil, Platipus underground guying system; tree pits 1500 x 1500 x 1500 mm deep inclusive of Platimats; excavated material not backfilled to tree pit spread to surrounding area		
16 - 18 cm girth - £69.00	each	213.63
18 - 20 cm girth - £77.00	each	236.34
20 - 25 cm girth - £130.00	each	313.00
25 - 30 cm girth - £179.00	each	383.15
30 - 35 cm girth - £220.00	each	552.08
As above but tree pits 2.00 x 2.00 x 1.5 m deep		
40 - 45 cm girth - £715.00	each	1000.38
45 - 50 cm girth - £805.00	each	1161.78
55 - 60 cm girth - £1265.00	each	1558.70
67 - 70 cm girth - £1800.00	each	2066.86
75 - 80 cm girth - £2400.00	each	3209.22
Extra to the above for imported topsoil moved 25 m from tipping area and disposal off site of excavated material		
Tree pits 1500 x 1500 x1500 m deep		
16 - 18 cm girth	each	132.53
18 - 20 cm girth	each	126.05
20 - 25 cm girth	each	120.15
25 - 30 cm girth	each	115.00
30 - 35 cm girth	each	110.44
TREE PLANTING WITH MOBILE CRANES		
Excavate tree pit 1.50 x 1.50 x 1.00 m deep; supply and plant semi mature trees delivered in full loads; trees lifted by crane; inclusive of backfilling tree pit with imported topsoil, compost, fertilizers and underground guying using "Platipus" anchors		
Self Managed lift; local authority applications, health and safety, traffic management or road closures not included; tree size and distance of lift; 35 tonne crane		
25 - 30 cm; max 25 m distance	each	475.65
30 - 35 cm; max 25 m distance	each	628.83
35 - 40 cm; max 25 m distance	each	937.95
55 - 60 cm; max 15 m distance	each	1517.47
80 - 90 cm; max 10 m distance	each	4139.40
Managed lift; inclusive of all local authority applications, health and safety, traffic management or road closures all by crane hire company; tree size and distance of lift; 35 tonne crane		
25 - 30 cm; max 25 m distance	each	481.90
30 - 35 cm; max 25 m distance	each	640.42
35 - 40 cm; max 25 m distance	each	949.54
55 - 60 cm; max 15 m distance	each	1546.02
80 - 90 cm; max 10 m distance	each	4237.52

PLANTING

Item Excluding site overheads and profit	Unit	Total rate £
Self Managed lift; local authority applications, health and safety, traffic management or road closures not included; tree size and distance of lift; 80 tonne crane		
25 - 30 cm; max 40 m distance	each	480.82
30 - 35 cm; max 40 m distance	each	638.43
35 - 40 cm; max 40 m distance	each	951.26
55 - 60 cm; max 33 m distance	each	1541.10
80 - 90 cm; max 23 m distance	each	4220.64
Managed lift; inclusive of all local authority applications, health and safety, traffic management or road closures all by crane hire company; tree size and distance of lift; 35 tonne crane		
25 - 30 cm; max 40 m distance	each	488.09
30 - 35 cm; max 40 m distance	each	651.92
35 - 40 cm; max 40 m distance	each	966.83
55 - 60 cm; max 33 m distance	each	1574.32
80 - 90 cm; max 23 m distance	each	4334.82
SHRUBS GROUND COVERS AND BULBS		
Excavate planting holes on 250 x 250 mm x 300 deep to area previously ripped and rotavated; excavated material left alongside planting hole by mechanical auger		
250 mm centres (16 plants per m²)	m²	10.56
300 mm centres (11.11 plants per m²)	m²	7.33
400 mm centres (6.26 plants per m²)	m²	4.13
450 mm centres (4.93 plants per m²)	m²	3.25
500 mm centres (4 plants per m²)	m²	2.64
600 mm centres (2.77 plants per m²)	m²	1.83
750 mm centres (1.77 plants per m²)	m²	1.17
900 mm centres (1.23 plants per m²)	m²	0.81
1.00 m centres (1 plants per m²)	m²	0.66
1.50 m centres (0.44 plants per m²)	m²	0.29
As above but excavation by hand		
250 mm centres (16 plants per m²)	m²	10.56
300 mm centres (11.11 plants per m²)	m²	7.33
400 mm centres (6.26 plants per m²)	m²	4.13
450 mm centres (4.93 plants per m²)	m²	3.25
500 mm centres (4 plants per m²)	m²	2.64
600 mm centres (2.77 plants per m²)	m²	1.83
750 mm centres (1.77 plants per m²)	m²	1.17
900 mm centres (1.23 plants per m²)	m²	0.81
1.00 m centres (1 plants per m²)	m²	0.66
1.50 m centres (0.44 plants per m²)	m²	0.29

PLANTING

Item Excluding site overheads and profit	Unit	Total rate £
SHRUBS GROUND COVERS AND BULBS - cont'd		
Clear light vegetation from planting area and remove to dump on site, dig planting holes, plant whips with roots well spread out, backfill with excavated topsoil, including one 38 x 38 treated softwood stake, two tree ties, and mesh guard 1.20 m high; planting matrix 1.5 m x 1.5 m; allow for beating up once at 10% of original planting, cleaning and weeding round whips once, applying weedkiller once at 35 gm/m², applying fertilizer once at 35 gm/m², using the following mix of whips, bare rooted plant bare root plants average price 0.28 p each to a required matrix		
plant mix as above	100 m²	212.98
plant bare root plants average price 0.75 p each to a required matrix		
plant mix as above	100 m²	221.17
Cultivate and grade shrub bed, bring top 300 mm of topsoil to a fine tilth, incorporating Mushroom compost at 50 mm and Enmag slow release fertilizer; rake and bring to given levels, remove all stones and debris over 50 mm, dig planting holes average 300 x 300x 300 mm deep; supply and plant specified shrubs in quantities as shown below, backfill with excavated material as above; water to field capacity and mulch 50 mm bark chips 20-40 mm size; water and weed regularly for 12 months and replace failed plants Shrubs -3L PC £2.50; Ground covers - 9 cm PC £1.50		
100 % shrub area 300 centres		
300 mm centres	100 m²	5775.32
400 mm centres	100 m²	3442.75
500 mm centres	100 m²	2363.08
600 mm centres	100 m²	1776.57
100 % groundcovers		
200 mm centres	100 m²	7255.40
300 mm centres	100 m²	3469.55
400 mm centres	100 m²	2146.66
500 mm centres	100 m²	1533.58
Groundcover 30% / Shrubs 70% at the distances shown below		
200mm / 300 mm	100 m²	6154.17
300 mm / 400 mm	100 m²	3420.11
300 mm / 500 mm	100 m²	2683.11
400 mm / 500 mm	100 m²	2277.86
Groundcover 50% / Shrubs 50% at the distances shown below		
200mm / 300 mm	100 m²	6520.07
300 mm / 400 mm	100 m²	3451.65
300 mm / 500 mm	100 m²	2924.48
400 mm / 500 mm	100 m²	2249.58

PLANTING

Item Excluding site overheads and profit	Unit	Total rate £
Cultivate ground by machine and rake to level; plant bulbs as shown; bulbs PC 24.50/100		
15 bulbs per m²	100 m²	660.18
25 bulbs per m²	100 m²	1080.68
50 bulbs per m²	100 m²	2131.93
Cultivate ground by machine and rake to level; plant bulbs as shown; bulbs PC 12.90/100		
15 bulbs per m²	100 m²	458.73
25 bulbs per m²	100 m²	744.93
50 bulbs per m²	100 m²	1460.43
Form holes in grass areas and plant bulbs using bulb planter, backfill with organic manure and turf plug; bulbs PC £13.00/100		
15 bulbs per m²	100 m²	663.75
25 bulbs per m²	100 m²	1106.25
50 bulbs per m²	100 m²	2212.50
BEDDING		
Spray surface with glyphosate; lift and dispose of turf when herbicide action is complete; cultivate new area for bedding plants to 400 mm deep; spread compost 100 deep and chemical fertilizer "Enmag" and rake to fine tilth to receive new bedding plants; remove all arisings to skip		
Existing turf area		
Disposal to skip	100 m²	811.05
Disposal to compost area on site; distance 25 m	100 m²	809.43
Plant bedding to existing planting area; bedding planting PC £0.25 each		
Clear existing bedding; cultivate soil to 230 mm deep; incorporate compost 75 mm and rake to fine tilth; Collect bedding from nursery and plant at 100 mm ccs; irrigate on completion; maintain weekly for 12 weeks		
mass planted 100 mm ccs	m²	32.60
to patterns; 100 mm ccs	m²	35.72
mass planted 150 mm ccs	m²	18.71
to patterns; 150 mm ccs	m²	21.83
mass planted 200 mm ccs	m²	18.71
to patterns; 200 mm ccs	m²	21.83
Extra for watering by hand held hose pipe		
Flow rate 25 litres / minute		
10 litres / m²	100 m²	0.11
15 litres / m²	100 m²	0.17
20 litres / m²	100 m²	0.22
25 litres / m²	100 m²	0.28
Flow rate 40 litres / minute		
10 litres / m²	100 m²	0.07
15 litres / m²	100 m²	0.10
20 litres / m²	100 m²	0.14
25 litres / m²	100 m²	0.17

Approximate Estimates

PLANTING

Item Excluding site overheads and profit	Unit	Total rate £
PLANTING PLANTERS		
To brick planter, coat insides with 2		
coats RIW liquid asphaltic composition;		
fill with 50 mm shingle and cover with		
geofabric; fill with screened topsoil		
incorporating 25% Topgrow compost and		
Enmag		
Planters 1.00 m deep		
1.00 x 1.00	each	96.78
1.00 x 2.00	each	166.80
1.00 x 3.00	each	237.35
Planters 1.50 m deep		
1.00 x 1.00	each	144.88
1.00 x 2.00	each	218.43
1.00 x 3.00	each	355.44
Container planting; fill with 50 mm		
shingle and cover with geofabric; fill		
with screened topsoil incorporating 25%		
Topgrow compost and Enmag		
Planters 1.00 m deep		
400 x 400 x 400 mm deep	each	6.96
400 x 400 x 600 mm deep	each	9.06
1.00 x 400 wide x 400 deep	each	15.92
1.00 x 600 wide x 600 deep	each	25.53
1.00 x 100 wide x 400 deep	each	26.79
1.00 x 100 wide x 600 deep	each	39.11
1.00 x 100 wide x 1.00 deep	each	63.75
1.00 m diameter x 400 deep	each	21.25
1.00 m diameter x 1.00 m deep	each	50.81
2.00 m diameter x 1.00 m deep	each	197.87

FENCING

Item Excluding site overheads and profit	Unit	Total rate £
CHAIN LINK AND WIRE FENCING		
Chain link fencing; supply and erect **chain link fencing; form post holes and** **erect concrete posts and straining posts** **with struts at 50 m centres all set in** **1:3:6 concrete; fix line wires and** 3 mm galvanized wire 50 mm chainlink fencing		
900 mm high	m	22.97
1200 mm high	m	24.06
1800 mm high	m	28.59
2400 mm high	m	52.37
plastic coated 3.15 gauge galvanized wire mesh		
900 mm high	m	20.46
1200 mm high	m	22.20
1800 mm high	m	25.87
Extra for additional concrete straining posts with		
1 strut set in concrete		
900 high	each	61.78
1200 high	each	64.68
1400 high	each	71.68
1800 high	each	76.48
2400 high	each	99.45
Extra for additional concrete straining posts with		
2 struts set in concrete		
900 high	each	88.99
1200 high	each	92.00
1400 high	each	108.69
1800 high	each	113.51
2400 high	each	112.73
Chain link fencing 3 mm galvanized as **above but with mild steel angle posts** **and straining posts instead of concrete** **posts**		
900 mm high	m	12.41
1200 mm high	m	14.08
1800 mm high	m	21.00
2400 mm high	m	23.71
Extra for additional angle iron straining posts with		
1 strut set in concrete		
900 high	each	42.41
1200 high	each	59.32
1400 high	each	68.95
1800 high	each	78.40
2400 high	each	93.60
Extra for additional angle iron straining posts with		
2 struts set in concrete		
900 high	each	63.46
1200 high	each	87.32
1400 high	each	103.38
1800 high	each	125.69
2400 high	each	156.00
TIMBER FENCING		
Erect chestnut pale fencing, cleft **chestnut pales, two lines galvanized** **wire, galvanized tying wire, treated** **softwood posts at 3.0 m centres and** **straining posts and struts at 50 m** **centres driven into firm ground**		
900 high, posts 75 dia. x 1200 long	m	7.77
1200 high, posts 75 dia. x 1500 long	m	9.34

FENCING

Item Excluding site overheads and profit	Unit	Total rate £
TIMBER FENCING - cont'd		
Construct timber rail, horizontal hit and miss type, rails 150 x 25, posts 100 x 100 at 1.8 m centres, twice stained with coloured wood preservative, including excavation for posts and concreting into ground (C7P)		
In treated softwood		
1800 mm high	m	48.25
In primed softwood		
1800 mm high	m	54.88
Erect close boarded timber fence in treated softwood, pales 89 x 19 lapped, 152 x 25 gravel boards, two 76 x 38 rectangular rails		
Concrete posts 100 x 100 at 3.0 m centres set into ground in 1:3:6 concrete		
900 mm high	m	20.74
1350 mm high	m	26.96
1800 mm high	m	28.45
As above but with oak posts		
1350 mm high	m	26.17
1800 mm high	m	28.54
As above but with softwood posts		
1350 mm high	m	23.32
1650 mm high	m	24.31
1800 mm high	m	24.75
Erect post and rail fence, three horizontal rails 100 x 38, fixed with galvanized nails to 100 x 100 posts driven into firm ground		
In treated softwood		
1200 mm high	m	16.47
In oak or chestnut		
1200 mm high	m	31.34
Construct post and rail fence to BS 1722: Part 7 Type MPR 13/4 1300 mm high with five rails 87 x 38 mm; rails morticed into treated softwood posts 2100 75 x 150 mm at 3.0 m centres set 700 mm into firm ground with intermediate prick posts 1800 x 87 x 38		
1300 high	m	17.61
Construct cleft oak rail fence, with rails 300 mm minimum girth tennoned both ends; 125 x 100 mm treated softwood posts double mortised for rails, corner posts 125 x 125 mm, driven into firm ground at 2.5 m centres		
3 rails	m	20.81
4 rails	m	24.78

FENCING

Item Excluding site overheads and profit	Unit	Total rate £
DEER STOCK RABBIT FENCING		
Construct rabbit-stop fencing, erect galvanized wire netting, mesh 31, 900 above ground, 150 below ground turned out and buried, on 75 dia. treated timber posts 1.8 m long driven 700 into firm ground at 4.0 m centres, netting clipped to top and bottom straining wires 2.63 mm diameter, straining post 150 mm diameter, x 2.3 m long driven into firm ground at 50 m intervals		
turned in 150 mm	100 m	871.62
buried 150 mm	100 m	952.62
Deer fence		
Construct deer-stop fencing, erect 5 no. 4 dia. plain galvanized wires and 5 no. 2 ply galvanized barbed wires at 150 spacing, on 45 x 45 x 5mm angle iron posts 2.4 m long driven into firm ground at 3.0 m centres, driven into firm ground at 3.0 m centres, with timber droppers 25 x 38 x 1.0 m long at 1.5 m centres	100 m	1874.57
Forestry fencing		
Supply and erect forestry fencing of three lines of 3 mm plain galvanized wire tied to 1700 x 65 mm dia angle iron posts at 2750 m centres with 1850 x 100 mm dia straining posts and 1600 x 80 mm dia struts at 50.0 m centres driven into firm ground		
1800 high; 3 wires	100 m	1221.38
1800 high; 3 wires including cattle fencing	100 m	1512.38
CONCRETE FENCES		
Supply and erect precast concrete post and panel fence in 2 m bays, panels to be shiplap profile, aggregate faced one side, posts set 600 mm in ground in concrete		
1500 mm high	m	29.37
1800 mm high	m	34.25
2100 mm high	m	39.24
Extra over for exposed aggregate faced panels	m²	8.20

FENCING

Item Excluding site overheads and profit	Unit	Total rate £
SECURITY FENCING		
Supply and erect chainlink fence, 51 mm x 3 mm mesh, with line wires and stretcher bars bolted to concrete posts at 3.0 m centres and straining posts at 10 m centres; posts set in concrete 450 x 450 x 33% of height of post deep; fit straight extension arms of 45 x 45 x 5 mm steel angle with three lines of barbed wire and droppers; all metalwork to be factory hot- dip galvanized for painting on site		
900 high	m	26.98
1200 high	m	29.24
1800 high	m	40.36
As above but with PVC coated 3.15 mm mesh (diameter of wire 2.5 mm)		
900 high	m	24.47
1200 high	m	27.38
1800 high	m	37.64
Add to fences above for base of fence to be fixed with hairpin staples cast into concrete ground beam 1:3:6 site mixed concrete; mechanical excavation disposal to on site spoil heaps		
125 x 225 mm deep	m	5.15
Add to fences above for straight extension arms of 45 x 45 x 5 mm steel angle with three lines of barbed wire and droppers	m	3.60
Supply and erect palisade security fence Jacksons "Barbican" 2500 mm high with rectangular hollow section steel pales at 150 mm centres on three 50 x 50 x 6 mm rails; rails bolted to 80 x 60 mm posts set in concrete 450 x 450 x 750 mm deep at 2750 mm centres, tops of pales to pointes and set at 45 degree angle; all metalwork to be hot-dip factory galvanized for painting on site	m	85.87
Supply and erect single gate to match above complete with welded hinges and lock		
1000 mm wide	each	778.03
4.0 mm wide	each	836.32
8.00 mm wide	pair	1704.77
Supply and erect Orsogril proprietary welded steel mesh panel fencing on steel posts set 750 mm deep In concrete foundations 600 x 600 mm; supply and erect proprietary single gate 2.0 m wide to match fencing		
930 high	100 m	10586.00
1326 high	100 m	14653.00
1722 high	100 m	17622.00

FENCING

Item Excluding site overheads and profit	Unit	Total rate £
RAILINGS		
Erect ms bar railing of 19 mm balusters at 115 mm centres welded to ms top and bottom rails 40 x 10, bays 2.0 m long, bolted to 51 x 51 ms hollow section posts set in C15P concrete, all metal work galvanized after fabrication		
900 mm high	m	57.00
1200 mm high	m	78.39
1500 mm high	m	86.07
Supply and erect mild steel pedestrian guard rail Class A to BS 3049, panels 1000 mm high x 2000 mm wide with 150 mm toe space and 200 mm visibility gap at top, rails to be rectangular hollow sealed section 50 x 30 x 2.5 mm, vertical support 25 x 19 mm central between intermediate and top rail; posts to be set 300 mm into paving base; all components factory welded and factory primed for painting on site	m	72.57
BALLSTOP FENCING		
Supply and erect plastic coated 30 x 30 mm netting fixed to 60.3 diameter 12 mm solid bar lattice galvanized dual posts, top, middle and bottom rails with 3 horizontal rails on 60.3 mm dia. nylon coated tubular steel posts at 3.0 m centres and 60.3 mm dia. straining posts with struts at 50 m centres set 750 mm into FND2 concrete footings 300 x 300 x 600 mm deep; include framed chain link gate 900 x 1800 mm high to match complete with hinges and locking latch		
4500 high	100 m	7419.00
5000 high	100 m	8499.00
6000 high	100 m	9355.00
CATTLE GRID		
Cattle Grid Excavate pit 4.0 m x 3.0 m x 500, dispose of spoil on site, lay concrete base (C7P) 100 thick on 150 hardcore, excavate trench and lay 100 dia. clay agricultural drain outlet, form concrete sides to pit 150 thick (C15P) including all formwork, construct supporting walls one brick thick in engineering brick laid in cement mortar (1:3) at 400 mm centres, install cattle grid by H S Jackson & Son (Fencing), erect side panels on steel posts set in concrete (1:3:6), erect 2 nr warning signs standard DoT pattern		
3.66 x 2.55 m	each	2745.46
Erect stock gate, ms tubular field gate, diamond braced 1.8 m high hung on tubular steel posts set in concrete (C7P), complete with ironmongery, all galvanized		
width 3.00 m	each	255.41
width 4.20 m	each	276.35

FENCING

Item Excluding site overheads and profit	Unit	Total rate £
TRIP RAILS		
Steel trip rail Erect trip rail of galvanized steel tube 38 internal dia. with sleeved joint fixed to 38 dia. steel posts as above 700 long, set in C7P concrete at 1.20 m centres, metalwork primed and painted two coats metal preservative paint	m	115.48
Birdsmouth fencing; Timber 600 high 900 high	m m	17.56 18.46

STREET FURNITURE

Item Excluding site overheads and profit	Unit	Total rate £
BENCHES SEATS BOLLARDS		
Bollards and Access restriction Supply and install powder coated steel parking posts and bases to specified colour; fold down top locking type, complete with keys and instructions; posts and bases set in concrete footing 200 x 200 x 300 mm deep;	each	171.74
supply and install 10 no. Cast iron "Doric" bollards 920 mm high above ground x 170 diameter, bedded in concrete base 400 dia, x 400 mm deep	10 nr	1321.90
Benches and seating In grassed area excavate for base 2500 x 1575 mm and lay 100 mm hardcore, 100 mm concrete, brick pavers in stack bond bedded in 25 mm cement:lime:sand mortar; supply and fix where shown on drawing proprietary seat, hardwood slats on black powder coated steel frame, bolted down with 4 no. 24 x 90 mm recessed hex-head stainless steel anchor bolts set into concrete	set	1309.33
Litter bins Supply and fix to wall where shown on drawing wall-mounted proprietary litter bins powder coated timber slatted, open top with metal liner, fixed with 2 no. stainless steel recessed hex-head anchor bolts 16 x 60 mm	each	489.98
Cycle stand Supply and fix cycle stand 1250 m long x 550 mm long of 60.3 mm black powder coated hollow steel sections to BS 4948: Part 2, one-piece with rounded top corners, set 250 mm into paving	each	283.13

Street planters
Supply and locate in position precast concrete
planters, fill with topsoil placed over 50 mm
shingle and terram; plant with assorted 5 litre and
3 litre shrubs at to provide instant effect
  970 diameter x 470 high; white exposed
  aggregate finish
	each	37.19

DRAINAGE

Item Excluding site overheads and profit	Unit	Total rate £
AGRICULTURAL DRAINAGE		
Excavate and form ditch and bank with **45 degree sides in light to medium soils;** **all widths taken at bottom of ditch**		
300 wide x 600 deep	100 m	102.67
600 wide x 900 deep	100 m	132.82
1.20 m wide x 900 deep	100 m	735.00
1.50 m wide x 1.20 m deep	100 m	1237.50
Clear and bottom existing ditch average **1.50 m deep, trim back vegetation and** **remove debris to licensed tip not** **exceeding 13 km, lay jointed concrete** **pipes; to BS 5911 pt.1 class S; including** **bedding, haunching and topping with** **150 mm concrete; 11.50 N/mm² - 40 mm** **aggregate; back fill with approved spoil** **from site**		
pipes 300 dia.	100 m	5378.31
pipes 450 dia.	100 m	6877.07
pipes 600 dia.	100 m	8837.54
Clay Land Drain Excavate trench by excavator to 450 deep, lay 100 vitrified clay drain with butt joints, bedding Class B, backfill with excavated material screened to remove stones over 40, backfill to be laid in layers not exceeding 150, top with 150 mm topsoil remove surplus material to approved dump on site not exceeding 100m, final level of fill to allow for settlement	100 m	1501.18
SUB SOIL DRAINAGE. BY MACHINE		
Main drain; remove 150 topsoil and **deposit alongside trench, excavate drain** **trench by machine and lay flexible** **perforated drain, lay bed of gravel** **rejects 100 mm; backfill with gravel** **rejects or similar to within 150 of finished** **ground level, complete fill with topsoil,** **remove surplus spoil to approved dump** **on site**		
Main drain 160 mm in supplied in 35 m lengths		
450 mm deep	100 m	752.11
600 mm deep	100 m	849.12
900 mm deep	100 m	1236.06
Extra for couplings	each	2.74
As above but with 100 mm main drain supplied in 100 m lengths		
450 mm deep	100 m	526.67
600 mm deep	100 m	618.86
900 mm deep	100 m	778.66
Extra for couplings	each	2.19
Laterals to mains above; herringbone **pattern; excavation and back filling as** **above; inclusive of connecting lateral to** **main drain**		
160 mm pipe to 450 mm deep trench		
laterals at 1.0 m centres	100 m²	752.11
laterals at 2.0 m centres	100 m²	376.07
laterals at 3.0 m centres	100 m²	248.19
laterals at 5.0 m centres	100 m²	150.42
laterals at 10.0 m centres	100 m²	75.21

DRAINAGE

Item Excluding site overheads and profit	Unit	Total rate £
160 mm pipe to 600 mm deep trench		
laterals at 1.0 m centres	100 m²	849.12
laterals at 2.0 m centres	100 m²	424.57
laterals at 3.0 m centres	100 m²	280.21
laterals at 5.0 m centres	100 m²	169.82
laterals at 10.0 m centres	100 m²	84.91
160 mm pipe to 900 mm deep trench		
laterals at 1.0 m centres	100 m²	1236.06
laterals at 2.0 m centres	100 m²	618.03
laterals at 3.0 m centres	100 m²	407.90
laterals at 5.0 m centres	100 m²	247.22
laterals at 10.0 m centres	100 m²	123.60
Extra for 160/160 mm couplings connecting laterals to main drain		
laterals at 1.0 m centres	10 m	108.80
laterals at 2.0 m centres	10 m	54.40
laterals at 3.0 m centres	10 m	36.23
laterals at 5.0 m centres	10 m	21.76
laterals at 10.0 m centres	10 m	10.88
100 mm pipe to 450 mm deep trench		
laterals at 1.0 m centres	100 m²	526.67
laterals at 2.0 m centres	100 m²	263.34
laterals at 3.0 m centres	100 m²	173.80
laterals at 5.0 m centres	100 m²	105.34
laterals at 10.0 m centres	100 m²	52.67
100 mm pipe to 600 mm deep trench		
laterals at 1.0 m centres	100 m²	618.86
laterals at 2.0 m centres	100 m²	309.43
laterals at 3.0 m centres	100 m²	204.23
laterals at 5.0 m centres	100 m²	123.77
laterals at 10.0 m centres	100 m²	61.88
100 mm pipe to 900 mm deep trench		
laterals at 1.0 m centres	100 m²	778.66
laterals at 2.0 m centres	100 m²	389.33
laterals at 3.0 m centres	100 m²	256.96
laterals at 5.0 m centres	100 m²	155.73
laterals at 10.0 m centres	100 m²	77.87
80 mm pipe to 450 mm deep trench		
laterals at 1.0 m centres	100 m²	462.29
laterals at 2.0 m centres	100 m²	231.15
laterals at 3.0 m centres	100 m²	152.55
laterals at 5.0 m centres	100 m²	92.47
laterals at 10.0 m centres	100 m²	46.23
80 mm pipe to 600 mm deep trench		
laterals at 1.0 m centres	100 m²	554.48
laterals at 2.0 m centres	100 m²	277.24
laterals at 3.0 m centres	100 m²	182.98
laterals at 5.0 m centres	100 m²	110.90
laterals at 10.0 m centres	100 m²	55.44
80 mm pipe to 900 mm deep trench		
laterals at 1.0 m centres	100 m²	714.28
laterals at 2.0 m centres	100 m²	357.14
laterals at 3.0 m centres	100 m²	235.71
laterals at 5.0 m centres	100 m²	142.86
laterals at 10.0 m centres	100 m²	71.43
Extra for 100/80 mm junctions connecting laterals to main drain		
laterals at 1.0 m centres	10 m	46.10
laterals at 2.0 m centres	10 m	23.05
laterals at 3.0 m centres	10 m	15.35
laterals at 5.0 m centres	10m	9.22
laterals at 10.0 m centres	10m	4.61

DRAINAGE

Item Excluding site overheads and profit	Unit	Total rate £
SUB SOIL DRAINAGE; BY HAND		
Main drain; remove 150 topsoil and **deposit alongside trench, excavate drain** **trench by machine and lay flexible** **perforated drain, lay bed of gravel** **rejects 100 mm; backfill with gravel** **rejects or similar to within 150 of finished** **ground level, complete fill with topsoil,** **remove surplus spoil to approved dump** **on site**		
Main drain 160 mm in supplied in 35 m lengths		
450 mm deep	100 m	1827.83
600 mm deep	100 m	2251.14
900 mm deep	100 m	2887.07
Extra for couplings	each	2.74
As above but with 100 mm main drain supplied in 100 m lengths		
450 mm deep	100 m	1181.64
600 mm deep	100 m	1472.34
900 mm deep	100 m	2034.01
Extra for couplings	each	2.19
Laterals to mains above; herringbone **pattern; excavation and back filling as** **above; inclusive of connecting lateral to** **main drain**		
160 mm pipe to 450 mm deep trench		
laterals at 1.0 m centres	100 m²	1827.83
laterals at 2.0 m centres	100 m²	913.92
laterals at 3.0 m centres	100 m²	603.18
laterals at 5.0 m centres	100 m²	365.56
laterals at 10.0 m centres	100 m²	182.78
160 mm pipe to 600 mm deep trench		
laterals at 1.0 m centres	100 m²	2251.14
laterals at 2.0 m centres	100 m²	1125.57
laterals at 3.0 m centres	100 m²	742.87
laterals at 5.0 m centres	100 m²	450.23
laterals at 10.0 m centres	100 m²	225.11
160 mm pipe to 900 mm deep trench		
laterals at 1.0 m centres	100 m²	2887.07
laterals at 2.0 m centres	100 m²	1443.53
laterals at 3.0 m centres	100 m²	952.74
laterals at 5.0 m centres	100 m²	577.42
laterals at 10.0 m centres	100 m²	288.71
Extra for 160/160 mm couplings connecting laterals to main drain		
laterals at 1.0 m centres	10 m	108.80
laterals at 2.0 m centres	10 m	54.40
laterals at 3.0 m centres	10 m	36.23
laterals at 5.0 m centres	10 m	21.76
laterals at 10.0 m centres	10 m	10.88
100 mm pipe to 450 mm deep trench		
laterals at 1.0 m centres	100 m²	1181.64
laterals at 2.0 m centres	100 m²	590.81
laterals at 3.0 m centres	100 m²	389.94
laterals at 5.0 m centres	100 m²	236.32
laterals at 10 m centres	100 m²	118.17
100 mm pipe to 600 mm deep trench		
laterals at 1.0 m centres	100 m²	1472.34
laterals at 2.0 m centres	100 m²	736.17
laterals at 3.0 m centres	100 m²	485.87
laterals at 5.0 m centres	100 m²	294.47
laterals at 10.0 m centres	100 m²	147.24

DRAINAGE

Item Excluding site overheads and profit	Unit	Total rate £
100 mm pipe to 900 mm deep trench		
laterals at 1.0 m centres	100 m²	2034.01
laterals at 2.0 m centres	100 m²	1017.00
laterals at 3.0 m centres	100 m²	671.22
laterals at 5.0 m centres	100 m²	406.80
laterals at 10.0 m centres	100 m²	203.40
80 mm pipe to 450 mm deep trench		
laterals at 1.0 m centres	100 m²	1117.26
laterals at 2.0 m centres	100 m²	558.62
laterals at 3.0 m centres	100 m²	368.69
laterals at 5.0 m centres	100 m²	223.45
laterals at 10.0 m centres	100 m²	111.73
80 mm pipe to 600 mm deep trench		
laterals at 1.0 m centres	100 m²	1407.96
laterals at 2.0 m centres	100 m²	703.98
laterals at 3.0 m centres	100 m²	464.62
laterals at 5.0 m centres	100 m²	281.60
laterals at 10.0 m centres	100 m²	140.80
80 mm pipe to 900 mm deep trench		
laterals at 1.0 m centres	100 m²	1969.63
laterals at 2.0 m centres	100 m²	984.81
laterals at 3.0 m centres	100 m²	649.97
laterals at 5.0 m centres	100 m²	393.93
laterals at 10.0 m centres	100 m²	196.96
Extra for 100/80 mm couplings connecting laterals to main drain		
laterals at 1.0 m centres	10 m	46.10
laterals at 2.0 m centres	10 m	23.05
laterals at 3.0 m centres	10 m	15.35
laterals at 5.0 m centres	10 m	9.22
laterals at 10.0 m centres	10 m	4.61
DRAIN BY MOLE PLOUGH		
50 dia. mole at 450 deep		
1.20 m centres	100 m²	11.58
1.50 m centres	100 m²	10.64
2.00 m centres	100 m²	9.68
2.50 m centres	100 m²	8.60
3.00 m centres	100 m²	7.52
75 dia. mole at 600 deep		
1.20 m centres	100 m²	13.83
1.50 m centres	100 m²	12.60
2.00 m centres	100 m²	12.17
2.50 m centres	100 m²	10.17
3.00 m centres	100 m²	10.64
SURFACE WATER DRAINAGE		
Clay Gully		
Excavate hole, supply and set in concrete (C10P) vitrified clay trapped mud (dirt) gully with rodding eye to BS 65, complete with galvanized bucket and cast iron hinged locking grate and frame, flexible joint to pipe; connect to drainage system with flexible joints	each	166.21

DRAINAGE

Item Excluding site overheads and profit	Unit	Total rate £
SURFACE WATER DRAINAGE – cont'd		
Concrete Road Gully Excavate hole and lay 100 concrete base (1:3:6) 150 x150 to suit given invert level of drain, supply and connect trapped precast concrete road gully 450 dia x 1.07 m deep with 160 outlet to BS 5911, set in concrete surround, connect to vitrified clay drainage system with flexible joints, supply and fix straight bar dished top cast iron grating and frame, bedded in cement:sand mortar (1:3)	each	323.53
Gullies PVC-u Excavate hole and lay 100 concrete (C20P) base 150 x 150 to suit given invert level of drain, connect to drainage system, backfill with DoT Type 1 granular fill; install gully; complete with cast iron grate and frame		
trapped PVC-U gully	each	76.92
bottle gully 228 x 228 x 317 deep	each	90.47
bottle gully 228 x 228 x 642 deep	each	106.48
yard gully 300 diameter x 600 deep	each	333.50
Inspection Chambers; brick manhole; **excavate pit for inspection chamber** **including earthwork support and disposal** **of spoil to dump on site not exceeding** **100m, lay concrete (1:2:4) base 1500** **dia. x 200 thick, 110 vitrified clay** **channels, benching in concrete (1:3:6)** **allowing one outlet and two inlets for** **110 dia. pipe, construct inspection** **chamber 1 brick thick walls of** **engineering brick Class B, backfill with** **excavated material, complete with 2 no.** **cast iron step irons** 1200 x 1200 x 1200 mm		
cover slab of precast concrete	each	863.86
Access cover; Group 2; 600 x 450 mm	each	859.67
1200 x 1200 x 1500 mm		
Access cover; Group 2; 600 x 450 mm	each	1029.96
Recessed cover 5 tonne load; 600 x 450 mm; filled with block paviors	each	1192.86
Cast iron Inspection chamber; **excavate pit for inspection chamber,** **supply and install cast iron inspection** **chamber unit, bedded in Type 1 granular** **material**		
650 mm deep; 100 x 150 mm; one branch each side	each	247.41
Interceptor Trap Supply and fix vitrified clay interceptor trap 100 mm inlet; 100 mm outlet; to manhole bedded in 10 aggregate concrete (C10P), complete with brass stopper and chain	each	134.33
Connect to drainage system with flexible joints	each	34.46
Pipe Laying Excavate trench by excavator 600 deep, lay Type 2 bedding, backfill to 150 mm above pipe with gravel rejects, lay non woven geofabric and fill with topsoil to ground level		
160 PVC-u drainpipe	100 m	3231.91
110 PVC-u drainpipe	100 m	1806.91
150 vitrified clay	100 m	1984.91
100 vitrified clay	100 m	1394.91

Item Excluding site overheads and profit	Unit	Total rate £
SOAKAWAYS		
Construct soakaway from perforated concrete rings; excavation, casting insitu concrete ring beam base; and filling with gravel 250 mm deep; placing perforated concrete rings; surrounding with geofabric and backfilling with 250 mm granular surround and excavated material; step irons and cover slab; inclusive of all earthwork retention and disposal offsite of surplus material		
900 mm diameter		
1.00 m deep	nr	**437.21**
2.00 m deep	nr	**723.55**
1200 mm diameter		
1.00 m deep	nr	**607.53**
2.00 m deep	nr	**1003.78**
2400 mm diameter		
1.00 m deep	nr	**1440.20**
2.00 m deep	nr	**2409.32**

IRRIGATION

Item Excluding site overheads and profit	Unit	Total rate £
LANDSCAPE IRRIGATION		
Automatic irrigation; Quality Irrigation Ltd		
Landscape Irrigation		
Large garden consisting of 24 stations; 7000 m²		
irrigated area		
Turf only	nr	20875.00
70/30 % - turf/shrub beds	nr	21200.00
Medium sized garden consisting of 12 stations;		
3500 m² irrigated area		
Turf only	nr	13230.60
70/30 % - turf/shrub beds	nr	15175.60
Smaller garden consisting of 6 stations of irrigated		
area		
Turf only	nr	7200.00
70/30 % - turf/shrub beds	nr	7726.00
50/50 % - turf/shrub beds	nr	7880.00
Leaky Pipe Irrigation		
Works by machine; main supply and		
connection to laterals; excavate trench		
for main or ring main 450 mm deep;		
supply and lay pipe; backfill and lightly		
compact trench		
20 mm LDPE	100 m	415.80
16 mm LDPE	100 m	386.80
Works by hand; main supply and		
connection to laterals; excavate trench		
for main or ring main 450 mm deep;		
supply and lay pipe; backfill and lightly		
compact trench		
20 mm LDPE	100 m	1330.77
16 mm LDPE	100 m	1301.77
Turf area irrigation; laterals to mains; to		
cultivated soil; excavate trench 150		
deep using hoe or mattock; lay moisture		
leaking pipe laid 150 mm below ground		
at centres of 350 mm		
low leak	100 m²	515.00
high leak	100 m²	466.00
Landscape area irrigation; laterals to		
mains; moisture leaking pipe laid to the		
surface of irrigated areas at 600 mm		
centres		
low leak	100 m²	265.00
high leak	100 m²	236.00
Landscape area irrigation; laterals to		
mains; moisture leaking pipe laid to the		
surface of irrigated areas at 900 mm		
centres		
low leak	100 m²	177.00
high leak	100 m²	158.00
Multistation controller	each	369.00
Solenoid valves connected to automatic		
controller	each	407.38

WATER FEATURES

Item Excluding site overheads and profit	Unit	Total rate £
LAKES AND PONDS		
FAIRWATER LTD		
Excavate for small pond or lake		
maximum depth 1.00 m; remove arisings		
off site; grade and trim to shape; lay 75		
mm sharp sand; line with 0.75 mm butyl		
liner 75 mm sharp sand and geofabric;		
cover over with sifted topsoil; anchor		
liner to anchor trench; install balancing		
tank and automatic top-up system		
Pond or lake of organic shape		
100 m²; perimeter 50 m	each	5364.81
250 m²; perimeter 90 m	each	8062.60
500 m²; perimeter 130 m	each	19609.45
1000 m²; perimeter 175 m	each	39505.10
Excavate for lake average depth 1.0 m,		
allow for bringing to specified levels,		
reserve topsoil, remove spoil to		
approved dump on site, remove all		
stones and debris over 75, lay polythene		
sheet including welding all joints and		
seams by specialist; screen and replace		
topsoil 200 mm thick		
Prices are for lakes of regular shape		
500 micron sheet	1000 m	8905.80
1000 micron sheet	1000 m	10273.80
Extra for removing spoil to tip	m³	16.88
Extra for 25 sand blinding to lake bed	100 m²	104.14
Extra for screening topsoil	m²	0.94
Extra for spreading imported topsoil	100 m²	515.24
Plant aquatic plants in lake topsoil		
Aponogeton distachyum - £210.00/100	100	350.00
Acorus calamus - £169.00/100	100	350.00
Butomus umbellatus - £169.00/100	100	350.00
Typha latifolia - £169.00/100	100	350.00
Nymphaea - £425.00/100	100	350.00
Formal water features		
Excavate and construct water feature of		
regular shape; lay 100 hardcore base		
and 150 concrete 1:2:4 site mixed; line		
base and vertical face with butyl liner		
0.75 micron and construct vertical sides		
of reinforced blockwork; rendering 2		
coats; anchor the liner behind		
blockwork; install pumps balancing		
tanks and all connections to mains		
supply		
1.00 x 1.00 x 1.00 m deep	nr	2159.67
2.00 x 1.00 x 1.00 m deep	nr	2742.97

TIMBER DECKING

Item Excluding site overheads and profit	Unit	Total rate £
WYCKHAM-BLACKWELL LTD		
Timber decking; support structure of **timber joists for decking laid on blinded** **base (measured separately)**		
joists 50 x 150 mm	10 m²	281.00
joists 50 x 200 mm	10 m²	319.00
joists 50 x 250 mm	10 m²	299.70
As above but inclusive of timber decking boards in yellow cedar 142 mm wide x 42 mm thick		
joists 50 x 150 mm	10 m²	711.50
joists 50 x 200 mm	10 m²	749.50
joists 50 x 250 mm	10 m²	730.20
As above but boards 141 x 26 mm thick		
joists 50 x 150 mm	10 m²	555.10
joists 50 x 200 mm	10 m²	319.00
joists 50 x 250 mm	10 m²	299.70
As above but decking boards in red cedar 131 mm wide x 42 mm thick		
joists 50 x 150 mm	10 m²	669.80
joists 50 x 200 mm	10 m²	707.80
joists 50 x 250 mm	10 m²	688.50
Add to all of the above for handrails fixed to posts 100 x 100 x 1370 high		
square balusters at 100 centres	m	69.97
square balusters at 300 centres	m	43.67
turned balusters at 100 centres	m	89.77
turned balusters at 300 centres	m	50.20

LIGHTING

Item Excluding site overheads and profit	Unit	Total rate £
LIGHTING		
Excavate trench 600 mm deep; supply **and install 3 core 2.5 mm armoured** **cable and backfill with excavated** **material**		
by machine	m	6.74
by hand	m	30.01
Precinct Lighting Supply and install Marlin Sphereline lighting units at locations shown on drawings; 3.0 m tapered aluminium columns to BS 5649, set 300 mm into paving, 500 mm dia. clear acrylic globe luminaires to IP54 complete with control gear and electrical supply, installed all in accordance with IEE regulations	each	801.75
Supply and install 30 no. illuminated bollards; 900 mm high blue powder coated steel bollard to set 450 mm into ground, louvered polycarbonate luminaire to IP54, complete with control gear and electrical supply, installed all in accordance with IEE regulations		
Floodlights Supply and install 10 no. floodlighting units at locations 20 m apart; 5.0 m galvanized painted tubular steel columns to BS 5649, set 450 mm into paving, low light pollution aluminium luminaires to BS 4533:102.5 to IP65 complete with control gear and electrical supply, installed all in accordance with IEE regulations inclusive of electrical connections	each	1039.17
PC of lighting unit £395.00	each	544.26
PC of lighting unit £610.00	each	803.61
Recessed Lighting Supply and install 4 no. 12 volt ground mounted recessed brick lights at locations 5 metres apart; black powder coated diecast aluminium casing 220 mm dia. to IP55 recessed to paving, polycarbonate diffuser, complete with control gear and electrical supply, installed all in accordance with IEE regulations	set	2448.46

The Presentation and Settlement of Contractors' Claims
Second Edition

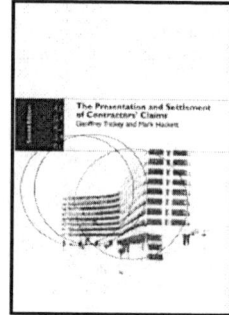

Geoffrey Trickey and Mark Hackett

Contractual disputes, often involving large sums of money, occur with increasing frequency in the construction industry. This book presents - in non-legal language - sound professional advice from a recognized expert in the field on the practical aspects of claims. This edition has been brought right up to date by taking into account legal decisions promulgated over the last 17 years, as well as reflecting the effect of current inflation on claims.

This new, fully updated edition of this practical guide is based on the 1998 JCT contract. The title contains numerous worked examples to support the advice offered, relating it to practitioners' experiences.

Contents: General. Introduction. 1998 edition of the Joint Contracts Tribunal standard form of building contract. Extensions of time. Variations and disruption. Ascertaining the loss or expense. Nominated sub-contractors and suppliers. Determination of the employment of the contractor. Comparison between the various editions of the JCT standard form of building contract. The JCT family of forms. The control of claims. A worked example of the ascertainment of direct loss and/or expense. Index.

November 2000: 234x156 mm: 512 pp.
5 line illustrations
HB: 0-419-20500-4: £85.00

Part 7

Tables and Memoranda

QUICK REFERENCE CONVERSION TABLES

Conversion Factors

	Imperial			Metric	
1. LINEAR					
0.039	in	1		mm	25.4
3.281	ft	1		metre	0.305
1.094	yd	1		metre	0.914
2. WEIGHT					
0.020	cwt	1		kg	50.802
0.984	ton	1		tonne	1.016
2.205	lb	1		kg	0.454
3. CAPACITY					
1.760	pint	1		litre	0.568
0.220	gal	1		litre	4.546
4. AREA					
0.002	in²	1		mm²	645.16
10.764	ft²	1		m²	0.093
1.196	yd²	1		m²	0.836
2.471	acre	1		ha	0.405
0.386	mile²	1		km²	2.59
5. VOLUME					
0.061	in³	1		cm³	16.387
35.315	ft³	1		m³	0.028
1.308	yd³	1		m³	0.765
6. POWER					
1.310	HP	1		kW	0.746

Conversion Factors – Metric to Imperial

Multiply Metric	Unit	By	To obtain Imperial	Unit
Length				
kilometre	km	0.6214	statute mile	ml
metre	m	1.0936	yard	yd
centimetre	cm	0.0328	foot	ft
millimetre	mm	0.0394	inch	in
Area				
hectare	ha	2.471	acre	
square kilometre	km²	0.3861	square mile	sq ml
square metre	m²	10.764	square foot	sq ft
square metre	m²	1550	square inch	sq in
square centimetre	cm²	0.155	square inch	sq in
Volume				
cubic metre	m³	35.336	cubic foot	cu ft
cubic metre	m³	1.308	cubic yard	cu yd
cubic centimetre	cm³	0.061	cubic inch	cu in
cubic centimetre	cm³	0.0338	fluid ounce	fl oz
Liquid volume				
litre	l	0.0013	cubic yard	cu yd
litre	l	61.02	cubic inch	cu in
litre	l	0.22	Imperial gallon	gal
litre	l	0.2642	US gallon	US gal
litre	l	1.7596	pint	pt
Mass				
metric tonne	t	0.984	long ton	lg ton
metric tonne	t	1.102	short ton	sh ton
kilogram	kg	2.205	pound, avoirdupois	lb
gram	g or gr	0.0353	ounce, avoirdupois	oz
Unit mass				
kilograms/cubic metre	kg/m³	0.062	pounds/cubic foot	lbs/cu ft
kilograms/cubic metre	kg/m³	1.686	pounds/cubic yard	lbs/cu yd
tonnes/cubic metre	t/m³	1692	pound/cubic yard	lbs/cu yd
kilograms/sq centimetre	kg/cm²	14.225	pounds/square inch	lbs/sq in
kilogram-metre	kg.m	7.233	foot-pound	ft-lb

Conversion Factors – Metric to Imperial - cont'd

Multiply Metric	Unit	By	To obtain Imperial	Unit
Force				
meganewton	MN	9.3197	tons force	tonf
kilonewton	kN	225	pounds force	lbf
newton	N	0.225	pounds force	lbf
Pressure and stress				
meganewton per square metre	MN/m^2	9.3197	tons force/square foot	$tonf/ft^2$
kilopascal	kPa	0.145	pounds/square inch	psi
bar		14.5	pounds/square inch	psi
kilogram metre	kgm	7.2307	foot-pound	ft-lb
Energy				
kilocalorie	kcal	3.968	British thermal unit	Btu
metric horsepower	CV	0.9863	horse power	hp
kilowatt	kW	1.341	horse power	hp
Speed				
kilometres/hour	km/h	0.621	miles/hour	mph

Conversion Factors – Imperial to Metric

Multiply Imperial	Unit	By	To obtain metric	Unit
Length				
statute mile	ml	1.609	kilometre	km
yard	yd	0.9144	metre	m
foot	ft	30.48	centimetre	cm
inch	in	25.4	millimetre	mm
Area				
acre	acre	0.4047	hectare	ha
square mile	sq ml	2.59	square kilometre	km^2
square foot	sq ft	0.0929	square metre	m^2
square inch	sq in	0.0006	square metre	m^2
square inch	sq in	6.4516	square centimetre	cm^2
Volume				
cubic foot	cu ft	0.0283	cubic metre	m^3
cubic yard	cu yd	0.7645	cubic metre	m^3
cubic inch	cu in	16.387	cubic centimetre	cm^3
fluid ounce	fl oz	29.57	cubic centimetre	cm^3
Liquid volume				
cubic yard	cu yd	764.55	litre	l
cubic inch	cu in	0.0164	litre	l
Imperial gallon	gal	4.5464	litre	l
US gallon	US gal	3.785	litre	l
US gallon	US gal	0.833	Imperial gallon	gal
pint	pt	0.5683	litre	l
Mass				
long ton	lg ton	1.016	metric tonne	tonne
short ton	sh ton	0.907	metric tonne	tonne
pound	lb	0.4536	kilogram	kg
ounce	oz	28.35	gram	g
Unit mass				
pounds/cubic foot	lb/ cu ft	16.018	kilogram's/cubic metre	kg/m^3
pounds/cubic yard	lb/cu yd	0.5933	cubic/cubic metre	kg/m^3
pounds/cubic yard	lb/cu yd	0.0006	tonnes/cubic metre	t/m^3
foot-pound	ft-lb	0.1383	kilogram-metre	kg.m

Conversion Factors – Imperial to Metric - cont'd

Multiply Imperial	Unit	By	To obtain metric	Unit
Force				
tons force	tonf	0.1073	meganewton	MN
pounds force	lbf	0.0045	kilonewton	kN
pounds force	lbf	4.45	newton	N
Pressure and stress				
pounds/square inch	psi	0.1073	kilogram/sq. centimetre	kg/cm²
pounds/square inch	psi	6.89	kilopascal	kPa
pounds/square inch	psi	0.0689	bar	
foot-pound	ft-lb	0.1383	kilogram metre	kgm
Energy				
British Thermal Unit	Btu	0.252	kilocalorie	kcal
horsepower (hp)	hp	1.014	metric horsepower	CV
horsepower (hp)	hp	0.7457	kilowatt	kW
Speed				
miles/hour	mph	1.61	kilometres/hour	km/h

Conversion Table

Length

Millimetre	mm	1 in	=	25.4 cm	1 mm	=	0.0394 in
Centimetre	cm	1 in	=	2.54 cm	1 cm	=	0.3937 in
Metre	m	1 ft	=	0.3048 m	1 m	=	3.2808 ft
		1 yd	=	0.9144 m	1 m	=	1.0936 yd
Kilometre	km	1 mile	=	1.6093 km	1 km	=	0.6214 mile

Note:

1 cm	=	10 mm	1 ft	=	12 in
1 m	=	100 cm	1 yd	=	3 ft
1 km	=	1,000 m	1 mile	=	1,760 yd

Area

Square millimetre	mm^2	$1\ in^2$	=	$645.2\ mm^2$	$1\ mm^2$	=	$0.0016\ in^2$
Square centimetre	cm^2	$1\ in^2$	=	$6.4516\ cm^2$	$1\ cm^2$	=	$1.1550\ in^2$
Square metre	m^2	$1\ ft^2$	=	$0.0929\ m^2$	$1\ m^2$	=	$10.764\ ft^2$
		$1\ yd^2$	=	$0.8361\ m^2$	$1\ m^2$	=	$1.1960\ yd^2$
Square Kilometre	km^2	1 mile2	=	$2.590\ km^2$	$1\ km^2$	=	0.3861 mile2

Note:

$1\ cm^2$	=	$100\ m^2$	$1\ ft^2$		=	$144\ in^2$
$1\ m^2$	=	$10,000\ cm^2$	$1\ yd^2$		=	$9\ ft^2$
$1\ km^2$	=	100 hectares	$1\ mile^2$		=	640 acres
			1 acre		=	$4,840\ yd^2$

Volume

Cubic Centimetre	cm^3	$1\ cm^3$	=	$0.0610\ in^3$	$1\ in^3$	=	$16.387\ cm^3$
Cubic Decimetre	dm^3	$1\ dm^3$	=	$0.0353\ ft^3$	$1\ ft^3$	=	$28.329\ dm^3$
Cubic metre	m^3	$1\ m^3$	=	$35.3147\ ft^3$	$1\ ft^3$	=	$0.0283\ m^3$
	$1\ m^3$	=		$1.3080\ yd^3$	$1\ yd^3$	=	$0.7646\ m^3$
Litre	L	1 L	=	1.76 pint	1 pint	=	0.5683 L
			=	2.113 US pt		=	0.4733 US L

Note:

$1\ dm^3$	=	$1,000\ cm^3$	$1\ ft^3$	=	$1.728\ in^3$
$1\ m^3$	=	$1,000\ dm^3$	$1\ yd^3$	=	$27\ ft^3$
1 L	=	$1\ dm^3$	1 pint	=	20 fl oz
1 HL	=	100 L	1 gal	=	8 pints

Conversion Table - cont'd

Mass

Milligram	mg	1 mg	=	0.0154 grain	1 grain	=	64.935 mg
Gram	g	1 g	=	0.0353 oz	1 oz	=	28.35 g
Kilogram	kg	1 kg	=	2.2046 lb	1 lb	=	0.4536 kg
Tonne	t	1 t	=	0.9842 ton	1 ton	=	1.016 t

Note:	1 g	=	1,000 mg	1 oz	=	437.5 grains
	1 kg	=	1000 g	1 lb	=	16 oz
	1 t	=	1,000 kg	1 stone	=	14 lb
				1 cwt	=	112 lb
				1 ton	=	20 cwt

Force

Newton	N	1 lbf	=	4.448 N	1 kgf	=	9.807 N
Kilonewton	kN	1 lbf	=	0.00448 kN	1 ton f	=	9.964 kN
Meganewton	MN	100 tonf	=	0.9964 MN			

Pressure and stress

Kilonewton per		1 lbf/in²	=	6.895 kN/m²
square metre	kN/m²	1 bar	=	100 kN/m²
Meganewton per		1 tonf/ft²	=	107.3 kN/m² = 0.1073 MN/m²
square metre	MN/m²	1 kgf/cm²	=	98.07 kN/m²
		1 lbf/ft²	=	0.04788 kN/m²

Temperature

Degree Celsius °C $\quad$ $^{\circ}C = \dfrac{5 \times (^{\circ}F - 32)}{9}$ $\qquad$ $^{\circ}F = \dfrac{(9 \times \,^{\circ}C) + 32}{5}$

Metric Equivalents

1 km	=	1000 m
1 m	=	100 cm
1 cm	=	10 mm
1 km²	=	100 ha
1 ha	=	10,000 m²
1 m²	=	10,000 cm²
1 cm²	=	100 mm²
1 m³	=	1,000 litres
1 litre	=	1,000 cm³
1 metric tonne	=	1,000 kg
1 quintal	=	100 kg
1 N	=	0.10197 kg
1 kg	=	1000 g
1 g	=	1000 mg
1 bar	=	14.504 psi
1 cal	=	427 kg.m
1 cal	=	0.0016 cv.h
torque unit	=	0.00116 kw.h
1 CV	=	75 kg.m/s
1 kg/cm²	=	0.97 atmosph

Imperial Unit Equivalents

1 mile	=	1760 yd
1 yd	=	3 ft
1 ft	=	12 in
1 sq mile	=	640 acres
1 acre	=	43,560 sq ft
1 sq ft	=	144 sq in
1 cu ft	=	7.48 gal liq
1 gal	=	231 cu in
	=	4 quarts liq
1 quart	=	32 fl oz
1 fl oz	=	1.80 cu in
	=	437.5 grains
1 stone	=	14 lb
1 cwt	=	112 lb
1 sh ton	=	2000 lb
1 lg ton	=	2240 lb
	=	20 cwt
1 lb	=	16 oz, avdp
1 Btu	=	778 ft lb
	=	0.000393 hph
	=	0.000293 kwh
1 hp	=	550 ft-lb/sec
1 atmosph	=	14.7 lb/in²

Power Units

kW	=	Kilowatt
HP	=	Horsepower
CV	=	Cheval Vapeur (Steam Horsepower)
	=	French designation for Metric Horsepower
PS	=	Pferderstarke (Horsepower)
	=	German designation of Metric Horsepower
1 HP	=	1.014 CV = 1.014 PS
	=	0.7457 kW
1 PS	=	1 CV = 0.9863 HP
	=	0.7355 kW
1 kW	=	1.341 HP
	=	1.359 CV
	=	1.359 PS

Speed Conversion

km/h	m/min	mph	fpm
1	16.7	0.6	54.7
2	33.3	1.2	109.4
3	50.0	1.9	164.0
4	66.7	2.5	218.7
5	83.3	3.1	273.4
6	100.0	3.7	328.1
7	116.7	4.3	382.8
8	133.3	5.0	437.4
9	150.0	5.6	492.1
10	166.7	6.2	546.8
11	183.3	6.8	601.5
12	200.0	7.5	656.2
13	216.7	8.1	710.8
14	233.3	8.7	765.5
15	250.0	9.3	820.2
16	266.7	9.9	874.9
17	283.3	10.6	929.6
18	300.0	11.2	984.3
19	316.7	11.8	1038.9
20	333.3	12.4	1093.6
21	350.0	13.0	1148.3
22	366.7	13.7	1203.0
23	383.3	14.3	1257.7
24	400.0	14.9	1312.3
25	416.7	15.5	1367.0
26	433.3	16.2	1421.7
27	450.0	16.8	1476.4
28	466.7	17.4	1531.1
29	483.3	18.0	1585.7
30	500.0	18.6	1640.4
31	516.7	19.3	1695.1
32	533.3	19.9	1749.8
33	550.0	20.5	1804.5
34	566.7	21.1	1859.1
35	583.3	21.7	1913.8

Speed Conversion - cont'd

km/h	m/min	mph	fpm
36	600.0	22.4	1968.5
37	616.7	23.0	2023.2
38	633.3	23.6	2077.9
39	650.0	24.2	2132.5
40	666.7	24.9	2187.2
41	683.3	25.5	2241.9
42	700.0	26.1	2296.6
43	716.7	26.7	2351.3
44	733.3	27.3	2405.9
45	750.0	28.0	2460.6
46	766.7	28.6	2515.3
47	783.3	29.2	2570.0
48	800.0	29.8	2624.7
49	816.7	30.4	2679.4
50	833.3	31.1	2734.0

FORMULAE

Two Dimensional Figures

Figure	Diagram of figure	Surface area	Perimeter
Square		a^2	$4a$
Rectangle		ab	$2(a + b)$
Triangle		$\frac{1}{2}ch$ $\frac{1}{2}ab \sin C$ $\sqrt{\{s(s-a)(s-b)(s-c)\}}$ where $s = \frac{1}{2}(a + b + c)$	$a + b + c = 2s$
Circle		πr^2 $\frac{1}{4}\pi d^2$ where $2r = d$	$2\pi r$ πd
Parallelogram		ah	$2(a + b)$
Trapezium		$\frac{1}{2}h(a + b)$	$a + b + c + d$
Ellipse		Approximately πab	$\pi(a + b)$
Hexagon		$2.6 \times a^2$	
Octagon		$4.83 \times a^2$	
Sector of circle		$\frac{1}{2}rb$ or $\frac{q}{360}\,\pi r^2$ note: $b = $ angle $\frac{q}{360} \times \pi 2r$	
Segment of a circle		$S - T$ where $S = $ area of sector $T = $ area of triangle	
Bellmouth		$\frac{3}{14} \times r^2$	

Three Dimensional Figures

Figure	Diagram of figure	Surface area	Volume
Cube		$6a^2$	a^3
Cuboid/ rectangular block		$2(ab + ac + bc)$	abc
Prism/ triangular block		$bd + hc + dc + ad$	$\frac{1}{2}hcd$ $\frac{1}{2}ab \sin C\, d$ $d\sqrt{\{s(s-a)(s-b)(s-c)\}}$ where $s = \frac{1}{2}(a + b + c)$
Cylinder		$2\pi rh + 2\pi r^2$ $\pi dh + \frac{1}{2}\pi d^2$	$\pi r^2 h$ $\frac{1}{4}\pi d^2 h$
Sphere		$4\pi r^2$	$\frac{4}{3}\pi r^3$
Segment of sphere		$2\pi Rh$	$\frac{1}{6}\pi h(3r^2 + h^2)$ $\frac{1}{3}\pi h^2(3R - h)$
Pyramid		$(a + b)l + ab$	$\frac{1}{3}abh$
Frustrum of a pyramid		$l(a+b+c+d) + \sqrt{(ab+cd)}$ [regular figure only]	$\frac{h}{3}(ab + cd + \sqrt{abcd})$
Cone		$\pi rl + \pi r^2$ $\frac{1}{2}\pi dh + \frac{1}{4}\pi d^2$	$\frac{1}{3}\pi r^2 h$ $\frac{1}{12}\pi d^2 h$
Frustrum of a cone		$\pi r^2 + \pi R^2 + \pi h(R + r)$	$\frac{1}{3}\pi h(R^2 + Rr + r^2)$

Geometric Formulae

Formula	Description
Pythagoras theorem	A2 = B2 + C2 where A is the hypotenuse of a right-angled triangle and B and C are the two adjacent sides
Simpsons Rule	The Area is divided into an even number of strips of equal width, and therefore has an odd number of ordinates at the division points area = $\dfrac{S (A + 2B + 4C)}{3}$ where S = common interval (strip width) A = sum of first and last ordinates B = sum of remaining odd ordinates C = sum of the even ordinates The Volume can be calculated by the same formula, but by substituting the area of each co-ordinate rather than its length.
Trapezoidal Rule	A given trench is divided into two equal sections, giving three ordinates, the first, the middle and the last. volume = $\dfrac{S \times (A + B + 2C)}{2}$ where S = width of the strips A = area of the first section B = area of the last section C = area of the rest of the sections
Prismoidal Rule	A given trench is divided into two equal sections, giving three ordinates, the first, the middle and the last. volume = $\dfrac{L \times (A + 4B + C)}{6}$ where L = total length of trench A = area of the first section B = area of the middle section C = area of the last section

EARTHWORK

Weights of Typical Materials Handled by Excavators

The weight of the material is that of the state in its natural bed and includes moisture. Adjustments should be made to allow for loose or compacted states.

Material	kg/m³	lb/cu yd
Adobe	1914	3230
Ashes	610	1030
Asphalt, rock	2400	4050
Basalt	2933	4950
Bauxite: alum ore	2619	4420
Borax	1730	2920
Caliche	1440	2430
Carnotite	2459	4150
Cement	1600	2700
Chalk (hard)	2406	4060
Cinders	759	1280
Clay: dry	1908	3220
Clay: wet	1985	3350
Coal: bituminous	1351	2280
Coke	510	860
Conglomerate	2204	3720
Dolomite	2886	4870
Earth: dry	1796	3030
Earth: moist	1997	3370
Earth: wet	1742	2940
Feldspar	2613	4410
Felsite	2495	4210
Fluorite	3093	5220
Gabbro	3093	5220
Gneiss	2696	4550
Granite	2690	4540
Gravel, dry	1790	3020
Gypsum	2418	4080
Hardcore (consolidated)	1928	120
Lignite broken	1244	2100

Weights of Typical Materials Handled by Excavators - cont'd

Material	kg/m³	lb/cu yd
Limestone	2596	4380
Magnesite, magnesium ore	2993	5050
Marble	2679	4520
Marl	2216	3740
Peat	700	1180
Potash	2193	3700
Pumice	640	1080
Quarry waste	1438	90
Quartz	2584	4360
Rhyolite	2400	4050
Sand: dry	1707	2880
Sand: wet	1831	3090
Sand and gravel - dry	1790	3020
- wet	2092	3530
Sandstone	2412	4070
Schist	2684	4530
Shale	2637	4450
Slag (blast)	2868	4840
Slate	2667	4500
Snow - dry	130	220
- wet	510	860
Taconite	3182	5370
Topsoil	1440	2430
Trachyte	2400	4050
Traprock	2791	4710
Water	1000	62

Transport Capacities

Type of vehicle	Capacity of vehicle	
	Payload	Heaped capacity
Wheelbarrow	150	0.10
1 tonne dumper	1250	1.00
2.5 tonne dumper	4000	2.50
Articulated dump truck (Volvo A20 6x4)	18500	11.00
Articulated dump truck (Volvo A35 6x6)	32000	19.00
Large capacity rear dumper (Euclid R35)	35000	22.00
Large capacity rear dumper (Euclid R85)	85000	50.00

Machine Volumes for Excavating and Filling

Machine Type	Cycles per minute	Volume per minute (m^3)
1.5 tonne excavator	1	0.04
	2	0.08
	3	0.12
3 tonne excavator	1	0.13
	2	0.26
	3	0.39
5 tonne excavator	1	0.28
	2	0.56
	3	0.84
7 tonne excavator	1	0.28
	2	0.56
	3	0.84
21 tonne excavator	1	1.21
	2	2.42
	3	3.63
Backhoe loader JCB3CX excavating Rear bucket capacity 0.28m3	1	0.28
	2	0.56
	3	0.84
Backhoe loader JCB3CX loading Front bucket capacity 1.00m3	1	1.00
	2	2.00

Machine Volumes for Excavating and Filling - cont'd

Machine Type	Loads per hour	Volume per minute (m^3)
1 tonne high tip skip loader	5	2.43
Volume 0.485m3	7	3.40
	10	4.85
3 tonne dumper	4	7.60
Max. volume 2.40m3	5	9.50
Available volume 1.9m3	7	13.30
	10	19.00
6 tonne dumper	4	15.08
Max. volume 3.40m3	5	18.85
Available volume 3.77m3	7	26.39
	10	37.70

Bulkage of Soils (After excavation)

Type of soil	Approximate bulking of 1 m^3 after excavation
Vegetable soil and loam	25 - 30%
Soft clay	30 - 40%
Stiff clay	10 - 20%
Gravel	20 - 25%
Sand	40 - 50%
Chalk	40 - 50%
Rock, weathered	30 - 40%
Rock, unweathered	50 - 60%

Shrinkage of Materials (On being deposited)

Type of soil	Approximate bulking of 1 m^3 after excavation
Clay	10%
Gravel	8%
Gravel and sand	9%
Loam and light sandy soils	12%
Loose vegetable soils	15%

Voids in Material Used as Sub-bases or Beddings

Material	m³ of voids/m³
Alluvium	0.37
River grit	0.29
Quarry sand	0.24
Shingle	0.37
Gravel	0.39
Broken stone	0.45
Broken bricks	0.42

Angles of Repose

Type of soil		degrees
Clay	- dry	30
	- damp, well drained	45
	- wet	15 - 20
Earth	- dry	30
	- damp	45
Gravel	- moist	48
Sand	- dry or moist	35
	- wet	25
Loam		40

Slopes and Angles

Ratio of base to height	Angle in degrees
5 : 1	11
4 : 1	14
3 : 1	18
2 : 1	27
1½ : 1	34
1 : 1	45
1 : 1½	56
1 : 2	63
1 : 3	72
1 : 4	76
1 : 5	79

Grades (In Degrees and Percents)

Degrees	Percent	Degrees	Percent
1	1.8	24	44.5
2	3.5	25	46.6
3	5.2	26	48.8
4	7.0	27	51.0
5	8.8	28	53.2
6	10.5	29	55.4
7	12.3	30	57.7
8	14.0	31	60.0
9	15.8	32	62.5
10	17.6	33	64.9
11	19.4	34	67.4
12	21.3	35	70.0
13	23.1	36	72.7
14	24.9	37	75.4
15	26.8	38	78.1
16	28.7	39	81.0
17	30.6	40	83.9
18	32.5	41	86.9
19	34.4	42	90.0
20	36.4	43	93.3
21	38.4	44	96.6
22	40.4	45	100.0

Bearing Powers

Ground conditions		Bearing Power		
		km/mý	lb/iný	metric t/m²
Rock (broken)		483	70	50
Rock (solid)		2,415	350	240
Clay,	dry or hard	380	55	40
	medium dry	190	27	20
	soft or wet	100	14	10
Gravel,	cemented	760	110	80
Sand,	compacted	380	55	40
	clean dry	190	27	20
Swamp and alluvial soils		48	7	5

Earthwork Support

Maximum depth of excavation in various soils without the use of earthwork support:

Ground conditions	Feet (ft)	Metres (m)
Compact soil	12	3.66
Drained loam	6	1.83
Dry sand	1	0.3
Gravelly earth	2	0.61
Ordinary earth	3	0.91
Stiff clay	10	3.05

It is important to note that the above table should only be used as a guide.
Each case must be taken on its merits and, as the limited distances given above are approached, careful watch must be kept for the slightest signs of caving in.

CONCRETE WORK

Weights of Concrete and Concrete Elements

Type of Material	kg/m³	lb/cu ft
Ordinary concrete (dense aggregates)		
Non-reinforced plain or mass concrete		
Nominal weight	2305	144
Aggregate - limestone	2162 to 2407	135 to 150
- gravel	2244 to 2407	140 to 150
- broken brick	2000 (av)	125 (av)
- other crushed stone	2326 to 2489	145 to 155
Reinforced concrete		
Nominal weight	2407	150
Reinforcement - 1%	2305 to 2468	144 to 154
- 2%	2356 to 2519	147 to 157
- 4%	2448 to 2703	153 to 163
Special concretes		
Heavy concrete		
Aggregates - barytes, magnetite	3210 (min)	200 (min)
steel shot, punchings	5280	330
Lean mixes		
Dry-lean (gravel aggregate)	2244	140
Soil-cement (normal mix)	1601	100

Weights of Concrete and Concrete Elements - cont'd

Type of material		kg/m² per mm thick	lb/sq ft per inch thick
Ordinary concrete (dense aggregates)			
Solid slabs (floors, walls etc.)			
Thickness:	75 mm or 3 in	184	37.5
	100 mm or 4 in	245	50
	150 mm or 6 in	378	75
	250 mm or 10 in	612	125
	300 mm or 12 in	734	150
Ribbed slabs			
Thickness:	125 mm or 5 in	204	42
	150 mm or 6 in	219	45
	225 mm or 9 in	281	57
	300 mm or 12 in	342	70
Special concretes			
Finishes etc			
Rendering, screed etc Granolithic, terrazzo		1928 to 2401	10 to 12.5
Glass-block (hollow) concrete		1734 (approx)	9 (approx)
Pre-stressed concrete		Weights as for reinforced concrete (upper limits)	
Air-entrained concrete		Weights as for plain or reinforced concrete	

Average Weight of Aggregates

Materials	Voids %	Weight kg/m³
Sand	39	1660
Gravel 10 - 20 mm	45	1440
Gravel 35 - 75 mm	42	1555
Crushed stone	50	1330
Crushed granite (over 15 mm)	50	1345
(n.e. 15 mm)	47	1440
'All-in' ballast	32	1800 - 2000

Material	kg/m³	lb/cu yd
Vermiculite (aggregate)	64-80	108-135

Material	kg/m³	lb/cu ft
All-in aggregate	1999	125

Common Mixes (per m³)

Recom- mended mix	Class of work suitable for: -	Cement (kg)	Sand (kg)	Coarse aggregate (kg)	No 50kg bags cement per m³ of combined aggregate
1:3:6	Roughest type of mass concrete such as footings, road haunching over 300 mm thick	208	905	1509	4.00
1:2.5:5	Mass concrete of better class than 1:3:6 such as bases for machinery, walls below ground etc.	249	881	1474	5.00
1:2:4	Most ordinary uses of concrete, such as mass walls above ground, road slabs etc. and general reinforced concrete work	304	889	1431	6.00
1:1.5:3	Watertight floors, pavements and walls, tanks, pits, steps, paths, surface of 2 course roads, reinforced concrete where extra strength is required	371	801	1336	7.50
1:1:2	Work of thin section such as fence posts and small precast work	511	720	1206	10.50

Prescribed Mixes for Ordinary Structural Concrete

Weights of cement and total dry aggregates in kg to produce approximately one cubic metre
of fully compacted concrete together with the percentages by weight of fine aggregate in total dry aggregates.

Conc. grade	Nominal max. size of aggregate (mm)	40		20		14		10	
	Workability	Med.	High	Med.	High	Med.	High	Med.	High
	Limits to slump that may be expected (mm)	50-100	100-150	25-75	75-125	10-50	50-100	10-25	25-50
7	Cement (kg)	180	200	210	230	-	-	-	-
	Total aggregate (kg)	1950	1850	1900	1800	-	-	-	-
	Fine aggregate (%)	30-45	30-45	35-50	35-50	-	-	-	-
10	Cement (kg)	210	230	240	260	-	-	-	-
	Total aggregate (kg)	1900	1850	1850	1800	-	-	-	-
	Fine aggregate (%)	30-45	30-45	35-50	35-50	-	-	-	-
15	Cement (kg)	250	270	280	310	-	-	-	-
	Total aggregate (kg)	1850	1800	1800	1750	-	-	-	-
	Fine aggregate (%)	30-45	30-45	35-50	35-50	-	-	-	-
20	Cement (kg)	300	320	320	350	340	380	360	410
	Total aggregate (kg)	1850	1750	1800	1750	1750	1700	1750	1650
	Sand								
	Zone 1 (%)	35	40	40	45	45	50	50	55
	Zone 2 (%)	30	35	35	40	40	45	45	50
	Zone 3 (%)	30	30	30	35	35	40	40	45
25	Cement (kg)	340	360	360	390	380	420	400	450
	Total aggregate (kg)	1800	1750	1750	1700	1700	1650	1700	1600
	Sand								
	Zone 1 (%)	35	40	40	45	45	50	50	55
	Zone 2 (%)	30	35	35	40	40	45	45	50
	Zone 3 (%)	30	30	30	35	35	40	40	45
30	Cement (kg)	370	390	400	430	430	470	460	510
	Total aggregate (kg)	1750	1700	1700	1650	1700	1600	1650	1550
	Sand								
	Zone 1 (%)	35	40	40	45	45	50	50	55
	Zone 2 (%)	30	35	35	40	40	45	45	50
	Zone 3 (%)	30	30	30	35	35	40	40	45

Weights of Bar Reinforcement

Nominal sizes (mm)	Cross-sectional area (mm²)	Mass kg/m	Length of bar m/tonne
6	28.27	0.222	4505
8	50.27	0.395	2534
10	78.54	0.617	1622
12	113.10	0.888	1126
16	201.06	1.578	634
20	314.16	2.466	405
25	490.87	3.853	260
32	804.25	6.313	158
40	1265.64	9.865	101
50	1963.50	15.413	65

Weights of Bars at Specific Spacings

Weights of metric bars in kilograms per square metre.

Size (mm)	Spacing of bars in millimetres									
	75	100	125	150	175	200	225	250	275	300
6	2.96	2.220	1.776	1.480	1.27	1.110	0.99	0.89	0.81	0.74
8	5.26	3.95	3.16	2.63	2.26	1.97	1.75	1.58	1.44	1.32
10	8.22	6.17	4.93	4.11	3.52	3.08	2.74	2.47	2.24	2.06
12	11.84	8.88	7.10	5.92	5.07	4.44	3.95	3.55	3.23	2.96
16	21.04	15.78	12.63	10.52	9.02	7.89	7.02	6.31	5.74	5.26
20	32.88	24.66	19.73	16.44	14.09	12.33	10.96	9.87	8.97	8.22
25	51.38	38.53	30.83	25.69	22.02	19.27	17.13	15.41	14.01	12.84
32	84.18	63.13	50.51	42.09	36.08	31.57	28.06	25.25	22.96	21.04
40	131.53	98.65	78.92	65.76	56.37	49.32	43.84	39.46	35.87	32.88
50	205.51	154.13	123.31	102.76	88.08	77.07	68.50	61.65	56.05	51.38

Basic weight of steelwork taken as 7850 kg/m³

Basic weight of bar reinforcement per metre run = 0.00785 kg/mm²

The value of PI has been taken as 3.141592654

Fabric Reinforcement

Fabric reference	Longitudinal wires			Cross wires			Mass
	Nominal wire size (mm)	Pitch (mm)	Area (mm/m²)	Nominal wire size (mm)	Pitch (mm)	Area (mm/m²)	(kg/m²)
Square mesh							
A393	10	200	393	10	200	393	6.16
A252	8	200	252	8	200	252	3.95
A193	7	200	193	7	200	193	3.02
A142	6	200	142	6	200	142	2.22
A98	5	200	98	5	200	98	1.54
Structural mesh							
B1131	12	100	1131	8	200	252	10.90
B785	10	100	785	8	200	252	8.14
B503	8	100	503	8	200	252	5.93
B385	7	100	385	7	200	193	4.53
B283	6	100	283	7	200	193	3.73
B196	5	100	196	7	200	193	3.05
Long mesh							
C785	10	100	785	6	400	70.8	6.72
C636	9	100	636	6	400	70.8	5.55
C503	8	100	503	5	400	49.00	4.34
C385	7	100	385	5	400	49.00	3.41
C283	6	100	283	5	400	49.00	2.61
Wrapping mesh							
D98	5	200	98	5	200	98	1.54
D49	2.5	100	49	2.5	100	49	0.77
Stock sheet size	Length 4.8 m		Width 2.4 m			Sheet area 11.52m²	

Wire

SWG	6g	5g	4g	3g	2g	1g	1/0g	2/0g	3/0g	4/0g	5/0g
diameter											
in	0.192	0.212	0.232	0.252	0.276	0.300	0.324	0.348	0.372	0.400	0.432
mm	4.9	5.4	5.9	6.4	7.0	7.6	8.2	8.8	9.5	0.2	1.0
area											
in²	0.029	0.035	0.042	0.050	0.060	0.071	0.082	0.095	0.109	0.126	0.146
mm²	19	23	27	32	39	46	53	61	70	81	95

Average Weight (kg/m³) of Steelwork Reinforcement in Concrete for Various Building Elements

	kg/m³ concrete
Substructure	
Pile caps	110 - 150
Tie beams	130 - 170
Ground beams	230 - 330
Bases	90 - 130
Footings	70 - 110
Retaining walls	110 - 150
Superstructure	
Slabs - one way	75 - 125
Slabs - two way	65 - 135
Plate slab	95 - 135
Cantilevered slab	90 - 130
Ribbed floors	80 - 120
Columns	200 - 300
Beams	250 - 350
Stairs	130 - 170
Walls - normal	30 - 70
Walls - wind	50 - 90

Note: For exposed elements add the following %: Walls 50%, Beams 100%, Columns 15%

Formwork Stripping Times – Normal Curing Periods

	Minimum periods of protection for different types of cement					
Conditions under which concrete is maturing	Number of days (where the average surface temperature of the concrete exceeds 10ºC during the whole period)			Equivalent maturity (degree hours) calculated as the age of the concrete in hours multiplied by the number of degrees Celsius by which the average surface temperature of the concrete exceeds - 10ºC		
	Other	SRPC	OPC or RHPC	Other	SRPC	OPC or RHPC
1. Hot weather or drying winds	7	4	3	3500	2000	1500
2. Conditions not covered by 1	4	3	2	2000	1500	1000

KEY

OPC - Ordinary Portland Cement

RHPC - Rapid-hardening Portland cement

SRPC - Sulphate-resisting Portland cement

Minimum period before striking formwork

	Minimum period before striking		
	Surface temperature of concrete		
	16 º C	17 º C	t º C(0-25)
Vertical formwork to columns, walls and large beams	12 hours	18 hours	$\frac{300}{t+10}$ hours
Soffit formwork to slabs	4 days	6 days	$\frac{100}{t+10}$ days
Props to slabs	10 days	15 days	$\frac{250}{t+10}$ days
Soffit formwork to beams	9 days	14 days	$\frac{230}{t+10}$ days
Props to beams	14 days	21 days	$\frac{360}{t+10}$ days

MASONRY

Weights of Bricks and Blocks

Walls and components of walls	kg/m² per mm thick	lb/sq ft per inch thick
Blockwork		
Hollow clay blocks; average)	1.15	6
Common clay blocks	1.90	10
Brickwork		
Engineering clay bricks	2.30	12
Refractory bricks	1.15	6
Sand-lime (and similar) bricks	2.02	10.5

Weights of Stones

Type of stone		kg/m3	lb/cu ft
Natural stone (solid)			
Granite		2560 to 2927	160 to 183
Limestone	- Bath stone	2081	130
	- Marble	2723	170
	- Portland stone	2244	140
Sandstone		2244 to 2407	140 to 150
Slate		2880	180
Stone rubble (packed)		2244	140

Quantities of Bricks and Mortar

Materials per m² of wall:		
Thickness	**No. of Bricks**	**Mortar m³**
Half brick (112.5 mm)	58	0.022
One brick (225 mm)	116	0.055
Cavity, both skins (275 mm)	116	0.045
1.5 brick (337 mm)	174	0.074
Mass brickwork per m³	464	0.36

Mortar Mixes: Quantities of Dry Materials

	Imperial cu yd			Metric m³		
Mix	Cement cwts	Lime cwts	Sand cu yds	Cement tonnes	Lime tonnes	Sand cu m
1:3	7.0	-	1.04	0.54	-	1.10
1:4	6.3	-	1.10	0.40	-	1.20
1:1:6	3.9	1.6	1.10	0.27	0.13	1.10
1:2:9	2.6	2.1	1.10	0.20	0.15	1.20
0:1:3	-	3.3	1.10	-	0.27	1.00

Mortar Mixes for Various Uses

Mix	Use
1:3	Construction designed to withstand heavy loads in all seasons
1:1:6	Normal construction not designed for heavy loads. Sheltered and moderate conditions in spring and summer. Work above d:p:c - sand, lime bricks, clay blocks etc.
1:2:9	Internal partitions with blocks which have high drying shrinkage, pumic blocks, etc. any periods
0:1:3	Hydraulic lime only should be used in this mix and may be used for construction not designed for heavy loads and above d:p:c spring and summer.

Quantities of Bricks and Mortar Required per m² of Walling

Description	Unit	Nr of bricks required	Mortar required (m³)		
			No frogs	Single frogs	Double frogs
Standard bricks					
Brick size					
215 x 102.5 x 50 mm					
half brick wall (103 mm) (103 mm)	m²	72	0.022	0.027	0.032
2 x half brick cavity wall (270 mm)	m²	144	0.044	0.054	0.064
one brick wall (215 mm)	m²	144	0.052	0.064	0.076
one and a half brick wall (328 mm)	m²	216	0.073	0.091	0.108
mass brickwork	m³	576	0.347	0.413	0.480
Brick size					
215 x 102.5 x 65 mm					
half brick wall (103 mm)	m²	58	0.019	0.022	0.026
2 x half brick cavity wall (270 mm)	m²	116	0.038	0.045	0.055
one brick wall (215 mm)	m²	116	0.046	0.055	0.064
one and a half brick wall (328 mm)	m²	174	0.063	0.074	0.088
mass brickwork	m³	464	0.307	0.360	0.413
Metric modular bricks - Perforated					
Brick co-ordinating size					
200 x 100 x 75 mm					
90 mm thick	m²	67	0.016	0.019	
190 mm thick	m²	133	0.042	0.048	
290 mm thick	m²	200	0.068	0.078	
Brick co-ordinating size					
200 x 100 x 100 mm					
90 mm thick	m²	50	0.013	0.016	
190 mm thick	m²	100	0.036	0.041	
290 mm thick	m²	150	0.059	0.067	
Brick co-ordinating size					
300 x 100 x 75 mm					
90 mm thick	m²	33	-	0.015	
Brick co-ordinating size					
00 x 100 x 100 mm					
90 mm thick	m²	44	0.015	0.018	

Note: Assuming 10 mm deep joints

Mortar Required per m^2 Blockwork (9.88 blocks/m²)

Wall thickness	75	90	100	125	140	190	215
Mortar m^3/m^2	0.005	0.006	0.007	0.008	0.009	0.013	0.014

Mortar Mixes

Mortar Group	Cement: lime: sand	Masonry cement: sand	Cement: sand with plasticiser
1	1 : 0-0.25:3		
2	1 : 0.5 :4-4.5	1 : 2.5-3.5	1 : 3-4
3	1 : 1:5-6	1 : 4-5	1 : 5-6
4	1 : 2:8-9	1 : 5.5-6.5	1 : 7-8
5	1 : 3:10-12	1 : 6.5-7	1 : 8

Group 1: strong inflexible mortar
Group 5: weak but flexible

All mixes within a group are of approximately similar strength.
Frost resistance increases with the use of plasticisers.
Cement:lime:sand mixes give the strongest bond and greatest resistance to rain penetration.
Masonry cement equals ordinary Portland cement plus a fine neutral mineral filler and an
air entraining agent.

Calcium Silicate Bricks

Type	Strength	Location
Class 2 crushing strength	14.0N/mm^2	not suitable for walls
Class 3	20.5N/mm2	walls above dpc
Class 4	27.5N/mm2	cappings and copings
Class 5	34.5N/mm^2	retaining walls
Class 6	41.5N/mm2	walls below ground
Class 7	48.5N/mm^2	walls below ground

The Class 7 calcium silicate bricks are therefore equal in strength to Class B bricks.
Calcium silicate bricks are not suitable for DPCs.

Durability of Bricks

FL	Frost resistant with low salt content
FN	Frost resistant with normal salt content
ML	Moderately frost resistant with low salt content
MN	Moderately frost resistant with normal salt content

Brickwork Dimensions

No. of Horizontal Bricks	Dimensions mm	No. of Vertical courses	No. of Vertical courses
1/2	112.5	1	75
1	225.0	2	150
1 1/2	337.5	3	225
2	450.0	4	300
2 1/2	562.5	5	375
3	675.0	6	450
3 1/2	787.5	7	525
4	900.0	8	600
4 1/2	1012.5	9	675
5	1125.0	10	750
5 1/2	1237.5	11	825
6	1350.0	12	900
6 1/2	1462.5	13	975
7	1575.0	14	1050
7 1/2	1687.5	15	1125
8	1800.0	16	1200
8 1/2	1912.5	17	1275
9	2025.0	18	1350
9 1/2	2137.5	19	1425
10	2250.0	20	1500
20	4500.0	24	1575
40	9000.0	28	2100
50	11250.0	32	2400
60	13500.0	36	2700
75	16875.0	40	3000

Standard Available Block Sizes

Block	Co-ordinating size Length x height (mm)	Work size Length x height (mm)	Thicknesses (work size) (mm)
A	400 x 100 400 x 200 450 x 225	390 x 90 440 x 190 440 x 215	75, 90, 100, 140, 190 75, 90, 100, 140, 190 75, 90, 100, 140, 190, 215
B	400 x 100 400 x 200 450 x 200 450 x 225 450 x 300 600 x 200 600 x 225	390 x 90 390 x 190 440 x 190 440 x 215 440 x 290 590 x 190 590 x 215	75, 90, 100, 140, 190 75, 90, 100, 140, 190 75, 90, 100, 140, 190, 215 75, 90, 100, 140, 190, 215 75, 90, 100, 140, 190, 215 75, 90, 100, 140, 190, 215 75, 90, 100, 140, 190, 215
C	400 x 200 450 x 200 450 x 225 450 x 300 600 x 200 600 x 225	390 x 190 440 x 190 440 x 215 440 x 290 590 x 190 590 x 215	60, 75 60, 75 60, 75 60, 75 60, 75 60, 75

TIMBER

Weights of Timber

Material	kg/m³	lb/cu ft
General	806 (avg)	50 (avg)
Douglas fir	479	30
Yellow pine, spruce	479	30
Pitch pine	673	42
Larch, elm	561	35
Oak (English)	724 to 959	45 to 60
Teak	643 to 877	40 to 55
Jarrah	959	60
Greenheart	1040 to 1204	65 to 75
Quebracho	1285	80

Material	kg/m² per mm thickness	lb/sq ft per inch thickness
Wooden boarding and blocks		
Softwood	0.48	2.5
Hardwood	0.76	4
Hardboard	1.06	5.5
Chipboard	0.76	4
Plywood	0.62	3.25
Blockboard	0.48	2.5
Fibreboard	0.29	1.5
Wood-wool	0.58	3
Plasterboard	0.96	5
Weather boarding	0.35	1.8

Conversion Tables (for sawn timber only)

Inches	>	Millimetres	Feet	>	Metres
1		25	1		0.300
2		50	2		0.600
3		75	3		0.900
4		100	4		1.200
5		125	5		1.500
6		150	6		1.800
7		175	7		2.100
8		200	8		2.400
9		225	9		2.700
10		250	10		3.000
11		275	11		3.300
12		300	12		3.600
13		325	13		3.900
14		350	14		4.200
15		375	15		4.500
16		400	16		4.800
17		425	17		5.100
18		450	18		5.400
19		475	19		5.700
20		500	20		6.000
21		525	21		6.300
22		550	22		6.600
23		575	23		6.900
24		600	24		7.200

Planed Softwood

The finished end section size of planed timber is usually 3/16" less than the original size from which it is produced. This however varies slightly dependant upon availability of material and origin of species used.

Standard (timber) to cubic metres and cubic metres to standards (timber)

m³	m³/Standards	Standard
4.672	1	0.214
9.344	2	0.428
14.017	3	0.642
18.689	4	0.856
23.361	5	1.070
28.033	6	1.284
32.706	7	1.498
37.378	8	1.712
42.05	9	1.926
46.722	10	2.140
93.445	20	4.281
140.167	30	6.421
186.890	40	8.561
233.612	50	10.702
280.335	60	12.842
327.057	70	14.982
373.779	80	17.122
420.502	90	19.263
467.224	100	21.403

Standards (timber) to Cubic Metres and Cubic Metres to Standards (timber)

1 cu metre	=	35.3148 cu ft	=	0.21403 std
1 cu ft	=	0.028317 cu metres		
1 std	=	4.67227 cu metres		

Basic Sizes of Sawn Softwood Available (cross sectional areas)

Thickness (mm)	Width (mm)								
	75	100	125	150	175	200	225	250	300
16	x	x	x	x					
19	x	x	x	x					
22	x	x	x	x					
25	x	x	x	x	x	x	x	x	x
32	x	x	x	x	x	x	x	x	x
36	x	x	x	x					
38	x	x	x	x	x	x	x		
44	x	x	x	x	x	x	x	x	x
47*	x	x	x	x	x	x	x	x	x
50	x	x	x	x	x	x	x	x	x
63	x	x	x	x	x	x	x		
75		x	x	x	x	x	x	x	x
100		x		x		x		x	x
150				x		x			x
200						x			
250								x	
300									x

* This range of widths for 47 mm thickness will usually be found to be available in construction quality only.

Note: The smaller sizes below 100 mm thick and 250 mm width are normally but not exclusively of European origin. Sizes beyond this are usually of North and South American origin.

Basic Lengths of Sawn Softwood Available (metres)

1.80	2.10	3.00	4.20	5.10	6.00	7.20
	2.40	3.30	4.50	5.40	6.30	
	2.70	3.60	4.80	5.70	6.60	
		3.90			6.90	

Note: Lengths of 6.00 m and over will generally only be available from North American species and may have to be re-cut from larger sizes.

Reductions From Basic Size to Finished Size of Timber By Planing of Two Opposed Faces

Purpose	15 – 35 mm	36 – 100 mm	101 – 150 mm	Over 150 mm
a) constructional timber	3 mm	3 mm	5 mm	6 mm
b) matching interlocking boards	4 mm	4 mm	6 mm	6 mm
c) wood trim not specified in BS 584	5 mm	7 mm	7 mm	9 mm
d) joinery and cabinet work	7 mm	9 mm	11 mm	13 mm

Note: The reduction of width or depth is overall the extreme size and is exclusive of any reduction of the face by the machining of a tongue or lap joints.

METAL

Weights of Metals

Material	kg/m³	lb/cu ft
Metals, steel construction, etc.		
Iron		
- cast	7207	450
- wrought	7687	480
- ore - general	2407	150
- (crushed) Swedish	3682	230
Steel	7854	490
Copper		
- cast	8731	545
- wrought	8945	558
Brass	8497	530
Bronze	8945	558
Aluminium	2774	173
Lead	11322	707
Zinc (rolled)	7140	446
	g/mm² per metre	**lb/sq ft per foot**
Steel bars	7.85	3.4

Structural steelwork	Net weight of member @ 7854 kg/m³
riveted	+ 10% for cleats, rivets, bolts, etc
welded	+ 1.25% to 2.5% for welds, etc
Rolled sections	
beams	+ 2.5%
stanchions	+ 5% (extra for caps and bases)
Plate	
web girders	+ 10% for rivets or welds, stiffeners, etc

	kg/m	lb/ft
Steel stairs : industrial type		
1 m or 3 ft wide	84	56
Steel tubes		
50 mm or 2 in bore	5 to 6	3 to 4
Gas piping		
20 mm or 3/4 in	2	1¼

KERBS/EDGINGS/CHANNELS

Precast Concrete Kerbs to BS 7263
Straight kerb units: length from 450 to 915 mm

150 mm high x 125 mm thick
   bullnosed   type BN
   half battered   type HB3

255 mm high x 125 mm thick
   45 degree splayed   type SP
   half battered   type HB2

305 mm high x 150 mm thick
   half battered   type HB1

Quadrant kerb units

150 mm high x 305 and 455 mm radius to match	type BN	type QBN
150 mm high x 305 and 455 mm radius to match	type HB2, HB3	type QHB
150 mm high x 305 and 455 mm radius to match	type SP	type QSP
255 mm high x 305 and 455 mm radius to match	type BN	type QBN
255 mm high x 305 and 455 mm radius to match	type HB2, HB3	type QHB
225 mm high x 305 and 455 mm radius to match	type SP	type QSP

Angle kerb units
   305 x 305 x 225 mm high x 125 mm thick
      bullnosed external angle   type XA
      splayed external angle to match type SP   type XA
      bullnosed internal angle   type IA
      splayed internal angle to match type SP   type IA

Channels
   255 mm wide x 125 mm high flat   type CS1
   150 mm wide x 125 mm high flat type   CS2
   255 mm wide x 125 mm high dished   type CD

Transition kerb units

from kerb type SP to HB	left handed	type TL
	right handed	type TR
from kerb type BN to HB	left handed	type DL1
	right handed	type DR1
from kerb type BN to SP	left handed	type DL2
	right handed	type DR2

Radial Kerbs and Channels

All profiles of kerbs and channels

External radius (mm)	Internal radius (mm)
1000	3000
2000	4500
3000	6000
4500	7500
6000	9000
7500	1050
9000	1200
1050	
1200	

Precast Concrete Edgings to BS 7263

Round top type ER	Flat top type EF	Bullnosed top type EBN
150 x 50 mm	150 x 50 mm	150 x 50 mm
200 x 50	200 x 50	200 x 50
250 x 50	250 x 50	250 x 50

BASES

Cement Bound Material for Bases and Sub-bases

CBM1:	very carefully graded aggregate from 37.5 - 75 ym, with a 7-day strength of 4.5N/mm2
CBM2:	same range of aggregate as CBM1 but with more tolerance in each size of aggregate with a 7-day strength of 7.0N/mm2
CBM3:	crushed natural aggregate or blast furnace slag, graded from 37.5 mm - 150 ym for 40 mm aggregate, and from 20 - 75 ym for 20 mm aggregate, with a 7-day strength of 10N/mm2
CBM4:	crushed natural aggregate or blast furnace slag, graded from 37.5 mm - 150 ym for 40 mm aggregate, and from 20 - 75 ym for 20 mm aggregate, with a 7-day strength of 15N/mm2

INTERLOCKING BRICK/BLOCK ROADS/PAVINGS

Sizes of Precast Concrete Paving Blocks to BS 6717: Part 1

Type R blocks
200 x 100 x 60 mm
200 x 100 x 65
200 x 100 x 80
200 x 100 x 100

Type S
Any shape within a 295 mm space

Sizes of Clay Brick Pavers to BS 6677: Part 1
200 x 100 x 50 mm thick
200 x 100 x 65
210 x 105 x 50
210 x 105 x 65
215 x 102.5 x 50
215 x 102.5 x 65

Type PA: 3 kN
Footpaths and pedestrian areas, private driveways, car parks, light vehicle traffic and over-run.

Type PB: 7 kN
Residential roads, lorry parks, factory yards, docks, petrol station forecourts, hardstandings, bus stations.

PAVING AND SURFACING

Weights and Sizes of Paving and Surfacing

Description of Item		Quantity per tonne
Paving 50 mm thick	900 x 600 mm	15
Paving 50 mm thick	750 x 600 mm	18
Paving 50 mm thick	600 x 600 mm	23
Paving 50 mm thick	450 x 600 mm	30
Paving 38 mm thick	600 x 600 mm	30
Path edging	914 x 50 x 150 mm	60
Kerb (including radius and tapers)	125 x 254 x 914 mm	15
Kerb (including radius and tapers)	125 x 150 x 914 mm	25
Square channel	125 x 254 x 914 mm	15
Dished channel	125 x 254 x 914 mm	15
Quadrants	300 x 300 x 254 mm	19
Quadrants	450 x 450 x 254 mm	12
Quadrants	300 x 300 x 150 mm	30
Internal angles	300 x 300 x 254 mm	30
Fluted pavement channel	255 x 75 x 914 mm	25
Corner stones	300 x 300 mm	80
Corner stones	360 x 360 mm	60
Cable covers	914 x 175 mm	55
Gulley kerbs	220 x 220 x 150 mm	60
Gulley kerbs	220 x 200 x 75 mm	120

Weights and Sizes of Paving and Surfacing

Material	kg/m³	lb/cu yd
Tarmacadam	2306	3891
Macadam (waterbound)	2563	4325
Vermiculite (aggregate)	64-80	108-135
Terracotta	2114	3568
Cork - compressed	388	24
	kg/m²	**lb/sq ft**
Clay floor tiles, 12.7 mm	27.3	5.6
Pavement lights	122	25
Damp proof course	5	1
	kg/m² per mm thickness	**lb/sq ft per inch thickness**
Paving Slabs (stone)	2.3	12
Granite setts	2.88	15
Asphalt	2.30	12
Rubber flooring	1.68	9
Poly-vinylchloride	1.94 (avg)	10 (avg)

Coverage (m²) Per Cubic Metre of Materials Used as Sub-bases or Capping Layers

Consolidated thickness laid in (mm)	Square metre coverage		
	Gravel	Sand	Hardcore
50	15.80	16.50	-
75	10.50	11.00	-
100	7.92	8.20	7.42
125	6.34	6.60	5.90
150	5.28	5.50	4.95
175	-	-	4.23
200	-	-	3.71
225	-	-	3.30
300	-	-	2.47

Approximate Rate of Spreads

Average thickness of course mm	Description	Approximate rate of spread			
		Open Textured		Dense, Medium & Fine Textured	
		kg/m²	m²/t	kg/m²	m²/t
35	14 mm open textured or dense wearing course	60-75	13-17	70-85	12-14
40	20 mm open textured or dense base course	70-85	12-14	80-100	10-12
45	20 mm open textured or dense base course	80-100	10-12	95-100	9-10
50	20 mm open textured or dense, or 28 mm dense base course	85-110	9-12	110-120	8-9
60	28 mm dense base course, 40 mm open textured of dense base course or 40 mm single course as base course		8-10	130-150	7-8
65	28 mm dense base course, 40 mm open textured or dense base course or 40 mm single course	100-135	7-10	140-160	6-7
75	40 mm single course, 40 mm open textured or dense base course, 40 mm dense roadbase	120-150	7-8	165-185	5-6
100	40 mm dense base course or roadbase	-	-	220-240	4-4.5

Surface Dressing Roads: Coverage (m²) per Tonne of Material

Size in mm	Sand	Granite chips	Gravel	Limestone Chips
Sand	168	-	-	-
3	-	148	152	165
6	-	130	133	144
9	-	111	114	123
13	-	85	87	95
19	-	68	71	78

Sizes of Flags to BS 7263

Reference	Nominal Size (mm)	Thickness (mm)
A	600 x 450	50 and 63
B	600 x 600	50 and 63
C	600 x 750	50 and 63
D	600 x 900	50 and 63
E	450 x 450	50 and 70 chamfered top surface
F	400 x 400	50 and 65 chamfered top surface
G	300 x 300	50 and 60 chamfered top surface

Sizes of Natural Stone Setts to BS 435

Width (mm)		Length (mm)		Depth (mm)
100	x	100	x	100
75	x	150 to 250	x	125
75	x	150 to 250	x	150
100	x	150 to 250	x	100
100	x	150 to 250	x	150

SPORTS

Sizes of Sports Areas
Sizes in metres given include clearances:

Association football	Senior		114	x	72
	Junior		108	x	58
	International		100 - 110	x	64 - 75
Football	American	Pitch	109.80	x	48.80
		Overall	118.94	x	57.94
	Australian Rules	Overall	135 - 185	x	110 - 155
	Canadian	Overall	145.74	x	59.47
	Gaelic		128 - 146.4	x	76.8 - 91.50
Handball			91 - 110	x	55 - 65
Hurling			137	x	82
Rugby	Union pitch		56	x	81
	League pitch		134	x	80
Hockey pitch			100.5	x	61
Men's lacrosse pitch			106	x	61
Women's lacrosse pitch			110	x	60
Target archery ground			150	x	50
Archery (Clout)			7.3m	firing area	
			Range 109.728 (Women), 146.304 (Men).		
			182.88 (Normal range)		
400m Running Track			115.61	bend length	x 2
6 lanes			84.39	straight length	x 2
		Overall	176.91 long	x	92.52 wide
Baseball		Overall	60m	x	70
Basketball			14.0	x	26.0

Sizes of Sports Areas - cont'd

Camogie			91 - 110	x	54 - 68
Discus and Hammer		Safety cage 2.74m square			
		Landing area 45 arc (65° safety) 70 m radius			
Javelin		Runway	36.5	x	4.27
		Landing area	80 - 95	x	48
Jump	High	Running area	38.8	x	19
		Landing area	5	x	4
	Long	Runway	45	x	1.22
		Landing area	9	x	2.750
	Triple	Runway	45	x	1.22
		Landing area	7.3	x	2.75
Korfball			90	x	40
Netball			15.25	x	30.48
Pole Vault		Runway	45	x	1.22
		Landing area	5	x	5
Polo			275	x	183
Rounders		Overall	19	x	17
Shot Putt		Base	2.135	dia	
		Landing area	65° arc,	25m radius	
Shinty			128 -183	x	64 - 91.5
Tennis		Court	23.77	x	10.97
		Overall minimum	36.27	x	18.29
Tug-of-war			46	x	5

SEEDING/TURFING AND PLANTING

BS 3882: 1994 Topsoil Quality

Topsoil grade	Properties
Premium	natural topsoil, high fertility, loamy texture, good soil structure, suitable for intensive cultivation
General Purpose	natural or manufactured topsoil of lesser quality than Premium, suitable for agriculture or amenity landscape, may need fertilizer or soil structure improvement.
Economy	selected subsoil, natural mineral deposit such as river silt or greensand. The grade comprises two sub-grades; "Low clay" and "High clay" which is more liable to compaction in handling. This grade is suitable for low production agricultural land and amenity woodland or conservation planting areas.

Forms of Trees to BS 3936: 1992

Standards:	shall be clear with substantially straight stems. Grafted and budded trees shall have no more than a slight bend at the union. Standards shall be designated as Half, Extra light, Light, Standard, Selected standard, Heavy, and Extra heavy.
Sizes of Standards	
Heavy standard	12-14 cm girth x 3.50 to 5.00 m high
Extra Heavy standard	14-16 cm girth x 4.25 to 5.00 m high
Extra Heavy standard	16-18 cm girth x 4.25 to 6.00 m high
Extra Heavy standard	18-20 cm girth x 5.00 to 6.00 m high
Semi-mature trees:	between 6.0 m and 12.0 m tall with a girth of 20 to 75 cm at 1.0 m above ground.
Feathered trees:	shall have a defined upright central leader, with stem furnished with evenly spread and balanced lateral shoots down to or near the ground.
Whips:	shall be without significant feather growth as determined by visual inspection.
Multi-stemmed trees:	shall have two or more main stems at, near, above or below ground.

Seedlings grown from seed and not transplanted shall be specified when ordered for sale as:

1+0 one year old seedling		
2+0 two year old seedling		
1+1 one year seed bed,	one year transplanted	= two year old seedling
1+2 one year seed bed,	two years transplanted	= three year old seedling
2+1 two year seed bed,	one year transplanted	= three year old seedling
1u1 two years seed bed,	undercut after 1 year	= two year old seedling
2u2 four years seed bed,	undercut after 2 years	= four year old seedling

Cuttings

The age of cuttings (plants grown from shoots, stems, or roots of the mother plant) shall be specified when ordered for sale. The height of transplants and undercut seedlings/cuttings (which have been transplanted or undercut at least once) shall be stated in centimetres. The number of growing seasons before and after transplanting or undercutting shall be stated.

0+1	one year cutting
0+2	two year cutting
0+1+1	one year cutting bed, one year transplanted = two year old seedling
0+1+2	one year cutting bed, two years transplanted = three year old seedling

Grass Cutting Capacities in m2 per Hour

Speed mph	Width Of Cut in metres												
	0.5	0.7	1.0	1.2	1.5	1.7	2.0	2.0	2.1	2.5	2.8	3.0	3.4
1.0	724	1127	1529	1931	2334	2736	3138	3219	3380	4023	4506	4828	5472
1.5	1086	1690	2293	2897	3500	4104	4707	4828	5069	6035	6759	7242	8208
2.0	1448	2253	3058	3862	4667	5472	6276	6437	6759	8047	9012	9656	10944
2.5	1811	2816	3822	4828	5834	6840	7846	8047	8449	10058	11265	12070	13679
3.0	2173	3380	4587	5794	7001	8208	9415	9656	10139	12070	13518	14484	16415
3.5	2535	3943	5351	6759	8167	9576	10984	11265	11829	14082	15772	16898	19151
4.0	2897	4506	6115	7725	9334	10944	12553	12875	13518	16093	18025	19312	21887
4.5	3259	5069	6880	8690	10501	12311	14122	14484	15208	18105	20278	21726	24623
5.0	3621	5633	7644	9656	11668	13679	15691	16093	16898	20117	22531	24140	27359
5.5	3983	6196	8409	10622	12834	15047	17260	17703	18588	22128	24784	26554	30095
6.0	4345	6759	9173	11587	14001	16415	18829	19312	20278	24140	27037	28968	32831
6.5	4707	7322	9938	12553	15168	17783	20398	20921	21967	26152	29290	31382	35566
7.0	5069	7886	10702	13518	16335	19151	21967	22531	23657	28163	31543	33796	38302

Number of Plants per m2: For Plants Planted on an Evenly Spaced Grid

Planting distances

mm	0.10	0.15	0.20	0.25	0.35	0.40	0.45	0.50	0.60	0.75	0.90	1.00	1.20	1.50
0.10	100.00	66.67	50.00	40.00	28.57	25.00	22.22	20.00	16.67	13.33	11.11	10.00	8.33	6.67
0.15	66.67	44.44	33.33	26.67	19.05	16.67	14.81	13.33	11.11	8.89	7.41	6.67	5.56	4.44
0.20	50.00	33.33	25.00	20.00	14.29	12.50	11.11	10.00	8.33	6.67	5.56	5.00	4.17	3.33
0.25	40.00	26.67	20.00	16.00	11.43	10.00	8.89	8.00	6.67	5.33	4.44	4.00	3.33	2.67
0.35	28.57	19.05	14.29	11.43	8.16	7.14	6.35	5.71	4.76	3.81	3.17	2.86	2.38	1.90
0.40	25.00	16.67	12.50	10.00	7.14	6.25	5.56	5.00	4.17	3.33	2.78	2.50	2.08	1.67
0.45	22.22	14.81	11.11	8.89	6.35	5.56	4.94	4.44	3.70	2.96	2.47	2.22	1.85	1.48
0.50	20.00	13.33	10.00	8.00	5.71	5.00	4.44	4.00	3.33	2.67	2.22	2.00	1.67	1.33
0.60	16.67	11.11	8.33	6.67	4.76	4.17	3.70	3.33	2.78	2.22	1.85	1.67	1.39	1.11
0.75	13.33	8.89	6.67	5.33	3.81	3.33	2.96	2.67	2.22	1.78	1.48	1.33	1.11	0.89
0.90	11.11	7.41	5.56	4.44	3.17	2.78	2.47	2.22	1.85	1.48	1.23	1.11	0.93	0.74
1.00	10.00	6.67	5.00	4.00	2.86	2.50	2.22	2.00	1.67	1.33	1.11	1.00	0.83	0.67
1.20	8.33	5.56	4.17	3.33	2.38	2.08	1.85	1.67	1.39	1.11	0.93	0.83	0.69	0.56
1.50	6.67	4.44	3.33	2.67	1.90	1.67	1.48	1.33	1.11	0.89	0.74	0.67	0.56	0.44

Grass Clippings Wet: Based on 3.5 m3 /tonne

Annual Kg/100 m2	Average 20 cuts Kg/100m2	m2 /tonne	m2 /m3
32.0	1.6	61162.1	214067.3

Nr of Cuts		22	20	18	16	12	4
Kg/cut		1.45	1.60	1.78	2.00	2.67	8.00
Area capacity of 3 tonne vehicle per load							
m2		206250	187500	168750	150000	112500	37500
Load m3		**100 m2 units / m3 of vehicle space**					
	1	196.4	178.6	160.7	142.9	107.1	35.7
	2	392.9	357.1	321.4	285.7	214.3	71.4
	3	589.3	535.7	482.1	428.6	321.4	107.1
	4	785.7	714.3	642.9	571.4	428.6	142.9
	5	982.1	892.9	803.6	714.3	535.7	178.6

Transportation of Trees

To unload large trees a machine with the necessary lifting strength is required. The weight of the trees must therefore be known in advance. The following table gives a rough overview. The additional columns with root ball dimensions and the number of plants per trailer provide additional information for example about preparing planting holes and calculating unloading times.

Girth in cm	Root Ball Diameter in cm	Ball Height in cm	Weight in kg	Numbers of Trees per Trailer
16 - 18	50 - 60	40	150	100 - 120
18 - 20	60 - 70	40 - 50	200	80 - 100
20 - 25	60 - 70	40 - 50	270	50 - 70
25 - 30	80	50 - 60	350	50
30 - 35	90 - 100	60 - 70	500	12 - 18
35 - 40	100 - 110	60 - 70	650	10 - 15
40 - 45	110 - 120	60 - 70	850	8 - 12
45 - 50	110 - 120	60 - 70	1100	5 - 7
50 - 60	130 - 140	60 - 70	1600	1 - 3
60 - 70	150 - 160	60 - 70	2500	1
70 - 80	180 - 200	70	4000	1
80 - 90	200 - 220	70 - 80	5500	1
90 - 100	230 - 250	80 - 90	7500	1
100 - 120	250 - 270	80 - 90	9500	1

Data supplied by Lorenz von Ehren GmbH

The information in the table is approximate; deviations depend on soil type, genus and weather.

FENCING AND GATES

Types of Preservative to BS 5589:1989

Creosote (tar oil) can be "factory" applied	by pressure to BS 144: pts 1&2
	by immersion to BS 144: pt 1
	by hot and cold open tank to BS 144: pts 1&2
Copper/chromium/arsenic (CCA)	by full cell process to BS 4072 pts 1&2
Organic solvent (OS)	by double vacuum (vacvac) to BS 5707 pts 1&3
	by immersion to BS 5057 pts 1&3
Pentachlorophenol (PCP)	by heavy oil double vacuum to BS 5705 pts 2&3

Boron diffusion process (treated with disodium octaborate to BWPA Manual 1986.

Note: Boron is used on green timber at source and the timber is supplied dry.

Cleft Chestnut Pale Fences to BS 1722: Part 4:1986

Pales	Pale spacing	Wire lines	
900 mm long	75 mm	2	temporary protection
1050	75 or 100	2	light protective fences
1200	75	3	perimeter fences
1350	75	3	perimeter fences
1500	50	3	narrow perimeter fences
1800	50	3	light security fences

Close-boarded Fences to BS 1722 :Pt 5: 1986

Close-boarded fences 1.05 to 1.8m high
Type BCR (recessed) or BCM (morticed) with concrete posts 140 x 115 mm tapered and Type BW with timber posts.

Palisade Fences to BS 1722: pt 6: 1986

Wooden palisade fences
Type WPC with concrete posts 140 x 115 mm tapered and Type WPW with timber posts.

For both types of fence:

Height of fence 1050 mm:	two rails
Height of fence 1200 mm:	two rails
Height of fence 1500 mm:	three rails
Height of fence 1650 mm:	three rails
Height of fence 1800 mm:	three rails

Post and Rail Fences to BS 1722: part 7

Wooden post and rail fences
Type MPR 11/3 morticed rails and Type SPR 11/3 nailed rails
Height to top of rail 1100 mm
Rails: three rails 87 mm 38 mm

Type MPR 11/4 morticed rails and Type SPR 11/4 nailed rails
Height to top of rail 1100 mm
Rails: four rails 87 mm 38 mm.

Type MPR 13/4 morticed rails and Type SPR 13/4 nailed rails
Height to top of rail 1300 mm
Rail spacing 250 mm, 250 mm, and 225 mm from top
Rails: four rails 87 mm 38 mm.

Steel Posts to BS 1722: Part 1

Rolled steel angle iron posts for chain link fencing:

Posts	Fence height	Strut	Straining post
1500 x 40 x 40 x 5 mm	900 mm	1500 x 40 x 40 x 5 mm	1500 x 50 x 50 x 6 mm
1800 x 40 x 40 x 5 mm	1200 mm	1800 x 40 x 40 x 5 mm	1800 x 50 x 50 x 6 mm
2000 x 45 x 45 x 5 mm	1400 mm	2000 x 45 x 45 x 5 mm	2000 x 60 x 60 x 6 mm
2600 x 45 x 45 x 5 mm	1800 mm	2600 x 45 x 45 x 5 mm	2600 x 60 x 60 x 6 mm
3000 x 50 x 50 x 6 mm with arms	1800 mm	2600 x 45 x 45 x 5 mm	3000 x 60 x 60 x 6 mm

Concrete Posts to BS 1722: Part 1

Concrete posts for chain link fencing:

Posts and straining posts	Fence height	Strut
1570 mm 100 x 100 mm	900 mm	1500 mm x 75 x 75 mm
1870 mm 125 x 125 mm	1200 mm	1830 mm x 100 x 75 mm
2070 mm 125 x 125 mm	1400 mm	1980 mm x 100 x 75 mm
2620 mm 125 x 125 mm	1800 mm	2590 mm x 100 x 85 mm
3040 mm 125 x 125 mm	1800 mm	2590 mm x 100 x 85 mm (with arms)

Rolled Steel Angle Posts to BS 1722: Part 2

Rolled steel angle posts for rectangular wire mesh (field) fencing

Posts	Fence height	Strut	Straining post
1200 x 40 x 40 x 5 mm	600 mm	1200 x 75 x 75 mm	1350 x 100 x 100 mm
1400 x 40 x 40 x 5 mm	800 mm	1400 x 75 x 75 mm	1550 x 100 x 100 mm
1500 x 40 x 40 x 5 mm	900 mm	1500 x 75 x 75 mm	1650 x 100 x 100 mm
1600 x 40 x 40 x 5 mm	1000 mm	1600 x 75 x 75 mm	1750 x 100 x 100 mm
1750 x 40 x 40 x 5 mm	1150 mm	1750 x 75 x 100 mm	1900 x 125 x 125 mm

Concrete Posts to BS 1722: Part 2

Concrete posts for rectangular wire mesh (field) fencing

Posts	Fence height	Strut	Straining post
1270 x 100 x 100 mm	600 mm	1200 x 75 x 75 mm	1420 x 100 x 100 mm
1470 x 100 x 100 mm	800 mm	1350 x 75 x 75 mm	1620 x 100 x 100 mm
1570 x 100 x 100 mm	900 mm	1500 x 75 x 75 mm	1720 x 100 x 100 mm
1670 x 100 x 100 mm	600 mm	1650 x 75 x 75 mm	1820 x 100 x 100 mm
1820 x 125 x 125 mm	1150 mm	1830 x 75 x 100 mm	1970 x 125 x 125 mm

Cleft Chestnut Pale Fences to BS 1722: part 4: 1986
Timber Posts to BS 1722: Part 2

Timber posts for wire mesh and hexagonal wire netting fences
Round timber for general fences

Posts	Fence height	Strut	Straining post
1300 x 65 mm dia.	600 mm	1200 x 80 mm dia	1450 x 100 mm dia
1500 x 65 mm dia	800 mm	1400 x 80 mm dia	1650 x 100 mm dia
1600 x 65 mm dia.	900 mm	1500 x 80 mm dia	1750 x 100 mm dia
1700 x 65 mm dia.	1050 mm	1600 x 80 mm dia	1850 x 100 mm dia
1800 x 65 mm dia.	1150 mm	1750 x 80 mm dia	2000 x 120 mm dia

Squared timber for general fences

Posts	Fence height	Strut	Straining post
1300 x 75 x 75 mm	600 mm	1200 x 75 x 75 mm	1450 x 100 x 100 mm
1500 x 75 x 75 mm	800 mm	1400 x 75 x 75 mm	1650 x 100 x 100 mm
1600 x 75 x 75 mm	900 mm	1500 x 75 x 75 mm	1750 x 100 x 100 mm
1700 x 75 x 75 mm	1050 mm	1600 x 75 x 75 mm	1850 x 100 x 100 mm
1800 x 75 x 75 mm	1150 mm	1750 x 75 x 75 mm	2000 x 125 x 100 mm

Steel Fences to BS 1722: Pt 9: 1992

	Fence height	Top/bottom rails and flat posts	Vertical bars
Light	1000 mm	40 x 10 mm 450 mm in ground	12 mm dia at 115 mm cs
	1200 mm	40 x 10 mm 550 mm in ground	12 mm dia at 115 mm cs
	1400 mm	40 x 10 mm 550 mm in ground	12 mm dia at 115 mm cs
Light	1000 mm	40 x 10 mm 450 mm in ground	16 mm dia at 120 mm cs
	1200 mm	40 x 10 mm 550 mm in ground	16 mm dia at 120 mm cs
	1400 mm	40 x 10 mm 550 mm in ground	16 mm dia at 120 mm cs
Medium	1200 mm	50 x 10 mm 550 mm in ground	20 mm dia at 125 mm cs
	1400 mm	50 x 10 mm 550 mm in ground	20 mm dia at 125 mm cs
	1600 mm	50 x 10 mm 600 mm in ground	22 mm dia at 145 mm cs
		50 x 10 mm 600 mm in ground	
	1800 mm	50 x 10 mm 600 mm in ground	22 mm dia at 145 mm cs
Heavy	1600 mm	50 x 10 mm 600 mm in ground	22 mm dia at 145 mm cs
		50 x 10 mm 600 mm in ground	
	1800 mm	50 x 10 mm 600 mm in ground	22 mm dia at 145 mm cs
	2000 mm	50 x 10 mm 600 mm in ground	22 mm dia at 145 mm cs
		50 x 10 mm 600 mm in ground	
	2200 mm	50 x 10 mm 600 mm in ground	22 mm dia at 145 mm cs

Notes:
Mild steel fences: round or square verticals; flat standards and horizontals.
Tops of vertical bars may be bow-top, blunt, or pointed.
Round or square bar railings.

Timber Field Gates to BS 3470: 1975

Gates made to this standard are designed to open one way only.
All timber gates are 1100 mm high.
Width over stiles 2400, 2700, 3000, 3300, 3600, and 4200 mm.
Gates over 4200 mm should be made in two leaves.

Steel Field Gates to BS 3470: 1975

All steel gates are 1100 mm high.
Heavy duty: width over stiles 2400, 3000, 3600 and 4500 mm
Light duty: width over stiles 2400, 3000, and 3600 mm

Domestic Front Entrance Gates to BS 4092: part 1: 1966

Metal gates: Single gates are 900 mm high minimum, 900 mm, 1000 mm and 1100 mm wide

Domestic Front Entrance Gates to BS 4092: part 2: 1966

Wooden gates: All rails shall be tenoned into the stiles
Single gates are 840 mm high minimum, 801 mm and 1020 mm wide
Double gates are 840 mm high minimum, 2130, 2340 and 2640 mm wide

Timber Bridle Gates to BS 5709:1979 (Horse Or Hunting Gates)

Gates open one way only
Minimum width between posts   1525 mm
Minimum height   1100 mm

Timber Kissing Gates to BS 5709:1979

Minimum width   700 mm
Minimum height   1000 mm
Minimum distance between shutting posts   600 mm
Minimum clearance at mid-point   600 mm

Metal Kissing Gates to BS 5709:1979

Sizes are the same as those for timber kissing gates.
Maximum gaps between rails 120 mm.

Categories of Pedestrian Guard Rail to BS 3049:1976

Class A for normal use.
Class B where vandalism is expected.
Class C where crowd pressure is likely.

DRAINAGE

Weights and Dimensions - Vitrified Clay Pipes

Product	Nominal diameter (mm)	Effective length (mm)	BS 65 limits of tolerance min (mm)	max (mm)	Crushing Strength (kN/m)	Weight kg/pipe	kg/m
Supersleve	100	1600	96	105	35.00	14.71	9.19
	150	1750	146	158	35.00	29.24	16.71
Hepsleve	225	1850	221	236	28.00	84.03	45.42
	300	2500	295	313	34.00	193.05	77.22
	150	1500	146	158	22.00	37.04	24.69
Hepseal	225	1750	221	236	28.00	85.47	48.84
	300	2500	295	313	34.00	204.08	81.63
	400	2500	394	414	44.00	357.14	142.86
	450	2500	444	464	44.00	454.55	181.63
	500	2500	494	514	48.00	555.56	222.22
	600	2500	591	615	57.00	796.23	307.69
	700	3000	689	719	67.00	1111.11	370.45
	800	3000	788	822	72.00	1351.35	450.45
Hepline	100	1600	95	107	22.00	14.71	9.19
	150	1750	145	160	22.00	29.24	16.71
	225	1850	219	239	28.00	84.03	45.42
	300	1850	292	317	34.00	142.86	77.22
Hepduct (Conduit)	90	1500	-	-	28.00	12.05	8.03
	100	1600	-	-	28.00	14.71	9.19
	125	1750	-	-	28.00	20.73	11.84
	150	1750	-	-	28.00	29.24	16.71
	225	1850	-	-	28.00	84.03	45.42
	300	1850	-	-	34.00	142.86	77.22

Weights and Dimensions - Vitrified Clay Pipes

Nominal internal diameter (mm)	Nominal wall thickness (mm)	Approximate weight kg/m
150	25	45
225	29	71
300	32	122
375	35	162
450	38	191
600	48	317
750	54	454
900	60	616
1200	76	912
1500	89	1458
1800	102	1884
2100	127	2619

Wall thickness, weights and pipe lengths vary, depending on type of pipe required.

The particulars shown above represent a selection of available diameters and are applicable to strength class 1 pipes with flexible rubber ring joints.

Tubes with Ogee joints are also available.

DRAINAGE – cont'd

Weights and Dimensions - PVC-U Pipes

	Nominal size	Mean outside diameter (mm)		Wall thickness	Weight
		min	max	(mm)	kg/m
Standard pipes					
	82.4	82.4	82.7	3.2	1.2
	110.0	110.0	110.4	3.2	1.6
	160.0	160.0	160.6	4.1	3.0
	200.0	200.0	200.6	4.9	4.6
	250.0	250.0	250.7	6.1	7.2
Perforated pipes					
- heavy grade	As above	As above	As above	As above	As above
- thin wall	82.4	82.4	82.7	1.7	-
	110.0	110.0	110.4	2.2	-
	160.0	160.0	160.6	3.2	-

Width of Trenches Required for Various Diameters of Pipes

Pipe diameter (mm)	Trench n.e. 1.5 m deep (mm)	Trench over 1.5 m deep (mm)
n.e. 100	450	600
100-150	500	650
150-225	600	750
225-300	650	800
300-400	750	900
400-450	900	1050
450-600	1100	1300

DRAINAGE BELOW GROUND AND LAND DRAINAGE

Flow of Water Which Can Be Carried by Various Sizes of Pipe

Clay or concrete pipes

	Gradient of pipeline							
	1:10	1:20	1:30	1:40	1:50	1:60	1:80	1:100
Pipe size	**Flow in litres per second**							
DN 100 15.0	8.5	6.8	5.8	5.2	4.7	4.0	3.5	
DN 150 28.0	19.0	16.0	14.0	12.0	11.0	9.1	8.0	
DN 225 140.0	95.0	76.0	66.0	58.0	53.0	46.0	40.0	

Plastic pipes

	Gradient of pipeline							
	1:10	1:20	1:30	1:40	1:50	1:60	1:80	1:100
Pipe size	**Flow in litres per second**							
82.4 mm i/dia	12.0	8.5	6.8	5.8	5.2	4.7	4.0	3.5
110 mm i/dia	28.0	19.0	16.0	14.0	12.0	11.0	9.1	8.0
160 mm i/dia	76.0	53.0	43.0	37.0	33.0	29.0	25.0	22.0
200 mm i/dia	140.0	95.0	76.0	66.0	58.0	53.0	46.0	40.0

Vitrified (Perforated) Clay Pipes and Fittings to BS En 295-5 1994

Length not specified		
75 mm bore	**250 mm bore**	**600 mm bore**
100	300	700
125	350	800
150	400	1000
200	450	1200
225	500	

Pre-cast Concrete Pipes: Pre-stressed Non-pressure Pipes and Fittings: Flexible Joints to BS 5911: Pt. 103: 1994

Rationalized metric nominal sizes: 450, 500			
Length:	500	-	1000 by 100 increments
	1000	-	2200 by 200 increments
	2200	-	2800 by 300 increments
Angles: length:	450 - 600 angles 45, 22.5, 11.25 °		
	600 or more angles 22.5, 11.25 °		

Pre-cast Concrete Pipes: Un-reinforced and Circular Manholes and Soakaways to BS 5911: Pt. 200: 1994

Nominal Sizes:	
Shafts:	675, 900 mm
Chambers:	900, 1050, 1200, 1350, 1500, 1800, 2100, 2400, 2700, 3000 mm.
Large chambers:	To have either tapered reducing rings or a flat reducing slab in order to accept the standard cover.
Ring depths:	1.　　　300 - 1200 mm by 300 mm increments except for bottom slab and rings　below cover slab, these are by 150 mm increments. 2.　　　250 - 1000 mm by 250 mm increments except for bottom slab and rings　below cover slab, these are by 125 mm increments.
Access hole:	750 x 750 mm for DN 1050 chamber 1200 x 675 mm for DN 1350 chamber

Calculation of Soakaway Depth

The following formula determines the depth of concrete ring soakaway that would be required for draining given amounts of water.

$$h = \frac{4ar}{3\pi D^2}$$

h　=　depth of the chamber below the invert pipe
A　=　The are to be drained
r　=　The hourly rate of rainfall (50 mm per hour)
π　=　pi
D　=　internal diameter of the soakaway

This table shows the depth of chambers in each ring size which would be required to contain the volume of water specified. These allow a recommended storage capacity of 1/3 (one third of the hourly rain fall figure).

Table Showing Required Depth of Concrete Ring Chambers in Metres

AREA m^2	50	100	150	200	300	400	500
Ring Size							
0.9	1.31	2.62	3.93	5.24	7.86	10.48	13.10
1.1	0.96	1.92	2.89	3.85	5.77	7.70	9.62
1.2	0.74	1.47	2.21	2.95	4.42	5.89	7.37
1.4	0.58	1.16	1.75	2.33	3.49	4.66	5.82
1.5	0.47	0.94	1.41	1.89	2.83	3.77	4.72
1.8	0.33	0.65	0.98	1.31	1.96	2.62	3.27
2.1	0.24	0.48	0.72	0.96	1.44	1.92	2.41
2.4	0.18	0.37	0.55	0.74	1.11	1.47	1.84
2.7	0.15	0.29	0.44	0.58	0.87	1.16	1.46
3.0	0.12	0.24	0.35	0.47	0.71	0.94	1.18

Pre-cast Concrete Inspection Chambers and Gullies to BS 5911: Pt. 230: 1994

Nominal sizes:	375 diameter, 750, 900 mm deep
	450 diameter, 750, 900, 1050, 1200 mm deep
Depths:	from the top for trapped or un-trapped units:
	centre of outlet 300 mm
	invert (bottom) of the outlet pipe 400 mm
Depth of water seal for trapped gullies:	
	85 mm, rodding eye int. diam. 100 mm
Cover slab:	65 mm min.

Ductile Iron Pipes to BS En 598: 1995

Type K9 with flexible joints should be used for surface water drainage.
5500 mm or 8000 mm long

80 mm bore	400 mm bore	1000 mm bore
100	450	1100
150	500	1200
200	600	1400
250	700	1600
300	800	
350	900	

Bedding Flexible Pipes: PVC-U Or Ductile Iron

Type 1	=	100 mm fill below pipe, 300 mm above pipe: single size material
Type 2	=	100 mm fill below pipe, 300 mm above pipe: single size or graded material
Type 3	=	100 mm fill below pipe, 75 mm above pipe with concrete protective slab over
Type 4	=	100 mm fill below pipe, fill laid level with top of pipe
Type 5	=	200 mm fill below pipe, fill laid level with top of pipe
Concrete	=	25 mm sand blinding to bottom of trench, pipe supported on chocks,
		100 mm concrete under the pipe, 150 mm concrete over the pipe.

Bedding Rigid Pipes: Clay Or Concrete

(for vitrified clay pipes the manufacturer should be consulted)

Class D:	Pipe laid on natural ground with cut-outs for joints, soil screened to remove stones over 40 mm and returned over pipe to 150 mm min depth. Suitable for firm ground with trenches trimmed by hand.
Class N:	Pipe laid on 50 mm granular material of graded aggregate to Table 4 of BS 882, or 10 mm aggregate to Table 6 of BS 882, or as dug light soil (not clay) screened to remove stones over 10 mm. Suitable for machine dug trenches.
Class B:	As Class N, but with granular bedding extending half way up the pipe diameter.
Class F:	Pipe laid on 100 mm granular fill to BS 882 below pipe, minimum 150 mm granular fill above pipe: single size material. Suitable for machine dug trenches.
Class A:	Concrete 100 mm thick under the pipe extending half way up the pipe, backfilled with the appropriate class of fill. Used where there is only a very shallow fall to the drain. Class A bedding allows the pipes to be laid to an exact gradient.
Concrete surround:	25 mm sand blinding to bottom of trench, pipe supported on chocks, 100 mm concrete under the pipe, 150 mm concrete over the pipe. It is preferable to bed pipes under slabs or wall in granular material.

PIPED SUPPLY SYSTEMS

Identification of Service Tubes From Utility to Dwellings

Utility	Colour	Size	Depth
British Telecom	grey	54 mm od	450 mm
Electricity	black	38 mm od	450 mm
Gas	yellow	42 mm od rigid	450 mm
		60 mm od convoluted	
Water	may be blue	(normally untubed)	750 mm

ELECTRICAL SUPPLY/POWER/LIGHTING SYSTEMS

Electrical Insulation Class En 60.598 BS 4533

Class 1:	luminaires comply with class 1 (I) earthed electrical requirements
Class 2:	luminaires comply with class 2 (II) double insulated electrical requirements
Class 3:	luminaires comply with class 3 (III) electrical requirements

Protection to Light Fittings

BS EN 60529:1992 Classification for degrees of protection provided by enclosures.
(IP Code - International or ingress Protection)

1st characteristic: against ingress of solid foreign objects		
The figure	2	indicates that fingers cannot enter
	3	that a 2.5 mm diameter probe cannot enter
	4	that a 1.0 mm diameter probe cannot enter
	5	the fitting is dust proof (no dust around live parts)
	6	the fitting is dust tight (no dust entry)

2nd characteristic: ingress of water with harmful effects		
The figure	0	indicates unprotected
	1	vertically dripping water cannot enter
	2	water dripping 15° (tilt) cannot enter
	3	spraying water cannot enter
	4	splashing water cannot enter
	5	jetting water cannot enter
	6	powerful jetting water cannot enter
	7	proof against temporary immersion
	8	proof against continuous immersion

Optional additional codes:		A-D protects against access to hazardous parts
	H	High voltage apparatus
	M	fitting was in motion during water test
	S	fitting was static during water test
	W	protects against weather

Marking code arrangement:	(example) IPX5S = IP (International or Ingress Protection)
	X (denotes omission of first characteristic);
	5 = jetting;
	S = static during water test.

LANDFILL TAX

Waste Liable at the Lower Rate

Group	Description of material	Conditions	
1	Rocks and soils	Naturally occurring	includes clay, sand, gravel, sandstone, limestone, crushed stone, china clay, construction stone, stone from the demolition of buildings or structures, slate, topsoil, peat, silt and dredgings glass includes fritted enamel, but excludes glass fibre and glass reinforced plastics.
2	Ceramic or concrete materials		ceramics includes bricks, bricks and mortar, tiles, clay ware, pottery, china and refractories concrete includes reinforced concrete, concrete blocks, breeze blocks and aircrete blocks, but excludes concrete plant washings
3	Minerals	Processed or prepared, not used	moulding sands excludes sands containing organic binders clays includes moulding clays and clay absorbents, including Fuller's earth and bentonite man-made mineral fibres includes glass fibres, but excludes glass-reinforced plastic and asbestos silica, mica and mineral abrasives
4	Furnace slags		vitrified wastes and residues from thermal processing of minerals where, in either case, the residue is both fused and insoluble slag from waste incineration
5	Ash		comprises only bottom ash and fly ash from wood, coal or waste combustion excludes fly ash from municipal, clinical, and hazardous waste incinerators and sewage sludge incinerators
6	Low activity inorganic compound		comprises only titanium dioxide, calcium carbonate, magnesium carbonate, magnesium oxide, magnesium hydroxide, iron oxide, ferric hydroxide, aluminium oxide, aluminium hydroxide & zirconium dioxide
7	Calcium sulphate	Disposed of either at a site not licensed to take putrescible waste or in a containment cell which takes only calcium sulphate	includes gypsum and calcium sulphate based plasters, but excludes plasterboard
8	Calcium hydroxide and brine	Deposited in brine cavity	
9	Water	Containing other qualifying material in suspension	

Volume to Weight Conversion Factors

Waste category	Typical waste types	Cubic metres to tonne - multiply by:	Cubic yards to tonne - multiply by:
Inactive or inert waste	Largely water insoluble and non or very slowly biodegradable: e.g. sand, subsoil, concrete, bricks, mineral fibres, fibreglass etc.	1.5	1.15

Visualization in Landscape and Environmental Planning
Technology and Applications

Ian Bishop and Eckart Lange

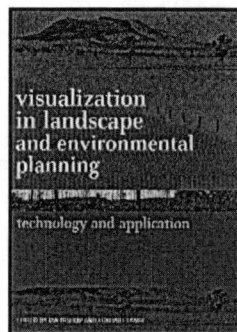

This major reference presents the challenges, issues and directions of computer-based visualization of the natural and built environment and the role of such visualization in landscape and environmental planning. It offers a uniquely systematic approach to the potential of visualization and the writers are acknowledged experts in their field of specialization. Case studies are presented to illustrate many aspects of landscape management including forestry, agriculture, ecology, mining and urban development.

Foreword: Stephen Ervin. Part One. 1. Communication, perception and visualization. 2. Visualization Classified. 3. Data sources for three-dimensional models. 4. Visualization Technology. 5. Validity, reliability, and ethics in visualization. Part Two - Applications. 6 - Application in the Forest Landscape. 7. Applications in the agricultural landscape. 8. Applications in Energy, Industry and Infrastructure. 9 Applications in the urban landscape Part Three - Prospects. 10. Visualization Prospects. Index.

May 2005: 246x189 mm: 340 pages
HB: 0-415-30510-1: £55.00

To Order: Tel: +44 (0) 1264 343071 Fax: +44 (0) 1264 343005, or
Post: Taylor and Francis Customer Services, Thomson Publishing Services, Cheriton House, Andover, Hants, SP10 5BE, UK Email: book.orders@tandf.co.uk

For a complete listing of all our titles visit:
www.sponpress.com

Taylor & Francis
Taylor & Francis Group plc

Index

Outdoor Lighting Guide

Institution of Lighting Engineers

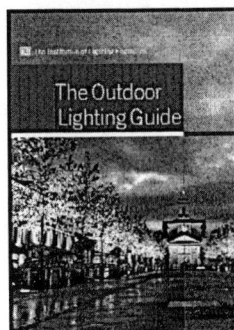

An all-inclusive guide to exterior lighting from The Institution of Lighting Engineers, recognised as the pre-eminent professional source in the UK for authoritative guidance on exterior lighting. As concern grows over environmental issues and light pollution, this book fills a need for a straightforward and accessible guide to the use, design and installation of outdoor lighting.

This book provides a comprehensive source of information and advice on all forms of exterior lighting, from floodlighting, buildings and road lighting to elaborate Christmas decorations, and will be useful to practitioners and non-experts alike. Specialists will value the dependable detail on standards and related design, installation and maintenance problems, and any user can find extensive practical guidance on safety issues, the lighting of hazardous areas, and avoiding potential difficulties.

August 2005: 234x156 mm: 320 pages
HB: 0-415-37007-8: £85.00

To Order: Tel: +44 (0) 1264 343071 Fax: +44 (0) 1264 343005, or
Post: Taylor and Francis Customer Services, Thomson Publishing Services, Cheriton House, Andover, Hants, SP10 5BE, UK Email: book.orders@tandf.co.uk

For a complete listing of all our titles visit:
www.sponpress.com

Taylor & Francis
Taylor & Francis Group plc

Garden History
Philosophy and Design
2000 BC-2000 AD

GARDEN HISTORY

Tom Turner

Highly illustrated to present and explain in a most appealing way, the historic styles of gardens, with particular emphasis on the philosophy of garden design. This carefully structured overview makes the large subject of garden history accessible to a wide range of readers.

The sections on history and philosophy are written as succinct essays, illustrated with photographs or perspective drawings. The essays deal with the ideas and historical conditions, which led to the making of particular types of gardens.

The section on styles will focus on plan analysis and is illustrated. Diagrams illustrate the key features of styles and representative garden plans with short descriptions explaining each style.

Tom Turner traces the evolution of gardens through four centuries, from Egypt and West Asia to Europe and America, relating designs to the philosophical, geographical, social and artistic context in which they developed.

September 2004: 250x250mm: 240 pages
360 colour photographs with 170 line drawings
HB: 0-415-31748-7: £50.00

To Order: Tel: +44 (0) 1264 343071 Fax: +44 (0) 1264 343005, or
Post: Taylor and Francis Customer Services, Thomson Publishing Services, Cheriton House, Andover, Hants, SP10 5BE, UK Email: book.orders@tandf.co.uk

For a complete listing of all our titles visit:
www.sponpress.com

Taylor & Francis
Taylor & Francis Group plc

CD-Rom Single-User Licence Agreement

We welcome you as a user of this Taylor & Francis CD-ROM and hope that you find it a useful and valuable tool. Please read this document carefully. **This is a legal agreement** between you (hereinafter referred to as the "Licensee") and Taylor and Francis Books Ltd. (the "Publisher"), which defines the terms under which you may use the Product. **By breaking the seal and opening the package containing the CD-ROM you agree to these terms and conditions outlined herein. If you do not agree to these terms you must return the Product to your supplier intact, with the seal on the CD case unbroken.**

1.   Definition of the Product
          The product which is the subject of this Agreement, *Spon's Landscape and External Works Price Book 2006 on CD-ROM* (the "Product") consists of:
1.1   Underlying data comprised in the product (the "Data")
1.2   A compilation of the Data (the "Database")
1.3   Software (the "Software") for accessing and using the Database
1.4   A CD-ROM disk (the "CD-ROM")

2.   Commencement and licence
2.1   This Agreement commences upon the breaking open of the package containing the CD-ROM by the Licensee (the "Commencement Date").
2.2   This is a licence agreement (the "Agreement") for the use of the Product by the Licensee, and not an agreement for sale.
2.3   The Publisher licenses the Licensee on a non-exclusive and non-transferable basis to use the Product on condition that the Licensee complies with this Agreement. The Licensee acknowledges that it is only permitted to use the Product in accordance with this Agreement.

3.   Multiple use
          For more than one user or for a wide area network or consortium, use is only permissible with the purchase from the Publisher of a multiple-user licence and adherence to the terms and conditions of that licence.

4.   Installation and Use
4.1   The Licensee may provide access to the Product for individual study in the following manner: The Licensee may install the Product on a secure local area network on a single site for use by one user.
4.2   The Licensee shall be responsible for installing the Product and for the effectiveness of such installation.
4.3   Text from the Product may be incorporated in a coursepack. Such use is only permissible with the express permission of the Publisher in writing and requires the payment of the appropriate fee as specified by the Publisher and signature of a separate licence agreement.
4.4   The CD-ROM is a free addition to the book and no technical support will be provided.

5.   Permitted Activities
5.1   The Licensee shall be entitled:
          5.1.1   to use the Product for its own internal purposes;
          5.1.2   to download onto electronic, magnetic, optical or similar storage medium reasonable portions of the Database provided that the purpose of the Licensee is to undertake internal research or study and provided that such storage is temporary;
          5.1.3   to make a copy of the Database and/or the Software for back-up/archival/disaster recovery purposes.
5.2   The Licensee acknowledges that its rights to use the Product are strictly set out in this Agreement, and all other uses (whether expressly mentioned in Clause 6 below or not) are prohibited.

6.   Prohibited Activities
          The following are prohibited without the express permission of the Publisher:
6.1   The commercial exploitation of any part of the Product.
6.2   The rental, loan, (free or for money or money's worth) or hire purchase of this product, save with the express consent of the Publisher.
6.3   Any activity which raises the reasonable prospect of impeding the Publisher's ability or opportunities to market the Product.
6.4   Any networking, physical or electronic distribution or dissemination of the product save as expressly permitted by this Agreement.
6.5   Any reverse engineering, decompilation, disassembly or other alteration of the Product save in accordance with applicable national laws.
6.6   The right to create any derivative product or service from the Product save as expressly provided for in this Agreement.
6.7   Any alteration, amendment, modification or deletion from the Product, whether for the purposes of error correction or otherwise.

7. General Responsibilities of the License

7.1 The Licensee will take all reasonable steps to ensure that the Product is used in accordance with the terms and conditions of this Agreement.

7.2 The Licensee acknowledges that damages may not be a sufficient remedy for the Publisher in the event of breach of this Agreement by the Licensee, and that an injunction may be appropriate.

7.3 The Licensee undertakes to keep the Product safe and to use its best endeavours to ensure that the product does not fall into the hands of third parties, whether as a result of theft or otherwise.

7.4 Where information of a confidential nature relating to the product of the business affairs of the Publisher comes into the possession of the Licensee pursuant to this Agreement (or otherwise), the Licensee agrees to use such information solely for the purposes of this Agreement, and under no circumstances to disclose any element of the information to any third party save strictly as permitted under this Agreement. For the avoidance of doubt, the Licensee's obligations under this sub-clause 7.4 shall survive the termination of this Agreement.

8. Warrant and Liability

8.1 The Publisher warrants that it has the authority to enter into this agreement and that it has secured all rights and permissions necessary to enable the Licensee to use the Product in accordance with this Agreement.

8.2 The Publisher warrants that the CD-ROM as supplied on the Commencement Date shall be free of defects in materials and workmanship, and undertakes to replace any defective CD-ROM within 28 days of notice of such defect being received provided such notice is received within 30 days of such supply. As an alternative to replacement, the Publisher agrees fully to refund the Licensee in such circumstances, if the Licensee so requests, provided that the Licensee returns the Product to the Publisher. The provisions of this sub-clause 8.2 do not apply where the defect results from an accident or from misuse of the product by the Licensee.

8.3 Sub-clause 8.2 sets out the sole and exclusive remedy of the Licensee in relation to defects in the CD-ROM.

8.4 The Publisher and the Licensee acknowledge that the Publisher supplies the Product on an "as is" basis. The Publisher gives no warranties:

    8.4.1 that the Product satisfies the individual requirements of the Licensee; or

    8.4.2 that the Product is otherwise fit for the Licensee's purpose; or

    8.4.3 that the Data are accurate or complete of free of errors or omissions; or

    8.4.4 that the Product is compatible with the Licensee's hardware equipment and software operating environment.

8.5 The Publisher hereby disclaims all warranties and conditions, express or implied, which are not stated above.

8.6 Nothing in this Clause 8 limits the Publisher's liability to the Licensee in the event of death or personal injury resulting from the Publisher's negligence.

8.7 The Publisher hereby excludes liability for loss of revenue, reputation, business, profits, or for indirect or consequential losses, irrespective of whether the Publisher was advised by the Licensee of the potential of such losses.

8.8 The Licensee acknowledges the merit of independently verifying Data prior to taking any decisions of material significance (commercial or otherwise) based on such data. It is agreed that the Publisher shall not be liable for any losses which result from the Licensee placing reliance on the Data or on the Database, under any circumstances.

8.9 Subject to sub-clause 8.6 above, the Publisher's liability under this Agreement shall be limited to the purchase price.

9. Intellectual Property Rights

9.1 Nothing in this Agreement affects the ownership of copyright or other intellectual property rights in the Data, the Database of the Software.

9.2 The Licensee agrees to display the Publishers' copyright notice in the manner described in the Product.

9.3 The Licensee hereby agrees to abide by copyright and similar notice requirements required by the Publisher, details of which are as follows:

"© 2006 Taylor & Francis. All rights reserved. All materials in *Spon's Landscape and External Works Price Book 2006* are copyright protected. © 2003 Adobe Systems Incorporated. All rights reserved. No such materials may be used, displayed, modified, adapted, distributed, transmitted, transferred, published or otherwise reproduced in any form or by any means now or hereafter developed other than strictly in accordance with the terms of the licence agreement enclosed with the CD-ROM. However, text and images may be printed and copied for research and private study within the preset program limitations. Please note the copyright notice above, and that any text or images printed or copied must credit the source."

9.4 This Product contains material proprietary to and copyedited by the Publisher and others. Except for the licence granted herein, all rights, title and interest in the Product, in all languages, formats and media throughout the world, including copyrights therein, are and remain the property of the Publisher or other copyright holders identified in the Product.

10.   Non-assignment

This Agreement and the licence contained within it may not be assigned to any other person or entity without the written consent of the Publisher.

11.   Termination and Consequences of Termination.

11.1   The Publisher shall have the right to terminate this Agreement if:

    11.1.1   the Licensee is in material breach of this Agreement and fails to remedy such breach (where capable of remedy) within 14 days of a written notice from the Publisher requiring it to do so; or

    11.1.2   the Licensee becomes insolvent, becomes subject to receivership, liquidation or similar external administration; or

    11.1.3   the Licensee ceases to operate in business.

11.2   The Licensee shall have the right to terminate this Agreement for any reason upon two month's written notice. The Licensee shall not be entitled to any refund for payments made under this Agreement prior to termination under this sub-clause 11.2.

11.3   Termination by either of the parties is without prejudice to any other rights or remedies under the general law to which they may be entitled, or which survive such termination (including rights of the Publisher under sub-clause 7.4 above).

11.4   Upon termination of this Agreement, or expiry of its terms, the Licensee must:

    11.4.1   destroy all back up copies of the product; and

    11.4.2   return the Product to the Publisher.

12.   General

12.1   **Compliance with export provisions**

The Publisher hereby agrees to comply fully with all relevant export laws and regulations of the United Kingdom to ensure that the Product is not exported, directly or indirectly, in violation of English law.

12.2   **Force majeure**

The parties accept no responsibility for breaches of this Agreement occurring as a result of circumstances beyond their control.

12.3   **No waiver**

Any failure or delay by either party to exercise or enforce any right conferred by this Agreement shall not be deemed to be a waiver of such right.

12.4   **Entire agreement**

This Agreement represents the entire agreement between the Publisher and the Licensee concerning the Product. The terms of this Agreement supersede all prior purchase orders, written terms and conditions, written or verbal representations, advertising or statements relating in any way to the Product.

12.5   **Severability**

If any provision of this Agreement is found to be invalid or unenforceable by a court of law of competent jurisdiction, such a finding shall not affect the other provisions of this Agreement and all provisions of this Agreement unaffected by such a finding shall remain in full force and effect.

12.6   **Variations**

This agreement may only be varied in writing by means of variation signed in writing by both parties.

12.7   **Notices**

All notices to be delivered to: Spon's Price Books, Taylor & Francis Books Ltd., 2 Park Square, Milton Park, Abingdon, Oxfordshire, OX14 4RN, UK.

12.8   **Governing law**

This Agreement is governed by English law and the parties hereby agree that any dispute arising under this Agreement shall be subject to the jurisdiction of the English courts.

If you have any queries about the terms of this licence, please contact:

Spon's Price Books
Taylor & Francis Books Ltd.
2 Park Square, Milton Park, Abingdon, Oxfordshire, OX14 4RN
Tel: +44 (0) 20 7017 6000
Fax: +44 (0) 20 7017 6702
www.sponpress.com

Spon Press
Taylor & Francis Group

CD-ROM Installation Instructions

System requirements

Minimum

- Pentium processor
- 32 MB of RAM
- 10 MB available hard disk space
- CD-ROM drive
- Microsoft Windows 95/98/2000/NT/ME/XP
- SVGA screen
- Internet connection

Recommended

- Pentium 266 MHz processor
- 256 MB of RAM
- 100 MB available hard disk space
- CD-ROM drive
- Microsoft Windows 2000/NT/XP
- XVGA screen
- Internet connection

Microsoft ® is a registered trademark and Windows™ is a trademark of the Microsoft Corporation.

Installation

How to install *Spon's Landscape and External Works Price Book 2006 CD-ROM*

Windows 95/98/2000/NT

Spon's Landscape and External Works Price Book 2006 CD-ROM should run automatically when inserted into the CD-ROM drive. If it fails to run, follow the instructions below.

- Click the **Start** button and choose **Run.**
- Click the **Browse** button.
- Select your CD-ROM drive.
- Select the Setup file (setup.exe) then click **Open.**
- Click the OK button.
- Follow the instructions on screen.
- The installation process will create a folder containing an icon for *Spon's Landscape and External Works Price Book 2006 CD-ROM* and also an icon on your desktop.

How to run the *Spon's Landscape and External Works Price Book 2006 CD-ROM*

- Double click the icon (from the folder or desktop) installed by the Setup program.
- Follow the instructions on screen.

The CD-ROM is a free addition to the book and no technical support will be provided. For help with the use of the CD-ROM please visit www.pricebooks.co.uk

Multiple-user use of the Spon Press CD–ROM

To buy a licence to install your Spon Press Price
Book CD–ROM on a secure local area network or a
wide area network, and for the supply of network key
files, for an agreed number of users please contact:

Spon's Price Books
Taylor & Francis Books Ltd.
2 Park Square, Milton Park, Abingdon, Oxfordshire, OX14 4RN
Tel: +44 (0) 207 7017 6000
Fax: +44 (0) 207 7017 6072
www.sponpress.com

Number of users	Licence cost
2–5	£360
6–10	£840
11–20	£1600
21–30	£2850
31–50	£4200
51–75	£6600
76–100	£9000
Over 100	Please contact Spon for details